Test Statistic

$$t = \frac{(M_1 - M_2) - k_0}{\sqrt{\left(\dfrac{N_1 S_1^2 + N_2 S_2^2}{N_1 + N_2 - 2}\right)\left(\dfrac{N_1 + N_2}{N_1 N_2}\right)}}$$

t with $\nu = N_1 + N_2 - 2$

$$t = \frac{(M_1 - M_2)}{\sqrt{s_D^2/N}} = \frac{(M_1 - M_2)}{\sqrt{S_D^2/(N - 1)}}$$

t with $\nu = N - 1$

$$F = \frac{\text{MSB}}{\text{MSE}} = \frac{\text{SSB}/(J - 1)}{\text{SSE}/(N - J)}$$

F with $\nu_1 = J - 1$
and $\nu_2 = N - J$

$$F = \frac{\text{MSR}}{\text{MSE}} = \frac{\text{SSR}/(J - 1)}{\text{SSE}/JK(n - 1)}$$

F with $\nu_1 = J - 1$
and $\nu_2 = JK(n - 1)$

$$F = \frac{\text{MSC}}{\text{MSE}} = \frac{\text{SSC}/(K - 1)}{\text{SSE}/JK(n - 1)}$$

F with $\nu_1 = K - 1$
and $\nu_2 = JK(n - 1)$

$$F = \frac{\text{MS}(R \times C)}{\text{MSE}} = \frac{\text{SS}(R \times C)/(J - 1)(K - 1)}{\text{SSE}/JK(n - 1)}$$

F with $\nu_1 = (J - 1)(K - 1)$
and $\nu_2 = JK(n - 1)$

$$F = \frac{\text{MSB}}{\text{MSE}} = \frac{\text{SSB}/(J - 1)}{\text{SSE}/(N - J)}$$

F with $\nu_1 = J - 1$
and $\nu_2 = N - J$

$$F = \frac{\text{MSR}}{\text{MS}(R \times C)} = \frac{\text{SSR}/(J - 1)}{\text{SS}(R \times C)/(J - 1)(K - 1)}$$

F with $\nu_1 = (J - 1)$
and $\nu_2 = (J - 1)(K - 1)$

$$F = \frac{\text{MSC}}{\text{MS}(R \times C)} = \frac{\text{SSC}/(K - 1)}{\text{SS}(R \times C)/(J - 1)(K - 1)}$$

F with $\nu_1 = K - 1$
and $\nu_2 = (J - 1)(K - 1)$

$$F = \frac{\text{MS}(R \times C)}{\text{MSE}} = \frac{\text{SS}(R \times C)/(J - 1)(K - 1)}{\text{SSE}/JK(n - 1)}$$

F with $\nu_1 = (J - 1)(K - 1)$
and $\nu_2 = JK(n - 1)$

$$F = \frac{\text{MSR}}{\text{MSE}} = \frac{\text{SSR}/(J - 1)}{\text{SSE}/JK(n - 1)}$$

F with $\nu_1 = J - 1$
and $\nu_2 = JK(n - 1)$

$$F = \frac{\text{MSC}}{\text{MS}(R \times C)} = \frac{\text{SSC}/(K - 1)}{\text{SS}(R \times C)/(J - 1)(K - 1)}$$

F with $\nu_1 = K - 1$
and $\nu_2 = (J - 1)(K - 1)$

$$F = \frac{\text{MS}(R \times C)}{\text{MSE}} = \frac{\text{SS}(R \times C)/(J - 1)(K - 1)}{\text{SSE}/JK(n - 1)}$$

F with $\nu_1 = (J - 1)(K - 1)$
and $\nu_2 = JK(n - 1)$

$$F = \frac{\text{MS treatments}}{\text{MS residuals}} = \frac{\text{SS treatment}/(K - 1)}{\text{SS residual}/(J - 1)(K - 1)}$$

F with $\nu_1 = K - 1$
and $\nu_2 = (J - 1)(K - 1)$

$$t = \frac{\hat{\psi}}{\sqrt{\text{MSE}\sum_j (c_j^2/N_j)}}$$

t with $\nu = N - J$

or

$$F = \frac{\text{SS}(\hat{\psi})}{\text{MSE}} = \frac{\dfrac{(\hat{\psi})^2}{\sum_j (c_j^2/N_j)}}{\text{MSE}}$$

F with $\nu_1 = 1$
and $\nu_2 = N - J$

THEORY AND APPLICATION OF STATISTICS

CECIL G. MISKEL
Consulting Editor

University of Michigan

THEORY AND APPLICATION OF STATISTICS

BRUCE E. WAMPOLD

University of Oregon

CLIFFORD J. DREW

University of Utah

McGRAW-HILL PUBLISHING COMPANY

New York   St. Louis   San Francisco   Auckland   Bogotá   Caracas
Hamburg   Lisbon   London   Madrid   Mexico   Milan   Montreal   New Delhi
Oklahoma City   Paris   San Juan   São Paulo   Singapore   Sydney
Tokyo   Toronto

This book was developed by Lane Akers, Inc.

THEORY AND APPLICATION OF STATISTICS

1 2 3 4 5 6 7 8 9 0   DOC DOC   8 9 4 3 2 1 0 9

ISBN 0-07-557239-7

This book was set in Meridien by Ruttle, Shaw & Wetherill, Inc.
The editors were Lane Akers and Eleanor Castellano;
the designer was Harry Rinehart;
the production supervisor was Valerie A. Sawyer.
R. R. Donnelley & Sons Company was printer and binder.

Library of Congress Cataloging-in-Publication Data

Wampold, Bruce E., (date).
    Theory and application of statistics/Bruce E. Wampold, Clifford J. Drew.
    p.   cm.
    Bibliography: p.
    Includes index.
    ISBN 0-07-557239-7
    1. Mathematical statistics.   I. Drew, Clifford J., (date).
II. Title.
QA276.W325   1990
519.5—dc20   89-8175

CONTENTS

PREFACE

T*heory and Application of Statistics* is designed to provide the beginning student with a comprehensive introduction to commonly used statistical methods in the social sciences. Our goal was to create a readable text that requires relatively little background knowledge (a course or two of high school algebra), but that still rigorously discusses important topics in statistics. The text is designed so that students will learn to apply statistical methods as well as to understand the rationale behind the statistics and issues related to those methods. Students mastering the material in this text will be able to read research reports and independently conduct research that uses the basic statistical techniques presented in this text. Before discussing the contents of this book more fully, we should answer a question that we are often asked by our students: Why learn statistics?

If we were to ask students to identify those factors that attracted them to the social sciences, we suspect that learning statistics would not be one of the most often cited factors. Realistically, very few students seek training in the social sciences to master the intricacies of statistical methods. Typically, students are interested in learning about behavior. The knowledge accumulated in the social sciences is fascinating; most of us are interested in some aspect of human or animal behavior. Consider the possibilities: How do humans solve problems? What are the characteristics of distressed marital partners? What factors are responsible for differences in intelligence? How do hormones affect behavior? What are the factors that increase the probability that an opinion will be changed? What are the most effective ways to instruct? To some students, answers provided by others regarding these questions are stimulating in and of themselves; to other students, answers from others are not enough. Instead, it is far more fascinating to explore, to investigate the world, and, in some small fashion, to find their own answers—discover knowledge rather than merely acquire it. For this latter group, learning about the application of research methods is particularly important, because it is through this process that students can launch their own investigations to find their own answers.

Regardless of the motivation for learning about behavior, we must recognize that it is the *research endeavor* that generates the knowledge that students find so fascinating. Without research, knowledge would be replaced by capricious speculation, and understanding behavior would be impossible. Because so much research in the social sciences involves quantifying characteristics related to behavior, procedures that make sense of numbers are critical to the research endeavor. Indeed, a strong case can be made that statistical methods are the basic research tool for social science researchers. Although researchers learn particular methods for a given area of inquiry (e.g., the neuropsychologist will certainly need to know how to measure brain activity), with few exceptions, an analysis of the numbers that result from these special methods is vital to understanding the phenomenon under investigation. The researcher skilled in statistics will be able to see an astounding image; the untrained will see only a huddle of numbers.

To further clarify why a student who is interested in knowledge should learn about a research tool such as statistics, we present a few reasons here. First, it is hoped that many of the students using this book want to (or will want to in the future) become skilled researchers. We feel that the only thing more exciting than acquiring knowledge is discovering knowledge. To those future researchers a fluency in statistics is essential. Some people may say that researchers can hire statistical consultants to "crunch" the numbers for them. Even though every researcher needs to consult a statistician from time to time, the researcher who doesn't understand the basic analysis will soon lose control of his or her research and eventually will not fully know what is transpiring. This will diminish the excitement as well as the interest in the research endeavor. We have seen it happen over and over again. Second, learning statistics is important because an understanding of statistics is necessary in order to acquire a comprehensive knowledge in the area of social science. The state of the science is such that there are few uncontested facts, and thus the inquiring student must be able to independently evaluate the evidence vis-à-vis knowledge in this area. Perusing a review of a substantive area in the social sciences will convince many about the complexity of knowledge and the apparent lack of concordance of research results. Often the argument about the "correct" view of the world becomes a statistical discussion. For example, the nature of intelligence (e.g., single factor versus multiple factor theories) is intricately intertwined with the esoteric statistical procedure factor analysis; knowledge of factor analysis is required to understand the issues involved in this area. To a large extent, knowledge in the social sciences is built on a statistical foundation. An understanding of statistics is necessary to appreciate fully the essence of the area.

There are many ways to approach teaching statistics. At one extreme is the "cook book" approach, which gives step-by-step instructions to conduct a statistical analysis; at the other extreme is the theoretical approach, which explicates the theory of statistical analyses. We feel there is merit to both approaches although the extremes are unlikely to be helpful to the typical student. Two components are involved in analyzing the numbers that result from research: a statistical procedure is applied, and the results of that procedure are interpreted. Knowledge of *applications* allows the student to perform a statistical analysis (either by hand or more likely with the aid of a computer). *Theoretical* knowledge allows the student to make sense of the analysis conducted. Both are necessary, but as authors, we believe it is not possible to dictate the exact balance between application and theory—the optimal balance depends on a number of factors—the objectives of the student, the nature of the training program, the background knowledge of the student, and/or the orientation of the instructor. Consequently, we have written this book on two levels, and the instructor and student can choose how to balance the application and theory. Applications, being more basic, should be understood by all readers. The degree to which attention is paid to the theoretical material is up to the discretion of the instructor and the student. Although the application and theory are integrated in the text for the most part, the reader should be able to "tune out" aspects of theory and concentrate on the more concrete applications without loss of continuity. In our teaching we find that this approach accommodates a wide range of student abilities, backgrounds, and interests.

The objective of this book is to provide an introduction to basic statistical methods in the social sciences. Although multiple regression is discussed, the topics presented in this text typically are referred to as univariate statistics. We have attempted to be comprehensive, but clearly many procedures contained in the univariate domain could not be included. However, we present the univariate procedures that are commonly used in social science research. With regard to theory, the presentation of it is at an introductory level, although we present a fairly heavy dose, as mentioned previously. The beginning student may have the distinct impression that the application of statistical methods is cut and dried. It is not; numerous complex issues are involved with the application of even the simplest procedure and many of these issues are unresolved by statisticians. The theory is presented to give the student an appreciation of the issues. However, we have only scratched the surface, and often we recommend further investigations of important topics. As well, many points discussed in *Theory and Application of Statistics* require that the reader take a particular result on faith.

We emphasize that the presentation is introductory. The material in this text is appropriate for a year's sequence in statistics for those with little background in the subject. A basic knowledge of algebra (e.g., one or two years of high school algebra) is recommended, although a richer background will allow the student to understand and appreciate some of the finer theoretical points. Mastery of the material will allow the student to progress to more advanced topics, such as multivariate techniques, although more advanced courses will likely require a more extensive background in mathematics. We suggest that this text be used in the first year of a graduate program in psychology or education or for more advanced students at the undergraduate level.

A number of features of this text, if recognized, may add to the readers' appreciation of the topics.

1. *Examples and problems are computationally simple,* so that most applications will not have to be performed with computer assistance. For this reason, many "computational formulas" have been omitted. Nevertheless, we want students to perform the statistical tests because it is important to know where the results of computer analyses "come from" and because hand calculations often provide understanding.

2. *Problems are aimed at applications.* The solutions to odd-numbered problems are provided as well as the answers to the even-numbered problems. Each even-numbered problem is a restatement of the odd-numbered problem (with different numbers, of course). At the end of the text is a "data set" designed to be used in a computer analysis, if the instructor should so desire.

3. *Important material is denoted in several ways.* Initial introduction of a term is denoted by printing the term in italics. Equations that are referred to later in the text or may be referred to by an external source are numbered. Important textual material or formulas are printed in bold type.

4. *An annotated bibliography is provided at the end of each chapter for those interested in a better understanding of various topics.* Clearly, we cannot mention every important book and article; nevertheless, we have tried to discuss those that we think are most valuable.

5. *The first five chapters provide the theoretical foundation for understanding the applications that are discussed in the remaining seven chapters.* Roughly, the foundations comprise about one-third of the material. Although the text is designed for one-year's sequence, we have found that a two-quarter sequence can be accomplished by presenting the foundation material (Chapters 1 through 5) and applications of t tests (Chapter 6), fixed-effects analysis of variance (Chapter 7), correlation (Sections 10.1 through 10.3), and univariate regression (Sections 11.1 through 11.3).

6. We have been careful to refer to hypothetical researchers in our text in a nonsexist way. Besides being socially appropriate at this point in time, it is our way to encourage women to become interested in statistics and participate in research.

We wish to thank the following persons for their insightful reading of the manuscript during various stages of the book's development: James Austin, University of Illinois; Richard Hasse, SUNY, Albany; Carl Huberty, University of Georgia; and Terrence Tracey, University of Illinois. Furthermore, we express our gratitude to Betsy Davis, Kay-Hyon Kim, and Michael Stoolmiller for their assistance, and to the other students who suffered through various typescript copies of this book.

Bruce E. Wampold
Clifford J. Drew

1 INTRODUCTION TO SET AND PROBABILITY THEORY

R esearch in education and psychology is characterized by uncertainty and ambiguity. Rarely can it be said that a result of an investigation leads with 100% certainty to a certain conclusion. Although the link between smoking and disease has clearly been established, any one study of smoking and disease can at best result in the conclusion that there is a strong **possibility** that a link is present. The researchers might be 90 or 95% certain that their conclusion is correct, but in research in the social sciences there is always doubt. Thus, we are left with having to qualify our results knowing that some chance exists that we have reached an incorrect conclusion. Before putting faith into our results, consumers of research will want (or should want) to know what is the probability that we have reached an incorrect conclusion. The task of the researcher then is not only to reach a conclusion but to inform the scientific community about the chances that the conclusion is incorrect (sort of a consumer's protection policy).

The need to state the chances that a conclusion might be incorrect necessitates the study of probability. Although probability theory is ubiquitous in statistics, an understanding of probability theory (at an elementary level) is absolutely necessary to understand the results of even the simplest statistical test. Failure to understand that the statement "statistical significance was .05" is a conditional probability will lead to much confusion.

Before discussing probability theory, we turn to a short discussion of set theory. Set theory is useful because we will characterize the various possible outcomes from an experiment as elements of a set. Set and probability theory are branches of mathematics in their own right; in this chapter a cursory introduction of these topics is presented and additional material on them is discussed as needed.

Set Theory

1.1 Set Theory

The definition of a set is quite simple. A *set* is a collection of objects. The objects are called *elements*. For example, the vowels of the English language constitute a set, the even numbers constitute a set, and so forth. To denote a set, the capital letters A, B, C will be used. So A could be the set that contains the vowels and B could be the set containing the even numbers. The elements are denoted by the lower-case letters a, b, c, ...

A set can be defined by either listing the elements of the set or by describing the property of the set. Accordingly, the set containing the vowels could be defined in the following ways:

List $A = \{a, e, i, o, u\}$ or

Property $A = \{a|a$ is a vowel$\}$

Two notations need explaining. Included in the brackets, { and }, are either the elements of a set or a description of the elements in the set. The symbol "|" can be read as "such that," so that the definition of A that uses the description of its properties would be read as "the set A contains all the elements a such that a is a vowel." Similarly, the set B could be defined in two ways:

List $B = \{2, 4, 6, . . .\}$ or

Property $B = \{b|b$ is an even positive integer$\}$

Sets A and B differ in a fundamental way. Set A is said to be a *finite* set because it contains a finite number of elements; similarly, set B is said to be an *infinite* set because it contains an infinite number of elements.

Sets can be manipulated in a number of ways. To illustrate these properties, we use three sets:

$$A = \{1, 3, 5, 7, 9\}, \quad B = \{1, 4, 9, 16\}, \text{ and } C = \{2, 4, 6, 8, 10\}$$

New sets can be formed by finding the union or the intersection of existing sets. The set D that is the *union* of sets A and B is denoted by $D = A \cup B$ and is defined as $D = \{x|x$ is in A or in B or in both$\}$. For the preceding sets A and B, then, $D = A \cup B = \{1, 3, 4, 5, 7, 9, 16\}$. The set E that is the *intersection* of sets A and B is denoted by $E = A \cap B$ and is defined as $E = \{x|x$ is in A and in $B\}$. Thus, for the sets A and B, $E = A \cap B = \{1, 9\}$. Because sets A and C contain no elements in common, $A \cap C$ is the set that contains no elements. A set that contains no elements is called the *empty set* and is denoted by $\emptyset$. If $A \cap C = \emptyset$, as in the preceding case, A and C are said to be *disjoint*.

The set F is said to be a *subset* of B provided every element of F is also an element of B. If F is a subset of B, then the notation $F \subseteq B$ is used. The empty set is a subset of all sets. The following sets are all the subsets of B:

$\emptyset$, $\{1\}$, $\{4\}$, $\{9\}$, $\{16\}$, $\{1, 4\}$, $\{1, 9\}$, $\{1, 16\}$, $\{4, 9\}$, $\{4, 16\}$, $\{9, 16\}$

$\{1, 4, 9\}$, $\{1, 4, 16\}$, $\{4, 9, 16\}$, $\{1, 9, 16\}$, and $\{1, 4, 9, 16\}$

The *complement* of A is denoted by $\overline{A}$ and defined as $\overline{A} = \{x|x$ is not an element of $A\}$. So, if $A = \{x|x$ is an odd whole number$\}$, then $\overline{A} = \{x|x$ is an even whole number$\}$. The compulsive reader might argue that $\overline{A}$ should also contain the letters of the English language and the rocks in his or her rock collection since they are not contained in A. However, implicit in forming the complement is that there is some universal set from which the elements of A and $\overline{A}$ are drawn. In this example the universal set would be the set containing all the whole numbers.

Although this concludes the basic material on set theory, the subsequent discussion of probability theory is related to sets. For example, in the next section the outcomes of experiments will be defined as sets.

Probability Theory

1.2   Experiments

Everyone who lives in the modern world develops an intuitive understanding of probability. When we hear that "there is a 60% chance of rain" or that "the probability of obtaining a head on a flip of a coin is .50," we have an understanding of what these phrases mean. Obviously, that the probability of obtaining a head is .50 does not mean that one-half head will occur on each toss; rather, we would expect that if we flipped the coin 100 times, about 50 heads would occur. However, it is doubtful that many of us determined that the probability of obtaining a head is .50 by tossing a coin 100 times. More likely we realized that there are two possible outcomes, each of which has the same chance of occurring and therefore the probability that each would occur is .50. Determining the probability that it will rain tomorrow is a more difficult task. Even though there are two outcomes, it will rain or it won't rain, they do not have an equal chance of occurring, and in fact this changes from one day to the next. So, what does it mean to say that "there is a 60% chance of rain?" An adequate explanation of this phrase would be to say that 60% of the days that have been similar to today (i.e., similar satellite picture configuration, similar temperature and barometric pressure, etc.) it has rained. Although these ideas may be familiar, a brief presentation of these concepts and others will be helpful to understand fully the discussions of statistics that follow.

A *simple experiment* is a process that leads to a single outcome from a specified set of outcomes. For example, tossing a die will lead to one outcome out of a possible six outcomes. The set of outcomes is called the *sample space* and is denoted by $\mathcal{S}$. For the example of the die $\mathcal{S} = \{1, 2, 3, 4, 5, 6\}$. Each of the elements of $\mathcal{S}$ is called an *elementary event* and is denoted by an uppercase letter. Thus, $A = \{3\}$ is an example of an elementary event. An *event*, which is also denoted by an upper-case letter, is *any* subset of $\mathcal{S}$. An event differs from an elementary event in that an event may contain more than one outcome. For example, the event B may be defined as $B = \{x|x$ is an even number$\}$. Thus, B contains three outcomes: 2, 4, and 6.

Often it is necessary to "split up" the sample space in certain ways. For instance, "nonoverlapping" events are of interest. Nonoverlapping events are called mutually exclusive. If $A \cap B = \emptyset$ then events A and B are said to be *mutually exclusive*. Another way of splitting up the sample space is to have two events contain all the elementary events. If this is so, then the two events are said to be exhaustive. If $A \cup B = \mathcal{S}$; then events A and B are said to be *exhaustive*. Finally, if two events are mutually exclusive and exhaustive, then they are said to form a *partition* of the sample space. To illustrate, again consider the experiment of tossing a die. Let $A = \{x|x$ is an even number$\}$ and $B = \{x|x$ is an odd number$\}$. A and B are mutually exclusive and exhaustive and thus form a partition of $\mathcal{S}$.

1.3   Calculation of Probabilities

The probability of an event A is denoted $P(A)$. **If A is an elementary event and all elementary events are equally likely, then**

$$P(A) = \frac{1}{N} \tag{1.1}$$

where N is the number of elementary events in the sample space $\mathcal{S}$. If A is the elementary event of obtaining a "5" on a roll of a die, then $P(A) = 1/N = \frac{1}{6}$. **If B is an event and all elementary events are equally likely, then**

$$P(B) = \frac{n(B)}{N} \tag{1.2}$$

where $n(B)$ is the number of elementary events contained in B. For example, if $B = \{x|x$ is an odd number$\}$, $P(B) = n(B)/N = \frac{3}{6} = \frac{1}{2}$, since there are three elementary events contained in B.

In the discussion of calculating probability the phrase "equally likely" was used. Elementary events are *equally likely* provided each elementary event has an equal chance of occurring. The six elementary events for a fair die are equally likely because each of the faces has an equal chance of occurring. However, instances may occur when the assumption that elementary events are equally likely is not justified. Consider a bag of red and black marbles where the red marbles are larger than the black marbles. If the experiment involves drawing out a marble and noting its color, then the assumption that each marble has an equal chance of being selected may not be valid because large marbles may be more (or less) easily selected. Often it is unknown whether or not the elementary events of an experiment are equally likely. For instance, one may want to know whether or not a coin is fair; that is, is the probability of obtaining a heads (or a tails) equal to .50? In a later chapter (Chapter 13, p. 415), the binomial test will be used to determine whether or not a coin is fair.

The rules for calculating probabilities can be related to the intuitive idea of probability that indicates that the probability of obtaining a head from the tossing of a coin is .50, because one-half of the times a coin is tossed a head will appear. No one would be surprised to find that one-half of the tosses did not result in heads if a fair coin were tossed 2 times. That is, it would not be surprising if both tosses resulted in heads or in tails. However, if the coin were tossed 20 times, it would be expected that about one-half of the times heads would appear, although slight deviations from ten heads, say, eight or nine heads, would not be too disturbing. If the coin were tossed a very large number of times, say, 1000 or more, it would be expected that very nearly one-half of the tosses would be heads. This notion is embodied in the intuitive idea that "in the long run" one-half of the tosses will be heads. This intuitive idea can be expressed as the *Law of Large Numbers*.

Before stating the Law of Large Numbers, we need to define the relative frequency of an event. When a simple experiment is replicated N times, the *relative frequency* of an event A is the number of times the event occurred divided by N. (Note that N is used differently here from how it was used in the rules for calculating probabilities.) For example, if a coin is tossed 100 times and 48 heads result, then the relative frequency of the event "heads" is .48. **The Law of Large Numbers states that the difference between the relative frequency of an event A and P(A) approaches zero as N becomes infinitely large.** So, if a fair coin is tossed a few number of times, say, 10, it would not be surprising to find that the relative frequency of heads differed considerably from .50, whereas if the coin were tossed many times, say, 1000, it would be very surprising to find that the relative frequency of heads differed much from .50.

COMMENTS

Comment 1.   Although the method to determine whether or not a coin is fair will be deferred to a later chapter, the Law of Large Numbers has intuitive implications in this context. Suppose a coin whose "fairness" is unknown is tossed five times and four heads were obtained. Although the relative frequency of "heads" is .80, which differs from .50, no one would give up the notion that the coin was fair; the Law of Large Numbers implies that relative frequencies will differ from the probabilities for small number of replications. Now suppose that the coin were tossed 10 times and 8 heads resulted. The evidence is now becoming stronger that the coin is not fair. However, if the coin were tossed 100 times and 80 heads resulted, it would be clear that the coin was not fair. So, although in each instance the relative frequency of "heads" was .80, as the number of replications increases, the stronger is the conclusion that can be reached.

Comment 2.   Often the phrase "beginner's luck" is heard. The phenomenon underlying this phrase can be understood by applying the Law of Large Numbers. Suppose that a beginner is playing blackjack and has a probability of .40 of winning each hand, which is considerably less than that of an expert. If event A is the event "winning," then $P(A) = .40$. Suppose that for the first 4 hands played by a beginner, he or she wins 3. An observer might attribute this performance to beginner's luck. However, for a small number of replications it is not surprising that the relative frequency of winning differs from the probability of winning. On the other hand, if a beginner plays 100 hands, we expect that he or she will win 40 and would be very surprised to find that the individual had won 75.

1.4   Properties of Probability

Now that the notation and rules for calculating probabilities have been given, a number of properties can be discussed. Many of these properties will be intuitive; those that are not will be illustrated with examples.

PROPERTY 1

  If A is any event, then $P(A) \geq 0$ **(1.3)**

This property states that the probability of any event cannot be negative, which should be intuitive.

PROPERTY 2

  $P(\mathscr{S}) = 1$ **(1.4)**

This property states that the probability of the entire sample space is 1. In the die-rolling experiment, $\mathcal{S} = \{1, 2, 3, 4, 5, 6\}$ and the probability of obtaining a "1," "2," "3," "4," "5," or "6" is, of course, 1.

PROPERTY 3

If $A_1, A_2, \ldots, A_N$ are mutually exclusive, then

$$P(A_1 \cup A_2 \cup \cdots \cup A_N) = P(A_1) + P(A_2) + \cdots + P(A_N) \quad \textbf{(1.5)}$$

The subscripts found in this property refer to different events; to indicate a series of events it is easier to use subscripts and ellipses rather than the letters A, B, C, and so on. Again, the example of rolling a die is used to illustrate this property. Let $A_1 = \{1\}$, $A_2 = \{2\}$, and $A_3 = \{3\}$. Property 3 indicates that $P(A_1 \cup A_2 \cup A_3) = P(A_1) + P(A_2) + P(A_3) = \frac{1}{6} + \frac{1}{6} + \frac{1}{6} = \frac{1}{2}$. That is, the probability of obtaining a "1," "2," or "3" is equal to .50.

PROPERTY 4

$$P(\overline{A}) = 1 - P(A) \qquad \textbf{(1.6)}$$

This property states that the probability of the complement of an event is equal to 1 minus the probability of the event. If the event A is the event of obtaining a "6" on the roll of a die, then $P(A) = \frac{1}{6}$. Hence, the probability of the complement of A, which is the event of obtaining "1," "2," "3," "4," or "5," is $1 - \frac{1}{6} = \frac{5}{6}$.

PROPERTY 5

$$P(A) \leq 1 \qquad \textbf{(1.7)}$$

This property states the intuitive idea that the probability of an event cannot exceed 1.

PROPERTY 6

$$P(\emptyset) = 0 \qquad \textbf{(1.8)}$$

This is another intuitive property: clearly, the probability of any event that contains no outcomes is zero.

PROPERTY 7

$$P(A \cup B) = P(A) + P(B) - P(A \cap B) \qquad (1.9)$$

This property can be illustrated by the following example. Suppose that the following experiment is conducted: draw a card at random from a standard deck of cards. Let A be the event of drawing a king and B be the event of drawing a heart. Then, by the rules of calculating probabilities, $P(A) = n(A)/N = 4/52$ and $P(B) = n(B)/N = 13/52$. The intersection of A and B is the event of obtaining the king of hearts; $P(A \cap B) = 1/52$. Thus, $P(A \cup B) = P(A) + P(B) - P(A \cap B) = 4/52 + 13/52 - 1/52 = 16/52$. This example can be further illustrated by Figure 1.1, which is called a Venn diagram. If the probability of $A \cup B$ was calculated by summing the probabilities of A and B, then $P(A \cap B)$ would be accounted for twice, necessitating the subtraction of $P(A \cap B)$. In terms of the example the probability of obtaining the king of hearts was included in both the probability of obtaining a king and in the probability of obtaining a heart. Thus, it was necessary to subtract the probability of obtaining the king of hearts from the union.

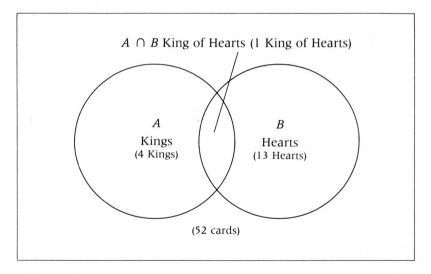

FIGURE 1.1   Venn Diagram for $A \cap B$

COMMENTS   **Comment 1.**   Technically, the material presented on probability is premature. The rules for calculating probability and the properties were given without defining the concept of probability. This seemed like an acceptable way to proceed because nearly everyone has an intuitive idea of probability and because the definition of probability is somewhat esoteric. It should be noted, however, that the first three properties of probability can serve as the definition of what is called a probability space. The remaining properties can be derived from the first three. To illustrate this, we derive property 4 from the first three properties. We want to show that $P(\overline{A}) = 1 - P(A)$. We know $\mathscr{S} = A \cup \overline{A}$. Hence, $P(\mathscr{S}) = P(A \cup \overline{A})$. Because A and $\overline{A}$ are mutually exclusive

$$P(A \cup \overline{A}) = P(A) + P(\overline{A})$$

and because $P(\mathscr{S}) = 1$, we now have

$$P(\mathscr{S}) = P(A \cup \overline{A}) = P(A) + P(\overline{A}) = 1$$

Taking the last two terms and manipulating, we have

$$P(\overline{A}) = 1 - P(A)$$

which is the desired result.

1.5   Conditional Probability

When calculating probabilities it is not uncommon to have additional information that impinges on the calculation of those probabilities. Consider the experiment of drawing a card at random from a standard deck. Let event B be the event of drawing a king: $P(B) = n(B)/N = 4/52 = 1/13$. However, the probability of obtaining a king would change if the additional information that the card drawn was a face card was provided. Clearly, the probability of drawing a king given that the card is a face card is greater than the probability of simply drawing a king. Conditional probability involves the calculation of probabilities in light of additional information. Continuing the previous example, let A be the event that a face card is obtained: $P(A) = n(A)/N = 12/52 = 3/13$. The situation described earlier is denoted as

$$P(B|A)$$

and is read as "the conditional probability of B given A," and in terms of the example it would be read as "the conditional probability of obtaining a king given that the card is a face card." In one sense conditional probabilities can be thought of as limiting the sample space. In the preceding example the sample space has been reduced from the entire deck of cards to the face cards. Thus

$$P(B|A) = \frac{\text{number of kings}}{\text{number of face cards}} = \frac{4}{12} = \frac{1}{3}$$

More formally, *conditional probability* can be defined in the following way:
   If A and B are two events, then the conditional probability of B given A, denoted by $P(B|A)$, is given by

$$P(B|A) = \frac{P(B \cap A)}{P(A)} \tag{1.10}$$

First, it should be noted that this definition does not make sense if $P(A) = 0$; if such is the case, the conditional probability is undefined. Although this definition seems unrelated to the calculation of the previous example, there is a clear rationale for it. Taking Equation (1.10), applying the rules of probability, and manipulating algebraically, we find that

$$P(B|A) = \frac{P(B \cap A)}{P(A)} = \frac{n(B \cap A)/N}{n(A)/N} = \frac{n(B \cap A)}{n(A)} = \frac{4}{12}$$

which is the result obtained earlier.

1.6   Independence

Before defining the independence of two events formally, we present an intuitive understanding and an example. Intuitively, two events are independent provided one does not give any information about the other. In other words, if A and B are independent, then the probability of B occurring will not be changed by knowing that event A has occurred. An example will help. Suppose that event B is the event of drawing a king from a standard deck; thus, $P(B) = 4/52 = 1/13$. Suppose that event A is the event of drawing a heart from the same deck; thus, $P(A) = 13/52 = 1/4$. The conditional probability of B given A is calculated as follows:

$$P(B|A) = \frac{P(B \cap A)}{P(A)} = \frac{1/52}{1/4} = \frac{1}{13}$$

since $B \cap A$ is the event of drawing the king of hearts. Of interest here is the fact that the conditional probability of B given A was exactly the same as the unconditional probability of B; that is, $P(B|A) = P(B)$. The fact that A occurred did not change the probability of B occurring. We can also verify that $P(A|B) = P(A)$. Thus, A and B are independent.

Now a more formal definition of probability can be formulated. If A and B are independent, then $P(B) = P(B|A) = P(B \cap A)/P(A)$. Multiplying the first and last terms by $P(A)$, we have

$$P(B \cap A) = P(A)P(B)$$

Hence, the following definition of independence is given:
  Two events A and B are *independent* if and only if

$$P(A \cap B) = P(A)P(B) \tag{1.11}$$

Using this definition, we can verify that the events A and B are independent. B was the event of drawing a king; thus, $P(B) = 1/13$. A was the event of drawing a heart; thus, $P(A) = 1/4$. The intersection of A and B is the event of drawing the king of hearts; thus, $P(A \cap B) = 1/52$. Thus, $P(A)P(B) = P(A \cap B)$ and A and B are independent.

The definition of independence has another important application. If it is known that two events are independent, then the probability of the inter-section can be calculated by multiplying the probabilities of the respective events. This can be illustrated by using the example of flipping a coin twice. Suppose that we are interested in finding the probability of obtaining a head on each flip. Because the coin has no "memory" of past events, it is reason-able to assume that the event of obtaining a head on the first flip is inde-pendent from the event of obtaining a head on the second flip. If we let A and B denote the events of obtaining a head on the first and second flips, respectively, then $A \cap B$ is the event of obtaining a head on both flips. By the definition of independence, $P(A \cap B) = P(A)P(B) = (\frac{1}{2})(\frac{1}{2}) = \frac{1}{4}$.

There is another way of obtaining the probability of obtaining two heads on two flips. In the preceding example the problem was conceptualized as two trials of a simple experiment. However, it can be conceptualized as a single experiment with four possible outcomes: $\{H, H\}, \{H, T\}, \{T, H\}, \{T, T\}$. These four outcomes become readily apparent if the following table is con-sidered:

	2nd Flip	
	H	T
1st Flip H	H, H	H, T
1st Flip T	T, H	T, T

Thus, the four outcomes can be thought of as elementary events even though they are composed of the results of two flips. Because there are four ele-mentary events, the probability of obtaining $\{H, H\}$ is equal to $\frac{1}{4}$, which is the result obtained earlier.

1.7 Bayes' Rule

Bayes' rule finds interesting applications to a number of problems that may be of interest to social scientists. Bayes' rule allows the calculation of $P(A|B)$ when $P(B|A)$ is known. *Bayes' rule* is given by the following formula:

$$P(A|B) = \frac{P(B|A)P(A)}{P(B|A)P(A) + P(B|\overline{A})P(\overline{A})} \tag{1.12}$$

An interesting example will illustrate the usefulness of this rule. Suppose that a researcher is seeking to develop a blood test that can be used to screen for schizophrenia in the general population. Let A be the event that a person will be considered schizophrenic at some point in his or her life, and let B be the event that a person exhibits a positive finding on the blood test. We know that the general incidence of schizophrenia in the population is about 1%; thus, $P(A) = .01$. Suppose that the researcher does a number of studies to test the efficacy of the blood test. First, a number of diagnosed schizophrenics are tested and 90% are found to have a positive test; thus, $P(B|A) = .90$. Furthermore, a number of normals are tested and only 5% have a positive test; thus, $P(B|\overline{A}) = .05$. It appears that the blood test is quite good. However, the probability that we want is $P(A|B)$; that is, the conditional probability that the person will be schizophrenic given the test was positive. This probability can be obtained by using Bayes' rule. First, the information needed for application of Bayes' rule is summarized:

A = "Schizophrenic"
B = "Positive test"
$P(A) = .01$

By property 4, $P(\overline{A}) = 1 - P(A) = .99$

$P(B|A) = .90$
$P(B|\overline{A}) = .05$

Therefore, by Bayes' rule

$$P(A|B) = \frac{P(B|A)P(A)}{P(B|A)P(A) + P(B|\overline{A})P(\overline{A})} = \frac{(.90)(.01)}{(.90)(.01) + (.05)(.99)}$$
$$= .154$$

So, although the test appeared to be useful, in fact, only 15% of the time will a person manifest schizophrenia given a positive test! Given the costs of misdiagnosing schizophrenia, one would be hesitant to use this test as a screening device. The reasoning behind Bayes' rule can be further illustrated by continuing the previous example with the hypothetical results of using the test with 2000 individuals. The results are presented in Figure 1.2. The entries in each cell are the number of individuals who possess the indicated characteristics. For instance, the entry in the upper left-hand cell, which is 18, is the number of individuals who are schizophrenic and who have a positive test. The row and column totals are given at the margins and are called the marginal totals. The grand total is given in the lower right-hand corner. $P(A)$ is thus found by dividing the marginal total for A, which is 20, by the grand total, which is 2000, yielding a result of $P(A) = .01$. (The calculations of probability use the relative frequency interpretation of probability, which is acceptable given the large N.) The conditional probabilities can also be found from this figure. For instance, to calculate $P(B|A)$, one is limited to those subjects in column A; thus, $P(B|A) = {}^{18}\!/_{20} = .90$. The other probabilities can be verified likewise. The desired probability $P(A|B)$ is clearly ${}^{18}\!/_{117} = .154$.

Schizophrenic

	A (Schizophrenic)	$\overline{A}$ (Normal)	

Blood Test

B (Positive)

18
Note:
$P(B|A) = 18/20 = .90$
$P(A|B) = 18/117 = .154$

99
Note:
$P(B|\overline{A}) = 99/1980 = .05$

117
Note:
$P(B) = 117/2000 = .0585$

$\overline{B}$ (Negative)

2

1881

1883
Note:
$P(\overline{B}) = 1883/2000 = .9415$

20
Note:
$P(A) = 20/2000 = .01$

1980
Note:
$P(\overline{A}) = 1980/2000 = .99$

Total = 2000

FIGURE 1.2   Table for Bayes' Rule

NOTES AND SUPPLEMENTARY READINGS

Feller, W. (1968). *An introduction to probability theory and its applications.* Vol. I (3rd ed.). New York: Wiley.

Feller's presentation of probability theory may be seen as somewhat mathematical for the uninitiated, but it is commonly viewed as the bible of mathematically based probability theory.

Hays, W. L. (1981). *Statistics* (3rd ed.). New York: Holt, Rinehart and Winston.

This is a very comprehensive statistics textbook that has had great durability over the years. Many statisticians can trace their early training to Hays' text in one form or another and for most it remains on their shelves as a valued reference. Some may find the presentation a bit difficult to read but it is thorough. Hays' presentation on probability and set theory may be of definite interest to the reader seeking further information.

PROBLEMS

1. Given the sets $A = \{2, 4, 6, 8, 10\}$, $B = \{1, 3, 5, 7, 9\}$, and $C = \{1, 2, 5\}$
   a. Find $A \cup B$.
   b. Find $A \cap C$.
   c. Find $A \cap B$.

13

    d. List all of the subsets of C.

    e. If the universal set is the set $\{x|x$ is an integer and $1 \leq x \leq 10\}$, find $\overline{A}$.

2. Given the sets $A = \{a, e, i, o, u\}$, $B = \{b, c, d, f, g, h, j, k, l, m, n, p, q, r, s, t, v, w, x, y, z\}$ and $C = \{a, b, c\}$

    a. Find $A \cup C$.

    b. Find $A \cap C$.

    c. Find $A \cap B$.

    d. List all of the subsets of C.

    e. If the universal set is the set of lower-case letters of the English alphabet, find $\overline{B}$.

3. Find the probabilities of the following events:

    a. A "head" on the toss of a coin.

    b. A "1" on a roll of a die.

    c. A "1" or a "2" on a roll of a die.

    d. A "king" on a random draw from a standard deck of cards.

    e. An even numbered card or a heart on a random draw from a standard deck of cards.

    f. A "10," given the card drawn is an even numbered card, on a random draw from a standard deck of cards.

4. Find the probabilities of the following events:

    a. A "tail" on the toss of a coin.

    b. A "6" on the roll of a die.

    c. An even number on the roll of a die.

    d. A face card or a heart on a random draw from a standard deck of cards.

    e. A "jack," given the card drawn is a face card, on a random draw from a standard deck of cards.

5. Suppose that the following six objects are in a box and the probability of withdrawing any one of the six objects is 1/6 (i.e., equally likely events):

Object	Shape	Color	Material
1	Cube	Blue	Metal
2	Sphere	Blue	Metal
3	Cube	Red	Plastic
4	Sphere	White	Metal
5	Cube	Blue	Metal
6	Pyramid	White	Plastic

    a. Find the probability of withdrawing an object that is (i) a cube, (ii) a blue cube, (iii) a blue plastic cube, (iv) plastic.

    b. What is the probability of withdrawing a cube given that the object withdrawn is white?

c. If A is the event that a cube is withdrawn and B is the event that a blue object is withdrawn, are the events A and B independent?

d. If A is the event that a cube is withdrawn and C is the event that a plastic object is withdrawn, are the events A and C independent?

6. Suppose that the following eight people belong to an organization and the probability of selecting any of the eight people to be president is ⅛ (i.e., equally likely events):

Person	Gender	Age	Hair Color
1	Male	32	Blond
2	Female	35	Black
3	Female	35	Blond
4	Female	32	Red
5	Male	32	Brown
6	Male	37	Brown
7	Female	32	Black
8	Male	37	Brown

a. Find the probability of selecting a president who (i) is female, (ii) is a 35-year-old female, (iii) is a 35-year-old female with blond hair, (iv) has brown hair.

b. What is the probability that the person selected to be president is a male given that the person has brown hair?

c. If A is the event that a female is selected to be president and B is the event that the person selected is 32-years-old, are the events A and B independent?

d. If A is the event that a female is selected to be president and C is the event that the person has blond hair, are the events A and C independent?

7. Factory Forgery produces widgets. The probability that any widget is defective is .05. The manager of Factory Forgery wishes to implement a test to identify defective widgets. The probability that the test is positive given a widget is defective is .99; the probability that the test was positive given the widget was not defective is .03. The manager wishes to know the probability that a widget randomly selected from the assembly line is defective given that the test of this widget is positive. Calculate this probability.

8. Put yourself in the place of a judge at a murder trial. Suppose that you know the probability that the suspect's fingerprints are found on the murder weapon given the suspect is the murderer is .90. However, the probability that the suspect's fingerprints are found on the murder weapon given the suspect is not the murderer is .05. If, in general, 5 out of every 1000 people are murderers, what is the probability that a suspect is guilty of murder, given his or her fingerprints are found on the weapon?

2 RANDOM VARIABLES AND OBTAINED DATA

Often in the social sciences it is assumed that some underlying, unobservable process is occurring. Although that process cannot be observed directly, the social scientist's task is to make as many inferences as possible about that process. An experiment is designed and conducted, data are obtained and analyzed statistically, and inferences are made. If the experiment was well designed and conducted and the data appropriately analyzed, then knowledge of some aspect of the social sciences might be furthered.

When an experiment involves the tossing of a die, not much excitement is generated (unless, of course, a wager was placed on the outcome) because the probabilities of obtaining various outcomes are well known. However, if the experiment involves determining whether some drug improved the memory of brain damaged individuals, the outcome is very interesting and extremely important. The probability that the drug will facilitate memory is unknown; nevertheless, a well-designed and executed study will allow the social scientist an opportunity to make inferences about the drug's effectiveness.

In this chapter underlying processes and obtaining data from those processes are explained further, although the relation between the underlying process and obtained data are not formalized until Chapter 4. Underlying processes are approached through random variables; a random variable is a convenient statistical concept for characterizing the "real" world.

Random Variables

2.1   Definition of Random Variables

Underlying processes can be better understood by discussing random variables. Typically, numerical values are assigned to the outcomes of an experiment. If each and every elementary event in $\mathscr{S}$ is assigned a numerical value x, then the rule that assigns the values is said to be a *random variable*. Random variables usually are denoted by upper-case letters toward the end of the alphabet (e.g., X, Y, and Z). In general, the values that a random variable can assume are denoted by the corresponding lower-case letter (e.g., the random variable X assumes the value x). Rather specific values that the random variable can assume frequently are denoted by lower-case letters at the beginning of the alphabet. Although it might seem that the a's, b's, and c's will be confused with the x's, y's and z's, it becomes clear as the discussion progresses.

When a die is rolled, there are six elementary events. The simplest way in which to assign the numbers to these outcomes is to "count" the number of spots that appear. Thus, the random variable X could be defined as the number of spots that appear on a roll of a die. If a "1" appears, then $x = 1$, if a "2" appears, then $x = 2$, and so on. However, more elaborate schemes are possible and useful. For instance, one might want to define a random variable Y as the square of the number of spots that appear (i.e., $Y = X^2$). If a "1" appeared, then $y = x^2 = 1$, if a "2" appeared, then $y = x^2 = 4$, and so on.

Random variables can be classified into two types, discrete random variables and continuous random variables. The notion of continuous and discrete random variables depends on mathematical concepts that are beyond the scope of this book and are not essential in order to understand their properties. Hence, only limited definitions are given. A *continuous random variable* is a random variable that takes on an infinite number of values; furthermore, between any two specified values the random variable assumes a value. In a reaction time study the experimenter studies the time that it takes for the subject to press a button after a stimulus is presented. If the random variable is defined as the time (measured in seconds) until a subject presses the button, then there are an infinite number of values that the random variable can assume. In addition, between any two specified values, say, 2 seconds and 3 seconds, the random variable can assume another value, say, 2.5 seconds. Between 2 and 2.5 seconds the random variable can assume the value 2.25. Intuitively, all the "wholes" are filled in and the random variable becomes continuous.

Discrete random variables are of two subtypes, finite and countably infinite. The *finite discrete random variable* assumes only a finite number of values. For instance, if X is the number of spots that appear when a die is rolled, then X is a finite discrete random variable, because it can take on only six values. A *countably infinite random variable* assumes an infinite number of values but may not assume a value between two other values. Suppose that an experiment involves tossing a coin. If the random variable Y is defined as the number of tosses that occur before a head appears, then Y is a countably infinite discrete random variable. Y can assume the values 1 through infinity, but cannot assume a number between, say, 5 and 6. The "wholes" are not filled in. Although countably infinite random variables are of interest to statisticians and are useful to understand some phenomena in the social sciences, they will not be discussed further here.

2.2   Probability Distribution Functions of Discrete Random Variables

Because each value of a random variable is associated with an elementary event, probabilities can be associated with the various values that the random variable can assume. For instance, if X is defined as the number of spots that appear when a die is rolled, the probability that X equals 1 is $\frac{1}{6}$. Notationally, we write

$$P(X = 1) = \frac{1}{6}$$

Or, in general, $P(X = a)$ indicates the probability that the random variable X assumes the value a. More complex probabilities can be formed such as $P(a \le X \le b)$.

The possible values of X and their associated probabilities form a *theoretical probability distribution function* or simply a *probability distribution* or *distribution*. The probability distribution for Example 2.1 is given in the first two columns of Table 2.2. The probability distribution often has attached the word "function." *Function* implies that for each value of X there is one and only one probability associated with it.

The sum of the probabilities for the random variable X in Table 2.2 is equal to 1. That is, $P(X = 2) + P(X = 3) + \cdots + P(X = 12) = 1$. This should be intuitively clear if one remembers that a random variable assumes a value for each and every elementary event, that a probability is associated with each possible value of X, and that $P(\mathcal{S}) = 1$ (Property 2 of Section 1.4, p. 6).

Often it is useful to find the probability that a random variable is equal to or less than a specified value. For instance, a school psychologist might want to know the probability that an IQ score would be less than or equal

EXAMPLE 2.1   To illustrate the probability of a discrete random variable, the following example will be useful. Suppose that an experiment involves tossing two dice and that the random variable X is the sum of the spots that appear on the dice. X can take on the values 2, 3, 4, . . . , 12. To determine the probabilities of these 11 values, Table 2.1 is useful. Clearly, there are 36 elementary events although the random variable X can only assume 11 values. $P(X = 2) = \frac{1}{36}$ since for only one elementary event (viz., the first die is a "1" and the second die is a "1") does X assume the value of 2. $P(X = 9) = \frac{4}{36}$ since for four elementary events X assumes the value of 9.

TABLE 2.1

Sum of Spots on Two Dice

		First Die					
		1	2	3	4	5	6
	1	2	3	4	5	6	7
	2	3	4	5	6	7	8
Second Die	3	4	5	6	7	8	9
	4	5	6	7	8	9	10
	5	6	7	8	9	10	11
	6	7	8	9	10	11	12

TABLE 2.2

Probability Distribution and Cumulative Probability Distribution for X Where X Is the "Sum of the Spots on Two Dice"

x	$P(X = x)$	$F(x)$
2	$\frac{1}{36}$	$\frac{1}{36}$
3	$\frac{2}{36}$	$\frac{3}{36}$
4	$\frac{3}{36}$	$\frac{6}{36}$
5	$\frac{4}{36}$	$\frac{10}{36}$
6	$\frac{5}{36}$	$\frac{15}{36}$
7	$\frac{6}{36}$	$\frac{21}{36}$
8	$\frac{5}{36}$	$\frac{26}{36}$
9	$\frac{4}{36}$	$\frac{30}{36}$
10	$\frac{3}{36}$	$\frac{33}{36}$
11	$\frac{2}{36}$	$\frac{35}{36}$
12	$\frac{1}{36}$	$\frac{36}{36}$
	$\frac{36}{36} = 1$	

to 90. Because this idea will be used frequently, the following notation is used.

$$P(X \leq a) = F(a) \tag{2.1}$$

Descriptively, $F(a)$ is the probability that the random variable X is less than or equal to a and is called the *cumulative probability* of a. It should be noticed that when this notation is used, the name of the random variable (e.g., X) does not appear but should be clear by the context. In the continuing Example 2.1 (p. 19)

$$F(4) = P(X \leq 4) = P(X = 2) + P(X = 3) + P(X = 4) = 6/36$$

Because each value of the random variable is associated with a cumulative probability, another function can be defined. The *cumulative probability function* associates with each value x of the random variable a cumulative probability $F(x)$. The cumulative probabilities for the continuing example (2.1, p. 19) are given in the third column of Table 2.2 (p. 19).

Given the notation for discrete random variables, the continuing example (2.1, p. 19), and the general properties of probability, some problems involving notation can be solved:

$$P(7 < X \leq 10) = P(X = 8) + P(X = 9) + P(X = 10) = 12/36$$

$$\begin{aligned} F(10) &= P(X \leq 10) = 1 - P(X > 10) \\ &= 1 - P(X = 11) - P(X = 12) \\ &= 33/36 \quad \text{(recall Property 4 Section 1.4, p. 7)} \end{aligned}$$

$$P(X = 2\tfrac{1}{2}) = 0 \quad \text{(recall that } X \text{ is a discrete random variable)}$$

$$F(2\tfrac{1}{2}) = P(X \leq 2\tfrac{1}{2}) = P(X = 2) = 1/36 \quad \text{(note this example!)}$$

Often it will be useful to graph the probability distribution and the cumulative distribution. Figure 2.1 is a histogram for our continuing example (2.1, p. 19). Each "bar" extends to the height that indicates the probability

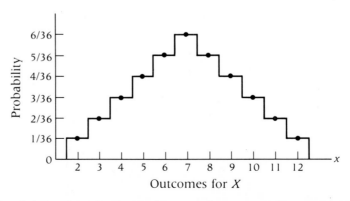

FIGURE 2.1   Graph of Probability Function for X Where X Is the "Sum of the Spots on Two Dice"

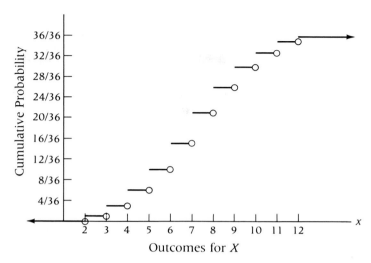

FIGURE 2.2   Graph of Cumulative Distribution Function for X Where X Is the "Sum of the Spots on Two Dice"

for each value of the random variable found on the horizontal axis. Although probabilities for values of X are zero for values other than 2, 3, ..., 12, the "bar" extends one-half unit to each side of these values. The "bars" are extended in this fashion so that an important property is present—**the area underneath the histogram is equal to 1.**

The graph of the cumulative distribution is a little more complex. The graph for our continuing example (2.1, p. 19), which is found in Figure 2.2, probably needs some explanation. First, notice that $F(x)$ for values of X less than 2 is equal to zero. However, $F(2) = \frac{1}{36}$. So the graph "jumps" from zero to $\frac{1}{36}$ at the point 2 on the horizontal axis. Similarly, $F(2) = F(2\frac{1}{2}) = F(2\frac{3}{4}) = F(2\frac{7}{8}) = \frac{1}{36}$. At 3 on the horizontal axis, the graph "jumps" to $\frac{3}{36}$. The open circles indicate that at that point on the horizontal axis, the graph has "jumped" to the next step. For obvious reasons graphs of this type are called step functions and characterize the graphs of cumulative distribution functions of discrete random variables.

2.3   Probability Distribution Functions of Continuous Random Variables

When the notion of the probability of obtaining a particular value of a random variable is extended to continuous random variables, a problem is encountered. Recall that the probability of an elementary event is equal to $1/N$, where N is the number of elementary events. In the continuous case N is infinite, so the probability of obtaining any specified outcome is zero.

(Several rigorous mathematical ideas have been glossed over here, but the general principle is all that is needed.) An amusing example will illustrate the point. Suppose that the instructor of a class informs the class that he or she is thinking of a number between 1 and 10 (including 1 and 10); if a student guesses the number, he or she can have an "A" without taking the examinations. However, if the guess is incorrect, the student will have to attend special tutorial groups. Thinking that the payoff is great and the loss minimal (and maybe beneficial), and that the probability of obtaining the correct number is $\frac{1}{10}$, the student decides to guess and selects the number 3. Unfortunately, the student made the assumption that the instructor had limited the numbers to the whole numbers (i.e., 1, 2, . . .). When the instructor said a number between 1 and 10, he meant *all* the numbers (including fractions and numbers like π, that is, the *real* numbers). The number that the instructor had selected was 2.37521. Because there are an infinite number of real numbers between 1 and 10, the probability of selecting a given number is zero, much to the instructor's delight. All is not lost, however, because probabilities for values of continuous random variables still can be assigned, as illustrated in Example 2.2.

Although the probability of any particular value of a continuous random variable is zero, a probability distribution function for a continuous random variable, often referred to as a *probability density function*, can be formed. The concept of a probability density function is approached backward in that first the graph of a probability density function is discussed.

The graph of the probability density function is similar to that of the graph of the probability distribution function of a discrete random variable. On the horizontal axis are the possible values for the random variable, but probabilities are not on the vertical axis. The area under the entire graph is required to equal 1, since $P(\mathscr{S}) = 1$. The graph is formed so that the area under any portion of the graph is equal to the probability for obtaining values of the random variable in that interval. The graph of the probability density function for the spinner, which appears in Figure 2.4, illustrates this seemingly confusing statement. The area under the graph between 0 and 1 on the horizontal axis is $\frac{1}{2}$, which is also $P(0 \leq X \leq 1)$. Similarly, the area under the graph between 1 and 2 is $\frac{1}{8}$, which is $P(1 \leq X \leq 2)$. This relationship between the area and probability is not limited to integers. For instance, the area under the graph between 3 and $3\frac{1}{4}$ is equal to the probability that X takes on a value in this region (viz., $\frac{1}{16}$). Suppose that the previous interval (3 to $3\frac{1}{4}$) is allowed to become smaller (say, 3 to $3\frac{1}{8}$) and smaller (say, 3 to $3\frac{1}{16}$), the height of the graph in this region will always be $\frac{1}{4}$. When the interval becomes infinitesimally small, the following statement makes sense: Even though the probability that a continuous random variable assumes a particular value is zero, each value x of the random variable has a height, denoted by $f(x)$ assigned to it (as depicted in the graph). This assignment is what is meant by the probability density function. The notation $f(x)$ is the typical notation for a function. For example, $f(2) = \frac{1}{8}$, $f(3\frac{1}{2}) = \frac{2}{8}$. However, remember that for a continuous random variable, probabilities

EXAMPLE 2.2   A spinner is arranged as shown in Figure 2.3. That is, the real numbers between 0 and 1 are evenly spaced in the right semicircle as shown. As well, the real numbers between 1 and 2 are evenly spaced in an eighth-circle, the real numbers between 2 and 3 are evenly spaced in an eighth-circle, and the real numbers between 3 and 4 are evenly spaced in a quarter-circle, as shown. The experiment consists of flicking the spinner so that it points to some random place on the circumference of the circle. Let the random variable X be the real number assigned (according to the preceding scheme) to the place where the spinner lands.

As indicated earlier, the probability that the random variable assumes any particular value is zero; for example, $P(X = 2\frac{1}{2}) = 0$. However, clearly, the probability of obtaining a value between 0 and 1 is $\frac{1}{2}$; that is, $P(0 \leq X \leq 1) = \frac{1}{2}$. Similarly, $P(3 \leq X \leq 4) = \frac{1}{4}$. Also, $P(3 \leq X \leq 3\frac{1}{2}) = \frac{1}{8}$.

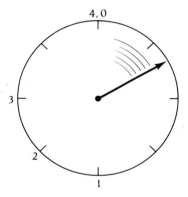

FIGURE 2.3   Spinner for Example 2.2

are only assigned by considering the area under the graph. That is, the probability that the continuous random variable assumes values in an interval is equal to the area under the graph for that interval. For instance, $P(1 \leq X \leq 3) = \frac{1}{4}$. However, note that $P(X = 1) = 0$ and that $P(X = 3) = 0$. So, unlike the case for the discrete random variable, the "or equal to" is irrelevant. That is, $P(1 \leq X \leq 3) = P(1 < X < 3)$.

The shape of the graph in Figure 2.4 is rather atypical of probability density functions that interest statisticians. The graph was horizontal between the values 0 and 1 because the real numbers were evenly spaced on the spinner between 0 and 1.

The cumulative probability distribution for a continuous random vari-

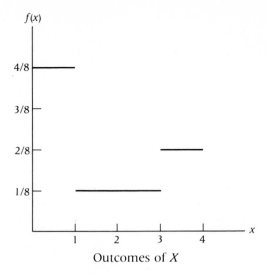

FIGURE 2.4   Probability Density Function for Spinner

able is formed in the same manner as it was for a discrete random variable. The graph of the cumulative probability distribution for the spinner can easily be graphed by calculating a few values: $F(0) = 0$, $F(\frac{1}{2}) = \frac{1}{4}$, $F(1) = \frac{1}{2}$, $F(2) = \frac{5}{8}$, $F(3) = \frac{3}{4}$, and $F(4) = 1$. The entire distribution is shown in Figure 2.5.

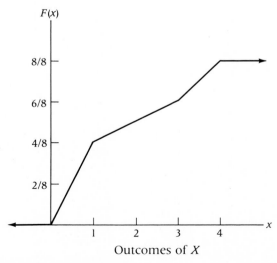

FIGURE 2.5   Cumulative Distribution Function for Spinner

2.4   A Very Special Probability Density Function— The Standard Normal

Although the basis for discussing particular probability density functions has not been built, the standard normal distribution is so very important to statistics that it is discussed here. If the random variable Z has a standard normal distribution, then the formula for the probability density function is

$$f(z) = \frac{1}{\sqrt{2\pi}} e^{-z^2/2}$$

(2.2)

where π is the familiar(?) symbol from high school geometry (the circumference divided by the diameter, which is approximately equal to 3.1416), and e is a number that is approximately equal to 2.718. This is the first instance where the probability density function has been defined mathematically; previously, the random variable was defined descriptively. Nevertheless, the principle is the same. It should be noted that the letter Z typically is reserved for any random variable that has been standardized (a term that has not been defined yet).

It is not important to memorize, or, for that matter, to recognize the formula for the probability density function of the standard normal distribution. However, the standard normal has several important properties that are discussed here and used throughout this book.

The graphs of the probability density function and the cumulative density function of the standard normal are presented in Figure 2.6. The graph of the probability density function is formed simply by substituting values of z into Equation (2.2) (p. 25). For instance, if $z = 1$, then $f(z) = (1/\sqrt{2\pi})e^{-1/2} = .242$. The graph of the probability density function should look familiar; it is sometimes referred to as the "bell-shaped distribution," although distributions other than normal distributions also look like a bell. The following characteristics of the standard normal distribution are important. First, because the standard normal is a probability density function, the area under the curve is equal to 1. Second, although the "tails" of the graph approach the horizontal axis as they are extended in each direction, they do not touch the axis. Third, the standard normal distribution is symmetric around zero and reaches a maximum at zero.

Given either the graph of the probability density function or the cumulative density function of the standard normal distribution, the probabilities of obtaining values in certain regions are quite apparent:

$$P(-\infty < Z < \infty) = 1$$
$$P(Z < 0) = F(0) = .5$$

Remember that the "or equal to" sign is irrelevant in the case of a continuous random variable. Because the standard normal is so vitally important, tables of probabilities have been derived for other intervals. Before learning to use

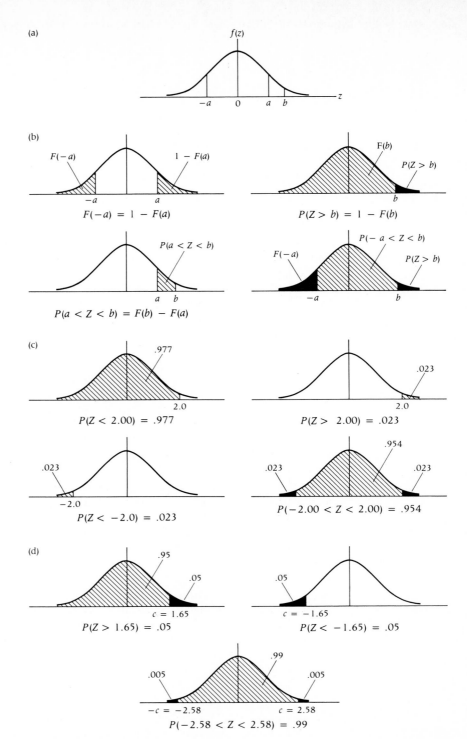

FIGURE 2.6   Standard Normal Distribution with Examples

a standard normal table, several observations are useful. If a and b are positive numbers, as illustrated in panel (b) of Figure 2.6, then

$$(i)\ \ F(-a) = 1 - F(a) \tag{2.3}$$
$$(ii)\ \ P(Z > b) = 1 - F(b) \tag{2.4}$$
$$(iii)\ \ P(a < Z < b) = F(b) - F(a) \tag{2.5}$$
$$(iv)\ \ P(-a < Z < b) = F(b) - F(-a) = F(b) - (1 - F(a))$$
$$= F(b) + F(a) - 1 = 1 - F(-a) - P(Z > b) \tag{2.6}$$

From these few examples it should be apparent that the probability that the random variable Z falls in some interval can be written in terms of the cumulative distribution function for positive values. Therefore, probabilities for the standard normal distribution are typically tabled as cumulative standard normal probabilities for values z greater than or equal to zero, as shown in Table B.1 of the Appendix (p. 449). Using Table B.1 we can calculate the probability of any interval, as illustrated by the following examples:

$$(i)\ \ P(Z \leq 2.00) = F(2.00) = .977$$
$$(ii)\ \ P(Z > 2.00) = 1 - F(2.00) = .023$$
$$(iii)\ \ P(Z < -2.00) = 1 - F(2.00) = 1 - .977 = .023$$
$$(iv)\ \ P(-2.00 < Z < 2.00) = F(2.00) - F(-2.00)$$
$$= F(2.00) - (1 - F(2.00))$$
$$= 2F(2.00) - 1 = 1.954 - 1 = .954$$

As the student becomes familiar with the standard normal distribution, much of the algebra can be omitted in favor of a graphical approach, as illustrated in panel (c) of Figure 2.6. For instance, in the last example the area under the graph is the total area minus two tails that have equal areas; that is, $1 - 2$(area in a tail for $Z > 2$), which is $1 - 2(.023) = .954$.

Often in the course of significance testing, tables of cumulative distributions will be used in the "reverse direction." That is, an area under the curve will be known and the value or values that define the region will be determined by reference to the tables of the cumulative distribution. For example, situations will arise where the value of c such that $P(Z > c) = .05$ needs to be determined (c need not be positive). The following list presents several examples of this type of problem that are solved algebraically (and, of course, by reference to Table B.1); panels (d) of Figure 2.6 illustrates the solutions graphically.

(i) $P(Z > c) = .05$ implies that $F(c) = .95$, and looking under the column $F(z)$ for .95, we see that, to the nearest hundredths, $c = 1.65$.

(ii) $P(Z < c) = .05$ implies that $P(Z > -c) = .05$ and further implies that $F(-c) = .95$, and therefore $-c = 1.65$ and $c = -1.65$. (The easier way to solve this is to note the symmetry with case (*i*), as illustrated in Figure 2.6.)

(*iii*) $P(-c < Z < c) = .99$, which by using symmetry implies that $P(Z > c) = .005$ and further implies that $F(c) = .995$, and therefore $c = 2.58$ and $-c = -2.58$.

2.5   Transformations of Random Variables

Frequently, occasions will arise when it is necessary to transform random variables in some manner. For example, the achievement of elementary schoolchildren might be measured in two different ways and it would be important to transform the achievement variables so that the results using the two different methods would be comparable. Many times during the development of statistical techniques, the need to transform random variables arises. In this section an introduction to the topic of transformations is presented and the discussion is continued in succeeding sections where applications of this topic are made to the development of related topics.

Generally, a *transformation* of a random variable is a function of the variable so that values of the original random variables are mapped onto values of the transformed random variable. That is

$$y = f(x)$$

where x represents values of the original random variable X and y represents values of the transformed random variable Y. Each of the following are examples of transformations:

$$y = 2x + 3$$
$$y = 4x^2 - 2x + 1$$
$$y = \log x$$

Although a transformation is a function rule that maps each value of one random variable onto a value of another transformed random variable, the transformation is often written in terms of capital letters, so that it is permissible to write $Y = 2X + 3$ rather than $y = 2x + 3$. In either case it should be realized that the values of X are multiplied by 2 and added to 3 to yield the corresponding values of the transformed random variable. For example, if $x = 5$, $y = 2(5) + 3 = 13$.

One type of transformation, the *linear transformation*, is particularly important in subsequent discussions. A linear transformation is a transformation of the type

$$Y = BX + A \qquad (2.7)$$

where B and A are constants and B is not equal to zero.

Obtained Data

As was mentioned previously, random variables characterize some underlying process, which usually is not observable directly. To make inferences about the underlying process, data are obtained from measurement of the behavior (in the broad sense of the word) of subjects. The following sections will discuss the measurement and the distributions that occur from repeated measurement.

2.6 Measurement

Measurement is the assignment of numbers to outcomes according to some rule. The numbers that result from measurement are the "raw material" that are subjected to statistical analysis and that lead to some conclusions about our world. To lead to valid conclusions, however, the numbers must reflect reality and that is where the rule comes into play. The rule used by the researcher may or may not lead to numbers that indicate the degree to which an object possesses a certain property. Suppose that a researcher wishes to measure the intelligence of several children. There are several ways (i.e., rules) in which this could be accomplished. The researcher might administer an individual test of intelligence; the IQ score for each child that resulted would be the number assigned to that child. Another means to assign numbers related to intelligence would be to look up the child's IQ score in his or her cumulative file in the school records. A third way to assign numbers would be for the child's teacher to rate the child's intelligence on a 10-point scale. A fourth means would be for the student to rate his or her own intelligence on a 10-point scale. Clearly, the rule used to assign the numbers is important. Better rules result in numbers that more closely reflect the characteristic or property being measured.

Determining the quality of the rule that is used is not a simple task. When the characteristic being measured has physical manifestations, such as the gender of a subject, then assigning numbers to the objects is relatively simple (e.g., $0 =$ male, $1 =$ female). However, when the characteristic is more psychological and less physical, then measurement is more problematic. Consider aggressiveness. We cannot take a biopsy to determine the degree to which a subject is aggressive, so we resort to more indirect methods, such as observing the subject and recording the frequency of aggressive behaviors, asking significant others to rate the aggressiveness of the subject, or having the subject rate his or her own aggressiveness. Apparently, because we don't have a purely objective or direct means to measure aggressiveness, it would be difficult to determine the degree to which the indirect methods reflect reality. Although making this determination is difficult, there are many ways to estimate the "goodness" of measurements, and an entire discipline called measurement is devoted to this purpose. The concepts "reliability" and "validity'' of measurements, which might be familiar to students, are central to a discussion of this topic but are beyond the scope of this book. Anyone conducting research in education or psychology should have a good understanding of measurement as well as of statistics. However, it should be noted that the boundary between measurement and statistics is not clear and there are many issues, some of which are unresolved, that are raised by the relation between measurement and statistics.

For our purposes a few considerations about measurement are needed for our discussion of statistics. The numbers that result from a measurement rule are often called observations, scores, or measurements. Further, measurements themselves have properties and are classified on the basis of these properties. Generally, there are four types of measurements: nominal or categorical, ordinal, interval, and ratio. We say that a nominal scale leads to nominal measurements, an ordinal scale to ordinal measurements, and so forth. The following example and the discussion describe the four scales.

EXAMPLE 2.3

Suppose that a 100-meter race is run, with the results summarized in Table 2.3. The results of the race can be characterized in four ways: (1) the elapsed time for the race, (2) the time behind the winner of the race, (3) the place of finish, and (4) whether or not the participant qualified for the finals.

TABLE 2.3

Results of 100-Meter Race

Name	Elapsed Time	Time Behind Winner	Place of Finish	Qualification Status
John	12.2	0.0	1	Qualifies
Sally	12.5	0.3	2	Qualifies
Susan	15.0	2.8	3	Qualifies
Peter	17.0	4.8	4	Does not qualify
Josh	24.4	12.2	5	Does not qualify

Each of the ways of characterizing the outcome of the race in Example 2.3 corresponds to a measurement scale. The crudest measurement scale is the *nominal or categorical scale*. In a nominal scale an outcome is classified into a category so that each category is different. In the example outcomes are classified as ''qualifies'' or ''does not qualify.'' Although this categorization could imply that one category is "better" than another (qualifiers are faster than nonqualifiers), nominal measurements only indicate that outcomes in two different categories are different. Other examples of nominal scales include gender (male vs. female), treatment group (treatment *A* vs. treatment *B* vs. control), and temperature (hot vs. cold). To form a nominal scale, each outcome must be classified into one and only one category; thus, a nominal scale consists of a set of mutually exclusive and exhaustive categories. When the outcomes refer to characteristics of subjects, the set of mutually exclusive and exhaustive categories is called an *attribute*. For example, gender is an attribute of organisms.

At this point the reader might object that the nominal scale does not reflect measurement because numbers were not assigned. However, numbers can be assigned to categories; for instance, a 0 for males and a 1 for females. It should be noted that 1 does not signify that the females possess one of some quality while males do not possess the quality at all. Rather, it means that on some basis males are different from females (see Section 12.7, p. 398 for a more complete discussion of the assignment of numbers to categorical measurements).

An *ordinal scale* gives a finer measurement than the nominal scale in that it describes direction as well as difference. In the 100-meter race example the place of finish is measured on an ordinal scale. Although we know that the third place finisher finished before the fourth place finisher (direction), we do not know how much before. One's position in the graduating class is another example of an ordinal scale. The numbers assigned to outcomes for ordinal scales are often called *ranks* and outcomes arranged by ranks are said to be *rank ordered*.

The next finer measurement is the *interval scale*. The interval scale describes the difference, direction, and amount of the difference. The time behind the winner constitutes an interval scale. Temperature on the Fahrenheit scale is a good example of an interval scale. If the temperature today was 80° and yesterday it was 70°, we know that the temperatures for the two days were not the same (difference); today was hotter (direction) by 10° (amount of difference). Intelligence test scores (IQ scores) form an interval scale in that it makes sense to say that one's IQ score is 10 points above another's IQ score. Furthermore, the difference between IQ scores of 120 and 110 is equivalent to the difference of IQ scores of 90 and 80.

The most sophisticated measurement scale is the *ratio scale*. Ratio scales characterize the difference, direction, amount of the difference, and have an absolute zero. For the elapsed time there is an absolute zero (zero seconds); therefore, elapsed time constitutes a ratio scale. Because absolute zero (no temperature at all) does not correspond to 0° on the Fahrenheit scale, the Fahrenheit scale is not a ratio scale. However, in the Kelvin scale the absolute zero corresponds to 0°, thus constituting a ratio scale. In the social sciences reaction time is an example of a ratio scale.

The presence of an absolute zero permits meaningful statements to be made about ratios of scores. The ratio of Josh's elapsed time to John's elapsed time was 2.0, which means that John ran twice as fast as Josh. On the Kelvin scale a temperature of 4000° represents four times the energy represented by a temperature of 1000°. However, because intelligence test scores do not form a ratio scale, it is incorrect to say that someone with an intelligence test score of 160 is twice as smart as someone with a score of 80.

It should be realized that the scores on a particular scale have all the properties of scores on the lower-level scales. Ratio scale scores have the properties of interval scale scores, interval scale scores have the properties of ordinal scales scores, and so forth. It is not uncommon to express scores on a lower-level scale so that a particular statistical technique can be used. In several instances in this book scores on an interval level will be converted to ranks, which, when there are no ties in the scores, is accomplished by simply rank ordering the scores. This is illustrated in the following scores, where lower scores are given lower ranks:

Scores	Ranks
5	2
9	5
8	4
6	3
2	1
10	6

When there are ties in the scores, assignment of ranks becomes slightly more complicated. In this text ranks will be assigned to tied scores by calculating the average of the ranks for the tied scores, as illustrated for the following scores:

Scores	Ranks
8	8
7	6
6	4
7	6
5	2.5
11	9
7	6
5	2.5
2	1

Ranks determined for tied scores in this way are called *midranks*.

Because interval and ratio scales refer to magnitude, we refer to these scales as *quantitative* scales. Some authors also include the ordinal scale as a quantitative scale.

2.7   Obtaining Data from a Process Characterized by a Discrete Random Variable

Suppose that the experiment described in Section 2.2 (p. 18) (rolling two dice) was undertaken and that scores for this experiment consisted of the sum of the spots on the two dice. Although the probabilities of the possible outcomes of this experiment are easily determined, it is possible to estimate those probabilities by repeated trials of the experiment. Table 2.4 represents the outcome of this experiment replicated 36 times. In the context of obtained data the symbol f denotes the *frequency* of each outcome, which is simply the number of times that the outcome occurs. Although related to the probability density function, f is the frequency with which an outcome occurs in repeated trials of an experiment, whereas $f(a)$ is the height of the graph associated with the value a of a random variable. The possible outcomes and the associated frequencies form a *frequency distribution*. The relative frequency of the outcomes (f/number of trials), which is denoted by relative f, is also found in Table 2.4.

The law that connects the frequency distribution to the probability distribution is, recall, the Law of Large Numbers (see Section 1.3, p. 4). Because 36 replications of this experiment are relatively few, the discrepancy between the relative frequencies (Table 2.4) and the theoretical probabilities (Table 2.2, p. 19) is relatively large.

The histogram for the results of the 36 trials of this experiment is presented in Figure 2.7. This figure is constructed similarly to a probability distribution (see Figure 2.1, p. 20), except that on the vertical axis there are frequencies or relative frequencies rather than probabilities.

TABLE 2.4

Frequency Distribution for "Two-Dice Experiment" (36 Trials)

Outcome	f	Relative f
2	1	$1/36$
3	3	$3/36$
4	4	$4/36$
5	5	$5/36$
6	4	$4/36$
7	7	$7/36$
8	3	$3/36$
9	3	$3/36$
10	4	$4/36$
11	2	$2/36$
12	0	$0/36$

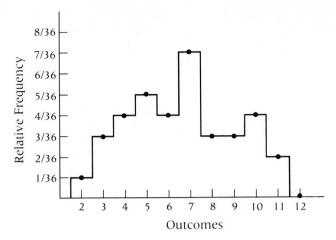

FIGURE 2.7   Graph of Relative Frequency Distribution for "Two-Dice Experiment"

2.8   Obtaining Data from a Process Characterized by a Continuous Random Variable

The process of obtaining data from a continuous random variable is similar to that of a discrete random variable. However, because the probability of any particular outcome is zero, frequencies for outcomes must be determined for intervals. For the example of the spinner (Example 2.2, p. 23), relative frequencies for an interval of length ½ might be found.

Somewhat more interesting is the experiment where a sample, say, of size 100, is drawn from a standard normal distribution. Suppose that intervals of size ¼ are selected and that the outcome is graphed as points drawn above the midpoints of the interval, as shown in Figure 2.8. Although this graph

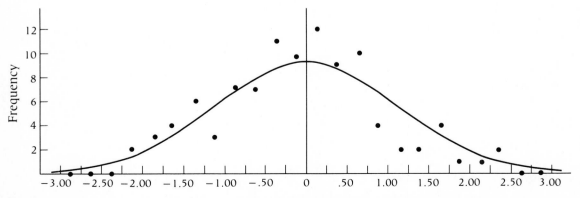

FIGURE 2.8   Graph of Frequency Distribution for Sample from Standard Normal Distribution (100 Trials)

consists of a number of discrete points, a "smooth" curve can be drawn through the points. If the intervals are small and the sample size is large, then the "smoothed" curve will resemble the theoretical standard normal distribution.

NOTES AND SUPPLEMENTARY READINGS

Hays, W. L. (1981). *Statistic* (3rd ed.). New York: Holt, Rinehart and Winston.

Comments pertaining to Hays' text in Chapter 1 are also relevant here. His coverage includes random variables and obtained data.

Kerlinger, F. N. (1986). *Foundations of behavioral research* (3rd ed.). New York: Holt, Rinehart and Winston.

This book contains a straightforward discussion of measurement as it relates to research and to statistics. Also, concise and clear discussions of many other issues related to research and statistical analysis, as well as conceptual discussions of several advanced techniques, are included in this text.

Nunnally, J. C. (1967). *Psychometric theory*. New York: McGraw-Hill.

This source contains a comprehensive discussion of classical topics in measurement and scaling.

Stevens, S. S. (1951). Mathematics, measurement, and psychophysics. In S. S. Stevens (Ed.), *Handbook of experimental psychology*, pp. 1–49.

An often-cited chapter delineating Stevens' view of problems pertaining to performing certain types of mathematical computations, and thereby certain analyses, on data of nominal and ordinal data scales. This classic seems to have prompted a debate that has continued to date with strong arguments on both sides. Evidence of the disagreement abounds, differing authors have different views, and one has only to read the *Psychological Bulletin* on a current basis to see that the controversy is alive and well.

PROBLEMS

1. Consider the following discrete probability distribution function:

x	$P(X = x)$
2	$1/10$
3	$2/10$
4	$5/10$
5	$1/10$
6	$1/10$

    a. Draw a histogram of the probability distribution function.
    b. Draw a graph of the cumulative probability distribution function.
    c. Find the following probabilities:
       (*i*) $P(2 \leq X < 5)$
      (*ii*) $F(3)$
    (*iii*) $P(X > 4)$
    (*iv*) $1 - F(5)$
     (*v*) $P(X \leq 4)$

2. Consider the following discrete probability distribution function:

x	$P(X = x)$
4	$1/12$
5	$3/12$
6	$5/12$
7	$1/12$
8	$1/12$
9	$1/12$

    a. Draw a histogram of the probability distribution function.
    b. Draw a graph of the cumulative probability distribution function.
    c. Find the following probabilities:
       (*i*) $P(6 \leq X < 8)$
      (*ii*) $F(5)$
    (*iii*) $P(X > 6)$
    (*iv*) $1 - F(8)$
     (*v*) $P(X \leq 6)$

3. Consider the following spinner (with the same conditions as the spinner in Example 2.2, p. 23):

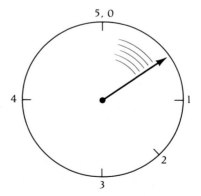

    a. Draw a graph of the probability density function for the spinner.
    b. Draw a graph of the cumulative probability distribution.

c. Find the following probabilities:
   (*i*) $P(X > 3)$
   (*ii*) $F(2)$
   (*iii*) $P(3 \leq X \leq 4)$
   (*iv*) $F(3.5)$

4. Consider the following spinner (with the same conditions as the spinner in Example 2.2, p. 23):

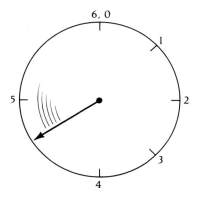

a. Draw a graph of the probability density function for the spinner.

b. Draw a graph of the cumulative probability distribution.

c. Find the following probabilities:
   (*i*) $P(X > 3)$.
   (*ii*) $F(2)$
   (*iii*) $P(3 \leq X \leq 4)$
   (*iv*) $F(3.5)$

5. Suppose that the random variable Z has a standard normal distribution.

a. Find the following probabilities:
   (*i*) $P(Z \leq 2.5)$
   (*ii*) $P(Z > 2.25)$
   (*iii*) $P(-1.00 < Z < 1.00)$
   (*iv*) $P(Z < -3.00)$
   (*v*) $P(Z > -1.50)$

b. Find the value of *a* such that
   (*i*) $P(Z < a) = .850$
   (*ii*) $P(Z > a) = .100$
   (*iii*) $P(Z < a) = .025$
   (*iv*) $P(-a < Z < a) = .900$

6. Suppose that the random variable Z has a standard normal distribution.

   a. Find the following probabilities:
      (*i*) $P(Z \leq 2.75)$
      (*ii*) $P(Z > 2.40)$
      (*iii*) $P(-1.50 < Z < 1.50)$
      (*iv*) $P(Z < -2.50)$
      (*v*) $P(Z > -1.00)$

   b. Find the value of a such that
      (*i*) $P(Z < a) = .750$
      (*ii*) $P(Z > a) = .050$
      (*iii*) $P(Z < a) = .05$
      (*iv*) $P(-a < Z < a) = .800$

3 MEASURES OF CENTRAL TENDENCY AND DISPERSION

I n response to a question about the probability distribution of a discrete random variable, one may, for example, list all possible values that the random variable can assume as well as the corresponding probabilities. Similarly, in response to an inquiry about the distribution of obtained data, one may present the frequency distribution. For instance, one may be interested in the intelligence tests scores of persons living in New York State. A frequency distribution for these intelligence tests scores would exceed the ability of anyone to make sense of the distribution without summarizing it in some way.

To characterize a distribution in one and only one way, most people would select the "average" as the measure that best characterizes a distribution. Although there are a number of ways to define average, all of them characterize the distribution by referring to the central tendency of the distribution. If the average intelligence test score for residents of New York was 103, then much is known about the intelligence test scores of these persons. Of course, much is also unknown. The residents of Maine might also have average intelligence tests score of 103 but have a very different distribution.

To characterize a distribution in two and only two ways, one would probably select the average and some measure of how the scores were spread out. The intelligence tests scores of residents of New York may be more "spread out" than the intelligence tests scores of the residents of Maine. Measures of the "spread" of scores are called measures of dispersion. Again, there is more than one such measure. Although other measures characterize a distribution (e.g., skewness and kurtosis, which will be mentioned in Section 3.16), measures of central tendency and dispersion are of paramount importance and are the primary focus of this chapter.

In the previous chapter we discussed the dichotomy of underlying and unobservable processes, characterized by a random variable, and data obtained from an underlying process. Thus, it will be necessary to speak of measures of central tendency and dispersion for probability distributions of random variables and for frequency distributions of obtained data. The measure of central tendency of a random variable and the corresponding measure for data obtained from that underlying process have an intimate relationship; the nature of that relationship will be explored in the following chapter. Because an intuitive feeling for measures of central tendency for obtained data typically is more easily developed, in this chapter obtained data are considered prior to random variables. However, before discussing any of the measures of central tendency or dispersion, we need to introduce a new (for some readers) notation—summation notation.

Summation   3.1 Summation Notation

In the discussion of statistics we frequently need to sum a set of numbers. Although this sum could be written out, it is parsimonious to use summation notation. To illustrate summation notation, consider a set of scores obtained by sampling some underlying process:

$$\{1, 2, 2, 4, 6, 8, 8, 9\}$$

$$\{x_1, x_2, x_3, x_4, x_5, x_6, x_7, x_8\}$$

The scores have been arranged, for convenience, in ascending order and labeled as x_1, x_2, and so on. The subscripts are used to distinguish one observation from another. For instance, x_2 simply refers to the second observation. (The notation "x" is used because each score might be thought of as a value of a random variable obtained from sampling from a population that is characterized by the random variable X. This concept, which was alluded to in Chapter 2 will be developed further in Chapter 4.)

If the sum of the eight preceding observations were desired, it would be sufficient to write

$$x_1 + x_2 + x_3 + x_4 + x_5 + x_6 + x_7 + x_8$$
$$= 1 + 2 + 2 + 4 + 6 + 8 + 8 + 9 = 40$$

The summation notation is used to indicate the sum in the following way:

$$\sum_{i=1}^{8} x_i = x_1 + x_2 + x_3 + x_4 + x_5 + x_6 + x_7 + x_8$$

The symbol "Σ" (sigma) indicates a sum. The quantity immediately to the right of Σ indicates the quantity to be summed, in this case values of x will be summed. The subscript "i," which assumes integer values, is called the index and is used to indicate which of the values of the quantity will be summed. The "$i = 1$" below Σ indicates that the sum begins with the first value (e.g., x_1); the "8" above the summation sign indicates that the sum ends with the eighth value (e.g., x_8). Formally, summation notation can be defined in the following way:

$$\sum_{i=1}^{N} x_i = x_1 + x_2 + \cdots + x_N \tag{3.1}$$

A few examples will help.

$$\sum_{i=2}^{4} x_i = x_2 + x_3 + x_4 = 8$$

$$\sum_{i=1}^{2} x_i = x_1 + x_2 = 3$$

When referring to a data set that contains N observations, the notation $\sum_{i=1}^{N} x_i$ indicates that the sum of all the observations is computed. For the eight preceding observations

$$\sum_{i=1}^{N} x_i = x_1 + x_2 + \cdots + x_N = 40$$

Because the sum of all the observations is used so frequently, it is not unusual to omit part or all of the indexing. The following "shorthand" notations are interchangeable:

$$\sum_{i=1}^{N} x_i = \sum_{i} x_i = \sum x_i = x_1 + x_2 + \cdots + x_N$$

Summation can become more complicated by changing the quantity to be summed. For instance, the square of x could be summed:

$$\sum_{i=1}^{3} x_i^2 = x_1^2 + x_2^2 + x_3^2 = 1^2 + 2^2 + 2^2 = 9$$

Or, a constant could be added to the quantity to be summed:

$$\sum_{i=1}^{4} (x_i + 2) = (x_1 + 2) + (x_2 + 2) + (x_3 + 2) + (x_4 + 2)$$

$$= 3 + 4 + 4 + 6 = 17$$

Note that this is different from

$$\sum_{i=1}^{4} x_i + 2 = x_1 + x_2 + x_3 + x_4 + 2 = 11$$

Remember that Σ indicates that only the quantity immediately to the right will be summed.

3.2   Rules of Summation

In several instances some common rules of summation are needed. The first rule indicates that the sum of N constants is N times the constant:

$$\sum_{i=1}^{N} a = Na, \quad \textbf{where } a \textbf{ is a constant} \tag{3.2}$$

The second rule refers to the multiplication of the quantity to be summed by a constant:

$$\sum_{i} a x_i = a \sum_{i} x_i, \quad \textbf{where } a \textbf{ is a constant} \tag{3.3}$$

The third rule refers to the "sum of sums:"

$$\sum_{i=1}^{N} (x_i + y_i) = \sum_{i=1}^{N} x_i + \sum_{i=1}^{N} y_i \tag{3.4}$$

The fourth rule refers to the addition of a constant to the quantity to be summed:

$$\sum_{i=1}^{N} (x_i + a) = \sum_{i=1}^{N} x_i + \sum_{i=1}^{N} a = \sum_{i=1}^{N} x_i + Na$$

$$\textbf{where } a \textbf{ is a constant} \tag{3.5}$$

It will be necessary to use these rules to simplify expressions. When such is the case, the justification will consist of a statement such as "by the rules of summation."

An Important Measure of Central Tendency— The Mean

3.3 The Mean of Obtained Data

Although there are several measures of central tendency, the *arithmetic mean*, which will be referred to as simply the *mean*, is the most important and widely used measure and is discussed here. Consider the following set of observations:

$$\{x_1, x_2, ..., x_N\}$$

The mean of the observation is simply the sum of the observations divided by the number of observations. If M is used to denote the mean of the observations,

$$M = \frac{\sum_{i=1}^{N} x_i}{N} = \frac{1}{N}\sum_{i=1}^{N} x_i \tag{3.6}$$

Clearly, the mean is that measure of central tendency that is often referred to as the "average." Although the mean has many important properties that will be discussed subsequently, several characteristics of the mean are illustrated here. Each of these characteristics becomes apparent by the use of the rules of summation, which is not surprising since the mean is defined as a sum. The first characteristic is that if all the observations of a data set are equal to some constant a, then the mean is that constant a. This is easily verified by using the rules of summation:

$$\frac{1}{N}\sum a = \frac{1}{N}(Na) = a$$

The second characteristic of the mean is that if each of the observations is multiplied by a constant a, the mean of the new set is equal to the mean of the original set multiplied by a:

$$\frac{1}{N}\sum ax_i = a\frac{1}{N}\sum x_i = aM$$

The third characteristic of the mean is that if a constant a is added to each of the observations, the mean of the new set is equal to the constant added to the mean of the original set:

$$\frac{1}{N}\sum (x_i + a) = \frac{1}{N}\sum x_i + \frac{1}{N}(Na) = M + a$$

Finally, consider two sets of observations that would be derived from observing each of N subjects twice:

$$\{x_1, x_2, ..., x_N\} \quad \text{and} \quad \{y_1, y_2, ..., y_N\}$$

with means M_x and M_y, respectively. (The subscripts x and y are used to differentiate the means for the two sets.) A new set of observations can be formed by adding the corresponding observations, yielding $\{(x_1 + y_1), (x_2 + y_2), ..., (x_N + y_N)\}$. In other words, for each subject there is a new observation, which is the sum of two other observations for that subject. A relevant example would be scores on the GRE (Graduate Record Examination), where an x would represent a score on the verbal section, a y would represent a score on the quantitative section, and the sum $(x_i + y_i)$ would represent a total score. The mean of the total scores (i.e., the mean of the set of observations formed by adding the corresponding observations) is the sum of the means of the two original sets of observations:

$$\frac{1}{N}\sum (x_i + y_i) = \frac{1}{N}\sum x_i + \frac{1}{N}\sum y_i = M_x + M_y$$

If the mean verbal and mean quantitative GRE for students in a graduate program were 580 and 620, respectively, then the mean total score for the students would be 1200.

3.4   The Mean of a Random Variable—Expectation

The mean of a random variable, which will also be referred to as the expectation of the random variable, is defined in a manner that makes it the analogue of the mean of a set of observations. This definition is motivated by Example 3.1.

To find the mean of a discrete random variable, such as the one in Example 3.1, it might be tempting to calculate the mean of the possible outcomes. This strategy ignores the fact, however, that certain outcomes are more probable than others. To give each value of the random variable its "fair share," each possible value is weighted by its probability. **If the mean of a discrete random variable X is denoted by either $E(X)$ or μ (*mu*), then the mean of X is defined as follows:**

$$E(X) = \mu = \sum_i x_i P(X = x_i) \qquad (3.7)$$

In other words, for a discrete random variable X, "the mean of X" or "the expectation of X" or "the expected value of X" is the product of each value of X and its associated probability summed over the possible values of X. The rationale for using the symbol "μ" will become apparent in the following chapter.

To illustrate the formula for the expectation, we use the random variable from Example 3.1:

$$E(X) = \sum x_i P(X = x_i) = 12(\tfrac{1}{4}) + (-6)(\tfrac{1}{2}) + (-4)(\tfrac{1}{4}) = -1$$

Several comments about the expectation of a discrete random variable will help make the idea of an expectation of a random variable more intuitive. The formula for the mean of an obtained data set and the expectation of a discrete random variable are more similar than might first appear. The mean of obtained data can be modified in the following way:

$$M_x = \frac{1}{N}\sum x_i = \sum \frac{1}{N} x_i$$

The essential difference between the term on the right of this equation and the definition of the expectation of a discrete random variable is that $1/N$ is replaced by $P(X = x_i)$. If each of the values of the random variable were equally likely, the formulas would be identical.

EXAMPLE 3.1   Suppose that a spinner is arranged as shown in Figure 3.1, with the usual stipulation that the real numbers are "evenly" spaced in each of the intervals. Further, suppose that payoffs are contingent on the outcome of a flick of the spinner, specifically, $12 is won if the spinner lands between zero and 1, $6 is lost if the spinner lands between 1 and 2, and $4 is lost if the spinner lands between 2 and 3. Let the random variable X be defined as the amount of money won or lost (loss is assigned a negative number). Although the outcomes of X were generated by a process that involved a continuous set of numbers (the spinner), X is a discrete random variable because there are only three distinct outcomes (viz., $+12$, -6, -4). The assignment of probabilities to these three outcomes, as shown in Figure 3.1, constitutes a probability distribution function.

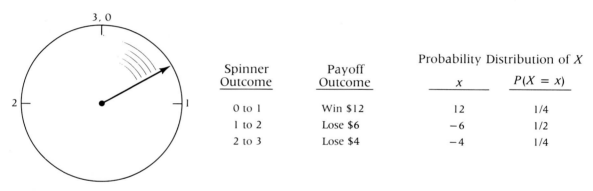

Spinner Outcome	Payoff Outcome	Probability Distribution of X	
		x	$P(X = x)$
0 to 1	Win $12	12	1/4
1 to 2	Lose $6	-6	1/2
2 to 3	Lose $4	-4	1/4

FIGURE 3.1   Spinner with Payoff Outcomes

There is another connection between the mean of obtained data and the expectation of a discrete random variable. The expectation of the random variable derived from the spinner was -1. What is the intuitive meaning of this value? Certainly, one does not expect to lose $1.00 on each spin; this is an impossibility. However, it would be reasonable to say that *in the long run*, the player would expect to lose $1.00 per spin. That is, if the game were played an infinite number of times, the player would lose $1.00 per spin. The motivation for the word "expectation" should now be clear.

Suppose that an unsuspecting player was not aware of the probability distribution associated with this game. The player decided to conduct an experiment to determine what payoff was expected from the gaming device. Having an intuitive idea of statistics, the player decides to conduct a relatively large number of trials, say, 100. Suppose that the player lost $80.00, for an average or mean loss of $0.80 per trial. The mean loss was not $1.00 because the obtained distribution is not identical to the underlying distribution. However, in theory, if the player had conducted an infinite number of trials, the mean loss would have been $1.00. So, clearly, there is a strong and intimate relationship between the expectation of a random variable and the mean of data obtained from that random variable; the exact nature of the relationship will be a major focus of the next chapter.

The notion of the expectation of a continuous random variable is not nearly so intuitive as is the expectation of a discrete random variable. The brief discussion that follows gives a sketch of how the expectation of a continuous random variable is calculated. Consider the continuous random variable presented in Figure 3.2. If the range of the random variable X is broken into many small intervals, the probability associated with each interval is the area under the curve of that interval. This area can be approximated by calculating the area of the rectangle that is formed as shown in Figure 3.2. An approximation of the expectation of X can be found by using the formula for the discrete random variable [Equation (3.7), p. 44], where x_i would be the midpoint of each interval and $P(X = x_i)$ would be the area of the rectangle. That is

$$E(X) \cong \sum_{\substack{\text{all} \\ \text{rectangles}}} x_i P(X \text{ lies in the interval})$$

$$\cong \sum_{\substack{\text{all} \\ \text{rectangles}}} x_i \text{ (area of the rectangle)}$$

(The symbol "$\cong$" means "approximately equal to.") To obtain better approximations, the size of the intervals (i.e., the width of the rectangles) would

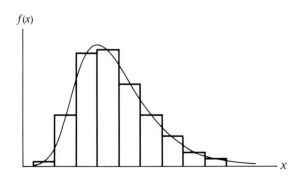

$$E(X) \cong \sum_{\text{all intervals}} x_i P(x \text{ lies in interval}) \cong \sum x_i \text{ (area of rectangle)},$$

where x_i is the midpoint of interval

FIGURE 3.2   Approximation to the Expectation of a Continuous Random Variable

be decreased. To those students who are familiar with the calculus, it is clear that **if X is a continuous random variable, then the expectation of X, $E(X)$ or μ, is given by**

$$E(X) = \int_{-\infty}^{\infty} xf(x)\, dx \tag{3.8}$$

Those unfamiliar with the calculus will not be at a disadvantage; the expectation of the continuous distributions that are important in this book have been derived by statisticians. Incidentally, the mean of the standard normal distribution is zero, which should not be surprising if one recalls the graph of the distribution of the standard normal.

3.5 Properties of Expectation

Because the expectation is such an important concept, and because it is used so frequently in subsequent derivations, several properties of expectations are considered here. Before discussing these properties, however, we need to clarify a sticky little detail. Unless we understand this detail, several critical steps in derivations will seem like magic. Failure to understand the detail only implies that derivations that depend on the detail will need to be taken on faith; don't be alarmed, there are many instances in this book when the reader will be asked to accept something on faith. Remember, though, that this is not blind faith, for statistical "gnomes" will have done the work behind the scenes.

Back to the detail. Consider a discrete random variable X and transformations of X, such as

$X + 4$
$2X + 5$
X^2
and $(X - 3)^2$

An example of a random variable X and the transformed random variables is given in Table 3.1. From this table it should be clear that the probabilities associated with the transformed random variables are related to the probabilities associated with X. For instance, consider the random variables X and $X + 4$. Clearly, $P(X = 2) = P(X + 4 = 6)$. To find the probability for some value of a random variable expressed in terms of X, we need only refer to the probabilities associated with X. Thus

$$E(X + 4) = \sum (x_i + 4)P(X = x_i) = 6(\tfrac{1}{2}) + 8(\tfrac{1}{4}) + 12(\tfrac{1}{4}) = 8$$

$$E(2X + 5) = \sum (2x_i + 5)P(X = x_i) = 9(\tfrac{1}{2}) + 13(\tfrac{1}{4}) + 21(\tfrac{1}{4}) = 13$$

$$E(X^2) = \sum x_i^2\, P(X = x_i) = 4(\tfrac{1}{2}) + 16(\tfrac{1}{4}) + 64(\tfrac{1}{4}) = 22$$

$$E(X - 3)^2 = \sum (x_i - 3)^2 P(X = x_i)$$
$$= 1(\tfrac{1}{2}) + 1(\tfrac{1}{4}) + 25(\tfrac{1}{4}) = 7 \tag{3.9}$$

Pay close attention to the last example because something quite similar to this will appear shortly. Before returning to the properties of expectation, consider Example 3.2.

TABLE 3.1

Distribution of X and Several Variants

x_i	$P(X = x_i)$	$x_i + 4$	$2x_i + 5$	x_i^2	$(x_i - 3)^2$
2	$\tfrac{1}{2}$	6	9	4	1
4	$\tfrac{1}{4}$	8	13	16	1
8	$\tfrac{1}{4}$	12	21	64	25

EXAMPLE 3.2

Example. Suppose that an experiment consists of rolling two dice. Define the random variable X as the number of spots on the first die and Y as the number of spots on the second die. Define three new random variables, U, V, and W in terms of X and Y:

$$U = X + Y, \qquad V = X - Y, \qquad \text{and} \qquad W = XY$$

The probability distributions of U, V and W appear in Table 3.2. The following expectations are derived directly from the probability distributions.

$$E(X) = \sum x_i P(X = x_i)$$
$$= 1(\tfrac{1}{6}) + 2(\tfrac{1}{6}) + 3(\tfrac{1}{6}) + 4(\tfrac{1}{6}) + 5(\tfrac{1}{6}) + 6(\tfrac{1}{6})$$
$$= 3\tfrac{1}{2}$$

Similarly, $E(Y) = 3\tfrac{1}{2}$

$$E(U) = \sum u_i P(U = u_i) = 2(\tfrac{1}{36}) + 3(\tfrac{2}{36}) + 4(\tfrac{3}{36}) + 5(\tfrac{4}{36})$$
$$+ 6(\tfrac{5}{36}) + 7(\tfrac{6}{36}) + 8(\tfrac{5}{36}) + 9(\tfrac{4}{36}) + 10(\tfrac{3}{36}) + 11(\tfrac{2}{36})$$
$$+ 12(\tfrac{1}{36})$$
$$= 7$$

$$E(V) = \sum v_i P(V = v_i) = -5(\tfrac{1}{36}) - 4(\tfrac{2}{36}) - 3(\tfrac{3}{36}) - 2(\tfrac{4}{36})$$
$$- 1(\tfrac{5}{36}) + 0(\tfrac{6}{36}) + 1(\tfrac{5}{36}) + 2(\tfrac{4}{36}) + 3(\tfrac{3}{36}) + 4(\tfrac{2}{36})$$
$$+ 5(\tfrac{1}{36})$$
$$= 0$$

$$E(W) = \sum w_i P(W = w_i) = 1(\tfrac{1}{36}) + 2(\tfrac{2}{36}) + \cdots + 35(0) + 36(\tfrac{1}{36})$$
$$= 12\tfrac{1}{4}$$

TABLE 3.2

Probability Distributions for Sum, Difference, and Product of Rolls of Two Dice (X = "spots on 1st Die, Y = "spots on 2nd Die)

$U = X + Y$		$V = X - Y$		$W = XY$	
u	$P(U=u)$	v	$P(V=v)$	w	$P(W=w)$
2	1/36	−5	1/36	1	1/36
3	2/36	−4	2/36	2	2/36
4	3/36	−3	3/36	3	2/36
5	4/36	−2	4/36	4	3/36
6	5/36	−1	5/36	5	2/36
7	6/36	0	6/36	6	4/36
8	5/36	1	5/36	7	0
9	4/36	2	4/36	8	2/36
10	3/36	3	3/36	9	1/36
11	2/36	4	2/36	10	2/36
12	1/36	5	1/36	11	0
				12	4/36
				13	0
				14	0
				15	2/36
				16	1/36
				17	0
				18	2/36
				19	0
				20	2/36
				21	0
				22	0
				23	0
				24	2/36
				25	1/36
				26	0
				27	0
				28	0
				29	0
				30	2/36
				31	0
				32	0
				33	0
				34	0
				35	0
				36	1/36

X

	1	2	3	4	5	6
1	2	3	4	5	6	7
2	3	4	5	6	7	8
Y 3	4	5	6	7	8	9
4	5	6	7	8	9	10
5	6	7	8	9	10	11
6	7	8	9	10	11	12

$X + Y$

X

	1	2	3	4	5	6
1	0	1	2	3	4	5
2	−1	0	1	2	3	4
Y 3	−2	−1	0	1	2	3
4	−3	−2	−1	0	1	2
5	−4	−3	−2	−1	0	1
6	−5	−4	−3	−2	−1	0

$X - Y$

X

	1	2	3	4	5	6
1	1	2	3	4	5	6
2	2	4	6	8	10	12
Y 3	3	6	9	12	15	18
4	4	8	12	16	20	24
5	5	10	15	20	25	30
6	6	12	18	24	30	36

XY

Some of the properties to be presented are intuitive; those that are not will be illustrated in some depth. For these properties a is constant and X and Y are random variables.

Property 1

$$E(a) = a \qquad\qquad (3.10)$$

Simply, this property states that if a random variable assumes only one value a (with probability 1), then the expectation of the random variable is a.

Property 2

$$E(aX) = aE(X) \qquad\qquad (3.11)$$

This is a direct analogue of one of the characteristics of the mean of a set of observations and should be intuitive.

Property 3

$$E(X + a) = E(x) + a \qquad\qquad (3.12)$$

This is also a direct analogue of one of the characteristics of the mean of a set of observations and incidentally was illustrated earlier for $X + 4$.

Property 4

$$E(X + Y) = E(X) + E(Y) \qquad\qquad (3.13)$$

Although this is also a direct analogue to one of the characteristics of the mean of a set of observations, illustrating it will be helpful. This is done in Example 3.2, which should be familiar. Property 4 is easily verified by this example since $E(X) = 3\frac{1}{2}$, $E(Y) = 3\frac{1}{2}$, and $E(X + Y) = E(U) = 7$.

COMMENT

Using the rules of summation, we can easily prove each of the four properties. To illustrate, property 3 will be proved for discrete random variables. We want to show that $E(X + a) = E(X) + a$. Taking the term on the left, we can derive the term on the right (note the use of the detail explained earlier):

$$E(X + a) = \sum (x_i + a)P(X = x_i) = \sum x_i P(X = x_i) + \sum a P(X = x_i)$$

By the rules of summation and the definition of expectation the quantity reduces to

$$\sum x_i P(X = x_i) + a \sum P(X = x_i) = E(X) + a \sum P(X = x_i)$$

But $\Sigma P(X = x_i)$ is simply the probabilities summed over the range of the distribution and thus equals 1. Hence, the last expression reduces to $E(X) + a$, the desired result.

Important Measures of Dispersion— Variance and Standard Deviation

3.6 Variance of Obtained Data—The Sample Variance

Just as there are several measures of central tendency, there are several measures of dispersion. The measure of dispersion that corresponds to the mean is the variance. Essentially, variance is a measure that describes how spread out the data are *from* the mean.

Consider a set of observations $\{x_1, x_2, ..., x_N\}$. Recall that the mean $M = (1/N) \Sigma x_i$. As the variance is discussed and defined, it will be helpful to follow along with the data presented in Table 3.3. For this set of observations, the mean is easily calculated:

$$M = \frac{1}{N} \sum x_i = \frac{1}{5}(20) = 4$$

A first attempt to define variance might involve determining the difference between each observation and the mean, that is, $(x_i - M)$. This difference is called the *deviation* for the observation i. Certainly, the deviation describes how far each observation is from the mean. However, as illustrated in Table 3.3, when the deviations for a set of observations are summed, the sum is equal to zero. This is always true and is worth restating: **For a set of observations the sum of the deviations is equal to zero; that is**

$$\sum_{i=1}^{N} (x_i - M) = 0 \tag{3.14}$$

This is easily proved:

$$\sum (x_i - M) = \sum x_i - \sum M$$

but remember that M is a constant for a particular set of N observations. Thus

$$\sum x_i - \sum M = \sum x_i - NM = N\frac{1}{N}\sum x_i - NM = NM - NM = 0$$

This result is reasonable if it is remembered that the observations are spread out from the mean; some are above and some are below. Thus, some of the deviations are greater than zero and others are less than zero, thereby "canceling each other out."

TABLE 3.3

Calculation of the Sample Variance

Observation	x	$x - M$	$(x - M)^2$	x^2
1	7	3	9	49
2	4	0	0	16
3	0	−4	16	0
4	6	2	4	36
5	3	−1	1	9
	$\sum x_i = 20$	$\sum (x_i - M) = 0$	$\sum (x_i - M)^2 = 30$	$\sum x_i^2 = 110$

$M_i = (1/N) \sum x_i = (\frac{1}{5})(20) = 4$

$S^2 = (1/N) \sum (x_i - M)^2 = (\frac{1}{5})(30) = 6$

$S^2 = (1/N) \sum x_i^2 - M^2 = (\frac{1}{5})(110) - 4^2 = 6$ (computational formula)

To avoid the difficulty that the positive deviations are balanced by the negative deviations, we use the square of the deviations in the definition of the variance. The square of the deviations for each observation are presented in Table 3.3. To make a measure of variation that is comparable from one set of observations to another, the number of observations must be taken into account. Therefore, variation is defined as the average of the squared deviations: **The variance of a set of observations, which will be called the** *sample variance* **and denoted by** S^2, **is given by**

$$S^2 = \frac{1}{N} \sum_{i=1}^{N} (x_i - M)^2 \tag{3.15}$$

Using the sums of columns in Table (3.3), we can easily calculate the sample variance of that data set:

$$S^2 = \frac{1}{N} \sum (x_i - M)^2 = \frac{1}{5}(30) = 6$$

The adjective "sample" is appended to variance to indicate that the variance was calculated from a set of observations; these observations typically are sampled from a population, a concept that will be discussed in the next chapter. Whenever it is clear that S^2 is calculated from a set of observations, it is permissible to drop the word "sample." However, the student should always keep in mind that S^2 is the *sample* variance. The sample variance is a descriptive statistic. **A statistic is a rule that pairs any set of observations with one and only one value.** The formula for the sample variance of a set of observations [Equation (3.15), p. 52] is the rule.

The sample variance describes the dispersion of the scores in the set of observations, thus the term descriptive statistic. The mean of a set of observations is also a descriptive statistic; it would be permissible, even stylish, to call such a mean the sample mean. However, sample is appended to variance because there will be another, although related, index of variance calculated on a set of observations. No such confusion exists for the mean of a set of observations. Although this may seem confusing, the following chapter will discuss these concepts in detail.

There is another source of confusion about the sample variance. This confusion is related to the question, How large is large? The fact that the mean annual income of a sample of 100 Saudi Sheiks is $1,000,000 has intuitive meaning. Clearly, on the basis of income one might desire to be a Sheik. However, if the variance of the income was $100,000, it is not clear what meaning this has. This confusion can be attacked in three ways. First, the mean has more meaning because we have many referents to which to compare the mean. Certainly, we know our own income. Probably, we know the income of several others (e.g., friends, parents, and well-publicized business figures). We might even know the average income of persons in the United States. So, when the mean income of the Sheiks is heard, we instantly compare it to these referents. Typically, no such referents exist for the variance.

The second issue is that the size of the variance often makes sense only in certain contexts. One of the common strategies in statistics is to compare a statistic to the square root of the variance of that statistic. By itself, the size of this variance is meaningless; only when it is used in comparisons does its meaning become apparent. Hopefully, this will become clearer in future chapters. Finally, the variance is affected by the scale of the observations. The mean is affected by the scale also, but in a more direct way. If the mean income of the Sheiks was given in francs, this mean could be converted to dollars simply by dividing by 8 (8 francs = $1.00; recall the characteristics of the mean!). However, if the variance is given in francs, conversion is more complicated, as we will see. It takes time for the variance to become familiar; for now, keep in mind that the more spread out the scores are, the larger is the variance.

COMMENT

Before computers and calculators were accessible, many formulas were altered to ease the burden of computations. However, because computing devices are commonly available, computational formulas are omitted in this book for the most part. Partly for nostalgic reasons, partly because the derivation of the computational formula illustrates some of the rules that have been discussed, and partly because it appears in some subsequent derivations, the computational formula for the variance is derived here.

Remember that $S^2 = (1/N) \sum (x_i - M)^2$. Expanding the right-hand side and using the rules of summation gives

$$S^2 = \frac{1}{N} \sum (x_i^2 - 2Mx_i + M^2) = \frac{1}{N} \sum x_i^2 - \frac{1}{N} \sum 2Mx_i + \frac{1}{N} \sum M^2$$

Recalling that M (and thus M^2) are constants, and again using the rules of summation, we have

$$S^2 = \frac{1}{N} \sum x_i^2 - (2M) \frac{1}{N} \sum x_i + \frac{1}{N}(NM^2)$$

But notice that the middle term contains an expression for the mean; thus

$$S^2 = \frac{1}{N} \sum x_i^2 - 2M^2 + M^2$$

Finally, the computational formula for the sample variance is

$$S^2 = \frac{1}{N} \sum x_i^2 - M^2 \tag{3.16}$$

By the use of the sums in Table 3.3 (p. 52) the sample variance is easily calculated:

$$S^2 = \frac{1}{N} \sum x_i^2 - M^2 = \frac{1}{5}(110) - 4^2 = 6$$

which agrees with the result obtained with the definitional formula.

3.7　Standard Deviation of Obtained Data— The Sample Standard Deviation

Recall that when the sample variance was defined, the deviations were squared. Although squaring the deviations was desirable to avoid the problem that the deviations "canceled each other out," the result was that variance was defined in terms of square units, such as square inches. Frequently, it is undesirable to have a measure of dispersion that is expressed in square units. A simple solution is to calculate the square root of the sample variance. The resulting quantity is called the *sample standard deviation* and is denoted by S. Thus

$$S = \sqrt{\frac{1}{N} \sum (x_i - M)^2} \tag{3.17}$$

For the observations in Table 3.3 (p. 52) the sample standard deviation is found simply:

$$S = \sqrt{6} = 2.45$$

The comments about the variance can be extended to the standard deviation.

The sample standard deviation is a descriptive statistic because every set of observations has a unique standard deviation. The term sample may be dropped if the context is clear. In some sense the interpretation of the standard deviation is clearer than that for the variance because the standard deviation is expressed in the original measurement units. That is, if the standard deviation of the Sheiks is expressed in francs, dividing by 8 gives the standard deviation in dollars.

3.8   Variance and Standard Deviation of a Random Variable

Just as the expectation of a random variable was the analogue of the mean of a set of observations, the variance of a random variable is the analogue to the sample variance. To facilitate the development of the definition of the variance of a random variable, let us examine the formula for the sample variance:

$$S^2 = \frac{1}{N} \sum (x_i - M)^2$$

That is, the sample variance is the average of the squared deviations. The variance of a random variable will also be an average of the squared deviations. For a random variable the deviation is simply $(x_i - E(X))$, where the x_i are the possible values of the random variable. The average is found by weighting the deviations by the probability of each value of the random variable. Thus, the *variance of a discrete random variable X*, denoted by $V(X)$ or σ^2 (sigma squared), is defined as follows:

$$V(X) = \sigma^2 = \sum_i P(X = x_i)(x_i - E(X))^2 \tag{3.18}$$

One should convince oneself that this formula is the analogue of the formula for the sample variance given earlier. The rationale for the use of σ^2 will become apparent in the next chapter. Consider the following probability distribution for the random variable X:

x_i	$P(X = x_i)$
0	$1/12$
4	$1/4$
6	$1/3$
9	$1/3$

The expectation of X is calculated in the usual fashion:

$$E(X) = \sum x_i P(X = x_i) = 0(1/12) + 4(1/4) + 6(1/3) + 9(1/3) = 6$$

Using Equation 3.18 (p. 55), we find the variance of X:

$$V(X) = \sum P(X = x_i)(x_i - E(X))^2$$
$$= (1/12)(0 - 6)^2 + (1/4)(4 - 6)^2 + (1/3)(6 - 6)^2 + (1/3)(9 - 6)^2$$
$$= 3 + 1 + 0 + 3 = 7$$

The formula for the variance of X can be manipulated to another form. This manipulation is fairly tricky; one is advised to review the "detail" explained in Section 3.5 (p. 47). Once again, examine the formula for the variance of a random variable X:

$$V(X) = \sum P(X = x_i)(x_i - E(X))^2$$

Notice the similarity of this formula to Equation (3.9) (p. 48), which is rewritten here with the terms reversed:

$$E(X - 3)^2 = \sum (x_i - 3)^2 P(X = x_i) = \sum P(X = x_i)(x_i - 3)^2$$

Of particular importance here is the term $(x_i - E(X))^2$ in the formula for the variance. Recall that $E(X)$ is a constant—for any random variable X, $E(X)$ assumes a particular value. Therefore, $(x_i - E(X))^2$ is a random variable, which is formed by subtracting the constant $E(X)$ from each value of the random variable and by squaring the difference. Now examine the equation for the variance once again. Notice that the variance is the sum of terms that are each the product of a random variable formed from X and the corresponding probability. This is an expectation. In fact, Equation (3.9) (p. 48) and the formula for the variance are identical, with the exception that "3" is replaced with $E(X)$. Therefore, the variance of X can be written as an expectation:

$$V(X) = E(X - E(X))^2 \tag{3.19}$$

In other words, the variance of a random variable is the mean (euphemistically, the average) of the squared deviations, which should not be surprising given our understanding of variance. By expanding Equation (3.19) and using the rules of expectation, we can further alter the formula for the variance of X to yield

$$V(X) = E(X^2) - [E(X)]^2 \tag{3.20}$$

This is the direct analogue of the computational formula for the sample variance of a set of observations that was discussed in the comment in Section 3.6 (p. 51). The computational formula can be illustrated with the random variable X presented earlier in this section. First, $E(X^2)$ needs to be calculated:

$$E(X^2) = \sum X^2 P(X = x_i) = 0^2(1/12) + 4^2(1/4) + 6^2(1/3) + 9^2(1/3) = 43$$

Then

$$V(X) = E(X^2) - [E(X)]^2 = 43 - 6^2 = 7$$

which is the result obtained by the definitional formula.

In this section the discussion has been limited to discrete random variables. Using the calculus, we can make the analogous definition for the variance of a continuous random variable:

$$V(X) = \int_{-\infty}^{\infty} f(x - E(X))^2 \, dx$$

(3.21)

where $f(x)$ is the probability density function for X.

From the variance of a set of observations the standard deviation is derived by taking the square root of the variance. Similarly, the *standard deviation of a random variable X*, which is denoted by σ, is defined:

$$\sqrt{V(X)} = \sigma$$

(3.22)

Recall from Section 3.6 (p. 51) that the sum of the deviations for obtained data was always equal to zero. Similarly, for a random variable X the sum of the deviations, each weighted by the corresponding probability, is equal to zero; that is

$$\sum P(X = x_i)[x_i - E(X)] = 0$$

(3.23)

3.9 Properties of Variances

A number of rules for variances of random variables will be useful in subsequent chapters. Because variance typically is a less intuitive concept than expectation, several examples are presented. For this discussion consider a random variable X with $V(X) = \sigma_x^2$ and $E(X) = \mu_x$ and the constants a and b.

Property 1

$$V(X + a) = V(X) = \sigma_x^2$$

(3.24)

That is, adding a constant to a random variable does not change the variance. Recall the $E(X + a) = E(X) + a = \mu_x + a$. Essentially, adding a constant to a random variable moves the distribution to the right or left (depending on the sign of a) but does not change the shape of the distribution, as illustrated in Figure 3.3. Clearly, when a constant is added to a random variable, the standard deviation also remains unchanged.

Property 2

$$V(aX) = a^2 V(X) = a^2 \sigma_x^2$$

(3.25)

It should be somewhat intuitive that multiplying a random variable by a constant will change the variability. If $a > 1$, it is not surprising that

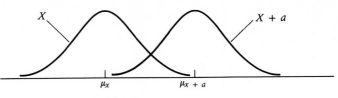

Adding a constant a

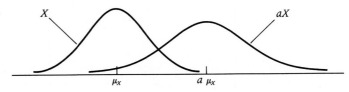

Multiplying by a constant a ($a > 1$)

FIGURE 3.3   Changing a Random Variable by Adding a Constant and Multiplying by a Constant

the variance would increase. But why is the variance changed by a factor of a^2? If we recall that the variance was expressed in terms of square units (the deviations were squared), then the fact that a is squared makes sense. Thus, when the variance of the Sheik's income was expressed in francs (recall 8 francs to $1.00), the variance was increased by a factor of 64. What happens when a is negative, say, -3? Because a is squared, the variance would also be increased by a factor of 9. Note that if $-1 < a < 1$, then the variance is decreased. Recall that $E(aX) = aE(X) = a\mu_x$. Thus, multiplying a random variable X moves the mean and changes the variance, as illustrated in Figure 3.3.

How is the standard deviation of X changed when it is multiplied by a constant? Algebraically, this can be derived easily:

$$\sqrt{V(aX)} = \sqrt{a^2 V(X)} = |a|V(X) \tag{3.26}$$

where $|a|$ is the absolute value of a. Thus, when a is positive, multiplying the random variable by a changes the standard deviation by a factor of a. Continuing with the Sheiks, when the standard deviation is expressed in terms of francs, the standard deviation is increased by a factor of 8, whereas the variance was increased by a factor of 64. Without the absolute value, one could obtain a negative standard deviation, which is an impossibility.

Property 3

$$V(aX + b) = a^2 V(X) = a^2 \sigma_x^2 \tag{3.27}$$

Actually, this property is a combination of properties 2 and 3.

Joint Distributions, Independence, and Covariance

3.10   Joint Distributions

Although joint distributions could have been discussed when random variables were introduced, it is helpful to have previously discussed summation notation and expectation. The discussion begins with discrete random variables. Continuing with Example 3.2 (p. 48) from a previous section where X was the number of spots on the first die and Y was the number of spots on the second die, we can form the joint distribution of X and Y. Considering the two random variables simultaneously, we can calculate probabilities that X will take on a given value **and** Y will take on a given value. For instance, $P(X = 1 \text{ and } Y = 1) = \frac{1}{36}$. The "and" typically is replaced by a comma; that is, $P(X = 1, Y = 1)$. All possible joint events and their associated probabilities form a *joint distribution* of the discrete random variables X and Y (or frequently referred to as the *bivariate distribution* of X and Y). The joint distribution for the continuing example of the two dice is presented in Table 3.4.

From a joint distribution two marginal distributions can be derived. Simply, the two marginal distributions are the distributions for X and Y. **More formally, the *marginal distribution* of X can be derived from the joint distribution of X and Y in the following way:**

$$P(X = a) = \sum_{y} P(X = a, Y = i) \qquad\qquad (3.28)$$

The previous statement is illustrated by the row totals in Table 3.4. The marginal distribution of Y is illustrated by the column totals.

As was the case of a single random variable, data can be obtained from joint random variables. In the case of joint random variables a pair of observations are obtained. Data obtained from the joint distribution of two random variables X and Y would take the form $\{(x_1, y_1), (x_2, y_2), ..., (x_N, y_N)\}$. Continuing with the example of the two dice, six rolls of the dice might yield $\{(1, 2), (5, 1), (4, 4), (3, 6), (5, 6), (4, 2)\}$.

Until now, the discussion of joint random variables has focused on discrete random variables. The principles that apply to the discrete case also apply to continuous random variables, although unfortunately the mathematics become more sophisticated. For these reasons only the essentials of the joint distribution of continuous random variables are presented here. Consider two continuous random variables X and Y such that $g(a)$ is associated with $X = a$ and $h(b)$ is associated with $Y = b$. The *joint density (or distribution)* of two continuous random variables X and Y is then the function f that associates a functional value $f(a, b)$ with $X = a$ and $Y = b$. Confusion at this point is natural; the concept of the joint distribution of two continuous random variables is a difficult one. Perhaps the joint probability density function f illustrated in Figure 3.4 will help. Recall that $g(a)$ was interpreted as the height of the graph of the probability density function for X when

TABLE 3.4

Joint Distribution of X and Y
(X = "spots" on first die, Y = "spots" on second die)

	$Y=1$	$Y=2$	$Y=3$	$Y=4$	$Y=5$	$Y=6$	Marginal Distribution of X
	$P(X=1, Y=1) = \frac{1}{36}$	$P(X=1, Y=2) = \frac{1}{36}$	$P(xX=1, Y=3) = \frac{1}{36}$	$P(X=1, Y=4) = \frac{1}{36}$	$P(X=1, Y=5) = \frac{1}{36}$	$P(X=1, Y=6) = \frac{1}{36}$	$P(X=1) = \frac{1}{6}$
	$P(X=2, Y=1) = \frac{1}{36}$	$P(X=2, Y=2) = \frac{1}{36}$	$P(X=2, Y=3) = \frac{1}{36}$	$P(X=2, Y=4) = \frac{1}{36}$	$P(X=2, Y=5) = \frac{1}{36}$	$P(X=2, Y=6) = \frac{1}{36}$	$P(X=2) = \frac{1}{6}$
	$P(X=3, Y=1) = \frac{1}{36}$	$P(X=3, Y=2) = \frac{1}{36}$	$P(X=3, Y=3) = \frac{1}{36}$	$P(X=3, Y=4) = \frac{1}{36}$	$P(X=3, Y=5) = \frac{1}{36}$	$P(X=3, Y=6) = \frac{1}{36}$	$P(X=3) = \frac{1}{6}$
	$P(X=4, Y=1) = \frac{1}{36}$	$P(X=4, Y=2) = \frac{1}{36}$	$P(X=4, Y=3) = \frac{1}{36}$	$P(X=4, Y=4) = \frac{1}{36}$	$P(X=4, Y=5) = \frac{1}{36}$	$P(X=4, Y=6) = \frac{1}{36}$	$P(X=4) = \frac{1}{6}$
	$P(X=5, Y=1) = \frac{1}{36}$	$P(X=5, Y=2) = \frac{1}{36}$	$P(X=5, Y=3) = \frac{1}{36}$	$P(X=5, Y=4) = \frac{1}{36}$	$P(X=5, Y=5) = \frac{1}{36}$	$P(X=5, Y=6) = \frac{1}{36}$	$P(X=5) = \frac{1}{6}$
	$P(X=6Y=1) = \frac{1}{36}$	$P(X=6Y=2) = \frac{1}{36}$	$P(X=6, Y=3) = \frac{1}{36}$	$P(X=6, Y=4) = \frac{1}{36}$	$P(X=6Yy=5) = \frac{1}{36}$	$P(X=6, Y=6) = \frac{1}{36}$	$P(X=6) = \frac{1}{6}$
Marginal Distribution of Y	$P(Y=1) = \frac{1}{6}$	$P(Y=2) = \frac{1}{6}$	$P(Y=3) = \frac{1}{6}$	$P(Y=4) = \frac{1}{6}$	$P(Y=5) = \frac{1}{6}$	$P(Y=6) = \frac{1}{6}$	

60

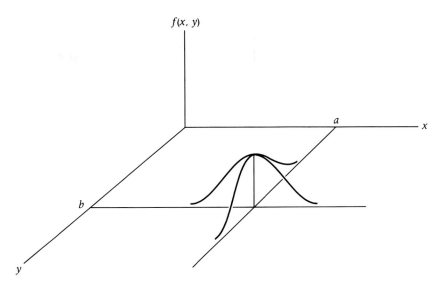

FIGURE 3.4   Joint Distribution of Two Continuous Random Variables

$X = a$; similarly, $h(b)$ was the height of the graph of the probability density function for Y when $Y = b$. The joint event $X = a$ and $Y = b$ has a probability density function f such that the height of the graph of the probability density function at the point $X = a$ and $Y = b$ is the value of the function $f(a, b)$. Instead of a curve, the function f is represented by a surface. The value $f(a, b)$ of the function $f(a, b)$ is the height of the surface from the X, Y plane.

The graph of the joint probability density function of X and Y has special properties. Just as the area under the graph of the probability density function of a single random variable was equal to 1, the *volume* under the surface of the graph of $f(a, b)$ is equal to 1. Hence, the probabilities of regions can be found by calculating volumes. For example, $P(a < X < c, b < Y < d)$ is equal to the volume under the surface for that region, which is a rectangle on the X, Y plane. As was the case for a single random variable, actual calculation of these volumes requires the calculus. Fortunately, calculating volumes are not required to understand the statistics presented in this text.

The similarities between the discrete and continuous cases continue. The marginal probability density functions of X and Y can be derived from the joint probability density function of X and Y. Specifically, the relationship between $g(a)$, the probability density function for X, and $f(a, b)$, the joint probability density function of X and Y, is given by the following expression, which is the analogue to that for the discrete case [Equation 3.28]:

$$g(a) = \int_y f(a, y) \, dy \qquad\qquad (3.29)$$

3.11　Independence of Random Variables

In Chapter 1 the independence of two events was defined in the following way (see Section 1.6, p. 10). Two events are independent if and only if

$$P(A \cap B) = P(A)P(B)$$

The rationale used for this definition can be applied to random variables, resulting in an analogous definition of independence. **Two discrete random variables X and Y are independent if and only if**

$$P(X = a, Y = b) = P(X = a)P(Y = b) \tag{3.30}$$

for all pairs of a and b. In other words, the criterion for independence is that the probability of all joint events is equal to the product of the probabilities of the corresponding marginal events. Criteria similar to this apply to a number of "cross-classified" tables, as we will see in subsequent chapters. Only a quick glance at Table 3.4 (p. 60) is needed to verify that X and Y (the spots on two dice) are independent, as would be expected.

The independence of two continuous random variables is defined in a similar manner. **Given that the continuous random variable X has a density $g(a)$ at $X = a$, the continuous random variable Y has a density $h(b)$ at $Y = b$, and the joint distribution of X and Y has a density $f(a, b)$ at $X = a$ and $Y = b$, X and Y are independent if and only if**

$$f(a, b) = g(a)h(b) \tag{3.31}$$

3.12　Properties of Independent Random Variables

When two random variables are independent, many special properties exist that will be of great help in understanding and deriving important results. The following properties of independent random variables are true regardless of whether they are discrete or continuous.

If X and Y are independent random variables

$$E(XY) = E(X)E(Y) \tag{3.32}$$

Although this statement may not be intuitive, it is easy to verify for the continuing example of the two dice (Example 3.2, p. 48). Recall that $W = XY$, $E(X) = E(Y) = 3\frac{1}{2}$, and $E(W) = 12\frac{1}{4}$. Clearly, $E(XY) = E(X)E(Y)$ in this case. Because this rule plays an important role, its truthfulness for discrete random variables will be demonstrated. By the definition of expectation

$$E(XY) = \sum_{x,y} (xy)P(X, Y = x, y)$$

But by the definition of independence, $P(X, Y = x, y) = P(X = x)P(Y = y)$. Thus

$$E(XY) = \sum_{x,y} xyP(X = x)P(Y = y) = \sum_{x} xP(X = x) \sum_{y} yP(Y = y)$$
$$= E(X)E(Y)$$

As is always the case, the contrapositive of this statement is also true. (The definition of the contrapositive is not needed, however, to understand the discussion.) The contrapositive states that **if**

$$E(XY) \neq E(X)E(Y) \tag{3.33}$$

then X and Y are not independent. Not surprisingly, if X and Y are not independent, then X and Y are said to be *dependent*. To illustrate, consider the joint distribution in Table 3.5. X and Y are not independent, since, for example, $P(X = 0, Y = 0) \neq P(X = 0)P(Y = 0)$. If it is noted that $P(XY = 0) = 1$, then the expectations are also easily calculated. $E(X) = \frac{3}{4}$, $E(Y) = \frac{1}{2}$, and $E(XY) = 0$. Clearly, Equation (3.33) is verified for this example. Because the inequality of $E(XY)$ and $E(X)E(Y)$ indicates that X and Y are dependent, the difference of these two quantities, $E(XY) - E(X)E(Y)$, is a measure of this dependence. In this case the difference is $-\frac{3}{8}$. This measure of dependence, which will be discussed in the next section, is closely related to the correlation coefficient, a concept with which most readers have some familiarity and one that will be discussed in a later chapter.

TABLE 3.5

Joint Distribution of Two Dependent Random Variables

			Marginal Distribution of X
$P(X = 0, Y = 0) = 0$	$P(X = 0, Y = 1) = \frac{1}{2}$	$P(X = 0) = \frac{1}{2}$
$P(X = 1, Y = 0) = \frac{1}{4}$	$P(X = 1, Y = 1) = 0$	$P(X = 1) = \frac{1}{4}$
$P(X = 2, Y = 0) = \frac{1}{4}$	$P(X = 2, Y = 1) = 0$	$P(X = 2) = \frac{1}{4}$
Marginal Distribution of Y	$P(Y = 0) = \frac{1}{2}$	$P(Y = 1) = \frac{1}{2}$

$E(X) = \frac{3}{4}$, $E(Y) = \frac{1}{2}$, $P(XY = 0) = 1$, $E(XY) = 0$

Let us now consider the converse of the statement embodied in Equation (3.32) (p. 62). The converse states that if $E(XY) = E(X)E(Y)$, then X and Y are independent random variables. The joint distribution presented in

Table 3.6 demonstrates that this statement is not true. Clearly, X and Y are not independent: for instance, $P(X = 0, Y = 6) \neq P(X = 0)P(Y = 6)$. (Intuitively, X and Y are not independent because when $Y = 6$ one knows with absolute certainty that $X = 0$.) Furthermore, $E(X) = 0$, $E(Y) = 3\frac{1}{2}$, and $E(XY) = 0$. Therefore, $E(XY) = E(X)E(Y)$, but X and Y are not independent. This is worth restating: **the statement $E(XY) = E(X)E(Y)$ does not imply that X and Y are independent.**

TABLE 3.6

Joint Distribution of Two Dependent Random Variables for Which $E(XY) = E(X)E(Y)$

		Marginal Distribution of X
$P(X = -1, Y = 1) = \frac{1}{4}$	$P(X = -1, Y = 6) = 0$	$P(X = -1) = \frac{1}{4}$
$P(X = 0, Y = 1) = 0$	$P(X = 0, Y = 6) = \frac{1}{2}$	$P(X = 0) = \frac{1}{2}$
$P(X = 1, Y = 1) = \frac{1}{4}$	$P(X = 1, Y = 6) = 0$	$P(X = 1) = \frac{1}{4}$
Marginal Distribution of Y	$P(Y = 1) = \frac{1}{2}$	$P(Y = 6) = \frac{1}{2}$

$E(X) = 0$, $E(Y) = 3\frac{1}{2}$, $P(XY = -1) = \frac{1}{4}$, $P(XY = 0) = \frac{1}{2}$, $P(XY = 1) = \frac{1}{4}$, $E(XY) = 0$

There are two properties of independent random variables related to variance that are important for our use. **If X and Y are independent**

$$V(X + Y) = V(X) + V(Y) = \sigma_x^2 + \sigma_y^2 \tag{3.34}$$

To illustrate this property, let us return to the example of the two dice (Example 3.2, p. 48). Recall that X and Y "counted" the number of spots on the first die and the second die, respectively, and were independent. As well, $U = X + Y$. Although the reader will be spared the arithmetic, the following variances were calculated:

$$V(X) = 2^{11}/_{12}$$
$$V(Y) = 2^{11}/_{12}$$
$$V(U) = 5^{10}/_{12}$$

So, $V(X + Y) = V(X) + V(Y)$ as expected.

The second property related to variance is the following: **If X and Y are independent**

$$V(X - Y) = V(X) + V(Y) = \sigma_x^2 + \sigma_y^2 \tag{3.35}$$

This property needs some explanation because it may seem counter-intuitive to many students. It should be noted that this property can be thought of as a restatement of the previous one with a negative number inserted:

$$V(X - Y) = V[X + (-1)Y] = V(X) + V[(-1)Y]$$

(Here we assume that if X and Y are independent, X and $(-1)Y$ are independent, which is true.) Because $V[(-1)Y] = (-1)^2 V(Y) = V(Y)$ (recall property 2 of variances, p. 57), we have $V(X - Y) = V(X) + V(Y)$.

Unfortunately, algebraic demonstrations do not always lead to increased intuition. Perhaps an example will help. Continuing with the die example, recall that V was the difference of X and Y (viz., $V = X - Y$). If the distributions of U (the sum of X and Y) and V are examined (see Table 3.2, p. 49), it can be seen that the variances of U and V are equal; that is, the dispersions about their respective means are equal. Thus

$$V(X + Y) = V(X - Y) = V(X) + V(Y) = 5^{10}/_{12}$$

Intuitively, one might want to subtract the variances; that is, $V(X - Y) = V(X) - V(Y)$. In this example that would lead to $V(X - Y) = 0$, which clearly is incorrect.

3.13 Covariance

Covariance is a measure that can be applied to joint random variables or to data obtained from these variables. We begin the discussion with obtained data. Consider the set of N observations from a joint distribution, that is, $\{(x_1, y_1), ..., (x_N, y_N)\}$. To illustrate, some fabricated data appear in Table 3.7. Covariance is a measure that indexes the degree to which the data covary or are dependent. Simply, if high scores on one variable are associated with high scores on another variable (and low scores are associated with low scores), the covariance should be large. If scores on the two variables are unrelated, the covariance should be zero. If high scores on one variable are associated with low scores on the other variable and vice versa, the covariance should be large in the negative direction. The covariance is defined to meet these criteria: **The *sample covariance*, S_{xy}, is given by**

$$S_{xy} = \frac{1}{N} \sum (x_i - M_x)(Y_i - M_y) \tag{3.36}$$

Before discussing this formula, we need to calculate the sample covariance for the data in Table 3.7. From the columns in this table

$$S_{xy} = \frac{1}{N} \sum (x_i - M_x)(Y_i - M_y) = \frac{1}{5}(21) = 4\frac{1}{5}$$

TABLE 3.7

Calculation of Sample Covariance

Observation	x	y	$x_i - M_x$	$y_i - M_y$	$(x_i - M_x)(y_i - M_y)$	xy
1	10	8	4	3	12	80
2	5	6	-1	1	-1	30
3	3	2	-3	-3	9	6
4	7	5	1	0	0	35
5	5	4	-1	-1	1	20
	$\Sigma x_i = 30$	$\Sigma y_i = 25$			$\Sigma(x_i - M_x)(y_i - M_y) = 21$	$\Sigma xy = 171$

$M_x = (1/N)\sum x_i = (\frac{1}{5})(30) = 6$

$M_y = (1/N)\sum y_i = (\frac{1}{5})(25) = 5$

$S_{xy} = (1/N)\sum (x_i - M_x)(y_i - M_y) = (\frac{1}{5})21 = 4\frac{1}{5}$

$S_{xy} = (1/N)\sum xy - M_x M_y = (\frac{1}{5})171 - (6)(5) = 4\frac{1}{5}$ (computational formula)

A few points about the formula for the sample covariance are in order. First, the formula for the sample covariance is similar to the definition of the sample variance; however, one deviation score $(x_i - M_x)$ is taken from one variable and the other deviation score $(Y_i - M_y)$ is taken from the other variable, rather than two deviation scores being taken from the same variable. (This should give some rationale for the notation S_{xy}.)

Second, the formula meets the criterion discussed earlier. From the example it can be seen that when high scores are associated with high scores, positive deviations are associated with positive deviations, and when low scores are associated with low scores, negative deviations are associated with negative deviations. Thus, the products of the deviations are positive and the sample covariance will be large in the positive direction. When high scores are associated with low scores (and vice versa), positive deviations will be paired with negative deviations, making the products of deviations negative and yielding a negative covariance.

Finally, the formula for the covariance can be altered in an important way:

$$S_{xy} = \frac{1}{N}\sum (x_i - M_x)(y_i - M_y)$$

$$= \frac{1}{N}\sum (x_i y_i - y_i M_x - x_i M_y + M_x M_y)$$

$$= \frac{1}{N}\sum (xy - M_x \sum y_i - M_y \sum x_i + \sum M_x M_y)$$

$$= \frac{1}{N} \sum xy - M_x \frac{1}{N} \sum y_i - M_y \frac{1}{N} \sum x_i + \frac{1}{N} \sum M_x M_y$$

$$= \frac{1}{N} \sum xy - M_x M_y - M_x M_y + M_x M_y$$

Thus

$$S_{xy} = \frac{1}{N} \sum xy - M_x M_y \qquad (3.37)$$

This formula is computationally simple, as can be seen by applying it to the data from Table 3.7 (p. 66).

$$S_{xy} = \frac{1}{N} \sum xy - M_x M_y = \frac{1}{5}(171) - (6)(5) = 4\frac{1}{5}$$

Although not much emphasis in this book is made on computational formulas, the analogue of this formula for the covariance of two random variables is extremely important.

The covariance of two discrete random variables is defined in the analogous manner. **Given two random variables X and Y with expectations $E(X)$ and $E(Y)$, respectively, the covariance of X and Y, Cov(X,Y) is given by**

$$\mathbf{Cov}(X, Y) = \sum P(X = x, Y = y)[x - E(X)][y - E(Y)] \qquad (3.38)$$

In general

$$\mathbf{Cov}(X, Y) = E\{[X - E(X)][Y - E(Y)]\} \qquad (3.39)$$

The covariance of continuous random variables involves the calculus and will not be given here. However, for both the continuous and discrete cases the formula for the covariance can be altered to the following formula:

$$\mathbf{Cov}(X, Y) = E(XY) - E(X)E(Y) \qquad (3.40)$$

First, note that this equation resembles the computational formula for the sample covariance given in Equation (3.37) (p. 67) (the derivation is similar to that for the obtained data). More importantly, this formula is identical to the measure of dependence that was discussed in Section 3.12 (p. 62). As was the case for obtained data, the covariance of two random variables is a measure of how the two variables are related. The statements about independence made in Section 3.12 (p. 62) can now be restated in terms of covariance. First, **if X and Y are independent, then**

$$\mathbf{Cov}(X, Y) = 0$$

However, **it is not true that Cov(X, Y) = 0 implies that X and Y are independent. But, if Cov(X, Y) $\neq$ 0, then X and Y are not independent.**

We can now consider the sum and difference of two random variables

X and Y in detail. We know that

$$E(X + Y) = E(X) + E(Y)$$

If X and Y are independent, then

$$V(X + Y) = V(X) + V(Y)$$

However, **for any two random variables (independent or not)**

$$V(X + Y) = V(X) + V(Y) + 2\text{Cov}(X, Y) \tag{3.41}$$

Clearly, if X and Y are independent, Equation (3.41) reduces to Equation (3.34) (p. 64). The proof of Equation (3.41) is within our grasp, but is not simple, and therefore is omitted here. However, illustration of this equation with an example will be helpful. Consider the joint distribution of X and Y as well as the distribution of $U = X + Y$ given in Table 3.8. From those distributions, $E(X)$, $V(X)$, $E(Y)$, $V(Y)$, $E(U)$, $V(U)$, and $\text{Cov}(X, Y)$ can be calculated:

$$E(X) = \sum xP(X = x) = 4(\tfrac{1}{8}) + 8(\tfrac{6}{8}) + 12(\tfrac{1}{8}) = 8$$

$$\begin{aligned}
V(X) &= \sum P(X = x)[x - E(X)]^2 \\
&= \tfrac{1}{8}(4 - 8)^2 + \tfrac{6}{8}(8 - 8)^2 + \tfrac{1}{8}(12 - 8)^2 \\
&= 4
\end{aligned}$$

$$\begin{aligned}
E(Y) &= \sum yP(Y = y) \\
&= 0(\tfrac{2}{8}) + 2(\tfrac{1}{4}) + 4(\tfrac{3}{8}) + 16(\tfrac{1}{8}) = 4
\end{aligned}$$

$$\begin{aligned}
V(Y) &= \sum P(Y = y)[y - E(Y)]^2 \\
&= \tfrac{2}{8}(0 - 4)^2 + \tfrac{2}{8}(2 - 4)^2 + \tfrac{3}{8}(4 - 4)^2 \\
&\quad + \tfrac{1}{8}(16 - 4)^2 \\
&= 23
\end{aligned}$$

$$\begin{aligned}
E(U) &= \sum uP(U = u) \\
&= 4(\tfrac{1}{8}) + 8(\tfrac{1}{8}) + 10(\tfrac{2}{8}) + 12(\tfrac{3}{8}) + 28(\tfrac{1}{8}) \\
&= 12
\end{aligned}$$

$$\begin{aligned}
V(U) &= \sum P(U = u)[u - E(U)]^2 \\
&= \tfrac{1}{8}(4 - 12)^2 + \tfrac{1}{8}(8 - 12)^2 \\
&\quad + \tfrac{2}{8}(10 - 12)^2 + \tfrac{3}{8}(12 - 12)^2 + \tfrac{1}{8}(28 - 12)^2 \\
&= 43
\end{aligned}$$

$$\begin{aligned}
\text{Cov}(X, Y) &= \sum P(X = x, Y = y)[X - E(X)][y - E(Y)] \\
&= \tfrac{1}{8}(4 - 8)(0 - 4) + \tfrac{1}{8}(8 - 8)(0 - 4) \\
&\quad + \tfrac{2}{8}(8 - 8)(2 - 4) + \tfrac{3}{8}(8 - 8)(4 - 4) \\
&\quad + \tfrac{1}{8}(12 - 8)(16 - 4) \\
&= 8
\end{aligned}$$

As expected, these quantities verify Equation (3.41):

$$V(X + Y) = 43$$

TABLE 3.8

Joint Distribution of X and Y and Distribution of X + Y
(Used to show $V(X + Y) = V(X) + V(Y) + 2\mathrm{Cov}(X, Y)$)

Joint Distribution of X and Y

				Marginal Distribution of X
$P(X=4,Y=0) = \frac{1}{8}$	$P(X=4,Y=2) = 0$	$P(X=4,Y=4) = 0$	$P(X=4,Y=16) = 0$	$P(X=4) = \frac{1}{8}$
$P(X=8,Y=0) = \frac{1}{8}$	$P(X=8,Y=2) = \frac{2}{8}$	$P(X=8,Y=4) = \frac{3}{8}$	$P(X=8,Y=16) = 0$	$P(X=8) = \frac{6}{8}$
$P(X=12,Y=0) = 0$	$P(X=12,Y=2) = 0$	$P(X=12,Y=4) = 0$	$P(X=12,Y=16) = \frac{1}{8}$	$P(X=12) = \frac{1}{8}$
Marginal Distribution of Y $P(Y=0) = \frac{2}{8}$	$P(Y=2) = \frac{2}{8}$	$P(Y=4) = \frac{3}{8}$	$P(Y=16) = \frac{1}{8}$	

Distribution of $U = X + Y$

$u_i = x_i + y_i$	$P(U = u)$
4	$\frac{1}{8}$
8	$\frac{1}{8}$
10	$\frac{2}{8}$
12	$\frac{3}{8}$
28	$\frac{1}{8}$

and

$$V(X) + V(Y) + 2\text{Cov}(X, Y) = 4 + 23 + 2(8) = 43$$

Because $\text{Cov}(X, -Y) = -\text{Cov}(X, Y)$ (this should be clear from the definition of covariance)

$$V(X - Y) = V(X) + V(Y) - 2\text{Cov}(X, Y) \qquad (3.42)$$

The analogous equation for sample data is important in future work:

$$S_D^2 = S_{x-y}^2 = S_x^2 + S_y^2 - 2S_{xy} \qquad (3.43)$$

where S_D^2 is the *sample variance of the difference.*

Another result that is useful for future work involves linear transformations of random variables. Suppose that X and Y are two random variables that are linearly transformed into two other random variables $AX + B$ and $CY + D$, respectively. How does the linear transformation affect the covariance? From examination of the definition of covariance [Equation (3.38)] it should be clear that covariance is not invariant under linear transformations. However, if X and Y are independent, the transformed random variables $AX + B$ and $CY + D$ are independent.

Other Topics Involving Measures of Central Tendency and Dispersion

3.14   Standardized Scores and Random Variables

When Johnny comes home from school proudly announcing that he received a score of 130 on a test, the statistically wise parent seeks some reference point to which to assign meaning to this score. One basic question involves whether the score is above or below the mean. Additional information may be sought about how far the score is above or below the mean. If the distance from the mean is expressed in standard deviation units, then the score from one test can be compared to the score on another test. Suppose that Johnny's score was derived from an intelligence test that had a mean of 100 and a standard deviation of 15. The obtained score of 130 was then two standard deviations above the mean. Suppose Johnny's sister Sally received a score of 750 on the SAT (with a mean of 500 and a standard deviation of 100), then Sally scored 2½ standard deviations above the mean, a score that was superior to Johnny's score. Thus, Johnny's score and Sally's score can be compared even though the scores were not originally expressed on a common scale.

Scores expressed as standard deviation units from the mean are called standardized scores. **Given a set of observations $\{x_1, x_2, ..., x_N\}$ with mean M and sample standard deviation S, then the** *standardized score* z_i **for the** i**th observation** x_i **is given by**

$$z_i = \frac{x_i - M}{S_x} \qquad (3.44)$$

TABLE 3.9

Standardized Scores

Observation	x	$x - M_x$	$z = \dfrac{x - M_x}{S_x}$
1	3	-2	-1.195
2	4	-1	-0.598
3	5	0	0.000
4	5	0	0.000
5	8	3	1.793
			$\sum z = 0$

$M_x = 5,\ S_x = 1.673$

Clearly, each standardized score represents the distance from the mean expressed in standard deviation units. The standardized scores for a set of observations are given in Table 3.9. Consider now the set of standardized scores $\{z_1, z_2, ..., z_N\}$. This set also has a mean M_z and a sample variance S_z^2, defined in the usual way. For the scores in Table 3.9, the mean and sample variance can be calculated:

$$M_z = \frac{1}{N} \sum z_i = \frac{1}{5}(-1.195 - 0.598 + 0.0 + 0.0 + 1.793) = 0$$

and

$$S_z^2 = \frac{1}{N} \sum (z_i - M_z)^2$$

$$= \frac{1}{5}[(-1.195)^2 + (-0.598)^2 + (0.0)^2 + (0.0)^2 + (1.793)^2]$$

$$= 1.00$$

The fact that the mean and variance of a set of standardized observations are equal to zero and 1, respectively, is not a coincidence. **A set of standardized scores always has a mean of zero and a sample variance of 1.** Although this is not difficult to prove, we will wait to prove the analogous statements for random variables.

For any random variable X with expectation $E(X) = \mu$ and standard deviation σ, a *standardized random variable Z* is defined in the following way:

$$Z = \frac{X - \mu}{\sigma} \tag{3.45}$$

Furthermore, $E(Z) = 0$ and $V(Z) = 1$.

This last statement is relatively simple to prove given the rules for expectation and variance. Clearly

$$E(Z) = E\left[\frac{X - \mu}{\sigma}\right] = \frac{1}{\sigma}E(X - \mu) = \frac{1}{\sigma}[E(X) - \mu] = 0$$

Moreover

$$V(Z) = V\left[\frac{X - \mu}{\sigma}\right] = \frac{1}{\sigma^2}V(X - \mu) = \frac{1}{\sigma^2}V(X) = \frac{1}{\sigma^2}\sigma^2 = 1$$

3.15   Normal Distribution Revisited

In Section 2.4 (p. 25) the standard normal density function was discussed. Although it was not mentioned at that point, the standard normal distribution is a special case of a set of distributions that are normal. In general, the formula for the probability density function of a normal distribution X is given by

$$f(x) = \frac{1}{\sqrt{2\pi\sigma^2}}e^{-(x-\mu)^2/2\sigma^2}$$

(3.46)

This formula differs from the one for the standard normal distribution in that it contains the parameters μ and σ^2 (for the moment forget that these symbols stand for the mean and variance). *Parameters*, which will be defined more rigorously in Chapter 4, may be thought of as constants whose values determine the various forms of the distribution. If the parameters μ and σ^2 are set to zero and 1, respectively, the formula for the standard normal is obtained [see Equation (2.2), p. 25]. When μ and σ^2 take on other values, other normal distributions are obtained. In other words, there exists a family of normal distributions, each one characterized by the parameters μ and σ^2. Three normal distributions are presented in Figure 3.5.

It is not a coincidence that the parameters for the normal distribution were assigned the symbols μ and σ^2. From the formula for the normal distribution [Equation (3.46), p. 72] it can be shown that if X is normally distributed with parameters μ and σ^2, then

$$E(X) = \mu \qquad \text{and} \qquad V(X) = \sigma^2$$

Frequently, we write

$$X \sim N(\mu, \sigma^2)$$

(The N indicates the normal distribution.) This is read as "X is normally distributed with mean μ and variance σ^2." The standard normal distribution

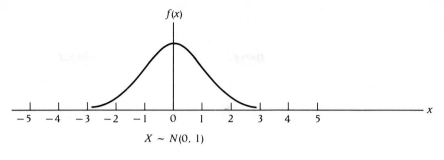

$X \sim N(0, 1)$

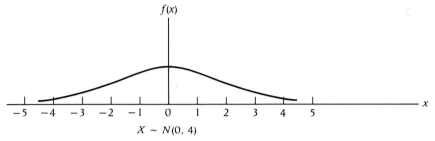

$X \sim N(0, 4)$

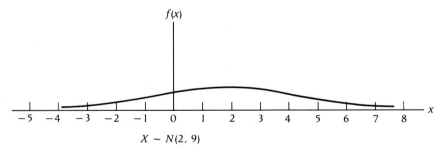

$X \sim N(2, 9)$

FIGURE 3.5   Three Normal Distributions

is the normal distribution that has a mean zero and a variance 1; thus, if Z is a standard normal random variable, $Z \sim N(0, 1)$.

Because the normal distribution is so important, we need to obtain cumulative probabilities associated with values of the various normal distributions. Of course, it is not possible to present tables for every normal distribution. However, any normal distribution can be easily converted to a standard normal. **If $X \sim N(\mu, \sigma^2)$ and $Z = (X - \mu)/\sigma$, then $Z \sim N(0, 1)$.** Note the following. **Any random variable can be standardized; however, the standardized random variable will be normal only if the original random variable was normal.**

We now have the background to find probabilities associated with the values of any normal distribution. For example, suppose that $X \sim N(10, 16)$

and we wish to find $P(X < 20)$. This is accomplished by standardizing X (noting that the expectation of X is 10 and the standard deviation is 4) and by using the standard normal tables:

$$P(X < 20) = P\left(\frac{X - 10}{4} < \frac{20 - 10}{4}\right) = P(Z < 2.5) = F(2.5) = .9938$$

Continuing with the same example, we obtain

$$P(X < 6) = P\left(\frac{X - 10}{4} < \frac{6 - 10}{4}\right) = P(Z < -1) = 1 - P(Z < 1)$$
$$= 1 - F(1) = 1 - .8413 = .1587$$

As one becomes familiar with the distributions, much of the algebra can be omitted in favor of an intuitive approach. For instance, in the latter example, as illustrated in Figure 3.6, it can be seen that 6 is one standard deviation ($\sigma = 4$) below the mean ($\mu = 10$). The desired area is to the left of a z-score (standardized score) of -1. This area is equal to the area to the right of a z-score of $+1$. But because the tables for the standard normal are cumulative, the area to the left of $+1$ can be found in Table B.1 (viz., .8413). Finally, the desired area is $1 - .8413 = .1587$.

A final algebraic example should be sufficient to illustrate the procedure (those who feel comfortable with the graphical approach should confirm the algebraic results). Suppose that $Y \sim N(6, 9)$, then

$$P(3 < Y < 12) = P\left(\frac{3 - 6}{3} < \frac{Y - 6}{3} < \frac{12 - 6}{3}\right) = P(-1 < Z < 2)$$
$$= F(2) - F(-1) = F(2) - [1 - F(1)]$$
$$= .9772 - [1 - .8413] = .8185$$

As was the case with the standard normal distribution, frequently, it will be necessary to reverse the process so that intervals for the random variable are

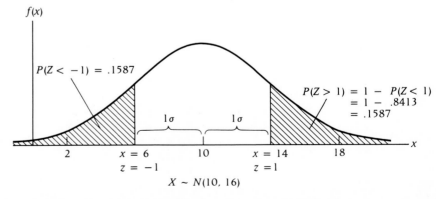

FIGURE 3.6   Graphical Solution to Finding Probabilities for Normal Distribution

found from the probabilities. For example, suppose that $X \sim N(10, 16)$ and the value of c is desired such that $P(X < c) = .95$. Algebraically

$$P(X < c) = P\left(\frac{X - 10}{4} < \frac{c - 10}{4}\right) = P\left(Z < \frac{c - 10}{4}\right) = .95$$

and by Table B.1 (p. 449)

$$\frac{c - 10}{4} = 1.65 \quad \text{and} \quad c = 16.6$$

Again, the reader might prefer a graphical approach where we know that $P(Z < 1.65) = .95$, and therefore for $X \sim N(10, 16)$, c must be four standard deviations above the mean of 10 [i.e., $10 + 4(1.65) = 16.6)$].

At various times it will be necessary to find the distributions of the sum of two or more independent random variables. Suppose that X and Y are independent and $X \sim N(\mu_X, \sigma_X^2)$ and $Y \sim N(\mu_Y, \sigma_Y^2)$. From the rules for expectation and variance we know that the sum, $X + Y$, will have the following mean and variance:

$$E(X + Y) = E(X) + E(Y) = \mu_X + \mu_Y,$$

and

$$V(X + Y) = V(X) + V(Y) = \sigma_X^2 + \sigma_Y^2$$

It is also true that **the sum (or difference) of two independent normal random variables is normal. Thus, if X and Y are independent and $X \sim N(\mu_X, \sigma_X^2)$ and $Y \sim N(\mu_Y, \sigma_Y^2)$, then**

$$(X + Y) \sim N(\mu_X + \mu_Y, \sigma_X^2 + \sigma_Y^2) \tag{3.47}$$

For instance, if X and Y are independent and $X \sim N(3, 16)$ and $Y \sim N(2, 9)$, then $(X + Y) \sim N(5, 25)$.

Another topic that relies on the normal distribution and that will be vital to understanding aspects of correlation is the *bivariate normal distribution*. A bivariate normal distribution is a continuous joint distribution with a particular probability density function. The probability density function for two standardized random variables Z_X and Z_Y is given by

$$f(z_X, z_Y) = \frac{e^{-(z_x + z_y - 2\rho_{xy} z_x z_y)/2(1 - \rho_{XY}^2)}}{\sqrt{2\pi(1 - \rho_{XY}^2)}} \tag{3.48}$$

where all the symbols are familiar with the exception of ρ_{XY} (rho), which is another parameter. As we will see, the parameter ρ_{XY} indexes the degree of linear relation between the two random variables X and Y (see Section 10.1, p. 287). Although the probability density function for the bivariate normal distribution of X and Y is much more complicated than that of the standardized variables, it also contains the parameter ρ_{XY} (as well as the parameters μ_X, μ_Y, σ_X, and σ_Y). It should be recognized that if X and Y have a bivariate normal distribution, the marginal distributions X and Y are normal. However, the normality of X and Y does not imply that X and Y have a bivariate normal distribution.

3.16   Other Parameters That Characterize a Distribution

Recall, as was mentioned previously, a normal distribution is completely characterized by the parameters μ and σ^2, which are the mean and variance of the distribution. That is, if a distribution is normal and the mean and variance is known, then the exact shape of the distribution is known. However, other distributions with the same mean and variance exist but are not normal. For example, three distributions with the same mean and variance, but very different forms, are shown in Figure 3.7. Distributions that are not symmetric are said to be *skewed*. Those distributions with a "long tail" on the right are said to be skewed to the right or positively skewed (see positively skewed distribution in Figure 3.7). Those distributions with a long tail on the left are said to be skewed to the left or negatively skewed. Symmetric distributions, such as the normal distribution, have zero skew.

Distributions can also be characterized by their *kurtosis*. Although kurtosis sounds like a dreaded disease, it also refers to the shape of a distribution. A *platykurtic* distribution is "flatter" and has "thicker tails" than does the normal distribution. One of the most important distributions in statistics (the *t* distribution) is platykurtic, and we will devote much time to it. A distribution that is "skinnier" than the normal distribution is *leptokurtic*.

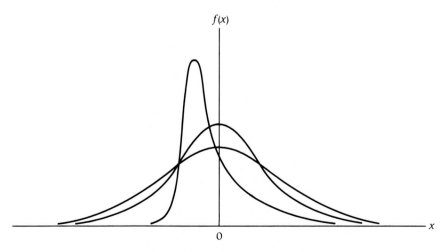

FIGURE 3.7   Three Distributions with Expectation Zero and Variance Two

COMMENT   This chapter began with a discussion of measures that adequately described distributions. The case was made that measures of central tendency and dispersion (usually the mean and variance) were good descriptors of a distribution. If one wants to describe the distribution further, skewness and kurtosis would be the next choices. These four measures—mean, variance, skewness, and kurtosis—are intimately connected. Their relationship can be examined by discussing *moments*. Moments about the origin are of the form $E[(X - 0)]^a$, and moments about the mean are defined as $E[X - E(X)]^a$, where a is a whole number. If a equals 1, then we speak of the first moment about the origin or about the mean, and so on. It is helpful to write the first four moments of each type.

	Origin	Mean
First	$E(X)$	$E[X - E(X)]$
Second	$E(X^2)$	$E[X - E(X)]^2$
Third	$E(X^3)$	$E[X - E(X)]^3$
Fourth	$E(X^4)$	$E[X - E(X)]^4$

Clearly, the first moment about the origin is the mean and the first moment about the mean is zero (remember that the deviations about the mean cancel each other out). The second moment about the mean is the variance. It can be shown that the third moment about the mean can be altered to yield a measure of skewness. Similarly, the fourth moment about the mean can be altered to yield a measure of kurtosis. Those with training in physics should be familiar with the notion of moments.

3.17   Other Measures of Central Tendency and Dispersion

Up to this point the discussion of central tendency has been restricted to the mean. Two other measures of central tendency are the mode and the median. The discussion of these measures are illustrated for obtained data by two sets of observations:

$$\{1, 3, 3, 3, 5, 6, 7, 9, 12\} \quad \text{and} \quad \{10, 12, 12, 12, 13, 15, 15, 15\}$$

and by the theoretical distributions graphed in Figure 3.7. **For obtained data, the *mode* is simply the score that occurs most often. If the data are grouped into intervals, the mode is the midpoint of the interval containing the largest number of observations.** For example, for the first set of observations, the mode is 3. Notice that the mode need not be unique. The second set of observations has two modes, 12 and 15. Such a distribution is called bimodal. **For a theoretical distribution, the mode**

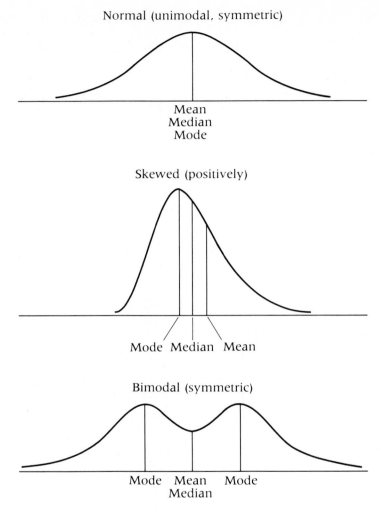

FIGURE 3.8   Measures of Central Tendency of Several Distributions

is the value or values for which the function is maximized, as illustrated in Figure 3.8.

The median, as denoted by *Md*, of a set of observations is the value that divides the distribution equally in the sense that one-half of the scores are less than this value (and hence, one-half are greater). If the number of observations N is odd, then the median is the score of observation $(N + 1)/2$, when the observations are arranged in order. For the first set of observations $N = 9$, thus the median corresponds to the fifth observation (i.e., $(9 + 1)/2$). If N is even, the median is the midpoint of observation $N/2$ and observation $[(N/2) + 1]$, when the observations are arranged in order. For the second set of observations the median is the midpoint of fourth and fifth observations, which is the value 12.5. Actually, by the definition of the median, any value between 12 and 13 would suffice. For instance, one-half the scores are above 12.8 and one-half are below. When obtained data are arranged by intervals, computing the median becomes somewhat laborious and the discussion of this method is omitted here. When the median is used, dispersion may be calculated using the *average deviation*, denoted by AD, and defined in the following way:

$$AD = \frac{1}{N} \sum |x_i - Md|$$

For theoretical distributions the median is that value a such that $F(a) = \frac{1}{2}$. The median for several probability distributions are shown in Figure 3.8 (p. 78).

There are several properties of the mean that make it more desirable than either the median or the mode. Remember that the underlying distributions have means (also called the expectation), medians, and modes and that sets of data also have medians, means, and modes. The first advantage is that the mean of a set of data (or the mean of an underlying distribution, for that matter) is unique, whereas the mode may not be. The second advantage of the mean is that it is relatively stable in the sense that changing a few scores will not alter its value greatly. If two different scores in a set of data have almost equal frequencies, then the mode can be changed drastically by the increase or decrease in frequencies for one of these scores. Similarly, the median can be made to fluctuate. The third advantage of the mean is related to its "quality" as a guess of any score in the set of data. Suppose that one were asked to guess an unknown score from a set of data. The mean is the guess that minimizes the square of the errors in guessing; that is

$$(guess - score)^2$$

is at its minimum when the guess is the mean. This property will play an important role in the development of many of the statistics in this text. It should be realized that the mean is a very poor guess if the criterion of "quality" is the probability that the guess is correct. It would not be unusual for the mean to be equal to a number that is not contained in the data, and therefore the probability that the mean was equal to any score would be zero. Under this criterion the mode is the best guess. The final advantage is of primary importance. The mean of a set of data gives a good estimate of the mean of the underlying distribution. Because this is so important, a major portion of the subsequent chapter will discuss concepts related to estimation. The sample median and mode are not so useful in estimating the median and mode of the underlying distribution.

However, before going off to the prom with the mean, several desirable characteristics of the median and mode should be mentioned. For data measured on a nominal scale the mode is the only measure that makes sense, since the nominal groups are not ordered. Similarly, for ordinal data the median is the appropriate measure.

NOTES AND SUPPLEMENTARY READINGS

It is rather safe to say that almost all textbooks on statistical methods discuss measures of central tendency and dispersion (high probability). Differences exist largely in style and the clarity with which authors present the material. Consequently, few recommended readings seem needed. However, the first chapter of Glass and Stanley (1970), noted here, may be of interest because of the delightful manner in which the authors return to a basic question, "Why employ statistics?" This amusing presentation reminds us of the views that some hold regarding statistics and statisticians as well as the seemingly simple but fundamental reasons that statistical procedures are helpful. It is one of the few chapters of "light" reading available in this area and helps us to focus on the trees as well as the forest.

Glass, G. V., & Stanley, J. C. (1970). *Statistical methods in education and psychology.* Englewood Cliffs, NJ: Prentice-Hall.

PROBLEMS

1. Suppose that the random variable X has the following distribution:

x	$P(X = x)$
9	$1/3$
4	$1/4$
0	$1/6$
8	$1/4$

   a. Find $E(X)$.
   b. Find $V(X)$.
   c. Find $E(2X + 3)$ and $V(2X + 3)$.
   d. Find $E(X^2 - 1)$.

2. Suppose that the random variable X has the following distribution:

x	$P(X = x)$
3	$1/3$
8	$1/2$
12	$1/12$
0	$1/12$

    a. Find $E(X)$.
    b. Find $V(X)$.
    c. Find $E(3X + 4)$ and $V(3X + 4)$.
    d. Find $E(X - 3)^2$.

3. Consider the following set of observations: $\{5, 3, 6, 9, 2, 11\}$.
    a. Find M.
    b. Find S^2 and S.

4. Consider the following observations: $\{4, 7, 8, 11, 15\}$.
    a. Find M.
    b. Find S^2 and S.

5. Consider the following joint distribution:

$P(X=4, Y=1) = \frac{1}{9}$ $P(X=4, Y=2) = \frac{1}{12}$ $P(X=4, Y=1) = \frac{1}{18}$

$P(X=6, Y=1) = \frac{1}{9}$ $P(X=6, Y=2) = \frac{1}{6}$ $P(X=6, Y=3) = \frac{2}{9}$

$P(X=8, Y=1) = \frac{1}{9}$ $P(X=8, Y=2) = \frac{1}{12}$ $P(X=8, Y=3) = \frac{1}{18}$

    a. Find the marginal distributions of X and Y.
    b. Find $E(X)$ and $V(X)$.
    c. Find $E(Y)$ and $V(Y)$.
    d. Are X and Y independent? Why or why not?
    e. Find $\text{Cov}(X, Y)$. How does this answer relate to (d)?
    f. Find the distribution of the random variable $U = X + Y$.
    g. Find $E(X + Y)$ and $V(X + Y)$.
    h. Verify Equation (3.41) [i.e., $V(X + Y) = V(X) + V(Y) + 2\text{Cov}(X, Y)$] in this instance.

6. Consider the following joint distribution:

$P(X=0, Y=0) = 0$ $P(X=0, Y=1) = \frac{1}{8}$ $P(X=0, Y=2) = \frac{1}{8}$ $P(X=0, Y=3) = \frac{1}{12}$

$P(X=3, Y=0) = \frac{1}{8}$ $P(X=3, Y=1) = \frac{1}{8}$ $P(X=3, Y=2) = 0$ $P(X=3, Y=3) = \frac{5}{12}$

    a. Find the marginal distributions of X and Y.
    b. Find $E(X)$ and $V(Y)$.
    c. Find $E(Y)$ and $V(Y)$.
    d. Are X and Y independent? Why or why not?
    e. Find $\text{Cov}(X, Y)$. How does this answer relate to (d)?
    f. Find the distribution of the random variable $U = X + Y$.
    g. Find $E(X + Y)$ and $V(X + Y)$.
    h. Verify Equation (3.41) [i.e., $V(X + Y) = V(X) + V(Y) + 2\text{Cov}(X, Y)$] in this instance.

7. Calculate the sample covariance for the following data:

Observation	x	y
1	3	5
2	7	13
3	8	11
4	1	12
5	5	9
6	6	10

8. Calculate the sample covariance for the following data:

Observation	x	y
1	9	9
2	3	10
3	10	11
4	2	14
5	9	7
6	10	10
7	6	16

9. Find the indicated probabilities for the following distributions:
   a. If $X \sim N(9, 16)$, find $P(X < 1)$.
   b. If $X \sim N(10, 25)$, find $P(X > 25)$.
   c. If $X \sim N(0, 9)$, find $P(X < 6)$.
   d. If $X \sim N(5, 9)$, find $P(2 < X < 8)$.

10. Find the indicated probabilities for the following distributions:
   a. If $X \sim N(12, 25)$, find $P(X < 7)$.
   b. If $X \sim N(4, 9)$, find $P(X > 10)$.
   c. If $X \sim N(4, 4)$, find $P(X < 10)$.
   d. If $X \sim N(12, 36)$, find $P(3 < X < 21)$.

11. Given each distribution, find the value of a such that
   a. $P(X < a) = .01$ given $X \sim N(14, 9)$
   b. $P(X > a) = .025$ given $X \sim N(3, 16)$
   c. $P(X > a) = .99$ given $X \sim N(24, 9)$

12. Given each distribution, find the value of a such that
   a. $P(X < a) = .025$ given $X \sim N(20, 4)$
   b. $P(X > a = .01$ given $X \sim N(6, 25)$
   c. $P(X < a) = .90$ given $X \sim N(15, 9)$

4 SAMPLING, SAMPLING DISTRIBUTIONS, AND ESTIMATION

I n the previous chapters the terms *underlying distribution, unobservable process, random variables, obtained data,* and *samples* were used loosely. In this chapter these terms, as well as many others, are defined more precisely and the intimate connections between them are explored extensively. Understanding this chapter is vital to having an intuitive feel for the essence of parametric statistical testing, the most widely used inferential strategy in the social sciences.

It is important for the results of investigations in the social sciences to be generalizable to organisms other than the ones actually studied. For instance, if a researcher wished to determine that an innovative reading program was superior to a traditional reading program for retarded elementary schoolchildren, clearly, it would be unreasonable to test all retarded elementary schoolchildren. Intuitively, it would seem reasonable to test a subset of these children. However, to make a statement about a sample is one thing, to make inferences about others that are not observed in the study is another thing. A number of factors are involved in the confidence with which one makes inferences about organisms that are not studied directly—the manner in which the sample was selected, the size of the sample, and the distribution of the characteristic under study, to name a few. This chapter begins to explore the theory behind making inferences from a limited sample.

Populations and Samples

4.1 Populations and Population Parameters

The set to which a researcher wishes to generalize is called the *population*. More technically, the population is a sample space of elementary events. Another way to think of the population is a set of units from which the researcher will sample. These units need not be restricted to people; researchers may be interested in populations of animals or objects. Populations can be *finite* (all retarded elementary schoolchildren at this instance) or *infinite* (all retarded elementary schoolchildren present and future). Although it is quite possible to consider the statistical properties of sampling from a finite population, we will limit our discussion to infinite populations for a number of reasons. First, finite populations are of limited usefulness. If a population is limited to a set of units that exist at any given moment, we cannot make a statement about that set of units at any other time. Second, often it is useful to investigate theoretical concepts that have no place in time or space; that is, there may be universal laws that govern the world. Third, the mathematics involved with infinite populations are much "nicer." However, a word of caution is needed. We must be careful when generalizing to a population other than the one from which the sample was drawn—the population might well be different at some point in the future than it was at the time of the investigation. Many psychological processes are affected by cultural milieus, which change over time.

Recall that a random variable is a rule that assigns a numerical value to each elementary event in a sample space (Section 2.1, p. 17). Thus, the numerical value assigned to each unit of the population is a value of a random variable. Typically, the distribution of the random variable is referred to as the *population distribution*. Often, we will say that the population is *characterized* by a random variable. The form of this population distribution is unknown to the researcher, although he or she may have good reasons to make assumptions about this distribution.

Suppose for the moment that the random variable X assigns numerical values of interest to the researcher who is studying a certain population. The values of the mean μ and the variance σ^2 of X have been defined previously. The mean μ and variance σ^2 for a particular population distribution are constants and are called *parameters of the population*. Technically, **a parameter of the population is a constant that is derived by some rule from the distribution of the random variable underlying the population.** Parameters other than the mean μ and variance σ^2 exist, such as a parameter that indicates the amount of skew in a distribution. The values of the parameters for a particular population distribution describe the population distribution. For instance, a population distribution might have a mean of 4 and a variance of 6, although, as discussed previously, there are many different population distributions that have a mean of 4 and a variance of 6. Let us return for a moment to the normal distribution, whose form is described by Equation (3.46) (p. 72).

$$f(x) = \frac{1}{\sqrt{2\pi\sigma^2}} e^{-(x-\mu)^2/2\sigma^2}$$

Note that the parameters μ and σ^2 appear in the formula, but no others. This is because the normal distribution is *completely* characterized by the parameters μ and σ^2. That is, if we know the population distribution is normally distributed with parameters μ and σ^2, then there is no wiggle room—the form and location of the distribution is completely determined.

The purpose of taking samples from a population is to make inferences about the form of the population distribution. Often we assume that the population distribution is of a certain type and then, based on the sample, make a statement about the parameters. For instance, it might be assumed that the population distribution is normal and then the values of the parameters μ and σ^2 would be estimated. The assumption that the population distribution is normal is not made arbitrarily, as we will see.

4.2 Samples and Sample Statistics

In actual practice a *sample* typically is formed by repeatedly selecting a unit from a population and noting its numerical value. If the units are animals or persons, typically they are referred to as *subjects*; the numerical values are often called *observations* or *scores*. Inasmuch as the examples in this book mostly involve animals or persons, the units frequently are referred to as subjects and the numerical values as observations or scores. There is an alternate means to understand sampling, which technically is more accurate. Every unit in the population has a numerical value assigned to it via the random variable and the sample simply is a subset of these numerical values. Some confusion may result here because the numerical values associated with each unit in the population are also referred to as observations or scores. However, in this text the term "scores" will be reserved for the numerical values in the sample. Ambiguity around the term "observations" will be clear from the context. A word about notation: The symbol x_1 represents the first score (or observation) in the sample (e.g., $x_1 = 6$). The symbol x_i represents the ith observation in the sample but the value represented depends on i. Two conceptualizations of sampling have been presented; some concepts to be discussed subsequently are best understood with the first approach, whereas others are best understood with the second approach.

Samples can be formed with replacement or without replacement. Sampling *with replacement* indicates that the unit sampled is replaced in the population before the next unit is selected, implying that a particular unit could be sampled more than once. Sampling *without replacement* indicates that the unit is removed from the population after it has been selected. Of course, it is not typically beneficial to observe a subject more than once, so it would seem that sampling without replacement would be desirable. Nevertheless, the mathematics is much simplified when sampling with replacement is used. However, because we are limiting ourselves to infinite populations, the distinction between sampling with and without replacement disappears, as the probability of selecting a particular unit twice is zero for infinite populations.

There is another aspect of sampling that must be considered—random sampling. A *random sample* is a sample in which the probability of selecting a particular unit is equal to the probability of selecting any other unit. **Simple random sampling implies that each sample of size N has the same probability of being selected as any other sample of size N.** It is a misconception to think that random samples are "representative" of the population. A random sample is not necessarily representative of the population; possibly the sample selected contains subjects who are very different from the typical subject in the population. How likely is it that the sample will be unrepresentative? Answers to questions of this type are the subject of this chapter.

We will assume, unless otherwise specified, that the observations obtained from sampling will be independent from each other. Recall that two events are independent if the probability of one of the events occurring has no influence on the probability of the occurrence of the other. In the same way, **observations are independent if the inclusion of one observation has no impact on the inclusion of another observation.** There are a number of ways that observations in a study inadvertently may not be independent. Suppose that a randomly selected subject appears at a laboratory with his or her friend. The laboratory assistant does not check the list of subjects and administers the tasks to both the subject and his or her friend. Because friends may well have many similar attributes, their performances will likely be related. Another instance in which observations may not be independent involves group administrations of tasks. If one subject leans over to view another subject's responses (i.e., "cheats"), clearly, the responses are no longer independent. Similarly, if a study includes some treatment administered to a group, such as instruction in reading, it is possible that one subject may influence the performance of many of the other subjects. For instance, the subject may be disruptive and this behavior would likely cause a decrement in not only his or her performance, but in that of the others as well. Independence is not black or white. Examination of the research procedures will be needed to make this determination. If a treatment is administered to a group of subjects but the subjects have little interaction, then it is probably safe to assume that the observations are independent. However, if the treatment involves the interaction of the subjects, which would be the case in the administration of group psychotherapy, there is precious little reason to believe that the observations are independent. **Strong violation of the assumption of independent observations can have dramatic and deleterious effects on the results of statistical tests.** The researcher must make a decision whether or not it is reasonable to assume that the observations are independent.

One issue should be clarified. The fact that a sample of observations is random and independent does not mean that the probability that the sample contains one particular value is equal to the probability that the sample contains another value. For instance, consider ten persons selected at random from the population comprised of individuals in the United States. It is much more likely that this sample contains a person who is 32-years-old than the sample contains a person who is 99-years-old, simply because there are more 32-year-olds in the population than 99-year-olds.

For each sample selected from a population we may calculate several values, such as the sample mean or the sample variance. These values associated with the sample are obtained by a rule (i.e., a formula). **A rule that associates a value with a particular sample is called a *statistic*.** For example, the sample mean is a statistic and the rule is that the scores in a sample are summed and divided by the total number of scores. Typically, the rule is expressed as a formula (i.e., $M = \Sigma\, x_i/N$). Because values of statistics are associated with samples, they are often referred to (redundantly) as *sample statistics*. Although we have discussed a number of important statistics such as the mean, median, variance, and standard deviation, we could stipulate any number of others. One such statistic that we will discuss later, but which has little value, is the mean of the first and last observations in a sample: $(x_1 + x_N)/2$. This is a crude measure of central tendency. What makes the mean superior to this measure? Later in this chapter criteria for determining whether a statistic is "good" will be discussed.

The ideas of population, population parameters, samples, and statistics are illustrated in Figure 4.1.

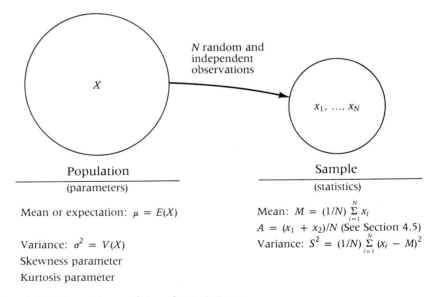

FIGURE 4.1   Population Parameters and Sample Statistics

Sampling Distributions and Estimation

4.3   The Concept of a Sampling Distribution

So far the following scheme has been described. An infinite population has a distribution that is characterized by various parameters. N random and independent observations are sampled from the population and various statistics are calculated from that sample. However, it should be recalled that under this scheme each and every sample of size N has an equal probability of being selected. Typically, the sample statistics calculated for one sample will differ from those calculated for another sample. It is not clear, then, how inferences about the population can be made if values of the sample statistics vary. The idea of a sampling distribution will supply the missing link.

The concept of a sampling distribution is approached by presenting an example, Example 4.1.

EXAMPLE 4.1

Consider the following experiment. Suppose that a researcher has an experimental task and wants to determine the difficulty of the task. Subjects attempt to complete the task and if a mistake is made, the subject must begin the task again. Difficulty is assessed by noting the number of the trial on which the task was successfully completed. Further, suppose that one-quarter of the population can successfully complete the task on the first trial, one-half on the second trial, and the remaining one-fourth on the third trial. Let the random variable X be the number of the trial at which the subject successfully completes the task; the distribution of X is found in Table 4.1. This distribution has parameters $\mu = E(X) = 2$ and $\sigma^2 = V(X) = \frac{1}{2}$.

TABLE 4.1

Population Distribution for Trial Number at Which Success Is Achieved

Number of Trial	
x	$P(X = x)$
1	$\frac{1}{4}$
2	$\frac{1}{2}$
3	$\frac{1}{4}$
$E(X) = 2, V(X) = \frac{1}{2}$	

Suppose that a sample of size 2 was selected; that is, two subjects were randomly selected, and they attempted the task until successful completion was accomplished, and the trial at which success was achieved was noted. There are many different samples of size 2. For instance, one sample might be composed in such a way that the first subject is able to complete the task successfully on the first trial and

TABLE 4.2

**Possible Samples
for N = 2**

Sample			
x_1	x_2	P(Sample)	M
1	1	$(\frac{1}{4})(\frac{1}{4}) = \frac{1}{16}$	1
1	2	$(\frac{1}{4})(\frac{1}{2}) = \frac{1}{8}$	$\frac{3}{2}$
1	3	$(\frac{1}{4})(\frac{1}{4}) = \frac{1}{16}$	2
2	1	$(\frac{1}{2})(\frac{1}{4}) = \frac{1}{8}$	$\frac{3}{2}$
2	2	$(\frac{1}{2})(\frac{1}{2}) = \frac{1}{4}$	2
2	3	$(\frac{1}{2})(\frac{1}{4}) = \frac{1}{8}$	$\frac{5}{2}$
3	1	$(\frac{1}{4})(\frac{1}{4}) = \frac{1}{16}$	2
3	2	$(\frac{1}{4})(\frac{1}{2}) = \frac{1}{8}$	$\frac{5}{2}$
3	3	$(\frac{1}{4})(\frac{1}{4}) = \frac{1}{16}$	3

the second subject is able to complete the task on the second trial. Table 4.2 lists the possible samples. Assuming that the performance of the first subject is independent of the performance of the second subject, the probability of obtaining samples with these values are the product of the probabilities of obtaining each of the values (see the section on independent events, Section 1.6, p. 10). For each possibility the value of the mean also is easily calculated and appears in Table 4.2. Thus, for each value of the mean M a probability can be assigned. For instance, a mean of 1 will be obtained only if the sample contains two subjects who can successfully complete the task on the first trial and this occurs with the probability of $\frac{1}{16}$. However, a mean of $\frac{3}{2}$ can be obtained in two ways; the first subject is able to complete the task successfully on the first trial and the second subject is able to complete the task successfully on the second trial or vice versa. Thus, the probability of obtaining a mean of $\frac{3}{2}$ is $\frac{1}{8} + \frac{1}{8}$. Table 4.3 summarizes the probabilities for the range of values of M. This distribution is called the sampling distribution of the mean.

TABLE 4.3

**Sampling
Distribution
of Mean for
Number of Trials
at Which Success
Is Achieved
(N = 2)**

m	P(M = m)
1	$\frac{1}{16}$
$\frac{3}{2}$	$\frac{1}{4}$
2	$\frac{3}{8}$
$\frac{5}{2}$	$\frac{1}{4}$
3	$\frac{1}{16}$
$E(M) = 2, V(M) = \frac{1}{4}$	

As illustrated in Example 4.1, the sampling distribution of the mean is a theoretical distribution. Precisely, **the theoretical distribution M that associates the values of the sample mean with the probability of each value for all possible samples of size N is called the** *sampling distribution of the mean.* A few comments about this definition are needed. First, the sampling distribution of the mean is a theoretical distribution. Just as the population distribution typically is unknown to the researcher, so is the sampling distribution of the mean. The sampling distribution is unknown because the population distribution from which it is formed is unknown. Second, although the sampling distribution of the mean is clearly related to the population distribution, it is different. Third, because the sampling distribution is a probability distribution, the sampling distribution has parameters associated with it. For Example 4.1 the value of the expectation $E(M) = \mu_M$ can be calculated:

$$E(M) = \sum mP(M=m)$$
$$= 1(\tfrac{1}{16}) + (\tfrac{3}{2})(\tfrac{1}{4}) + 2(\tfrac{3}{8}) + (\tfrac{5}{2})(\tfrac{1}{4}) + 3(\tfrac{1}{16})$$
$$= 2$$

(Not by coincidence, this value equals the value of the expectation of the population distribution.) As well, $V(M) = \sigma_M^2 = \tfrac{1}{4}$ (which is not equal to the variance of the population distribution). Fourth, from the example and the definition it is clear that the distribution for the mean depends on the size of the sample. That is, the sampling distribution of the mean for samples of size 5 will differ from the distribution obtained from samples of another size (say, 10). Fifth, a sampling distribution of other statistics could be formed, such as the sampling distribution of the median or the sampling distribution of the variance. However, the sampling distribution of the mean is most important to our development of inferential statistics. Finally, the term "sampling" is often omitted, so that one speaks of the distribution of the mean. Later in this chapter the sampling distribution of the mean will be examined more closely.

At this point it will be helpful to review the relationship among the population distribution, the sample, and the sampling distribution of a statistic. The distribution of the random variable that defines the underlying process in the population is called the population distribution. The parameters that characterize the population distribution are denoted by Greek letters. Although the mean μ and the variance σ^2 (or standard deviation σ) are the parameters of primary interest, we let θ (theta) denote a *parameter in general.* From the population a sample of size N is drawn. From these N observations statistics are defined and denoted by upper-case letters. Again, the mean M and the variance S^2 (or the standard deviation S) are of primary interest, although any number of other statistics could be defined. We will use G to denote a *statistic in general.* The statistic G has a sampling distribution. The sampling distribution of G may be thought of in this way. Take a sample of size N and calculate G from that sample. Take another sample of size N and calculate G. Continue this process for all possible random samples. Associated with each value of G is a probability (or density for continuous distributions) that each value is obtained. This probability distribution of G is the sampling distribution of G. Of course, we will not go through any such process; statistical theory will allow us to derive the sampling distributions that are needed.

Because the sampling distribution is a probability distribution, it has parameters associated with it. If the distribution of G is discrete, then the expectation of G is defined in the usual way:

$$E(G) = \mu_G = \sum gP(G = g) \tag{4.1}$$

The subscript G for μ indicates that this is the mean or expectation for the distribution G. If sampling distribution of G is continuous

$$E(G) = \mu_G = \int gf(g)\, dg \tag{4.2}$$

Similarly, the variance of the sampling distribution is defined:

$$V(G) = \sigma_G^2 = E(G - \mu_G)^2 = E(G^2) - \mu_G^2 \tag{4.3}$$

The standard deviation $\sigma_G = \sqrt{\sigma_G^2}$ of a sampling distribution is a parameter to which we will refer often; consequently, it has a special name, the *standard error of the statistic* G. To reiterate, **the standard deviation σ_G of the sampling distribution of G is called the standard error of the statistic G.**

4.4 Estimation

One of the primary reasons for sampling is that the information contained in the sample is useful to estimate the population parameters. Suppose that we desire to estimate the mean of the population. The obvious choice of an estimator would be the sample mean. However, another reasonable estimator could be $(x_1 + x_N)/2$, the mean of the first and last observations. What makes one estimator better than another? Obviously, we would hope that the estimator yields values that are near the value of the parameter. However, as we will see in the next section, characterizing the goodness of estimators is fairly complex. Before discussing properties of estimators, we need to define the concept of estimation more precisely.

An *estimator* is a rule for calculating a value from a sample. Thus, the sample mean M is an estimator; the formula for M is the rule. For any particular sample, the value obtained by the rule is the *point estimate* (or simply estimate) for the parameter being estimated. If the parameter θ is estimated by G, then we write $G = \hat{\theta}$, where the "$\hat{\ }$" is read as "estimator for." For example, if the sample mean is used to estimate the population mean, then $M = \hat{\mu}$. It should be noted that the definition of an estimator is identical to the definition of a statistic (see Section 3.6, p. 51). It is the purpose of the value obtained by the rule that determines whether the term "statistic" or "estimator" is used. When M is used to describe the central tendency of a set of observations, M is called a statistic; when M is used to estimate the population mean, M is called an estimator.

Before examining the properties of estimators, we must explain one important detail. Consider a random variable X with a particular distribution and suppose that we draw a sample of size 1. What is the theoretical distribution of this one observation? To answer this question, perform the hypothetical experiment of drawing a sample of one repeated an infinite number of times. That is, randomly select a subject and note the score for this subject, replace the subject in the population and randomly select another subject and note this score, and so on. Clearly, the resulting distribution will be identical to the population distribution. For example, if $X \sim N(6, 9)$, then $x_i \sim N(6, 9)$.

4.5   Properties of Estimators

To illustrate the properties of estimators, consider the following example.

EXAMPLE 4.2

The concept of estimating from a sample is illustrated by sampling from a normal distribution with a mean 6 and a variance 9; that is, $X \sim N(6, 9)$. The two samples, one of size 10 and one of size 20, that appear in Table 4.4 were randomly selected from this distribution. For each sample two estimators of μ were calculated: the sample mean M and the mean of the first and last scores $A = (x_1 + x_N)/2$.

TABLE 4.4

Two Random Samples ($N = 10$, $N = 20$) from $X \sim N(6, 9)$

Sample 1	Sample 2	
$x_1 = 4.608$	$x_1 = 5.727$	$x_{11} = 10.458$
$x_2 = 8.976$	$x_2 = 4.908$	$x_{12} = 1.524$
$x_3 = 7.929$	$x_3 = 6.027$	$x_{13} = 6.966$
$x_4 = 5.763$	$x_4 = 3.612$	$x_{14} = 3.615$
$x_5 = 4.569$	$x_5 = 5.328$	$x_{15} = 7.302$
$x_6 = 9.993$	$x_6 = 5.211$	$x_{16} = 9.825$
$x_7 = 2.850$	$x_7 = 5.907$	$x_{17} = 9.441$
$x_8 = 0.036$	$x_8 = 0.036$	$x_{18} = 8.400$
$x_9 = 2.487$	$x_9 = 7.254$	$x_{19} = 6.516$
$x_{10} = 2.574$	$x_{10} = 3.081$	$x_{20} = 0.015$
$M = 5.31$	$M = 5.56$	
$A = (x_1 + x_N) = 3.59$	$A = (x_1 + x_N) = 2.87$	

Let θ be the population parameter to be estimated and $G = \hat{\theta}$ be the point estimator for θ. Each property of estimators is defined in terms of this notation, but is illustrated by Example 4.2.

Obviously, it is unlikely that the estimator for a parameter will equal the parameter of any sample. Clearly, none of the estimates formed in Example 4.2 equals the mean of the population. However, if one took many samples of size 10, it would be expected that the mean of the estimates would equal the parameter to be estimated. Although one does not have the luxury of drawing many samples, it is still a desirable property that the mean of the estimates of a large (infinite) number of samples will equal the population parameter. When an estimator has this property, it is said to be *unbiased*, and is defined in the following way: **$G = \hat{\theta}$ is an unbiased estimator of θ provided $E(G) = \theta$.** In words, G is unbiased provided the expectation of the sampling distribution for G is equal to θ. In light of this property, let us put our two estimators M and A to the test. If M is an unbiased estimator of μ, then $E(M) = \mu$. By the rules of expectation and the definition of the sample mean M for N independent observations

$$E(M) = E\left[\frac{1}{N}\sum x_i\right] = \frac{1}{N}E\left(\sum x_i\right) = \frac{1}{N}\sum E(x_i)$$

Because a single observation has the same distribution as the population (see Section 4.4, p. 91), $E(x_i) = \mu$. Thus, we have

$$E(M) = \frac{1}{N}\sum E(x_i) = \frac{1}{N}\sum \mu = \frac{1}{N}N\mu = \mu$$

To summarize,

$$E(M) = \mu \tag{4.4}$$

Therefore, M is an unbiased estimator of μ.

However, A is also an unbiased estimator of μ, since the expectation of A is also equal to μ:

$$E(A) = E\left[\frac{(x_1 + x_N)}{2}\right] = \frac{1}{2}[E(x_1) + E(x_N)] = \frac{1}{2}(2\mu) = \mu$$

With regard to bias, M and A pass the test. Unfortunately, not all sample statistics are unbiased estimators of their corresponding population parameters. It so happens that $E(S^2) \neq \sigma^2$; that is, if one takes the mean of the sample variance for an infinite number of samples of size N, the result will not be the population variance. In the next section an unbiased estimator of σ^2 will be discussed.

In Example 4.2 samples of sizes 10 and 20 were drawn. Hopefully, larger samples would yield better estimates. **If as the size of the sample increases, the probability that the estimator is close to the parameter increases, and the estimator is said to be *consistent*.** Because M is a consistent estimator of μ, a sample of size 20 would **probably** yield a better estimate than would a sample of size 10, which was the case in Example 4.2. It should be clear that statements about samples, such as the definition of consistency, must be written with reference to probability. There is no guarantee that the mean of a sample of size 20 will be closer to μ than the mean of a sample of size 10. In Example 4.2 there is a probability (although somewhat small) that a sample of size 20 would be uncharacteristic of the population and have a mean, say, 4.5, that was far from the population mean of 6.0. With regard to A, which is the mean of the first and last observations, clearly, this estimator is not consistent inasmuch as increasing the size of the sample does not improve the estimate.

A third desirable property of an estimator is called *relative efficiency*. Even when an estimator is unbiased and consistent, there will be error in the estimate; that is, for a sample of size N the estimator G will not likely equal the parameter θ. A measure of the degree to which the estimator G is dispersed around the parameter θ is given by $(1/N)\Sigma(G - \theta)^2$ for discrete distributions. However, if G is unbiased, this may be rewritten as $(1/N)\Sigma[G - E(G)]^2$, which is the variance of the sampling distribution of G, σ_G^2. Therefore, if σ_G^2 is small, the estimator is efficient. However, because the size of variances typically have meaning when compared to some other quantity (see Section 3.6, p. 51), efficiency is expressed as the comparison of two rival estimators. **Given a sample of size N and two estimators G and H with variances of the sampling distributions σ_G^2 and σ_H^2, respectively, then the efficiency of G relative to H is given by**

$$\frac{\sigma_H^2}{\sigma_G^2}$$

If G is efficient relative to H, then this ratio will be greater than 1, since more efficient estimators have smaller variances.

There are two other properties of estimators that we have not discussed; sufficiency and maximum likelihood. Although they are of importance to statisticians, they are not central to our discussion.

4.6   Unbiased Estimator for the Population Variance

As mentioned in the previous section, the sample variance was a biased estimator for the population variance; that is, $E(S^2) \neq \sigma^2$. It can be shown that

$$E(S^2) = \frac{N - 1}{N} \sigma^2$$

Clearly, because $(N - 1)/N$ is less than 1, S^2 underestimates σ^2. This bias can be corrected by defining an unbiased estimator for σ^2 (denoted by s^2), formed by increasing S^2 by a factor of $N/(N - 1)$. Thus, **the unbiased**

estimator for the population variance for N independent observation is

$$s^2 = \frac{N}{N-1} S^2 \qquad\qquad (4.5)$$

Of course, s^2 can be calculated directly from the sample.

$$s^2 = \frac{N}{N-1} S^2 = \left(\frac{N}{N-1}\right)\frac{1}{N}\sum (x_i - M)^2 = \frac{1}{N-1}\sum (x_i - M)^2 \quad (4.6)$$

Notation is half the battle. S^2 is the uncorrected variance of the sample and S is the standard deviation based on S^2. S^2 is a statistic that describes the dispersion in a set of observations. On the other hand, s^2 is the unbiased or corrected variance, and s is the standard deviation based on s^2. We use s^2 as the estimator of the population variance. Many formulas depend on the sample variance; typically, we write two equivalent versions of the formulas, one with S^2 (or S) and one with s^2 (or s) [For example, see Equation (4.10), p. 99]. Thus, a number of different strategies can be chosen. One can calculate S^2 from the sample and use the formula with S^2; one can calculate S^2 from the sample, convert to s^2 [Formula (4.5)] and use the formula with s^2; one can calculate s^2 and use the formula with s^2; or one can calculate s^2, convert to S^2, and use the formula with S^2. Students frequently object to using a formula with S^2 because S^2 is biased; rest assured that any formula for an unbiased estimator that is related to the variance and contains the term S^2 has a correction factor built in.

You might have noticed that we did not speak of s as the unbiased estimator of σ. In fact, s is a biased estimator of σ. Unfortunately, the correction of s depends on the form of the population distribution. Because the correction is small for moderately large samples, s is an adequate estimator of σ.

Sampling Distribution of the Mean

4.7   Mean and Variance of the Sampling Distribution of the Mean

Because the sampling distribution of the mean is so vitally important, we want to look at this distribution in more depth. Suppose that the random variable X that underlies some population has a mean μ and a variance σ^2. Remember that the sampling distribution of the sample mean is the theoretical distribution of M calculated for all possible samples of size N. In the derivations it will be required that the N observations are independent.

We have already proven that $E(M) = \mu$ [see Equation (4.4), p. 93] in the context of estimators. However, in this context the result indicates that **the mean of the sampling distribution of means is the same as the mean of the population.** That is, $\mu_M = \mu$.

Derivation of the variance of the sampling distribution of the mean is more complex. In the comment at the end of this section it is shown that

$$V(M) = \sigma_M^2 = \frac{\sigma^2}{N} \tag{4.7}$$

In other words, the variance of the sampling distribution of the mean (for independent samples of size N) is the variance of the population divided by the sample size.

Quite clearly, the greater the sample size is, the smaller is the variance of the sampling distribution of the mean. To illustrate the effect of the sample size on the variance of the sampling distribution of the mean, we present sampling distributions of the mean for several sample sizes in Figure 4.2. Each of the distributions are derived from a population that is normally distributed with a mean of 20 and a variance of 100. When the sample size is 1, the sampling distribution of the mean is identical to the population distribution, which is reasonable since the mean of a sample of 1 is equal to the score itself. When the sample size is increased to 10, the variance of the sampling distribution of the mean is 10; when the sample size is increased by a factor of 10 (to 100), the variance of the sampling distribution is reduced by a factor of 10 (to 1). In the extreme case, when the sample size is infinite (i.e., encompasses the entire population), the variance of the sampling distribution is zero; that is, the probability that M equals μ is 1 and the population mean has been estimated perfectly, which makes perfect sense. It should be clear that larger samples give better estimates of the population mean (i.e., the sample mean is a consistent estimator of the population mean).

The standard deviation of the sampling distribution of the mean is called the standard error of the mean and is given by

$$\sigma_M = \sqrt{\sigma_M^2} = \frac{\sigma}{\sqrt{N}} \tag{4.8}$$

COMMENT

We now show that

$$V(M) = \sigma_M^2 = \frac{\sigma^2}{N}$$

To derive the variance, we need at one point to manipulate M^2, which we examine closely now:

$$M^2 = \left(\frac{1}{N}\right)^2 \sum (x_1 + x_2 + \cdots + x_N)^2$$

From the rules of algebra

$$(x_1 + x_2)^2 = x_1^2 + x_2^2 + 2x_1 x_2$$

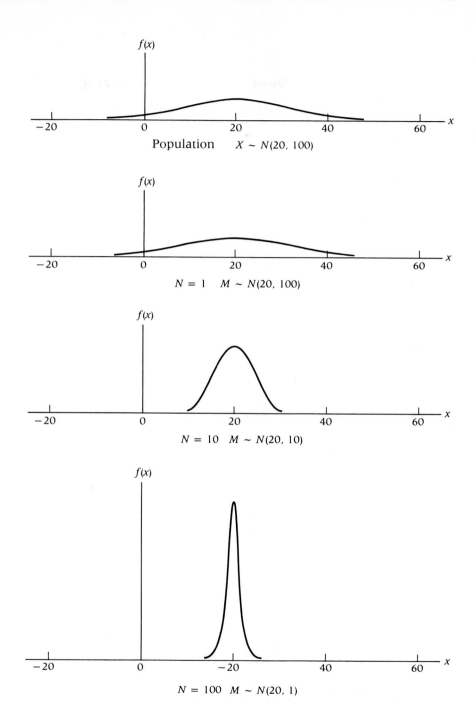

FIGURE 4.2   Sampling Distribution of M for Samples of Various Sizes

and

$$(x_1 + x_2 + x_3)^2 = x_1^2 + x_2^2 + x_3^2 + 2x_1x_2 + 2x_1x_3 + 2x_2x_3$$

With a little effort, similarly it can be shown that

$$(x_1 + x_2 + \cdots + x_N)^2 = x_1^2 + x_2^2 + \cdots + x_N^2 + 2\sum_{i<j} x_i x_j$$

Therefore

$$M^2 = \left(\frac{1}{N}\right)^2 (x_1^2 + x_2^2 + \cdots + x_N^2 + 2\sum_{i<j} x_i x_j)$$

Note that for a sample of size N, there are $(N)(N-1)/2$ pairs $x_i x_j$, where $i < j$. Because x_i and x_j are independent

$$E(2\sum_{i<j} x_i x_j) = 2\sum_{i<j} E(x_i x_j) = 2\sum_{i<j} \mu^2 = \frac{2(N)(N-1)\mu^2}{2} = N(N-1)\mu^2$$

Also, because $\sigma^2 = E(x_i^2) - \mu^2$ [see Equation (3.20), p. 56, and recall that the distribution of x_i is the same as X]

$$E(x_i^2) = \sigma^2 + \mu^2$$

So

$$
\begin{aligned}
E(M^2) &= \frac{E(x_1^2 + x_2^2 + \cdots + x_N^2 + 2\sum_{i<j} x_i x_j)}{N^2} \\
&= \frac{E(x_1^2 + x_2^2 + \cdots + x_N^2) + E(2\sum_{i<j} x_i x_j)}{N^2} \\
&= \frac{N(\sigma^2 + \mu^2) + N(N-1)\mu^2}{N^2} \\
&= \frac{N\sigma^2 + N^2\mu^2}{N^2} \\
&= \frac{\sigma^2}{N} + \mu^2
\end{aligned}
$$

Finally

$$V(M) = \sigma_M^2 = E(M^2) - \mu_M^2 = \frac{\sigma^2}{N} + \mu^2 - \mu_M^2$$

But since $\mu = \mu_M$

$$V(M) = \sigma_M^2 = \frac{\sigma^2}{N}$$

which is the desired result.

4.8   Estimating the Parameters of the Sampling Distribution of the Mean

In the previous section we showed that the mean μ_M and variance σ^2_M of the sampling distribution of the mean were given by the following formulas:

$$\mu_M = \mu \quad \text{and} \quad \sigma^2_M = \frac{\sigma^2}{N}$$

Obviously, the mean and variance of the sampling distribution of the mean can be calculated from these formulas only if the mean and variance of the population are known, which typically is not the case. In the same way that a sample was used to estimate the population parameters, a sample can be used to estimate the parameters of the sampling distribution of the mean. Because the mean of the sampling distribution of the mean is equal to the population mean,

$$\hat{\mu}_M = \hat{\mu} = M \tag{4.9}$$

Estimating the variance of the sample distribution of the mean is also straightforward:

$$\sigma^2_M = \frac{\hat{\sigma}^2}{N} = \frac{s^2}{N} = \frac{S^2}{N-1} \tag{4.10}$$

Similarly, we estimate the standard error of the mean:

$$\hat{\sigma}_M = \frac{\hat{\sigma}}{\sqrt{N}} = \frac{s}{\sqrt{N}} = \frac{S}{\sqrt{N-1}} \tag{4.11}$$

4.9   The Form of the Sampling Distribution of the Mean

In the previous sections we discussed the mean and variance of the sampling distribution of the mean. Earlier, we also noted that there are important aspects of a distribution other than the mean and variance, such as skewness (see Section 3.16, p. 76). That is, the mean and variance do not completely characterize the distribution. Thus, knowing the mean and variance of a sampling distribution of the mean does **not** imply that this distribution is known. However, when the population distribution is normal, the sampling distribution of the mean has a particularly nice form: **The sampling distribution of the mean derived from a normally distributed population distribution is itself normally distributed; that is,**

if

$$X \sim N(\mu, \sigma^2)$$

then

$$M \sim N(\mu, \frac{\sigma^2}{N}) \hspace{4cm} \text{(4.12)}$$

However, it should be remembered that typically the form of the population distribution is unknown to the researcher. Often we believe that the distribution of some traits is approximately normally distributed in the general population, such as intelligence or height. However, it is not reasonable that other distributions would be normal. Intelligence scores of applicants to elite graduate programs and height of professional basketball players are skewed to the left. Similarly, prices of homes in the United States are skewed to the right. For a number of reasons, not the least of which is simplicity, it would be nice if the population distribution were normal. Fortunately, even if the population is not normal, the sampling distribution of the mean may be very close to a normal distribution. This property is embodied in the following theorem, which is called the *central limit theorem*: **If a population has a finite variance σ^2 and a mean μ, then the sampling distribution of the mean for samples of N independent observations approaches a normal distribution with a variance σ_M^2 and a mean μ as the sample size increases.** Let us not minimize the importance of this theorem. It says that regardless of the form of the population distribution, the sampling distribution of the mean can be made arbitrarily close to a normal distribution by increasing sufficiently the size of the sample.

The obvious question is, "How large a sample size is large enough?" The answer to this question depends on how similar the population distribution is to a normal distribution. If the distribution resembles closely a normal distribution (perhaps, symmetric, unimodal, and slightly platykurtic), then moderate sample sizes (say, 5 or 10) will be sufficient to say that the sampling distribution of the mean is "approximately" normal. For other distributions, 30 observations might be required.

Some cautions should be expressed. If a researcher has good reason to believe that the population distribution is not normally distributed, then he or she is well advised to "transform" the distribution to a distribution that more closely resembles the normal distribution (see Section 6.9, p. 152). One should also be aware that the central limit theorem applies **only** to the sampling distribution of the mean. It does **not** apply to the sampling distribution of other statistics, such as the sample variance.

Frequently, the naive criticize statistical analyses of data on the grounds that it depends on the assumption of normality and that this assumption is not realistic in most instances. When confronted by such an argument, the reader needs only to cite the central limit theorem!

4.10 Standardized Scores for Sample Means

The mean obtained from a particular sample can be standardized in the usual way:

$$z_M = \frac{M - \mu}{\sigma_M} = \frac{M - \mu}{\sigma/\sqrt{N}} \qquad (4.13)$$

When the population distribution is normal, the sampling distribution of the mean is normal; thus, in this case

$$z_M \sim N(0, 1)$$

Notice that a sample mean is standardized by using "its own distribution," the sampling distribution of the mean. The following example, Example 4.3, illustrates a common statistical error that is made when this is ignored.

EXAMPLE 4.3   To make this illustration concrete, we use the construct of intelligence. Frequently, intelligence tests are constructed so that the mean is 100 and the standard deviation is 15. So, for this example, let us assume that the random variable X represents the score on such an intelligence test and that X is normally distributed. That is, $X \sim N(100, 225)$.

Suppose that a child was drawn at random and the administration of this intelligence test yielded a score of 115. To calculate the standardized score for this child, one uses the mean and standard deviation of the population distribution:

$$z = \frac{x - \mu}{\sigma} = \frac{115 - 100}{15} = 1.00$$

To see how "unusual" this child was, one would calculate the probability that the child would achieve a score of 115 or greater. This is easily accomplished by reference to the standard normal table (Table B.1, p. 449) because standardized scores are normally distributed if the population is normally distributed (see Section 3.15, p. 72).

$$P(Z \geq 1.00) = .159$$

That is, the probability of obtaining a child with a score of 115 or greater by chance would be .159. In other words, this child was in the top 16% of children, unusual but by no means extraordinary.

Now suppose that four children were drawn at random and that the *mean* of their scores was 115. Equation (4.13) yields a standardized score for this mean of

$$z_M = \frac{M - \mu}{\sigma/\sqrt{N}} = \frac{115 - 100}{15/\sqrt{4}} = 2.00$$

Because the sampling distribution of the mean derived from a normally distributed population is normal, the "unusualness" of this sample may be calculated by reference to the standard normal table:

$$P(Z_M \geq 2.00) = .023$$

Clearly, the probability of obtaining a sample of four children whose mean score was 115 or greater is unusual; it would happen by chance only about 2 times out of 100.

Now suppose that 25 children were randomly selected and the mean of this sample was 115. In this case

$$z_M = \frac{M - \mu}{\sigma/\sqrt{N}} = \frac{115 - 100}{15/\sqrt{25}} = 5.00$$

and

$$P(z_M \geq 5.00) < .0000001$$

a very unusual finding. Only 1 time in 10 million would you expect to find a sample of 25 children who had a mean intelligence score of 115 or greater!

Administrators rarely are aware of the sampling distribution of the mean. A school official who finds that the mean verbal SAT score for a sample of 400 students in his or her district was 485 is likely to say that the students in his or her district are about average because they are only 15 points below the mean (assume a population mean of 500 and a standard deviation of 100). However, note that $z_M = -3.00$ and the probability of obtaining a mean of 485 or smaller for a sample of this size is .001. It would have to be concluded that the achievement of students in this district was abominable.

NOTES AND SUPPLEMENTARY READINGS

Cochran, W. G. (1963). *Sampling techniques* (2nd ed.). New York: Wiley.

Sudman, S. (1976). *Applied sampling*. New York: Academic Press.

These volumes comprehensively examine various theoretical and applied aspects of sampling and provide very useful references for those with nearly any question pertaining to this aspect of planning or executing an investigation. Both discuss procedures that are appropriate for widely varying design configurations.

PROBLEMS

1. Suppose that the following set of observations were obtained by sampling from a normal distribution with mean μ and variance σ^2: {5, 7, 8, 10, 13, 17}.
   a. Estimate μ.
   b. Estimate σ^2.
   c. What is the best guess of the sampling distribution of the mean?

2. Suppose that the following set of observations were obtained by sampling from a normal distribution with mean μ and variance σ^2: {3, 5, 7, 10, 12, 12, 14}.
   a. Estimate μ.
   b. Estimate σ^2.
   c. What is the best guess of the sampling distribution of the mean?

3. Suppose that a population is characterized by the following distribution:

x	$P(X = x)$
2	$\frac{1}{4}$
4	$\frac{1}{2}$
6	$\frac{1}{4}$

   a. Find $E(X) = \mu$ and $V(X) = \sigma^2$.
   b. Find the mean and variance of the sampling distribution of M for samples of size 3 (i.e., $N = 3$) in two ways; (i) by knowing the parameters of the sampling distribution discussed in Section 4.7 and (ii) by listing all possible samples of size 3 (there are 27 different possibilities) as illustrated in Example 4.1.

4. Suppose that a population is characterized by the following distribution:

x	$P(X = x)$
0	$\frac{1}{8}$
2	$\frac{1}{2}$
4	$\frac{1}{8}$
6	$\frac{1}{4}$

   a. Find $E(X) = \mu$ and $V(X) = \sigma^2$.
   b. Find the mean and variance of the sampling distribution of M for samples of size 2 (i.e., $N = 2$) in two ways; (i) by knowing the parameters of the sampling distribution discussed in Section 4.7 and (ii) by listing all possible samples of size 2 (there are 16 different possibilities) as illustrated in Example 4.1.

CHAPTER **5** HYPOTHESIS TESTING

S cientifically naïve peple often think that the answers to scientific
problems are so clear that the answers literally hit the scientist
over the head, as was the case in the story of Sir Isaac Newton
and the apple. Although instances of scientific discovery by serendipity do
exist in the history of science, the usual course is for the scientist to make
conjectures about the nature of the world and then to put those conjectures
to some kind of test. The inspiration for such conjectures may originate from
careful observation, from a thorough understanding of previous research,
from someone's creative thought, from a suggestion of or discussion with a
colleague, or more likely, from a combination of these sources. In the aes-
thetic world of mathematics, for a statement to be true, it must **always** be
true. For example, in Euclidean geometry the statement "The sum of the
angles of a triangle equals 180°" is true for all triangles. For a mathematician
the fact that the sum of the angles of each of 100 randomly selected triangles
equals 180° is insufficient evidence to prove the statement. There is always
the possibility that the sum of the angles of the 101st triangle would not
equal 180°. In the complex "real" world rarely can an empirical test of a
conjecture about the mechanisms that govern the way "objects" behave yield
data that will prove the conjecture with absolute certainty. The evidence may

strongly support the conjecture or it may tend to disprove the conjecture. Even in the world of physics, in which the objects have no mind of their own, there is uncertainty. Einstein's theory of relativity is only a theory, in spite of the mathematical and empirical evidence that supports this theory. When the objects are people or even animals, the uncertainty is greater.

In the social and behavioral sciences the scientific method is complicated. A useful analogy with criminal law exists. Suppose that a defendant is charged with a crime. Two true states of affairs are possible: either the defendant is innocent or the defendant is guilty. Lawyers will present evidence that relates to the innocence or guilt of the defendant. The evidence may strongly indicate that the defendant is guilty (e.g., eyewitness testimony), may weakly indicate the defendant is guilty (e.g., circumstantial evidence), or may indicate that the defendant is innocent (e.g., alibi). However strong the evidence, it is impossible to know whether the defendant is innocent or guilty; that is, whatever judgment is made, there is room for error (uncertainty). Even if a jury convicts the defendant, there is the possibility that the defendant is innocent; similarly, if the defendant is found innocent, there is the possibility that he or she is guilty. Our jurisprudence system is conservative in the sense that we want the probability to be small that the defendant is found guilty when he or she is truly innocent. The criterion is "reasonableness"; the defendant is found guilty only if the evidence is sufficiently strong that the jury **believes** that the defendant is truly guilty without a reasonable doubt (i.e., "innocent until proven guilty"). This conservatism limits the probability that innocent people will be convicted of a crime. On the other hand, the probability that guilty people go free (i.e., are not convicted) is determined by the nature of the case. If the evidence is strong, the case is presented by a skilled prosecutor, and so forth, then this probability will be small. However, there is no guarantee in the legal system that this probability will be small.

Statistical inference utilizes a similar strategy. A researcher has some conjecture about psychological processes or constructs. He or she collects evidence that will provide useful information about the truth of that conjecture. A statistical test is conducted to indicate if the evidence is sufficiently supportive of the conjecture to "accept" it. Keep in mind, we have not "proven" that the conjecture is true; we only have strong evidence that it is true. It might be said that the conjecture is "supported" or "corroborated." As is the case in jurisprudence, science is conservative. The evidence must be very strong to accept the conjecture, which limits the probability that conjectures will be accepted as true when they are not. For instance, we do not want to be told that drugs are effective when they are not or that criminal behavior is genetic when it is not or that schizophrenia is a result of deviant family systems when it is not. Although individuals may not operate from this conservative stance (e.g., are willing to use a drug that has not been shown to be effective), generally science does.

In this chapter the process by which statistical hypotheses, which are related to the conjectures, are tested. The approach that is presented is a traditional one; students should be aware that other models of inference, such as Bayesian inference, can be adopted.

Hypothesis-Testing Strategies

5.1   Statistical Hypotheses

The term "conjecture" was used in the introduction to this chapter because the terms "scientific" and "statistical hypotheses" had not been defined. A *scientific hypothesis* is a conjecture about the nature of things; that is, a scientific hypothesis is a statement about relationships among constructs that may or may not be true. For instance, it might be hypothesized that hyperactive children have normal intelligence. A *statistical hypothesis* is a statement about one or more population distributions. Because statistical hypotheses make statements about population distributions, the characteristics of which are unknown, the truth of a statistical hypothesis is unknown (hence the word hypothesis!). Typically, statistical hypotheses make a statement about the form of the population distribution or about the parameters of the population distribution. For instance, it might be hypothesized that the population of hyperactive children has a mean intelligence quotient of 100. Clearly, scientific hypotheses and statistical hypotheses are related. Hopefully, statistical hypotheses are generated from scientific hypotheses and tests of statistical hypotheses provide evidence that is relevant to the determination of the truthfulness of scientific hypotheses. Regardless of the relationship between statistical and scientific hypotheses, statistical inference relies only on statistical hypotheses. In this text hypothesis refers to a statistical hypothesis.

We write a statistical hypothesis by an "H" followed by a statement about one or more population distributions. For example, the hypothesis about intelligence would be written as

H: the mean of the population is 100

or more simply

H: $\mu = 100$

The preceding hypothesis was concerned only with one parameter. Hypotheses may specify more than one parameter, for example

H: $\mu = 100$ and $\sigma = 15$

Or, the hypothesis might completely characterize the distribution:

H: $X \sim N(100, 225)$

The hypothesis need not be exact; for instance, the following hypothesis is permissible, and as we will see, very useful:

H: $\mu > 100$

It must be emphasized that statistical hypotheses are concerned with population distributions. Statistical hypotheses are never stated in terms of samples. For instance, the statement

$H: M = 100$

is not permissible; in fact, it is not a hypothesis because we know with certainty whether or not it is true. If the mean of the sample of 50 hyperactive children was 101, then the statement that $H: M = 100$ would be false. Of course, the fact that the sample mean was 101 does not indicate that $H: \mu = 100$ is false.

5.2   Hypothesis-Testing Strategy

The first step in hypothesis-testing strategies is to select two hypotheses such that if the first hypothesis is true, the second is false and, conversely, if the first is false, then the second is true. The guilt or innocence of a defendant in the jurisprudence example is an example of two hypotheses that meet this criterion. A more relevant example is the following:

$H: \mu = 100$

versus

$H: \mu \neq 100$

The "versus" is used because the two hypotheses "compete" in the sense that they both cannot be true.

The second step in hypothesis testing is to assume that one of the hypotheses is true. This hypothesis is designated as the null hypothesis (denoted by H_0) and the other is called the alternative hypothesis (denoted by H_a). In jurisprudence the defendant is assumed innocent until proven guilty; hence, innocence is the null hypothesis and guilt is the alternative. Although choice of the null hypothesis for a statistical example may be magical at this point, it will become clearer as this section progresses. For the moment continue with our statistical example

$H_0: \mu = 100$

versus

$H_a: \mu \neq 100$

The third step in hypothesis testing is to collect evidence that clarifies the nature of the truthfulness of the hypotheses. In statistical inference the evidence is derived from a sample or samples from population distributions and a statistical test is conducted on that evidence. The evidence is tested under the assumption that the null hypothesis is true. However, if the evidence is so incompatible with the null hypothesis that the assumption of truth of the null hypothesis is no longer reasonable, then the null hypothesis is rejected. The conservative nature of science indicates that belief in the null hypothesis will not be overturned unless the evidence is extremely convincing.

The role of the scientist is to collect the best evidence possible so that decisions to accept or reject a null hypothesis minimizes the error of making the wrong decision. Typically, a scientist's belief about the nature of the world is framed as an alternative hypothesis. The evidence must be terribly convincing before the scientist's view of the world will be accepted. In this sense a scientist takes the role of the prosecutor; that is, the scientist wishes to present evidence that the null hypothesis is unreasonable and the scientific community (jury) should believe that the alternative is true. For example, with regard to the efficacy of a newly developed drug, a researcher's hypothesis that the drug is effective is tested against the null hypothesis that it is not effective. Generally, but not always, the null hypothesis indicates that there is no effect, or that there is no relation among variables, or that one treatment is not more effective than another. The alternative then is that there is an effect, or that there is a relationship among the variables, or that one treatment is more effective than another. There are times, however, when the researcher hopes that the evidence indicates that the null hypothesis is true (e.g., hyperactive children have average intelligence). At other times, rejection of the null hypothesis will indicate that an assumption necessary for another statistical test is unlikely to be true (see Section 5.4, p. 110).

5.3   Type I and Type II Errors

Suppose that the evidence against a null hypothesis is very strong and that the null hypothesis is rejected (i.e., the alternative is accepted). However, we are not certain that the null hypothesis is false; we may have made an incorrect decision to reject the null hypothesis. Such an error is called a Type I error. Formally, **a Type I error is made when it is decided to reject the null hypothesis (i.e., accept the alternative) when the null hypothesis is true.** The probability of making a Type I error is denoted by α. In conditional probabilities we have

$$\alpha = P(H_0 \text{ is rejected} \mid H_0 \text{ is true}) \tag{5.1}$$

In terms of the jurisprudence example, α is the probability that the defendant is found guilty given that he or she is truly innocent. As we mentioned earlier, science is conservative in the sense that the criteria for rejection are selected so that α is small.

There is another possible error that can be made. **A Type II error is made when it is decided that the null hypothesis will not be re-**

jected when the alternative is true. The probability of making a Type II error is denoted by β. In conditional probabilities we have

$$\beta = P(H_0 \text{ is not rejected} \mid H_a \text{ is true}) \qquad (5.2)$$

In terms of the jurisprudence example β is the probability that the defendant is not found guilty given that he or she is truly guilty.

The relationship between the decision made and the true state of affairs is shown here:

	True State of Affairs	
	H_0 true (H_a false)	H_a true (H_0 false)
Do not reject H_0	$1 - \alpha$	β
Reject H_0 (**Accept** H_a)	α	$1 - \beta$ (power)

(row label: **Decision**)

If α and β are the probabilities of making errors, then $1 - \alpha$ and $1 - \beta$ are the probabilities of making correct decisions. Indeed

$$1 - \alpha = P(H_0 \text{ is not rejected} \mid H_0 \text{ is true})$$

and

$$1 - \beta = P(H_0 \text{ is rejected} \mid H_0 \text{ is false})$$
$$= P(H_a \text{ is accepted} \mid H_a \text{ is true}) \qquad (5.3)$$

The probability $1 - \beta$ is called the *power* of the statistical test. Although we will spend considerable effort to explain power in detail later in this chapter, clearly, we desire that the power of a statistical test be high so that H_a can be accepted if indeed it is true.

One important misconception needs to be eliminated. Consider again the jurisprudence example. Suppose that after hearing the evidence the judge (or the jury foreperson) makes the following statement: "I am 95% certain that the defendant is guilty." This statement could be rewritten as

$$P(\text{guilt}) = .95$$

or in terms of hypotheses

$$P(H_a \text{ is true}) = .95$$

These statements are very different from those for α and β [Equations (5.1) and (5.2)], which addressed the probability of accepting or rejecting hypotheses given the true state of affairs. Is the judge's statement legitimate? It is not and some effort is necessary to understand why not. H_a (and H_0) is either true or false—there is no middle ground. So, either $P(H_a$ is true) $= 0$ or $P(H_0$ is true) $= 1$ (of course, we don't know which). How would we interpret $P(H_a$ is true) $= .95$? Interpretations such as "95% of the time H_a is true" is clearly absurd. Remember that in classical hypothesis testing, **we refer to probabilities of a decision conditional on the true state of affairs; we do not refer to the probability that a hypothesis is true or false.**

5.4   Other Assumptions

As mentioned previously, hypothesis testing involves assuming H_0 is true and collecting evidence about the reasonableness of that assumption. If the evidence is compelling that the assumption of the truth of the null hypothesis is unreasonable, then the decision to reject H_0 is made. Typically, the intricacies of the statistical test also involve making one or more other assumptions about the population distribution. These collateral assumptions are important to the validity of the statistical test.

Consider again the example

$$H_0: \mu = 100$$

versus

$$H_a: \mu \neq 100$$

The hypothesis-testing strategy requires that these hypotheses be tested under the assumption that the null hypothesis is true; that is, $\mu = 100$. Acting under that assumption, we can collect evidence and perform a statistical test. The statistical test will indicate whether the assumption that the null hypothesis is true is reasonable. However, the statistical test may require that the population distribution be normal. Although it is unknown whether or not the population distribution is normal, we often assume that it is. A statistical test of the hypothesis only gives information about the reasonableness of the assumption that μ equals 100 (i.e., H_0 is true); it does not give any information about the reasonableness of the assumption of normality.

If the collateral assumptions (e.g., normality) are not true, then the decision to accept or reject H_0 may be flawed; that is, the probability of making a Type I error may be greater than realized. It would not be surprising if a reader made the following inference: The validity of the statistical test used to accept or reject the null hypothesis depends on some collateral assumptions, the truth of which are unknown; thus, there is precious little reason to conduct the statistical test in the first place. Fortunately, all is not lost. One solution is to conduct a preliminary test to determine the reasonableness of the collateral assumption. The second, and more often employed strategy, involves knowing the degree to which violations of the collateral assumption invalidates the statistical test. **If the validity of a statistical test is insensitive to violation of a collateral assumption, then the test is said to be *robust* with respect to the violation of that assumption.** The robustness of a statistical test is determined by statistical theory or by simulations (known as "monte-carlo" experiments). Recall that the central limit theorem (see Section 4.9, p. 99) states that the sampling distribution of the mean approaches a normal distribution as the sample size increases regardless of the form of the population distribution. Therefore, violation of the assumption of normality of the population distribution has little effect, for reasonably large sample sizes, on the sampling distribution of the mean. The robustness of statistical tests will be discussed throughout this text (see especially Section 6.8, p. 150).

5.5   Hypothesis Testing Illustrated and Formalized

Although hypothesis testing has been discussed rather extensively, the presentation has been primarily intuitive. Because hypothesis testing is the basis for inferential statistics, it is important to present the process more formally. Example 5.1 will be useful in illustrating hypothesis testing and in setting the stage for a more formal exposition. At this point the example needs to be rather artificial; by Chapter 6 the remaining statistical theory needed to examine a realistic example will be presented.

The administrator in Example 5.1 is now ready to conduct a statistical test that will determine whether or not the null hypothesis will be rejected, provided the assumption is made that the population distribution of intelligence scores is normally distributed. Because the decision to accept or reject the null hypothesis is made on the basis of the sample mean, the sampling distribution of the mean must be examined. Given the collateral assumption of normality and our knowledge of the sampling distribution of the mean (see Section 4.9, p. 99), the hypotheses for this example can now be rewritten as follows:

$$H_0: M \sim N(100, {}^{225}\!/_{16})$$

versus

$$H_a: M \sim N(110, {}^{225}\!/_{16})$$

(Although hypotheses typically are not written in this form, they are perfectly legitimate, because the sampling distribution of the mean is a theoretical distribution). The hypothesized distributions for H_0 and H_a are illustrated in Figure 5.1. This figure provides a wealth of information about hypothesis testing and should be studied carefully as the discussion in this and other sections continues.

EXAMPLE 5.1

Suppose that the following dilemma is facing a school administrator. The administrator is being offered the position as principal of one of two schools, a typical public school or a prestigious preparatory school. However, she is not allowed to choose the school. Nevertheless, she will be able to visit the school, conduct an experiment, and decide whether or not she wants to accept the position, although she will not be told which school she is visiting. The administrator does know that the mean and variance of the intelligence test scores of students in public schools are 100 and 225, respectively. Similarly, she knows that the mean and variance of the intelligence test scores of students in preparatory schools are 110 and 225, respectively. The administrator wants very dearly to be the principal of the preparatory school, but she does not want to accept the position in the public school (she is an elitist). Therefore, the following hypotheses make sense:

$$H_0: \mu = 100$$

versus

$$H_a: \mu = 110$$

(Because there are only two choices for μ in this fabricated example, if H_0 is true, H_a is false, and conversely, if H_0 is false, H_a is true.) If she decides to accept H_0 (i.e., the school is the public school), then she will refuse to take the position; if she decides to reject H_0 (i.e., accept the alternative that the school is the preparatory school), she will accept the position. Because of her aversion to the public school position, she does not want to decide to take the position if the school is in fact the public school. In other words, she does not want to make a Type I error. She is willing to have a 5% chance of making the wrong decision, so she sets α, the probability of deciding to take the position when the school is the public school, at .05. She now conducts the following experiment. Sixteen students from the school are randomly selected and their intelligence test scores are obtained from the school records. The sample mean $M = (1/N)\Sigma \, x_i$ is calculated and found to be 107. Clearly, she thinks that she has sampled from the preparatory school, but is a sample mean of 107 sufficiently large to take the risk of accepting the position at the public school? It is possible that a random sample of 16 students from the public school would yield a sample mean of 107. Sufficient statistical theory has been presented so that we can conduct a test of the hypotheses and decide whether the administrator should accept the position given the sample mean of 107 and the willingness to have a 5% chance of making the wrong decision.

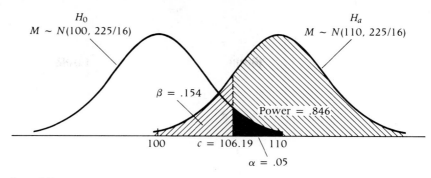

FIGURE 5.1 α, β, and Power

Our problem is this. We want to find a value c, called the critical value, such that if the sample mean M exceeds this value, the null hypothesis will be rejected. Of course, c must be selected so that the value of α is preserved. Given all the information for our example, the statement

$$\alpha = P(H_0 \text{ is rejected} \mid H_0 \text{ is true})$$

may now be rewritten as

$$.05 = P(M > c \mid M \sim N(100, {}^{225}\!/_{16})) \tag{5.4}$$

The meaning of c is clarified by referring to Figure 5.1. The value of c is selected so that the area under the distribution for the null hypothesis to the right of c is equal to .05, the α selected by the administrator. Clearly, in this example α is the area to the right since large values of the sample mean lead to rejection of the null hypothesis. The value of c is calculated from Equation (5.4) and if 107, the sample mean obtained, exceeds this value, the null hypothesis will be rejected and the administrator will take the principalship.

For clarity, Equation (5.4) is rewritten without the conditional statement, with the understanding that all calculations are now performed with the assumption that H_0 is true:

$$P(M > c) = .05$$

Clearly

$$P(M \leq c) = .95$$

Because we are assuming that H_0 is true (i.e., $M \sim N(100, {}^{225}\!/_{16})$), M may be standardized so that we now have

$$P\left(\frac{M - 100}{\sqrt{{}^{225}\!/_{16}}} \leq \frac{c - 100}{\sqrt{{}^{225}\!/_{16}}}\right) = .95$$

and

$$P\left(Z \le \frac{c - 100}{15/4}\right) = .95 \tag{5.5}$$

For the standard normal distribution $F(1.65) = P(Z \le 1.65) = .95$; consequently, from Equation (5.5) we can deduce that

$$1.65 = \frac{c - 100}{15/4}$$

and therefore

$$c = (15/4)(1.65) + 100 = 106.19$$

Because the value of c is so important, we reiterate its meaning. The value of 106.19 was selected so that the probability of observing a value of M greater than 106.19 is equal to .05 when the null hypothesis that the school is a public school is true (and the collateral assumption of normality is true). Because the observed value equaled 107, the administrator decides to reject the null hypothesis and hence agrees to accept the position.

Given the decision to reject the null hypothesis if the observed mean was greater than 106.19, β and $1 - \beta$ can be calculated easily. Recall that

$$\beta = P(H_0 \text{ is not rejected} \mid H_a \text{ is true})$$

Given the specifics of the continuing example,

$$\beta = P(M \le 106.19 \mid M \sim N(110, 225/16))$$

Standardizing M under the assumption of the alternative hypothesis, we obtain

$$\beta = P\left(\frac{M - 110}{\sqrt{225/16}} \le \frac{106.19 - 110}{\sqrt{225/16}}\right)$$
$$= P(Z \le -1.02)$$
$$= 1 - F(1.02)$$
$$= .154$$

Then

$$\text{Power} = 1 - \beta = P(H_a \text{ is accepted} \mid H_a \text{ is true}) = .846$$

That is, given the decision rule to accept the teaching position (reject H_0) if the obtained value of M was greater than 106.19, the probability of taking the position given the school was the private school was .846.

We may now state more formally the hypothesis-testing convention accepted by statisticians:

Set α, the probability of falsely rejecting H_0, equal to some small value. Then, considering the alternative hypothesis H_a, choose a region of rejection such that the probability of observing a sample value in that region is less than or equal to α

when H_0 is true. If the value of the sample statistic falls within the rejection region, the decision is made to reject the null hypothesis.

Typically, α is set at .05, .01, or .001, although these values are completely arbitrary. The alternative hypothesis needs to be considered to determine the direction of rejection; in the continuing example the obtained mean M needed to be sufficiently large to reject H_0 because the alternative hypothesis stipulated that the population mean was greater than that stipulated by the null hypotheses. In Sections 5.7 (p. 118) and 5.8 (p. 120) the direction of hypotheses will be discussed in more depth.

5.6 Test Statistics

Infrequently, one finds critical values for the rejection region presented in published research. This is so because there is a convenient shortcut to determining whether to reject the null hypothesis. Suppose that the following hypotheses are stipulated:

$$H_0: \mu = \mu_0$$

versus

$$H_a: \mu = \mu_a$$

where μ_0 and μ_a are hypothesized means for the null and alternative hypotheses, respectively (for the present assume $\mu_a > \mu_0$). Assume that the population distribution is normal and that the population standard deviation is known. Suppose that a sample of size N is randomly selected from this population. For an α of a given size the task is to identify a critical value c such that the null hypothesis will be rejected if the sample mean M exceeds c; simply, the rule may be stated as

Reject H_0 provided $M > c$

Recall that the particular value c for our example was found by standardizing M under the null hypothesis. However, this may be done more generally:

$$\text{Reject } H_0 \text{ provided } \frac{M - \mu_0}{\sigma/\sqrt{N}} > \frac{c - \mu_0}{\sigma/\sqrt{N}}$$

The right-hand side of this equation is a constant, which we call c'. The left-hand side of the equation is the random variable Z, which has a standard normal distribution (under the collateral assumption of normality). We may now say that H_0 will be rejected, provided

$$z = \frac{M - \mu_0}{\sigma/\sqrt{N}} > c'$$

or

$$z = \frac{M - \mu_0}{\sigma_M} > c'$$

since $\sigma/\sqrt{N} = \sigma_M$ (see Section 4.7, p. 95). Because Z has a standard normal distribution, determination of c' is straightforward, as illustrated in Figure 5.2. If α was set at .05, we want to find the value c' such that $F(c') = .95$. From the cumulative distribution for the standard normal (See Table B.1, p. 449), c' equals 1.65. Similarly, if α was set at .01, then c' would equal 2.33.

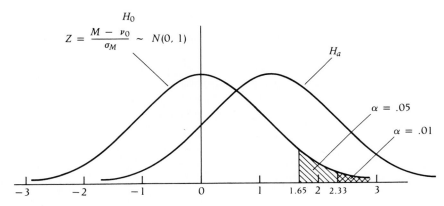

FIGURE 5.2   One-Tailed Rejection Regions for Z ($\mu_a > \mu_0$)

We can apply this shortcut to the problem of determining whether to select the principalship of the school. In this example

$$z = \frac{M - \mu_0}{\sigma/\sqrt{N}} = \frac{107 - 100}{\sqrt{225/16}} = 1.87$$

Thus, because this value exceeds 1.65, the null hypothesis may be rejected at the preset level of .05.

Because this "shortcut" represents the prototype for parametric statistical testing, it is useful to emphasize the method. As is the case in hypothesis testing, the null hypothesis and the alternative are specified as well as the level of α set by the experimenter. A sample is taken. From this sample a statistic is calculated. Here we used the statistic

$$z = \frac{M - \mu_0}{\sigma/\sqrt{N}} = \frac{M - \mu_0}{\sigma_M} \tag{5.6}$$

This statistic differs from previous statistics in that it is not used descriptively (as would be M or S^2) or as an estimator for a population parameter (as would be M or s^2). **Because this statistic is used to test hypotheses, it is called a *test statistic*.** Just as other statistics have sampling distributions, test statistics have sampling distributions. The distribution of the test statistic z described in Equation (5.6) is the standard normal, provided the population distribution is normal (see Figure 5.2). If the value of the test statistic (or as we will see, in some cases the absolute value of the test statistic) is sufficiently large, the null hypothesis is rejected.

It should be emphasized that the test statistic described in Equation (5.6) has no practical utility because σ_M is not known to the researcher. It would seem reasonable that one might estimate σ_M with $s/\sqrt{N}$, as described in Section 4.8 (p. 99). However, when this estimation is incorporated in the test statistic, the test statistic no longer has a normal distribution. Although for large N the effect on the decision to reject the null hypothesis is not drastic, for small or moderate N, it is necessary to be cognizant that the test statistic is not normally distributed. This change in distribution will be discussed in some detail in the next chapter.

COMMENT   Often one reads in articles of a "significance level" or a "p-value." These terms refer to the probability that the test statistic exceeds the obtained value of the test statistic under the assumption that the null hypothesis is true. If we represent this probability by p, then

$$p = P(Z > \text{obtained value} \mid H_0 \text{ is true})$$

In our continuing example

$$p = P(Z > 1.87) = .031$$

Clearly, because this value of p is less than the preset value of α of .05, the null hypothesis is rejected. However, there seems to be a widespread and inappropriate use of p-values. Instead of presetting α, the researcher calculates p. If $.05 > p \geq .01$, then the researcher says that the null hypothesis is rejected at the .05 level; if $.01 > p \geq .001$, then the researcher says that the null hypothesis is rejected at the .01 level, and so forth. Or, equally troublesome, the researcher says that the null hypothesis is rejected at the p level (in our case, at the .031 level). The convention to which we adhere stipulates that α be set **before** conducting the experiment, and the null hypothesis is either rejected or not rejected at that level. The level of α should be determined by how much the researcher wishes to avoid falsely rejecting the null hypothesis and not by sample data. Nevertheless, p has a nice heuristic interpretation. The p-value may be thought of as the probability of obtaining a sample value as or more deviant than the obtained one, given that the null hypothesis is true. In our continuing example we could say that about only 3 times in 100 would a sample mean of 107 be obtained if the population or true mean was 100.

5.7   Specification of the Alternative Hypothesis: One-Tailed Tests

The examples to this point have specified the null and alternative hypotheses exactly, as illustrated in Example 5.1 (p. 112). It is extremely unusual that a research problem would dictate that the hypotheses be formulated in an exact way. Typically, the research hypothesis would be more general, such as "It is hypothesized that the population of children being studied has below average intelligence." The null and alternative statistical hypotheses would then be written as follows:

$$H_0: \mu = 100$$

versus

$$H_a: \mu < 100$$

It should be noted that specification of the alternative inexactly (i.e., stated with an inequality) does not change the manner in which the hypothesis-testing strategy is conducted. This should be clear if we remember that determination of the rejection region is determined under the null hypothesis, which is stated exactly. These hypotheses also differ from those in Example 5.1 in another way. In this case the alternative states that the population mean is less than that stipulated in the null hypothesis. To reject the null hypothesis, therefore, an obtained mean of considerably **less** than 100 is needed. Figure 5.3 illustrates the rejection regions for the test statistic we used previously (see Equation (5.6), p. 116) when the alternative hypothesis specified a population mean greater than and less than that specified by the null hypothesis. **Because an alternative hypothesis that specifies a direction results in a rejection region that falls at one or the other extreme of the sampling distribution of the test statistic, the test of the hypothesis is called a one-tailed test.**

These concepts are illustrated by using the inexact alternative hypothesis discussed in this section. Suppose that the population is normally distributed with a variance of 225, a random sample of 16 was selected, and a mean of 97.5 was obtained. Would the null hypothesis be rejected at the .05 level? Using the test statistic of Equation (5.6), we calculate

$$z = \frac{M - \mu_0}{\sigma/\sqrt{N}} = \frac{97.5 - 100}{15/\sqrt{16}} = -.667$$

but because this value is not less than -1.65, the null hypothesis cannot be rejected.

The curious reader may be wondering what the researcher would do if a mean of 109 was obtained, a value that makes the null hypothesis appear unlikely. The answer to this problem lies in examining the hypotheses of this section further:

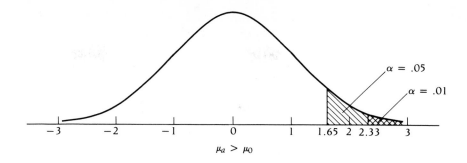

$\mu_a > \mu_0$

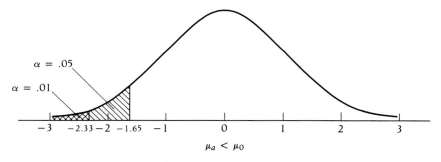

$\mu_a < \mu_0$

FIGURE 5.3   One-Tailed Rejection Regions for Standard Normal Distribution

$H_0: \mu = 100$

versus

$H_a: \mu < 100$

Clearly, these two hypotheses do not cover all the possibilities for μ. Although hypotheses are typically written in this form, the proper way to write them would be

$H_0: \mu \geq 100$

versus

$H_a: \mu < 100$

When written in this fashion, an obtained mean of 109 is included in the null hypothesis, and thus could not possibly provide evidence for its rejection. Because the researcher wanted to demonstrate that the true mean was less than 100, an obtained mean of 109 does not corroborate his or her claim in any way. Although it might be claimed that the evidence indicates that the true mean is greater than 100, the research was not designed with that purpose in mind. Nevertheless, such a finding is interesting, and the researcher way wish to verify this conjecture (viz., $H_a: \mu > 100$) with another experiment. To reiterate, **In a one-tailed test an obtained test statistic in the direction opposite from that predicted (i.e., as stated in the alternative hypothesis) provides no evidence for the rejection of the null hypothesis, and hence the null hypothesis is not rejected.** If the researcher has no notion about the direction of the test, a two-tailed test, as described in the next section, would be used.

5.8   Specification of the Alternative: Two-Tailed Tests

An example of hypotheses for a two-tailed test would be the following:

H_0: μ = 100

versus

H_a: μ ≠ 100

In other words, the alternative hypothesis is that the population mean is not equal to 100. In this case an obtained mean that was less than or greater than 100 would provide evidence that might result in the rejection of H_0; thus, there are rejection regions in each tail of the sampling distribution of the test statistic. Although not necessary, it is typical that the area in each of the tails of the distribution be equal to $\alpha/2$, thus making it equally likely that the null hypothesis be rejected in the two directions. Using the test statistic given in Equation (5.6) (p. 116), we illustrate the two rejection regions in Figure 5.4. Algebraically, the null will be rejected at the .05 level, provided

$$\left| \frac{M - \mu_0}{\sigma/\sqrt{N}} \right| > 1.96 \tag{5.7}$$

and at the .01 level, provided

$$\left| \frac{M - \mu_0}{\sigma/\sqrt{N}} \right| > 2.58 \tag{5.8}$$

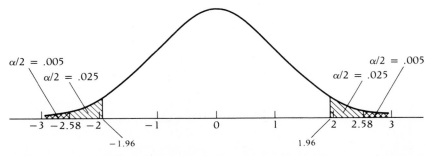

FIGURE 5.4   Two-Tailed Rejection Regions for Standard Normal Distribution

At first glance it might seem that a two-tailed test is preferable because the null hypothesis can be rejected regardless of the direction of the obtained mean from the population mean specified in the null hypothesis. However, as we will see over and over again, in the area of statistics there are "no free lunches." The price to be paid for choosing a two-tailed test can be illustrated easily. Returning for the moment to Example 5.1 (p. 112), recall that a mean of 107 was obtained, which was sufficiently large that the null hypothesis was rejected when a one-tailed test was used ($\alpha = .05$). For the two-tailed test calculation of the test statistic proceeds in the same fashion:

$$z = \frac{M - \mu_0}{\sigma/\sqrt{N}} = \frac{107 - 100}{15/\sqrt{16}} = 1.87$$

which is now **not** sufficiently large to reject the null hypothesis [cf. Equation (5.7)].

Although the issue of one- versus two-tailed tests will be discussed further in the next section, the following point needs to be made. When the researcher has a reason to specify the direction of the test, a one-tailed test is recommended. The rationale for a directional test may come from past research, theory, common sense, or utility. For example, when testing the efficacy of a drug, it would be predicted that subjects taking the drug would do "better" than those who are not receiving the drug. However, if the relative efficacy of two drugs was being tested and there was no reason to think one drug was superior to the other, a two-tailed test would be used.

Related Topics

5.9 Power

Recall that power was defined in the following way:

Power $= 1 - \beta = P(H_a \text{ is accepted} \mid H_a \text{ is true})$

Given the researcher's desire to accept the alternative hypothesis it is clear that the power of a statistical test should be large. There is little reason to conduct an experiment that lacks sufficient power to reject the null hypothesis when in fact the alternative is true.

In one sense power is related to the ease with which the null hypothesis is rejected (when the alternative is true). Because it is desirable to design a study so that the null hypothesis can be rejected when the alternative is true, it is useful to identify those factors that affect the power of a statistical test. By examining rejection regions for the test statistic found in Equation (5.7) (p. 120), we can identify some of the factors affecting power. Recall that the null hypothesis for a two-tailed test will be rejected, provided

$$\left| \frac{M - \mu_0}{\sigma/\sqrt{N}} \right| > c_{\alpha/2} \tag{5.9}$$

where $c_{\alpha/2}$ is the critical value for a given level of α. The ease with which the null hypothesis is rejected can be increased by decreasing the value of c

[i.e., the right-hand side of Inequality (5.9)] or by increasing the value of the test statistic [i.e., the left-hand side of Inequality (5.9)]. The value of c can be decreased in two ways, by increasing the value of α selected by the researcher or by changing from a nondirectional (two-tailed) test to a directional (one-tailed) test. The size of the test statistic is affected by the size of the sample and the standard deviation of the population. Reducing the standard deviation or increasing N or both will increase the size of the test statistic.

To determine how these and other factors affect the power of a statistical test, we return to Example 5.1 (p. 112). Given the hypotheses

$$H_0: \mu = 100$$

versus

$$H_a: \mu = 110$$

and the details described in Example 5.1, power was calculated (see Section 5.5, p. 111) and found to equal .846; that is, the probability that the alternative hypothesis will be correctly accepted is .846. This value is illustrated in Figure 5.5.

Clearly, changing the size of α affects the power of the statistical test. If α is decreased (say, from .05 to .01), it becomes more difficult to reject the null hypothesis. This is demonstrated by our continuing example (Example 5.1, p. 112) for which all aspects remain unchanged ($\sigma^2 = 225$, $N = 16$) with the exception that α is changed from .05 to .01. The change in the critical value for rejection and power is illustrated in Figure 5.5. Clearly, it is now more difficult to reject the null hypothesis (a sample mean of 108.74 is required, as opposed to 106.19) and power has been reduced (from .846 to .633). (A good exercise would be to calculate these values using the method of Section 5.5, p. 111.) As a rule, **reducing the size of α reduces the power of the statistical test.**

Another way to modify power is to change the size of the sample. Intuitively, increasing sample size increases the accuracy of estimation and decreases the variance of sampling distributions, thus increasing power. Indeed, generally such is the case. To illustrate, we will leave all the aspects of the Example 5.1 (p. 112) unchanged ($\alpha = .05$, $\sigma^2 = 225$), except that the sample size will be increased from 16 to 25. As illustrated in Figure 5.5, the critical value is more easily obtained (104.95 as opposed to 106.19) and power has been increased (from .846 to .954). Although there are rare instances when this is not true, generally **increasing the sample increases the power of a statistical test.**

Power is also changed by changing the standard deviation (and the variance) of the population distribution. Because reducing the variance of the population distribution reduces the variance of the sampling distribution, estimates become more accurate, improving the statistical test. To illustrate, we leave all aspects of Example 5.1 unchanged ($\alpha = .05$, $N = 16$), except that the variance is reduced from 225 to 169. As illustrated in Figure 5.5,

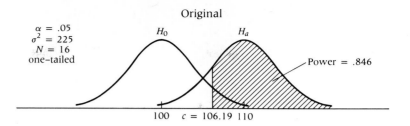

Original

$\alpha = .05$
$\sigma^2 = 225$
$N = 16$
one-tailed

H_0 H_a

Power = .846

100    $c = 106.19$ 110

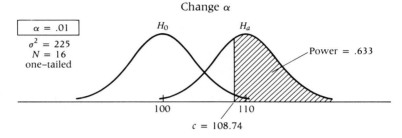

Change α

$\alpha = .01$
$\sigma^2 = 225$
$N = 16$
one-tailed

H_0 H_a

Power = .633

100   110

$c = 108.74$

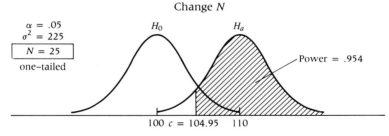

Change N

$\alpha = .05$
$\sigma^2 = 225$
$N = 25$
one-tailed

H_0 H_a

Power = .954

100 $c = 104.95$ 110

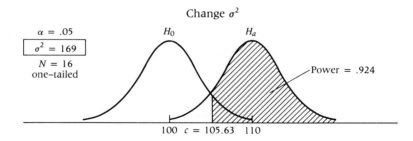

Change σ^2

$\alpha = .05$
$\sigma^2 = 169$
$N = 16$
one-tailed

H_0 H_a

Power = .924

100 $c = 105.63$ 110

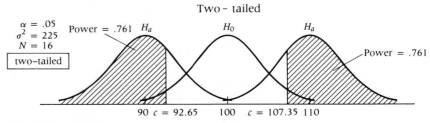

Two - tailed

$\alpha = .05$
$\sigma^2 = 225$
$N = 16$
two-tailed

Power = .761 H_a H_0 H_a

Power = .761

90 $c = 92.65$ 100 $c = 107.35$ 110

FIGURE 5.5   Factors Influencing Power

the critical value is again more easily obtained (105.36 as opposed to 106.19) and power has been increased (from .846 to .924). **Decreasing the variance of the population increases the power of the statistical test.** Admittedly, the criticism could be raised that changing a population parameter, such as the variance, is not within the prerogative of the researcher. However, there are two important ways in which investigators can have some influence over the variance. First, the researcher is able to define the population of interest; more homogeneous populations will have smaller variances. Of course, the more homogeneous the population is, the less generalizable are the results. Second, carefully designed and executed experiments reduce variance. Efficient control in experimentation minimizes the variance accounted for by extraneous factors. Compare the variances of "a population of subjects tested by an irresponsible research assistant" to "a population of subjects tested by a careful and conscientious research assistant."

As we have seen previously, it is more difficult to reject the null hypothesis with a two-tailed test than with a one-tailed test (subject to a caveat discussed subsequently). This is illustrated by Example 5.1 for which all factors remain unchanged ($\alpha = .05$, $\sigma^2 = 225$, $N = 16$), with the exception that a two-tailed test is conducted; that is

H_0: $\mu = 100$

versus

H_a: $\mu = 110$ or $\mu = 90$

As illustrated in Figure 5.5, there are critical values in both directions (107.35 and 92.65); the critical value to the right is now larger than was the case previously (107.35 as opposed to 106.19), thus reducing the power (from .846 to .761). However, it is vital to note that the power of the test when the true mean is less than 100 is also .761.

It is worthwhile to consider again the differences between one- and two-tailed tests. Consider the case where the true mean is 90 and a one-tailed test was used (H_a: $\mu = 110$). The null hypothesis may still be rejected, because, although highly unlikely, it is possible to select a sample with a mean greater than 107.35. However, to reject the null hypothesis and accept the alternative that $\mu = 110$ would be an error. On the other hand, if the obtained mean was less than 100, say, 88, the null hypothesis could not be rejected at all. In essence, the one-tailed test has no power to reject the null hypothesis in favor of the alternative that $\mu = 90$. To summarize, **if the true mean is in the hypothesized direction (as indicated in the one-tailed alternative hypothesis), a one-tailed test is more powerful than a two-tailed test; if the true mean is in the opposite direction (to that hypothesized), then the one-tailed test has no power.**

The final way to change power is to change the statistical test used. Up to this point we have no choice; only one test has been presented. Typically, for a particular research question the researcher will have a variety of statistical tests from which to choose. Depending on the statistical properties of the tests and on the degree to which assumptions for those tests are met, the tests will provide various levels of power. The relative power of statistical tests will be discussed in more detail when nonparametric tests are presented in Chapter 13. Furthermore, variations of statistical tests, such as the analysis of covariance (see Section 11.8, p. 343) have been designed that reduce unexplained variance, a concept that is not too different from reducing the population variance.

To this point we have considered power only in the context of alternative hypotheses that are stated exactly. Infrequently, however, is the research question so precise that this is the case. How is power calculated when the alternative is stated inexactly? The answer to this question involves the concept of effect size. In the case of population means effect size is some measure of the discrepancy of the true mean from the mean specified in the null hypothesis. An appropriate measure for effect size would be the absolute value of the difference between the true mean and the mean specified in the null hypothesis standardized by dividing by the standard deviation:

$$\text{Effect size} = \left| \frac{\mu - \mu_0}{\sigma} \right|$$

In other words, effect size is simply a measure of the "distance" from the true state of affairs to the state of affairs that is specified by the null hypothesis, which typically indicates no effect. So far effect size is not much help because μ and σ are unknown. The solution to this dilemma is to postulate the effect size. Evidence may be provided by a pilot study or by previous studies in the researcher's area. For instance, studies of programmed instruction for algebra skills might yield an effect size of .70, when this modality is compared to traditional instruction. If a researcher wishes to conduct a study to test the effectiveness of programmed instruction for geometry skills, the study should be designed to have sufficient power to detect an effect size of .70. In applied studies an investigator might postulate an effect size by determining the size of the effect that would be clinically significant. Finally, there have been attempts to create a taxonomy of effect sizes by defining effects as large effects, medium effects, and small effects. Although these attempts are justified in a number of ways, this classification is largely arbitrary, as effect sizes are likely to depend on the area of inquiry.

Typically, postulated effect sizes are used to determine sample sizes. First, the study is designed, the hypotheses are specified (one- or two-tailed), the statistical test is chosen, and the value of α is selected. Then, by whatever means, the desired effect size is postulated. The next step is to select an adequate level of power. Selection of a level of power is, however, also arbitrary, although the following solution has been proposed. If we consider a Type I error to be four times as serious as a Type II error, then power = $1 - \beta = 1 - 4\alpha$. Thus, if $\alpha = .05$, power = .80, and if $\alpha = .01$, power =

.96. In any event, after determination of the statistical test, α, the effect size, and power, sample size may be calculated. For most tests the calculations are not simple, however, and the researcher is advised to use tables that are prepared for this purpose. The most extensive tables have been compiled by Cohen (1977). To demonstrate how these tables would be used, we show an abridged version of a power table for the test performed in Example 5.1 in Table 5.1. Entering this table with α = .05, power = .80, effect size = .60, we find that 18 subjects are needed if a one-tailed test is used. If a two-tailed test is used, we enter the table at $\alpha/2$. In this case we enter at .025 and find that 22 subjects are needed.

TABLE 5.1

Number of Subjects Needed to Obtain Given Level of Power for Various α Levels and Effect Sizes (z test with σ known)

	α LEVEL											
	α = .05 Power			α = .025 Power			α = .01 Power			α = .005 Power		
Effect size	.70	.80	.90	.70	.80	.90	.70	.80	.90	.70	.80	.90
.2	119	156	217	155	196	264	205	251	328	242	293	375
.4	30	39	55	39	50	67	52	63	82	61	74	94
.6	14	18	25	18	22	30	23	28	37	27	33	42
.8	8	10	14	10	13	17	13	16	21	16	19	24

There is a tendency for beginning researchers to disregard determination of sample sizes. A total disregard of this issue often results in too few subjects to detect a reasonable effect size. How does the researcher interpret a nonsignificant result when sample size is small? One of two possible explanations for a nonsignificant finding when sample sizes are small is possible: either there was no effect or there was a reasonable sized effect but it was not detected. Because there is no evidence about which of these two explanations is true, nonsignificant findings with small sample sizes are essentially noninformative. A knee jerk reaction to this might be to use very large samples. However, very large samples are not desirous either. Besides being costly, it is relatively easy to detect very small effect sizes. If a small effect size is practically, theoretically, or clinically meaningless, a statistically significant finding may mislead the researcher and those who read about the study into believing that something important has been discovered. Statistical significance is not equivalent to importance. A very small effect, in some instances, may have an important implication. For instance, it might contradict an important theory. Often, however, small effects have little implications, and if found to be statistically significant, they may lead the researcher into a fruitless area of further experimentation or may lead to the adoption of programs with negligible clinical significance. Sample sizes should be selected so that a reasonable chance exists of detecting an effect size that is important. In this way a statistically significant finding will also be a worthwhile finding. Furthermore, a statistically nonsignificant finding will be informative, in that it can be said that there is a good chance that the size of the effect for which the researcher is searching was not present. It should be noted the studies presenting nonsignificant findings are subject to intense scrutiny as critics look for explanations for the nonsignificant findings other than a lack of an effect. The usual criticism is poor experimental method.

5.10   Confidence Intervals

The hypothesis-testing strategy presented in this chapter provides a method to choose between two hypotheses, the null and the alternative. Although this strategy provides important information about the population parameters, it does not in any way tell the whole story. Suppose that a study is undertaken to find the grade-level achievement of fifth graders in rural schools. Because it is unknown whether the children's achievement is below or above their grade level, the following hypotheses are formulated:

H_0: $\mu = 5.0$

versus

H_a: $\mu \neq 5.0$

Suppose that a random sample of 25 children is taken and the sample mean is equal to 5 years 10 months (5.833 years). Further, suppose that the population is normally distributed with a $\sigma^2 = 4$. Using Equation (5.6) (p. 116), we see that

$$z = \frac{M - \mu_0}{\sigma/\sqrt{N}} = \frac{5.833 - 5.0}{\text{²/₅}} = 2.083$$

which is sufficiently large to reject the null hypothesis when α is set at .05. The belief that the children achieve at the fifth-grade level is abandoned in favor of the belief that the children score above the fifth-grade level. How much above? If asked to give the best estimate of the true level of achievement, we would estimate the true mean by the sample mean 5.833 (see Section 4.5, p. 92). That is, our best guess is that the mean achievement level for the children is 5.833. It is extremely unlikely, however, that this estimate hits the bull's-eye; that is, due to sampling error, the true mean is not likely to equal 5.833, although this remains the best guess. To indicate the magnitude of the error, we would need to find an interval such that we were relatively confident that this interval included the true mean.

To understand the formal definition of confidence intervals, it will help to have already determined the interval for a particular example. Suppose that we want to construct an interval such that we are 95% certain that the interval covers the true value of μ where it is assumed that $X \sim N(\mu, \sigma^2)$ for a sample of size N. Essentially, we want to find the values of a and b such that

$$P(a \leq \mu \leq b) = .95$$

Under these conditions $M \sim N(\mu, \sigma^2/N)$ (see Section 4.9, p. 99). Standardizing M, we have $z_M \sim N(0, 1)$ (see Section 4.10, p. 101). For this latter distribution we know that

$$P(-1.96 \leq z_M \leq 1.96) = .95$$

But

$$z_M = \frac{M - \mu}{\sigma_M}$$

[see Equation (4.13), p. 101], and therefore

$$P\left(-1.96 \leq \frac{M - \mu}{\sigma_M} \leq 1.96\right) = .95$$

Isolating μ in the inequality within the parentheses, we have

$$P(M - 1.96\sigma_M \leq \mu \leq M + 1.96\sigma_M) = .95 \tag{5.10}$$

This is the 95% confidence interval for μ. Notice that the confidence interval is formed around the sample mean M. We know that M varies from one sample to another. Most of the time M is close to μ, but occasionally it will stray quite far from μ. When M is quite far from μ, the confidence interval will not cover μ, although most of the time the confidence interval will cover μ. This idea can be said more technically: we say that 95% of samples (of a given size) drawn from a population yield confidence intervals that include the true mean. More euphemistically, it is permissible to say that the 95% confidence interval has a 95% chance of covering the true mean μ. Notice that it is not permissible to say that there is a 95% chance that μ falls within the interval. This is not permissible for the same reason that we cannot say that the alternative hypothesis is true with 95% certainty (see Section 5.3, p. 108). Because μ is a parameter (i.e., a constant) and not a random variable, μ either lies or does not lie in the confidence interval. In the former case

$$P(M - 1.96\sigma_M \leq \mu \leq M + 1.96\sigma_M) = 1$$

and in the latter case

$$P(M - 1.96\sigma_M \leq \mu \leq M + 1.96\sigma_M) = 0$$

It should be reiterated that a confidence interval does not refer to the probability of μ. Instead, the confidence interval refers to the probability that a sample will yield a confidence interval that covers μ. For the purpose of this text, we define **a confidence interval as an estimated range that has a given probability of covering the true population value.** Confidence intervals are a form of estimation. Previously, we said that M was the point estimator for μ (see Section 4.4, p. 91). Confidence intervals constitute interval estimates. Confidence intervals typically are constructed for high probabilities, either .95 or .99.

To illustrate the use of a confidence interval, we calculate the 95% confidence interval for μ for the achievement data. Because $\sigma_M = \sigma/\sqrt{N}$, Equation (5.10) can be rewritten as

$$P\left(M - 1.96\frac{\sigma}{\sqrt{N}} \leq \mu \leq M + 1.96\frac{\sigma}{\sqrt{N}}\right) = .95 \tag{5.11}$$

and

$$P\left(5.833 - \frac{(1.96)(2)}{5} \leq \mu \leq 5.833 + \frac{(1.96)(2)}{5}\right)$$

$$= P(5.05 \leq \mu \leq 6.62) = .95$$

In words, there is a 95% chance that the interval from 5.05 to 6.62 covers the true mean μ.

Equation (5.10) is worth examining closer because it is similar to many other confidence intervals that we will consider. The equation for a $100(1 - \alpha)\%$ confidence interval for a parameter θ of a symmetric distribution is often written in the same form as Equation (5.10):

$$P(G - c_{\alpha/2}\hat{\sigma}_G \leq \theta \leq G + c_{\alpha/2}\hat{\sigma}_G) = 1 - \alpha \tag{5.12}$$

where θ is the parameter of interest, G is the point estimator for θ (i.e., $G = \hat{\theta}$), $\hat{\sigma}_G$ is the estimator for the standard error of the sampling distribution of G, and $c_{\alpha/2}$ is the value of the referent distribution such that $1 - F(c_{\alpha/2}) = \alpha/2$ (i.e., c is the value that cuts off an area of $\alpha/2$ in the upper tail). The referent distribution is dependent on the sampling distribution of G. Here we used the standard normal. As well, in the present case σ was known so it was not necessary to estimate $\hat{\sigma}_M$

Equation (5.11) can be used to determine the 99% confidence interval for μ:

$$P\left(M - \frac{2.58\sigma}{\sqrt{N}} \leq \mu \leq M + \frac{2.58\sigma}{\sqrt{N}}\right) = 1 - .01 \tag{5.13}$$

which for our achievement example is

$$P\left(5.833 - \frac{(2.58)(2)}{5} \leq \mu \leq 5.833 + \frac{(76.58)(2)}{5}\right)$$

$$= P(4.80 \leq \mu \leq 6.87)$$

$$= .99$$

Not surprisingly, the 99% confidence interval is larger than the corresponding 95% interval. In fact, changing any of the factors that affect the power of a statistical test will change the size of the corresponding confidence interval. Increasing power (e.g., increasing α, decreasing σ, increasing N) will decrease the size of the confidence interval.

As a shorthand notation, the confidence interval is often written with " $\pm$ " notation. For example, boundaries of the 99% confidence interval given in Equation (5.12), can be described as

$$M \pm \frac{2.58\sigma}{\sqrt{N}}$$

Although a confidence interval is a method of interval estimation, there is an intimate connection between confidence intervals and two-tailed statistical tests. Recall that for the achievement example the 95% confidence interval ranged from

5.05 to 6.62

That is, there is a 95% chance that this interval covers the true mean μ. Suppose that we adopt the following decision rule. If the value of μ under the null hypothesis falls outside the 95% confidence interval, then the null hypothesis will be rejected. For this example

H_0: $\mu = 5.0$

Consequently, we reject the null hypothesis. Because the probability is .95 that the interval 5.05 to 6.62 covers μ, the probability is at most .05 that the sample was selected from a population with a mean outside this interval. Thus, the probability of falsely rejecting H_0 is less than .05 and our decision rule was a reasonable one. In general, **the probability that the decision to reject H_0 when θ_0 does not fall within the $100(1 - \alpha)$% confidence interval for θ will be incorrect is equal to, at most, α.** Thus, for a two-tailed test the confidence interval may be used for hypothesis testing. If the value specified under the null hypothesis is not found within the $100(1 - \alpha)$ % confidence interval, the null hypothesis is rejected at the specified level of α.

NOTES AND SUPPLEMENTARY READINGS

Cohen, J. (1977). *Statistical power analysis for the behavioral sciences* (Revised ed.). New York: Academic Press.

This volume represents the main resource for researchers involved in determining statistical power. Cohen discusses the concept of power, examines power in relation to *t* tests, correlation coefficients, *F* tests, and others. He provides tables for determining power that, although a bit difficult to read in some circumstances, are incredibly useful.

Golding, S. L. (1975). Flies in the ointment: Methodological problems in the analysis of the percentage of variance due to persons and situations. *Psychological Bulletin, 82,* 278–288.

   Golding examines the use of ω^2 in determining the relative contributions of subjects and circumstances to variability, and proposes an alternative measure that is also based on variance components.

Kraemer, H. C., & Thiemann, S. (1987). *How many subjects? Statistical power analysis in research.* Newbury Park, CA: Sage.

   This slim volume presents an alternative to Cohen (1977) for finding the number of subjects needed to obtain a given level of power.

Serlin, R. C., & Lapsley, D. K. (1985). Rationality in psychological research: The good-enough principle. *American Psychologist, 40,* 73–83.

   Although there are a number of articles that discuss the relative merits of the hypothesis testing in psychological research, this article proposes an interesting alternative.

Tukey, J. W. (1960). Conclusions versus decisions. *Technometrics, 2,* 423–433.

   Although this article is somewhat dated, it presents a superb discussion regarding drawing inferences from data, an important part of hypothesis testing.

PROBLEMS

1. Given the following information, find the appropriate critical value(s) for rejecting H_0:
   a. $Y \sim N(\mu, 16)$, $N = 25$, H_0: $\mu = 12$ vs. H_a: $\mu < 12$, $\alpha = .05$
   b. $Y \sim N(\mu, 8)$, $N = 16$, H_0: $\mu = 0$ vs. H_a: $\mu \neq 0$, $\alpha = .01$
   c. $Y \sim N(\mu, 25)$, $N = 100$, H_0: $\mu = 10$ vs. H_a: $\mu = 12$, $\alpha = .05$
      Also, find the power given this critical value.

2. Given the following information, find the appropriate critical value(s) for rejecting H_0:
   a. $Y \sim N(\mu, 12)$, $N = 36$, H_0: $\mu = 10$ vs. H_a: $\mu < 10$, $\alpha = .01$
   b. $Y \sim N(\mu, 25)$, $N = 20$, H_0: $\mu = 100$ vs. H_a: $\mu \neq 100$, $\alpha = .05$
   c. $Y \sim N(\mu, 100)$, $N = 25$, H_0: $\mu = 6$ vs. H_a: $\mu = 2$, $\alpha = .05$
      Also, find the power given this critical value.

3. Test the following hypotheses with respect to the given information:
   a. H_0: $\mu = 100$ vs. H_a: $\mu \neq 100$ given $Y \sim N(\mu, 25)$, $N = 64$, $\alpha = .05$, $M = 98.50$

b. H_0: $\mu = 14$ vs. H_a: $\mu < 14$ given $Y \sim N(\mu, 4)$, $N = 30$, $\alpha = .01$, $M = 13.8$

c. H_0: $\mu = 27$ vs. H_a: $\mu > 27$ given $Y \sim N(\mu, 20)$, $N = 100$, $\alpha = .05$, $M = 21.2$

4. Test the following hypotheses with respect to the given information:
    a. H_0: $\mu = 0$ vs. H_a: $\mu \neq 0$ given $Y \sim N(\mu, 49)$, $N = 40$, $\alpha = .05$, $M = 1.90$
    b. H_0: $\mu = 30$ vs. H_a: $\mu < 30$ given $Y \sim N(\mu, 30)$, $N = 15$, $\alpha = .01$, $M = 25.6$
    c. H_0: $\mu = 100$ vs. H_a: $\mu > 100$ given $Y \sim N(\mu, 25)$, $N = 16$, $\alpha = .05$, $M = 102.88$

5. Find the indicated confidence intervals given the following information:
    a. 95% confidence interval given $Y \sim N(\mu, 20)$, $N = 100$, $M = 23.5$
    b. 99% confidence interval given $Y \sim N(\mu, 100)$, $N = 25$, $M = 3.0$
    c. 90% confidence interval given $Y \sim N(\mu, 8)$, $N = 10$, $M = 27.5$

6. Find the indicated confidence intervals given the following information:
    a. 95% confidence interval given $Y \sim N(\mu, 15)$, $N = 10$, $M = 153.2$
    b. 99% confidence interval given $Y \sim N(\mu, 36)$, $N = 18$, $M = 14.0$
    c. 90% confidence interval given $Y \sim N(\mu, 12)$, $N = 25$, $M = -2.3$

6 TESTING HYPOTHESES ABOUT MEANS— ONE AND TWO SAMPLES

I n the previous chapters we have presented statistical concepts related to probability, descriptive statistics, sampling distributions, estimation, and hypothesis testing. Although the necessity of learning about these concepts may have been somewhat obscure, it is now possible to use these concepts for the purpose of making inferences about populations in a realistic manner. In this chapter procedures are discussed for answering questions about the mean of a single population and about the difference between means of two populations. Inferences about more than two populations are deferred until later chapters. To make inferences about population means when the population variance is unknown, another distribution, the t distribution, must be used.

One-sample Problems

6.1 Back to Basics

It will be useful to review what we know about the one-sample problem. Suppose that we randomly select N observations $(y_1, y_2, \ldots, y_N)$ from a normally distributed population $Y \sim N(\mu, \sigma^2)$, for which μ is unknown but σ^2 is known. The sample mean M is easily calculated by using the formula

$$M = \frac{1}{N} \sum y_i$$

The sampling distribution of M has been derived earlier (see Section 4.9, p. 99).

$$M \sim N\left(\mu, \frac{\sigma^2}{N}\right)$$

Standardizing M, we have (see Section 4.10, p. 101)

$$z_M = \frac{M - \mu}{\sigma/\sqrt{N}}$$

and because the population distribution is normally distributed, $z_M \sim N(0, 1)$.

Suppose that we wish to test

$$H_0: \mu = \mu_0$$

versus

$$H_a: \mu \neq \mu_0$$

The test statistic for these hypotheses is (see Section 5.6, p. 115)

$$z = \frac{M - \mu_0}{\sigma/\sqrt{N}} = \frac{M - \mu_0}{\sigma_M}$$

and because the population is normally distributed, the test statistic is standard normally distributed; that is, $z \sim N(0, 1)$. If α is set at .05, the null hypothesis is rejected, provided

$$|z| = \left|\frac{M - \mu_0}{\sigma/\sqrt{N}}\right| \geq 1.96$$

The fact that σ is unknown to the researcher makes this procedure unrealistic. However, we should be tempted to estimate σ with s, the square root of the unbiased estimator of the population variance (see Section 4.6, p. 94), yielding a test statistic of the form

$$\text{Modified test statistic} = \frac{M - \mu_0}{s/\sqrt{N}} = \frac{M - \mu_0}{\hat{\sigma}_M} \qquad (6.1)$$

The temptation to use this test statistic is justified! However, although we will use this test statistic, it is not normally distributed. Decisions to reject the null hypothesis made by comparing the modified test statistic to the standard normal distribution would be flawed, unless the sample size was very large. All is not lost, though, because the distribution of the modified test statistic is well known; this distribution will be the topic of the next section.

6.2   The t Distribution

Suppose that we have the traditional one-sample problem. That is, a random sample of size N is drawn from a normally distributed population with unknown parameters μ and σ^2 [i.e., $Y \sim N(\mu, \sigma^2)$]. Given a null hypothesis that specifies a value of μ (i.e., H_0: $\mu = \mu_0$), then the appropriate test statistic is that given in previous section [Equation (6.1), p. 134].

$$t = \frac{M - \mu_0}{s/\sqrt{N}} = \frac{M - \mu_0}{\hat{\sigma}_M} \tag{6.2}$$

The distribution of the test statistic t was first discussed by W. S. Gosset under the pen name "Student," and is, therefore, sometimes referred to as "Student's t distribution." Actually, there are many t distributions, since it turns out that the form of the distribution of the t statistic is related to the size of the sample, as we will see.

In general, the probability density function for t is given by the rule

$$f(t) = G(\nu)\left[1 + \frac{t^2}{\nu}\right]^{-(\nu+1)/2} \tag{6.3}$$

Although the exact form of the density equation is not of interest to us, several aspects of it are important to understand. The Greek symbol ν (nu) is a parameter of the t distribution and $G(\nu)$ is simply a constant that depends on ν. The parameter ν must be positive and is called *the degrees of freedom*, a somewhat elusive concept that will be discussed in the next section. The important point to realize here is that *the t distribution is a family of distributions*, each member of which is determined by the value assigned to the degrees of freedom parameter. It is not unusual to find the degrees of freedom for the t distribution (and for other distributions that involve degrees of freedom) written as a subscript. Thus, the t distribution with ν degrees of freedom would be written as t_ν. In the context for which we use a t distribution, the degrees of freedom will always be related to sample sizes.

Although the probability density function of the t distribution does not seem to be similar to that of the standard normal (see Section 3.15, p. 72), they are closely related. This is easily seen in Figure 6.1, which presents the standard normal and three t distributions. First, note that as the degrees of freedom increases, the t distribution approaches that of a normal distribution. Thus

$$t \to N(0, 1) \qquad \text{as } \nu \to \infty \tag{6.4}$$

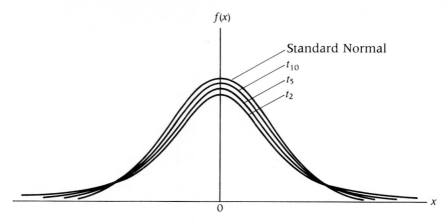

FIGURE 6.1

Standard Normal Distribution and t Distributions with 2, 5, and 10 Degrees of Freedom

The t distributions are also similar to the standard normal in that they are all unimodal, symmetric, and bell-shaped, and when the number of degrees of freedom is greater than 1, they have a mean of zero. However, of importance to hypothesis testing is the fact that a t distribution has a "fatter" tail than the standard normal (i.e., the t distribution is platykurtic); thus, for a given value of α, a greater value for the test statistic is required when the distribution of the test statistic is a t distribution rather than a standard normal.

The properties of a t distribution can be illustrated by examining Table B.2, p. 451, which presents the critical values of t distributions that create right-hand tails with various degrees of freedom. The critical value is found by entering the table at the row corresponding to the degrees of freedom and the column corresponding to the area in the right-hand tail. For instance, if we wish to know the value that cuts off a tail with area .05 for a t distribution with 10 degrees of freedom, we would enter the table with $\nu = 10$ and $Q = .05$ to find that the critical value is 1.812. Note that the corresponding critical value for a standard normal distribution is 1.645. As the degrees of freedom increase, the critical value for a tail with area .05 approaches 1.645. For example, for 40 and 120 degrees of freedom the critical values are 1.684 and 1.658, respectively. These changes in degrees of freedom and corresponding changes in critical values have practical importance, as we will see over and over again. As mentioned earlier, degrees of freedom are related to sample size. As the sample size increases, the degrees of freedom increase and the critical values decrease, making it "easier" to reject the null hypothesis.

Because the t distribution is symmetric, use of Table B.2 for two-tailed tests is straightforward. The column is selected so that $Q = \alpha/2$ and the critical values will be that value found in the table and the opposite of that

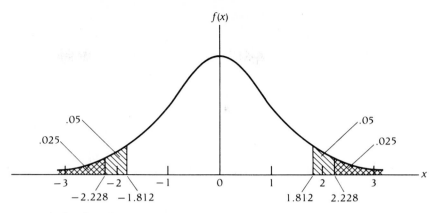

FIGURE 6.2   Some Critical Values for a *t* Distribution with $\nu = 10$

value. For example, if $\nu = 10$ and $\alpha = .05$, the table is entered with $Q = .025$ and the right-hand critical value is 2.228 and the left-hand critical value is -2.228. That is, if a test statistic is distributed as a *t* distribution with 10 degrees of freedom, then the null hypothesis will be rejected if the statistic is greater than 2.228 or less than -2.228. The properties of a *t* distribution can be seen by comparing Table B.2 to Figure 6.2, which presents some critical values for a *t* distribution with 10 degrees of freedom.

Recall that we began the discussion of the *t* distribution by noting that the test statistic

$$t = \frac{M - \mu_0}{s/\sqrt{N}} \tag{6.5}$$

has a *t* distribution, with $(N - 1)$ degrees of freedom, provided the population distribution is normal. That is

$$t \sim t_{N-1} \tag{6.6}$$

It is vital to note that "*t* " is used in two different ways in Statement (6.6). On the one hand, *t* is a test statistic as defined in Equation (6.5). This test statistic also has a *t* distribution with $(N - 1)$ degrees of freedom. Throughout this text the reader should be cognizant of the difference between a test statistic and the distribution of the test statistic.

To emphasize the difference between a test statistic and a distribution as well as another vital point, it should be repeated here that the *t* statistic has a *t* distribution only if the population has a normal distribution. **If the population distribution is not normal, then the *t* statistic [defined in Equation (6.5)] does not have a *t* distribution. However, for distributions that are reasonable approximations to the normal and for moderately large sample sizes, the *t* statistic is approximately distributed as a *t* distribution and we do not run much risk in using**

the *t* distribution to make decisions about rejecting the null hypothesis for the one-sample problem, as we will discuss later.

After discussing degrees of freedom in the next section, we will return to significance tests and confidence intervals based on the *t* distribution.

6.3   Degrees of Freedom

Degrees of freedom is a fairly ubiquitous concept in statistics. Except for the normal distribution, every continuous distribution discussed in this text is dependent on one or more degrees of freedom parameters. Although the concept of degrees of freedom is not a simple one, a discussion aimed at fostering an intuitive understanding is presented here.

In calculating the *t* statistic, we need to specify M in order to estimate σ. Given a value of the sample mean, and $N - 1$ of the N observations, the Nth observation is determined (i.e., not free to vary). Thus, the degrees of freedom associated with the *t* distribution involving one sample is equal to $N - 1$. To be precise, it is the degrees of freedom that determines the particular *t* distribution rather than the sample size, although clearly the two are closely related.

In this discussion the degrees of freedom of the *t* distribution for one sample was one less than the sample size. We often hear the statement that "a degree of freedom was lost." Although a degree of freedom is never "lost," the reason for the use of this term is not difficult to understand. Examination of Table B.2 (p. 451) demonstrates that as the degrees of freedom decrease, the critical value increases, making it more "difficult" to reject the null hypothesis. The use of the term "lose a degree of freedom" is unfortunate because it gives the impression that somehow if it could be found, the statistical test would be more powerful. The fact that the degrees of freedom for the statistical test we have been discussing is one less than the sample size is the price we pay—there is no way that this can be avoided. However, there are times, as we will see, that the researcher has a choice between two or more strategies for testing a hypothesis; the difference in the degrees of freedom among the different methods will be one, although minor, consideration in the selection of the procedure.

6.4   Significance Tests and Confidence Intervals Based on *t* Distribution—One Sample

In this section the procedures for the one-sample problem developed in the preceding sections are summarized and illustrated. As well, the confidence intervals using the *t* distribution are discussed.

Again, suppose that $Y \sim N(\mu, \sigma^2)$ and a random sample of size N is selected. The researcher stipulates some null hypothesis

$$H_0:\ \mu = \mu_0$$

and sets a level of α. Then the test statistic t is

$$t = \frac{M - \mu_0}{\hat{\sigma}_M} = \frac{M - \mu_0}{s/\sqrt{N}} = \frac{M - \mu_0}{S/\sqrt{N-1}} \qquad (6.7)$$

This test statistic is distributed as a t distribution with $\nu = (N - 1)$ degrees of freedom. For a one-tailed test the null hypothesis is rejected, provided

$$t > c_\alpha \qquad (\mu_a > \mu_0)$$

or

$$t < -c_\alpha \qquad (\mu_a < \mu_0)$$

and for a two-tailed test the null hypothesis is rejected, provided

$$|t| > c_{\alpha/2}$$

where c_α and $c_{\alpha/2}$ are the critical values for one- and two-tailed tests, respectively, found in the t distribution table (Table B.2, p. 451).

A minor bit of confusion might be caused by giving a version of the t statistic [Equation (6.7)] with the sample standard deviation S rather than the preferred estimator s. Although S^2 is a biased estimator of σ^2, this bias was corrected in Equation (6.7) by dividing S by $\sqrt{N-1}$ rather than by $\sqrt{N}$ (see Section 4.6, p. 94). In other words, the use of S is acceptable provided the appropriate correction is made elsewhere in the formula. If still in doubt, the reader should convince him- or herself of the equality of the versions of the t statistic given in Equation (6.7) by using the relationship between s and S given in Section 4.6 (p. 94).

The statistical test described in this section is often called the *one-sample t test*. We now have multiple terminology using "t": t statistic, t distribution, and t test. The t statistic is the statistic defined in Equation (6.7). When the population is normally distributed, the distribution of the t statistic is a t distribution. Finally, the procedure for making a decision of whether or not to reject the null hypothesis that is based on the t statistic and the t distribution is called the t test.

The only collateral assumption that is necessary for the validity of the t test is that the population distribution be normal. However, as we have discussed in Section 6.2 (p. 135), for moderately large N and reasonable approximation to the normal distribution, there is minimal risk in using the t test. That is, the t test is robust with regard to violation of the normality assumption. Students often ask how large is "moderately large" and how reasonable is a "reasonable approximation." Although there has been much written about these issues, there does not seem to be a simple rule for making decisions about the validity of the t test (or other tests, for that matter).

Generally, if the power of the test is sufficient to have a good chance to reject the null hypothesis, then the sample size is large enough so that the one-sample t test is a useful test provided the population distribution is not extremely strange (i.e., very different from the normal). When the evidence is clear that a population distribution is not normal, there are a variety of useful methods that are able to transform some distributions into a distribution that more closely approximates the normal distribution. The role of assumptions and methods to transform data will be discussed further when the two-sample t test is presented later in this chapter (see Section 6.8, p. 150).

The conceptualization of power presented in the context of the z test applies to the one-sample t test as well (see Section 5.9, p. 121). That is, power increases as α is increased, sample size is increased, variance of the population is decreased, and effect size is increased. Furthermore, when the result is in the expected direction, a one-tailed test is more powerful than a two-tailed test. Unfortunately, calculation of power is more complicated for a t test than for the tests based on the normal distribution. In the context of the t test, when the null hypothesis is not true, the distribution under the alternative is changed to a *noncentral t distribution*. The noncentral t distribution relies on two parameters—v, the degrees of freedom, and δ (delta), a noncentrality parameter. Calculation of power using the noncentral t distribution is not straightforward nor particularly instructive. Typically, applied researchers use tables to determine power (see Section 5.9, p. 121).

To illustrate the use of the t test, consider the following example. Suppose that the faculty of a graduate program wish to determine whether the applicants to their program are different in terms of their quantitative skills from applicants to similar programs. The measure chosen to make the comparison is the quantitative section of the Graduate Record Examination (GRE). Suppose that a comprehensive survey of graduate programs has indicated that the mean GRE score for applicants for the last three years was 620. The faculty decides to select 15 applicants randomly from those who have applied in the last three years. A priori, it is stipulated that the null hypothesis is that $\mu = 620$, $\alpha = .05$, and a two-tailed test will be used. The quantitative GRE scores for the 15 subjects are retrieved from the records and it is found that

$$M = 590 \quad \text{and} \quad s = 92$$

The value of the t statistic is calculated:

$$t = \frac{M - \mu_0}{s/\sqrt{N}} = \frac{590 - 620}{92/\sqrt{15}} = -1.26$$

When compared to a t distribution with 14 degrees of freedom, clearly, the absolute value of the t statistic is not sufficiently large to reject the null hypothesis, since a value of 2.145 is needed. The evidence is not sufficient to give up the belief that the applicants' quantitative skills are similar to those of other programs.

From our discussion of confidence intervals in the previous chapter (see Section 5.10, p. 127) the confidence interval for μ when σ is unknown is relatively easy to determine. Recall that Equation (5.12) (p. 129) was a general equation for confidence intervals that was applicable to many cases, including the present one:

$$P(G - c_{\alpha/2}\,\hat{\sigma}_G \le \theta \le G + c_{\alpha/2}\,\hat{\sigma}_G) = 1 - \alpha$$

In the present context the $100(1 - \alpha)\%$ confidence interval, provided the population distribution is normally distributed, is

$$P\left\{ M - t_{[\alpha/2;(N - 1)]}\frac{s}{\sqrt{N}} \le \mu \le M + t_{[\alpha/2;\,(N - 1)]}\frac{s}{\sqrt{N}} \right\} = 1 - \alpha \qquad (6.8)$$

where $t_{[\alpha/2;(N - 1)]}$ represents that value that cuts off an area of $\alpha/2$ in the upper tail of a t distribution with $(N - 1)$ degrees of freedom. For the GRE example, the limits for the 95% confidence interval are

$$M - t_{[\alpha/2;\,(N - 1)]}\frac{s}{\sqrt{N}} = 590 - 2.145\left(\frac{92}{\sqrt{15}}\right) = 590 - 50.95$$

$$= 539.05$$

and

$$M + t_{[\alpha/2;\,(N - 1)]}\frac{s}{\sqrt{N}} = 590 + 2.145\left(\frac{92}{\sqrt{15}}\right) = 590 + 50.95$$

$$= 640.95$$

Because μ_0 (viz., 620) falls inside the 95% confidence interval, the null hypothesis cannot be rejected, a result consistent with the test statistic procedure used earlier (see Section 5.10, p. 127).

Two-Independent-Samples Problem

Although the one-sample t test is an example of a practical parametric statistical test, it is not typical that the researcher has a problem for which the mean of one population is of interest. More interesting are problems for which the means of two populations are compared. For instance, it would be of interest to know whether the mean anxiety level of people receiving psychotherapy was less than that of untreated people. Or, we could compare the effectiveness of an innovative drug to the current drug of choice. The term ''independent'' is used to describe the samples because the observations of one sample are unrelated to the observations in the other sample.

6.5   Two Independent Samples When the Population Variance Is Known

The conditions for the two-independent-samples problem are as follows. Suppose that there are two population distributions Y_1 and Y_2 such that

$$Y_1 \sim N(\mu_1, \sigma_1^2) \qquad \text{and} \qquad Y_2 \sim N(\mu_2, \sigma_2^2)$$

The subscripts refer to the first and second populations. From the first population a sample of size N_1 is randomly selected, and from the second population a sample of size N_2 is randomly selected. From these samples the means are calculated:

$$M_1 = \frac{1}{N_1}\sum y_{i1} \qquad \text{and} \qquad M_2 = \frac{1}{N_2}\sum y_{i2}$$

(The double subscript is interpreted this way: y_{i1} is the ith observation in the first sample.) Inferences about the two populations are made by comparing the sample means M_1 and M_2.

In the two-sample problem the typical null hypothesis is that there are no differences between the means of the two populations. In this case the null hypothesis would be that the means are equal, or equivalently, that the difference between the means is zero:

$$H_0: \mu_1 = \mu_2 \qquad \text{or} \qquad H_0: \mu_1 - \mu_2 = 0$$

However, the test to be developed here can accommodate the hypothesis of a difference between the means that is not equal to zero. In general, for the two-sample problem the null hypothesis is

$$H_0: \mu_1 - \mu_2 = k_0$$

The alternative can be nondirectional (two-tailed)

$$H_a: \mu_1 - \mu_2 \neq k_0$$

or directional (one-tailed)

$$H_a: \mu_1 - \mu_2 > k_0 \qquad \text{or} \qquad H_a: \mu_1 - \mu_2 < k_0$$

Evidence against the null hypothesis is found by examining the difference between the sample means. If the difference between the sample means $(M_1 - M_2)$ is much smaller or larger than the hypothesized difference (i.e., k_0), then there is good reason to reject the null hypothesis. How much does $(M_1 - M_2)$ have to differ from k_0 before the null hypothesis can be rejected? We proceed in the same fashion as we did in the one-sample case. Recall that we examined the distribution of the sample mean under the null hypothesis and then standardized the sample mean to form a test statistic. Evidence to reject the null hypothesis was acquired by comparing the test statistic to the distribution of the test statistic.

The sample statistic of interest in the two-sample case is $(M_1 - M_2)$. The first task is to determine the sampling distribution of $(M_1 - M_2)$ under the null hypothesis. Using the properties of expectation (see Section 3.5, p. 47), we can easily determine $E(M_1 - M_2)$

$$E(M_1 - M_2) = E(M_1) - E(M_2) = \mu_1 - \mu_2 = k_0 \tag{6.9}$$

Because M_1 and M_2 are independent (i.e., determined from independent samples), the variance of $(M_1 - M_2)$ can be determined by using the properties of the variance (see Section 3.9, p. 57).

$$V(M_1 - M_2) = V(M_1) + V(M_2) = \frac{\sigma_1^2}{N_1} + \frac{\sigma_2^2}{N_2} \tag{6.10}$$

The square root of the variance of the difference is called the *standard error of the difference* $(\sigma_{(M_1 - M_2)})$:

$$\sigma_{(M_1 - M_2)} = \sqrt{\frac{\sigma_1^2}{N_1} + \frac{\sigma_2^2}{N_2}} \tag{6.11}$$

Finally, because the difference between two normally distributed random variables is also normally distributed (see Section 3.15, p. 72)

$$(M_1 - M_2) \sim N\left(\mu_1 - \mu_2, \frac{\sigma_1^2}{N_1} + \frac{\sigma_2^2}{N_2}\right) \tag{6.12}$$

which, when the null hypothesis is true, is equivalent to

$$(M_1 - M_2) \sim N\left(k_0, \frac{\sigma_1^2}{N_1} + \frac{\sigma_2^2}{N_2}\right) \tag{6.13}$$

We are now ready to form the test statistic for making decisions about rejecting H_0: $\mu_1 - \mu_2 = k_0$. The difference between the means $(M_1 - M_2)$ is now standardized by using Equations (6.9) and (6.11):

$$Z_{(M_1 - M_2)} = \frac{(M_1 - M_2) - (\mu_1 - \mu_2)}{\sigma_{M_1 - M_2}} = \frac{(M_1 - M_2) - (\mu_1 - \mu_2)}{\sqrt{(\sigma_1^2/N_1) + (\sigma_2^2/N_2)}} \tag{6.14}$$

Recall that the random variable formed by standardizing a normally distributed random variable is itself normally distributed (see Section 3.15, p. 72); consequently

$$Z_{(M_1 - M_2)} \sim N(0, 1)$$

Thus, when σ_1^2 and σ_2^2 are known, the test statistic for the two-sample problem is

$$z = \frac{(M_1 - M_2) - k_0}{\sigma_{(M_1 - M_2)}} = \frac{(M_1 - M_2) - k_0}{\sqrt{(\sigma_1^2/N_1) + (\sigma_2^2/N_2)}} \tag{6.15}$$

given the null hypothesis that $\mu_1 - \mu_2 = k_0$ (the subscript on Z has been deleted for convenience and a lower-case z is used to indicate that a particular value of z is obtained for the two samples). In the typical case in which the null hypothesis stipulates that $\mu_1 - \mu_2 = 0$, the test statistic reduces to

$$z = \frac{(M_1 - M_2)}{\sigma_{(M_1 - M_2)}} = \frac{(M_1 - M_2)}{\sqrt{(\sigma_1^2/N_1) + (\sigma_2^2/N_2)}} \tag{6.16}$$

In either case z is compared to the standard normal distribution. If the alternative hypothesis is directional (i.e., H_a: $\mu_1 - \mu_2 > k_0$ or $\mu_1 - \mu_2 < k_0$), and $\alpha = .05$, the null hypothesis is rejected provided $z > 1.65$ or $z < -1.65$ depending on the direction of the test. If the alternative is nondirectional (i.e., H_a: $\mu_1 - \mu_2 \neq k_0$) and $\alpha = .05$, then the null hypothesis is rejected provided $|z| > 1.96$.

This test will not be illustrated here because the z test is unrealistic in that it requires that σ_1^2 and σ_2^2 be known. As in the one sample case, the realistic solution is found by estimating unknown parameters. In this case the standard error of the difference is estimated and the resultant test statistic will be equal to Equation (6.15), except that $\sigma_{(M_1 - M_2)}$ will be replaced by $\hat{\sigma}_{(M_1 - M_2)}$, the estimate of the standard error of the difference. As was the case in the one-sample problem, the distribution of the test statistic involving an estimated parameter will have a t distribution rather than a standard normal. Because estimating the standard error of the difference involves making estimates based on the pooled samples, that topic will be discussed in the next section.

Using Equation (5.12) (p. 129) as a guide (although in this case, it is not necessary to estimate the standard error because σ is known), the limits for the 95% confidence interval for $(\mu_1 - \mu_2)$ are easily defined:

$$(M_1 - M_2) \pm 1.96\,\sigma_{(M_1 - M_2)} = (M_1 - M_2) \pm 1.96 \sqrt{\frac{\sigma_1^2}{N_1} + \frac{\sigma_2^2}{N_2}}$$

An important point needs to be made about the z test for two independent samples. The central limit theorem can be extended to the difference between two sample means so that **as both N_1 and N_2 become larger, the sampling distribution of the difference between means approaches a normal distribution, regardless of the form of the two population distributions.** Therefore, the z test is robust with regard to the assumption of normality.

6.6 Parameter Estimates Based on Pooled Samples

Estimating the standard error of the difference for the two-independent-sample problem involves estimating the population variance from the information contained in the two samples. An efficient way to make this estimate is to base the estimate on the data that result from combining the two samples. This strategy is used because estimates based on combining samples result in less sampling error than when estimates are based on a single sample. Intuitively, this should be clear because the sample size is increased by combining samples.

For the moment assume that two independent samples of size N_1 and N_2 are randomly selected from a population with mean μ and variance σ^2. Let the set $\{y_{11}, y_{21}, ..., y_{N_11}\}$ be the observations from the first sample and M_1, S_1^2, and s_1^2 be the sample mean, sample variance, and unbiased estimator for the variance, respectively, based on the first sample. Similarly, let the set $\{y_{12}, y_{22}, ..., y_{N_22}\}$ be the observations from the second sample and M_2, S_2^2, and s_2^2 be the sample mean, sample variance, and unbiased estimator for the variance, respectively, based on the second sample. Combining the two samples results in a *pooled sample*. **The pooled sample is the union of the two sets of observations;** that is

$$\text{Pooled sample} = \{y_{11}, y_{21}, ..., y_{N_11}\} \cup \{y_{12}, y_{22}, ..., y_{N_22}\}$$

and is of size $(N_1 + N_2)$. Let M_{pooled} be the sample mean based on the pooled sample. These concepts and notation are illustrated in Figure 6.3.

The mean of the pooled sample is easily calculated:

$$M_{\text{pooled}} = \frac{\sum_{i=1}^{N_1} y_{i1} + \sum_{i=1}^{N_2} y_{i2}}{N_1 + N_2}$$

Because $\sum_{i=1}^{N_1} y_{i1} = N_1 M_1$ and $\sum_{i=1}^{N_2} y_{i2} = N_2 M_2$

$$M_{\text{pooled}} = \frac{N_1 M_1 + N_2 M_2}{N_1 + N_2} \tag{6.17}$$

This result should not be surprising. The mean of the pooled sample is simply the weighted (by sample size) average of the means of the two individual samples. We now use the mean of the pooled sample as the estimator for the population mean:

$$\hat{\mu} = M_{\text{pooled}} = \frac{N_1 M_1 + N_2 M_2}{N_1 + N_2} \tag{6.18}$$

M_{pooled} is an unbiased estimator of μ and is more efficient than either M_1 or M_2 because it is based on a larger number of observations.

In a similar manner, the population variance can be estimated from the pooled sample:

$$\hat{\sigma}^2 = \frac{N_1 S_1^2 + N_2 S_2^2}{N_1 + N_2 - 2} = \frac{(N_1 - 1)s_1^2 + (N_2 - 1)s_2^2}{N_1 + N_2 - 2}$$

$$= \frac{\sum_{i=1}^{N_1} (y_{i1} - M_1)^2 + \sum_{i=1}^{N_2} (y_{i2} - M_2)^2}{N_1 + N_2 - 2} \tag{6.19}$$

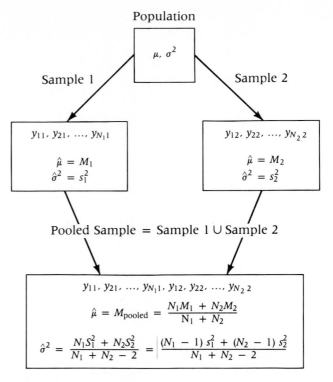

FIGURE 6.3   Pooled Sample and Population Estimators

This estimator is unbiased because the term $(N_1 + N_2 - 2)$ appears in the denominator rather than $(N_1 + N_2)$. Essentially, this estimator is a weighted average of the sample variances, except that the denominator is the sum of $(N_1 - 1)$ and $(N_2 - 1)$, which corrects for the bias in estimating σ^2 with S^2.

Estimating the population distribution based on the pooled sample is strategic in the development of the t test for the two-independent-sample problem. The two-independent-sample problem involves estimating the standard error of the difference, which is given by [see Equation (6.11), p. 143]

$$\sigma_{(M_1 - M_2)} = \sqrt{\frac{\sigma_1^2}{N_1} + \frac{\sigma_2^2}{N_2}}$$

To estimate the standard error of the difference, we assume that the variances of the two populations from which the two samples were drawn are equal; that is, $\sigma_1^2 = \sigma_2^2 = \sigma^2$ (i.e., we denote the variance of each of the populations by σ^2). The assumption that the variances of two or more populations are equal is called the *homogeneity of variance* assumption; when this assumption is violated, there is *heterogeneity of variance*. Because the homogeneity of variance assumption is made at this point, **we must keep in mind that**

one assumption of the two-independent-group _t_ test, developed in the next section, is that the variances of the two populations are equal. When this assumption is made

$$\sigma_{(M_1 - M_2)} = \sqrt{\frac{\sigma^2}{N_1} + \frac{\sigma^2}{N_2}} = \sqrt{\sigma^2 \left(\frac{1}{N_1} + \frac{1}{N_2} \right)} = \sqrt{\sigma^2 \left(\frac{N_1 + N_2}{N_1 N_2} \right)} \quad (6.20)$$

Because it is assumed that the variances of the two populations are equal, the best estimator for σ^2 is that based on the pooled sample [viz., Equation (6.19)]. Thus

$$\hat{\sigma}_{(M_1 - M_2)} = \sqrt{\hat{\sigma}^2 \left(\frac{N_1 + N_2}{N_1 N_2} \right)}$$

$$= \sqrt{\left(\frac{N_1 S_1^2 + N_2 S_2^2}{N_1 + N_2 - 2} \right) \left(\frac{N_1 + N_2}{N_1 N_2} \right)}$$

$$= \sqrt{\left(\frac{(N_1 - 1)s_1^2 + (N_2 - 1)s_2^2}{N_1 + N_2 - 2} \right) \left(\frac{N_1 + N_2}{N_1 N_2} \right)} \quad \textbf{(6.21)}$$

This estimator for the standard error of the difference will be used in the next section to derive the t test for two independent samples.

6.7   Significance Tests and Confidence Intervals Based on the _t_ Distribution—Two Independent Samples

The digression of the previous section was necessary in order to estimate the standard error of the difference when the population variance was not known. We can now develop the statistical test for two independent samples by using the estimate of the standard error of the difference in place of the known standard error of the difference, as found in the z test given in Section 6.5 (p. 142). Recall that for two independent samples the null hypothesis is (see Section 6.5, p. 142)

$$H_0\!: \mu_1 - \mu_2 = k_0$$

and that the test statistic, when σ_1^2 and σ_2^2 are known, is [Equation (6.15), p. 143]

$$z = \frac{(M_1 - M_2) - k_0}{\sigma_{(M_1 - M_2)}}$$

When σ_1^2 and σ_2^2 are unknown, this test statistic is altered by replacing the standard error of the difference by the estimate of the standard error of the difference given by Equation (6.21) (p. 147). Thus, the test statistic used for the two-independent-sample problem is

$$t = \frac{(M_1 - M_2) - k_0}{\sqrt{\left(\dfrac{N_1 S_1^2 + N_2 S_2^2}{N_1 + N_2 - 2}\right)\left(\dfrac{N_1 + N_2}{N_1 N_2}\right)}}$$

$$= \frac{(M_1 - M_2) - k_0}{\sqrt{\left(\dfrac{(N_1 - 1)s_1^2 + (N_2 - 1)s_2^2}{N_1 + N_2 - 2)}\right)\left(\dfrac{N_1 + N_2}{N_1 N_2}\right)}} \qquad (6.22)$$

When it is assumed that the variances of the two populations from which the samples were drawn are equal and that the two populations are normally distributed, the test statistic t has a t distribution with $\nu = (N_1 + N_2 - 2)$ degrees of freedom.

Several important points need to be made about the t statistic for two independent samples that is given by Equation (6.22). First, the t statistic is used in the typical fashion to make decisions about the null hypothesis. For a two-tailed test the null hypothesis is rejected, provided

$$|t| > c_{\alpha/2}$$

where $c_{\alpha/2}$ is the critical value of a t distribution with $(N_1 + N_2 - 2)$ degrees of freedom such that the area under the curve to the right of $c_{\alpha/2}$ is equal to $\alpha/2$. For a one-tailed test the null hypothesis is rejected, provided

$$t > c_\alpha \qquad \text{or} \qquad t < -c_\alpha$$

(depending on the direction of the test), where c_α is the critical value of a t distribution with $(N_1 + N_2 - 2)$ degrees of freedom such that the area under the curve to the right of c_α is equal to α. Second, the fact that the t distribution in this case has $(N_1 + N_2 - 2)$ degrees of freedom makes good sense. The first sample contributes N_1 observations but 1 degree of freedom is given up when s_1^2 is used to estimate σ^2. Similarly, the second sample contributes N_2 observations, but 1 degree of freedom is given up when s_2^2 is used to estimate σ^2. Thus

$$\nu = (N_1 - 1) + (N_2 - 1) = N_1 + N_2 - 2$$

Third, as was the case for the one-sample t test, the calculation of power for the two-sample t test relies on the noncentral t distribution. Again, the reader is referred to standard power tables. A useful measure of effect size for the two-independent-sample problem is given by

$$\text{Effect size} = \left| \frac{\mu_1 - \mu_2}{\sigma} \right| \qquad (6.23)$$

Finally, the importance of the homogeneity of variance and normality assumptions will be discussed in the next section.

A simple example will serve to illustrate the two-independent-sample *t* test. Suppose that the relative efficacy of two toothpastes is being investigated. Twenty children were randomly selected and ten were randomly assigned to a condition that used "Super-Smile Toothpaste" for two years and the other ten were assigned to a condition that used "Better-Breath Toothpaste" for two years. One child in the "Better-Breath" condition moved from the area during the two-year period. The number of cavities acquired during the two-year period was recorded and served as the outcome measure. Because the researcher has no preconceived idea about which toothpaste is more effective, the following hypotheses were used:

$$H_0: \mu_1 - \mu_2 = 0$$

versus

$$H_a: \mu_1 - \mu_2 \neq 0$$

The following sample results were obtained:

$$M_1 = 5 \qquad M_2 = 7$$
$$s_1^2 = 8 \qquad s_2^2 = 10$$
$$N_1 = 10 \qquad N_2 = 9$$

Thus, the *t* statistic is

$$t = \frac{(M_1 - M_2) - k_0}{\sqrt{\left(\dfrac{(N_1 - 1)s_1^2 + (N_2 - 1)s_2^2}{N_1 + N_2 - 2}\right)\left(\dfrac{N_1 + N_2}{N_1 N_2}\right)}}$$

$$= \frac{5 - 7}{\sqrt{\left(\dfrac{(10 - 1)(8) + (9 - 1)(10)}{10 + 9 - 2}\right)\left(\dfrac{(10 + 9)}{(10)(9)}\right)}}$$

$$= \frac{-2}{\sqrt{1.89}} = -1.45$$

When this value is compared to a *t* distribution with 17 degrees of freedom, it is clear that the null hypothesis cannot be rejected at conventional levels of significance and there is no basis to say that one of the toothpastes is more effective than the other.

When the sample sizes are equal, the formula for the *t* statistic reduces to a much simpler form. When $N_1 = N_2 = N$, the formula for the *t* statistic [Equation (6.22)] reduces to

$$t = \frac{(M_1 - M_2) - k_0}{\sqrt{\dfrac{S_1^2 + S_2^2}{N - 1}}} \tag{6.24}$$

which is distributed as a *t* distribution with $N_1 + N_2 - 2 = 2N - 2 = 2(N - 1)$ degrees of freedom. When the sample sizes are equal, this formula is much simpler; however, of course, the use of Equation (6.22) will result in exactly the same value of t.

In the typical manner, the confidence intervals for $(\mu_1 - \mu_2)$ are easily constructed. The limits for the $100(1 - \alpha)\%$ confidence intervals are

$$(M_1 - M_2) \pm t_{\alpha/2;\ (N_1 + N_2 - 2)} \hat{\sigma}_{(M_1 - M_2)}$$

$$(M_1 - M_2) \pm t_{\alpha/2;(N_1 + N_2 - 2)} \sqrt{\left(\frac{(N_1 - 1)s_1^2 + (N_2 - 1)s_2^2}{N_1 + N_2 - 2}\right)\left(\frac{N_1 + N_2}{N_1 N_2}\right)}$$

or

$$(M_1 - M_2) \pm t_{\alpha/2;\ (N_1 + N_2 - 2)} \sqrt{\left(\frac{(N_1 S_1^2 + N_2 S_2^2)}{N_1 + N_2 - 2}\right)\left(\frac{N_1 + N_2}{N_1 N_2}\right)} \qquad (6.25)$$

For the toothpaste example the limits for the 95% confidence interval are

$$(5 - 7) - 2.11 \sqrt{1.89} = -4.90$$

and

$$(5 - 7) + 2.11 \sqrt{1.89} = .90$$

Because this interval contains the value zero, the preceding null hypothesis could not be rejected in favor of the two-tailed alternative at the .05 level.

6.8 The Importance of the Assumptions in a t Test for Two Independent Samples

As noted before, the test statistic for the difference in means for two independent samples [Equation (6.22), p. 148] has a t distribution with $(N_1 + N_2 - 2)$ degrees of freedom provided (1) each population is normally distributed and (2) the variances of the two populations are equal. We briefly discussed violation of the first assumption in the context of the one-sample t test [see Section (6.4), p. 138]. It is now time to discuss in more depth the troublesome issues involved with violation of the assumptions in t tests. (The comments made here apply, for the most part, to the test statistics of the analysis of variance as well—see Chapter 7). We examine violations of each of the assumptions individually and then examine violation of both assumptions.

With regard to the first assumption that each of the populations is normally distributed, a variation of the central limit theorem [see Section (4.9), p. 99] establishes the robustness of the two-independent-sample t tests. Essentially, this variation stipulates that **the sampling distribution of the difference between means approaches a normal distribution as the size of N_1 and N_2 increases.** Because the central limit theorem is true regardless of the population distributions, decisions to reject the null hypothesis by comparing the test statistic to a t distribution with $(N_1 + N_2 - 2)$ degrees of freedom will not be affected much provided the sample sizes are moderately large and **provided the homogeneity of variance assumption is not violated.** Again, if the power of the statistical test is sufficient to reject the null hypothesis with a reasonable probability, then the sample size is sufficiently large to ignore the normality assumption in the case where the variances are homogeneous.

Just as large samples protect the validity of the t test from violations of the normality assumption, there is protection for violations of the homogeneity assumption. **When the population distribution is normally distributed and the sample sizes are equal, relatively large differences in the population variances have relatively small consequences on decisions to reject the null hypothesis.** As a rule, the experimenter should have equal and relatively large sample sizes, if at all possible.

It should be clear that the robustness of the t test for violation of each assumption was discussed in the context where the other assumption was true. What happens when both assumptions are violated? Unfortunately, deleterious and unpredictable things can happen. When distributions are highly skewed and multimodal **and** when variances are unequal, decisions to reject the null hypothesis by comparing the test statistic to a t distribution may be extremely flawed even when sample sizes are equal and extremely large. In fact, it has been found that in some altogether not unusual cases, decisions to reject the null hypothesis become more flawed as the sample sizes increase!

The fact that the t test for two independent samples is not robust with regard to serious violations of both assumptions leaves two important questions: (1) How does one know that the assumptions have been or will be violated? and (2) What does one do if one does know that the assumptions have been or will be violated? There are two ways in which the researcher can know that assumptions have been violated. First, the nature of the experimental task and/or the measurements may determine that the distribution of observations has certain characteristics. For example, instances in which there is a floor (point below which the measurements cannot fall), most observations are near the floor, and a few observations are very far from the floor, will result in positively skewed distributions. This is the case for the distribution of incomes and time to complete relatively simple tasks. Structural aspects of an experiment that create different types of distributions will be discussed further in the next section.

The second means by which violation of the assumptions can be detected is by examining the data obtained in the experiment. The most obvious way (and one of the best ways) to discover serious violations of assumptions is to examine graphs of the experimental data. If it is obvious that the data were not selected from a normal distribution or that the population variances were not equal, then one should question whether there are structural measurement reasons or theoretical reasons why this should be so. Clearly, this "eye-ball" approach has severe limitations. When sample sizes are small (and when determination of violations is most often needed!), it is most difficult to make inferences about the nature of the population distributions. Nevertheless, for relatively large samples a visual inspection of the graph of the frequency distributions can lead to a decision to transform the data (see the following section) or to throw out the data and redesign the study. There are also statistical tests that can be used to detect violations of the assumptions. Unfortunately, these tests also rely on assumptions and are most accurate when they are least needed (i.e., for large samples). These tests currently have little popularity in statistical practice. The reader looking for definitive rules here obviously will be disappointed. However, one point should be made crystal clear. **The researcher should never conduct an experiment if he or she is blind to clues about the distributional characteristics of the populations under investigation.** These clues may be derived from previous research, from theory, or from the nature of the experiment or the instruments used to obtain the observations. If these sources are insufficient to provide the information needed, then pilot studies should be conducted until such time that either it is reasonable to assume normality and homogeneity of variance, in which case the t test could be conducted, or until it is clear that the use of the t test is inappropriate and another remedy should be used.

This brings us to the question about what to do when there is good reason to believe that both the assumptions of normality and homogeneity of variance have been violated. Two alternatives are presented here. First, transformations of the data are often used to obtain normality or homogeneity of variance. Some of these transformations will be discussed in the following section. Second, the decision may be made not to use the t test and to use an alternative statistical test that does not rely on these assumptions. A nonparametric analogue to the t test, which does not depend on normality and homogeneity of variance, will be presented in Section 13.5 (p. 432).

6.9   Transformations

The goal of transformations in the two-independent-sample problem is either to make the distributions of the two groups more closely resemble normal distributions or to make the variances of the two groups more equal or both. Depending on the type of problem exhibited, different transformations will be needed.

Transformations of data are accomplished by forming a new random variable such that the new random variable is a function of an old random variable. As discussed earlier (Section 2.5, p. 28), a linear transformation is of the form

$$Y' = BY + A$$

where A and B are constants and Y' is the transformed random variable. Data are transformed analogously; the ith observation derived from the random variable Y is linearly transformed by using the equation

$$y'_i = B_{y_i} + A$$

The observation y'_i is now thought of as being sampled from Y'. Linear transformations are not useful with regard to violation of the assumptions of the two-independent-sample t test. Because linear transformations do not change the form of the distribution, the resulting distribution is no closer to a normal than the original (e.g., a skewed distribution remains skewed). Although clearly the mean and variance are changed by a linear transformation (see Section 3.9, p. 57), the variances of two populations would be changed proportionately and the heterogeneity of variance would remain unchanged. Incidentally, the standardization of a random variable is a linear transformation

$$Z = \frac{Y - \mu}{\sigma} = \frac{1}{\sigma} Y + \frac{-\mu}{\sigma}$$

There are a number of nonlinear transformations that attenuate the degree of violation of the normality and homogeneity of variance assumptions found in the nontransformed scale. We approach the different transformations by discussing different violations and the structural aspects of an experiment that might give rise to these patterns of violation.

The first violation discussed is heterogeneity of variance where the variance is a linear function of the means so that larger means have larger variances and vice versa. This might be the case when one is comparing retarded subjects to normals on some measure of functioning, in which case it might be expected that the variance for the retarded individuals is larger than the variance of the normals. In this instance a transformation that results in more homogeneous variances is

$$Y' = \sqrt{Y} \tag{6.26}$$

This transformation is called the *square root transformation*. A variation of the square root transformation that is appropriate when Y is a frequency measure (e.g., number of errors) and when Y is small is

$$Y' = \sqrt{Y} + \sqrt{Y + 1} \tag{6.27}$$

This latter square root transformation has the added advantage that it also tends to make the scores more normally distributed.

A second instance for which transformations seem to be effective is when the scores are proportions. The use of proportions occurs frequently in educational and psychological research. For instance, on a speeded test, the researcher may be interested in the rate of errors. Because some subjects are faster than others, the score of interest probably would be the proportion of the number of errors to the number of problems attempted. For proportions, the *arcsine transformation* is helpful in reducing heterogeneity of variances. This transformation is given by

$$Y' = 2 \text{ arcsine } \sqrt{Y} \tag{6.28}$$

The third instance for which transformations are effective is when the distributions are positively skewed, which often occurs when there is a floor below which the scores cannot fall. This is the case for incomes and time to complete tasks, which have floors of zero. Positive skewness often results from rating scales that do not adequately differentiate among individuals scoring at the lower end of the scale. For positively skewed distributions the *logarithmic transformation* helps to reduce the heterogeneity of variance as well as normalize the distribution. The logarithmic transformation is given by

$$Y' = \log Y \tag{6.29}$$

When a floor of zero exists, the following variation is preferred:

$$Y' = \log(Y + 1) \tag{6.30}$$

Several comments about transformations are needed. First, although the transformations involve relatively cumbersome functions to calculate (square root, arcsine, and logarithm), they are easily handled by the commonly used statistical packages. Typically, a control statement is inserted that specifies the transformation to be made. Second, it must be realized that a minimal treatment of the subject has been presented here and that the use of transformations is tricky. Unless the choice of a transformation is unambiguous, the researcher should seek expert help (see Winer, 1971, for a thorough introduction to this topic). Third, other transformations exist, one of which (Fisher's r to z transformation) will be used in the context of correlation (see Section 10.3, p. 291). Finally, there are distributions that appear to be intractable in the sense that transformations do not seem to improve the validity of the parametric tests, such as t tests or the analysis of variance, which will be the topic of the following chapter. An example of a distribution that is particularly resistant to transformations is presented in Figure 6.4. Such a distribution would be produced by an experiment for which the dependent variable was the time needed to complete successfully a relatively simple task, with the added condition that after a failure on the task the subject would begin that task anew. Most subjects would successfully complete the task on the first trial, resulting in the first "hump"; a small number of subjects would complete the task on the second trial, creating the second "hump"; and so forth. This type of distribution is particularly insidious in that decisions to reject the null hypothesis based on the t test (or the

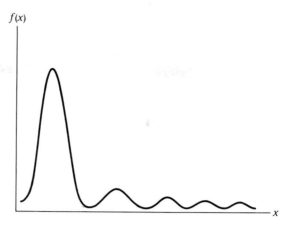

FIGURE 6.4   Distribution Resistant to Transformations

analysis of variance) can be extremely flawed. Sometimes the probability of a Type I error is smaller than one is lead to believe by the *t* test and sometimes it is larger—the direction of the flaw is unpredictable so that it is nearly impossible to correct decisions to reject the null hypothesis, and transformations do not improve the situation.

6.10 Inferences Made from Two-Independent-Sample *t* Tests

You may recall that the only structural requirement for the two-sample *t* test was that the observations be randomly selected from two populations (see Section 6.5, p. 142). In this context decisions are made with regard to the null hypothesis

$$H_0: \mu_1 - \mu_2 = k_0$$

Remember that information provided by a statistical test is restricted to the statistical hypotheses and not to the scientific hypotheses (see Section 5.1, p. 106). Invalidity of a research design may permit valid inferences about statistical hypotheses to result in extremely flawed inferences about research hypotheses. An example will illustrate the point. Suppose that a study was conducted to determine whether a series of films on parenting skills or a series of films on child development is more effective in helping mothers cope with newly born children. The researchers hypothesized that the parenting skills films were superior. Mothers of newly born children were randomly selected from public records to participate in the study. The first 20 mothers who agreed to participate in the study viewed the parenting skills films and the second 20 mothers who agreed to participate viewed the child development films. Suppose that the results of the study indicated that, based on an appropriate outcome measure, the mothers who viewed the parenting films scored significantly better; that is, the null hypothesis

$$H_0: \mu_1 - \mu_2 = 0$$

was rejected in favor of the alternative

$$H_a: \mu_1 - \mu_2 > 0$$

(here higher scores indicate ''better'' performance). It is not difficult to see that the inference that the parenting skills films were superior is flawed. Without some basis for saying that the first 20 mothers were comparable to the second 20 mothers, explanations for the statistically significant differences between the means other than superiority of parenting skills films exist. For example, it is quite likely that mothers who volunteered first were more motivated to acquire information and skills than were those mothers who volunteered later.

This example demonstrates the need to control extraneous factors in an experiment. The researcher wishes to keep constant all aspects of the experiment other than the experimental treatment or intervention. In that way it can be said that the results are due to the treatment or intervention and not to other factors. One source of factors that needs to be considered is characteristics of the units of observation that affect performance in the experimental task, such as motivation, intelligence, prior knowledge or skill, and so forth. It is desirable to form experimental groups such that they consist of units of observations that contain comparable amounts of these nuisance factors.

The most desirable method of establishing the comparability of experimental groups is *random assignment*. **When units of observation are randomly assigned, the probability that a unit is included in one group is the same as the probability that any other unit is included in that group.** This is typically accomplished by assigning numbers to the units and using a random numbers table to determine which units are included in which groups. Because it is desirable to have equal sample sizes, in the two-independent-sample problem, ideally one-half of the subjects would be randomly assigned to one group and the other half would be assigned to the other group. Random assignment ensures that whatever differences exist among groups are due to chance and not to systematic sources. Random assignment is the essence of experimental designs. The difference between an experimental design and nonexperimental designs is illustrated in Figure 6.5.

Several aspects of the designs illustrated in Figure 6.5 are worth discussing in more detail. First, examine the experimental design. The outcome measure is often referred to as the *dependent variable*. Typically, the dependent variable is denoted by Y and observed scores are denoted by y_i. The two

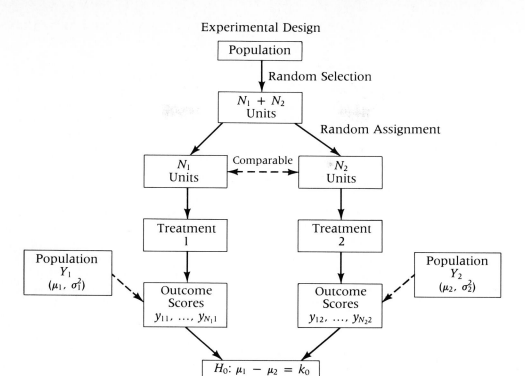

Experimental Design

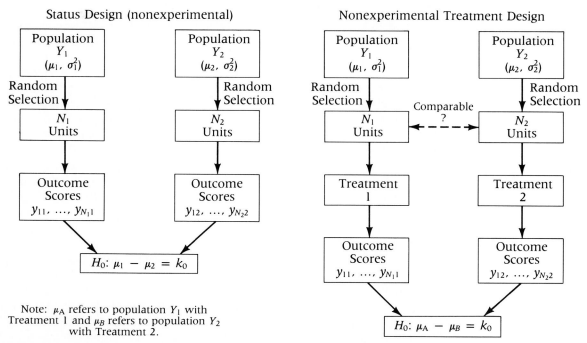

Status Design (nonexperimental)

Nonexperimental Treatment Design

Note: μ_A refers to population Y_1 with Treatment 1 and μ_B refers to population Y_2 with Treatment 2.

FIGURE 6.5   Experimental and Nonexperimental Designs

treatments 1 and 2 constitute the *independent variable*. "Treatments" is a generic term used to describe categories within an independent variable that is categorically measured. The treatments might well be therapeutic treatments, but they just as well could be control conditions (no treatment), placebo conditions, or any other manipulation planned by the researcher. The categories are also commonly referred to as *conditions, levels*, or *groups*. Frequently, the independent variable is denoted by X. Clearly, the experimenter controls the independent variable to determine the effect on the dependent variable. [To prevent some confusion, we should note that the terms "independent" and "dependent" have been used previously in four ways: the independence (or dependence) of random variables (see Section 3.11, p. 62), the independence of observations (see Section 4.2, p. 85), independent and dependent designs (which will be discussed more fully in Section 6.11, p. 160), and independent and dependent variables.] Another aspect of the experimental design is that although the subjects are selected from a single population, after the experimental manipulation has been accomplished, the observations in each of the experimental treatments are thought of as having been selected from a population of observations **that have been subjected to the treatment**, as illustrated by the dotted lines in Figure 6.5. In an experimental design it can be said that the independent variable is the **cause** of differences found between the populations. That is, if a statistically significant difference is found, then it is attributed to the relative efficacy of the treatments.

The two nonexperimental designs illustrated in Figure 6.5 are characterized by the fact that there is no random assignment of subjects. In the status design the differences between two populations are studied. Typically, the two populations are defined by some person variable, such as gender. This design provides information about differences between populations, but it does not indicate the cause of these differences. For example, differences between males and females may be due to heredity or to environment. In the nonexperimental treatment design the treatments are selected by the researcher, but the subjects are not randomly assigned to the groups. In this design the cause of differences between populations may be due either to the treatments or to other factors related to initial differences between the populations from which the subjects were selected, as was the case in the example of the mother's parenting skills. Unfortunately, in such a design there is no way to sort out the cause of the observed effect. Although in nonexperimental designs the researcher does not control the variables in the same way as in the experimental design, the terms "independent" and "dependent" variable still are used.

Frequently, neophyte researchers are concerned that random assignment might have "failed" and then desire to "check" the random assignment to determine whether, indeed, this was the case. This check usually consists of a statistical test such as the independent group t test on pretest scores or on measures such as motivation or intelligence, which are related to performance in the experimental task. **Checking to determine the adequacy of random assignment is unnecessary and in fact will likely lead to flawed judgments.** First, random assignment does not ensure that the composition of the groups are identical; in fact, from our knowledge of sampling, we know that it would be quite unlikely that this would be the case. Nevertheless, random assignment ensures that whatever differences exist occurred by chance. Although there is no guarantee that a particular result was not due to the quirks of random assignment (e.g., the more motivated subjects were significantly more often assigned to one group than to the other), repetitions of the experiment will reveal whether or not a true treatment effect exists. It should be emphasized that the results of a statistical test such as the independent group *t* test are phrased in terms of probabilities to account for chance occurrences. That is, if the null hypothesis was rejected at the .05 level, all we know is that the probability of observing differences as large as that obtained in the experiment by chance is less than .05, given the null hypothesis is true. The possibility that the differences observed were due to a particular random assignment and not to the treatment cannot be ruled out with certainty, although we are relatively certain that this was not the case. This uncertainty is the reason that we should always seek to replicate findings.

The second reason that a check on random assignment is inappropriate is that it is impossible to check all factors that might influence the outcome of an experiment. Comparability of motivation says nothing about the comparability of intelligence, prior knowledge or skill, or an infinite number of other factors. It is impossible to check the comparability of all factors. But even if it were possible to check all relevant factors, the results would not be particularly informative. We know that if we test a large number of factors, we will likely find that the differences between some of them are statistically significant by chance alone. What are we to do then? Should we conduct another random assignment and check the results again? A related problem here is that to check on random assignment, the researcher would want to accept the null hypothesis of no differences (i.e., comparability of the groups). Should differences that are large but not quite statistically significant (say, *p* = .06) be ignored? This is complicated by the fact that statistical significance is partly a function of sample size. Therefore, experiments with small sample sizes are unlikely to show statistically significant differences between groups, whereas experiments with large sample sizes are more likely to show statistically significant differences between groups. Hence, a "comparability check" for small samples would likely lead to a decision that the samples are comparable, whereas the large samples would lead to the opposite decision.

Occasionally, a researcher wishes to check the comparability of two nonrandomly assigned groups. For the same reasons this check is also flawed. It is not uncommon for researchers to try to "correct" statistically for initial differences found in nonrandomly assigned groups; the problems of such corrections will be discussed later (see Section 11.8, p. 343).

Three comments about the issues raised in this section need to be made. First, it is important to note that there are instances when recommendations will be made to measure and use nuisance factors in the design and analysis of experiments. We know that the power of a statistical test is increased by decreasing the variance. Variance can be decreased by extracting the variance due to the nuisance factors. This topic will be discussed in the next sections of this chapter (Sections 6.11 and 6.12, pp. 160 and 165) and in later chapters (see Sections 11.8, p. 343 and 12.9, p. 405). Second, the differences between random selection and random assignment should be reiterated. Random selection is technically required for the validity of many statistical tests, such as *t* tests, although, as discussed previously, true random selection is most often impossible to accomplish. Random assignment is not required for the validity of the two-independent-sample *t* test; nevertheless, it usually is vital to the validity of the research design. Without random assignment, one can say that the means of the two groups are significantly different, but it may be impossible to explain the ''cause'' of that difference. Even though research design is beyond the scope of this book, it is useful to note a rough similarity between random assignment and the *internal validity* of an experiment and random selection and the *external validity* of an experiment. Finally, the discussion of this section is applicable to independent variables with more than two treatments and should be kept in mind in the next chapters that examine the analysis of variance.

Dependent Sample Problem

6.11 Dependent Sample *t* Test

There are research designs for which the observations in two groups are paired. This pairing may be accomplished structurally when subjects in the experiment are associated in some way. For example, if twins were subjects, then for each set of twins, one twin would be assigned to one group and the other twin to the other group. Other structural matches include husbands and wives, siblings, classmates, and litter mates. A second way that the pairing may occur is when the experimenter *matches* the subjects on some criterion (or occasionally criteria), such as motivation, intelligence, or prior knowledge or skill (notice the similarity of these criteria to the nuisance factors discussed in Section 6.10, p. 155). Two subjects are matched if they have the same or nearly the same score on the matching variable. In either of these two ways, after pairing subjects, one of the matched subjects is assigned to one group and the other subject is assigned to the other group. To ensure the validity of the treatment studies, the paired subjects should be randomly assigned to the two groups; that is, for each pair, one subject is **randomly** assigned to one group (by default, the other subject is assigned to the other group). A third way to pair observations is to measure a subject at two different points in time. Typically, the two times are before treatment (pretest) and after treatment (posttest), although any two times could be used provided all subjects are tested at the same times. In this way the two observations for each subject are paired. A design in which subjects are observed two or more times is often called a repeated measures design.

The groups formed by any of these designs are called *dependent groups*. An important point to be made about the use of this term is that the groups are dependent in the *design* sense. That is, something about the design of the experiment (paired observations) makes the groups dependent. There is nothing about the design, however, that indicates that the scores in the two groups are *statistically* dependent. For example, if husbands and wives are paired, then it may be that the scores are positively related (intelligent husbands are married to intelligent wives), negatively related (passive husbands are married to dominant wives and vice versa), or unrelated (no relation between the blood type of husbands and wives).

Why would one prefer a dependent sample design to an independent sample design? The answer to this question, which will be discussed in more detail in the next section, is related to nuisance factors discussed previously (see Section 6.10, p. 155). If the nuisance factor is indeed related to the dependent measure of interest, then the variance in the dependent variable will be reduced by pairing observations on that nuisance factor, by either matching on the nuisance factor, using repeated observations, or using a structural pairing (in which it is assumed that the pairs have equal or nearly equal amounts of the nuisance factor). Essentially, the variance due to the nuisance factor is "removed" from the variance of the dependent measure, increasing the power of the statistical test.

The requirements for the dependent sample *t* test are as follows. A set of N observations $\{y_{11}, y_{21}, ..., y_{i1}, ..., y_{N1}\}$ are randomly selected from one population Y_1 and N observations $\{y_{12}, y_{22}, ..., y_{i2}, ..., y_{N2}\}$ are selected from a second population Y_2, with the provision that y_{i1} is paired in some way to y_{i2}. (In this notation the first subscript refers to the observation and the second to the population; it doesn't make any difference in the subsequent statistical test which population is designated as first and which as second.) The null hypothesis is

$$H_0: \mu_1 - \mu_2 = 0$$

(actually, any difference could be specified, but the need to test a difference other than zero rarely occurs). As was the case for the two-independent-sample *t* test, we calculate the means M_1 and M_2 for the two samples and examine the sampling distribution of the difference in the means $(M_1 - M_2)$. By the rules of expectation

$$E(M_1 - M_2) = E(M_1) - E(M_2) = \mu_1 - \mu_2 \tag{6.31}$$

Hence, the difference in the sample means is an unbiased estimator for the difference between the population means. However, because the scores in one sample are paired (i.e., are dependent) on the scores in the other sample, it is generally not true that

$$V(M_1 - M_2) = V(M_1) + V(M_2)$$

(see Section 3.12, p. 62). In fact, as will be discussed in more detail in the next section, when the paired observations are positively related (high scores are paired with high scores and vice versa)

$$V(M_1 - M_2) < V(M_1) + V(M_2)$$

This is desirable because it accomplishes the goal of reducing variance, which as we know, increases the power of the statistical test. Returning to the development of the statistical test, we note that because $V(M_1 - M_2)$ is unknown when the groups are dependent, a statistical test involving $V(M_1 - M_2)$ is precluded. The strategy used to avoid this is simply to take the difference between the respective pairs of observations creating a single set of differences

$$\{(y_{11} - y_{12}), (y_{21} - y_{22}), ..., (y_{i1} - y_{i2}), ..., (y_{N1} - y_{N2})\}$$

(It would not make any difference in the subsequent test if we had chosen to form the opposite difference; that is, $y_{i2} - y_{i1}$ rather than $y_{i1} - y_{i2}$.) Let the ith difference score be denoted by d_i; that is

$$d_i = y_{i1} - y_{i2} \tag{6.32}$$

Let the d_is be sampled from a population of difference scores whose distribution is

$$D = Y_1 - Y_2$$

The null hypothesis may now be rewritten as

$$H_0: \mu_1 - \mu_2 = \mu_D = 0$$

where μ_D is the mean of the population of difference scores. Essentially, we have one sample of N independent observations from a population of differences and we wish to determine whether there is sufficient evidence to reject the null hypothesis that the true mean of the difference scores is zero. This is the same paradigm as the one-sample t test!

The one-sample test on the differences is developed in this way. Let D denote the random variable that characterizes the population of differences and let σ_D^2 denote the variance of the population of differences. If we assume that the population of differences is normally distributed, then

$$D \sim N(\mu_D, \sigma_D^2)$$

or

$$D \sim N(\mu_1 - \mu_2, \sigma_D^2)$$

A sample of size N is randomly selected from the population of difference scores. The sampling distribution of the mean is thus (see Section 4.9, p. 99)

$$M_D \sim N\left(\mu_1 - \mu_2, \frac{\sigma_D^2}{N}\right) \tag{6.33}$$

Standardizing M_D, we have

$$\frac{(M_1 - M_2) - (\mu_1 - \mu_2)}{\sqrt{\sigma_D^2/N}} \sim N(0, 1) \tag{6.34}$$

Under the null hypothesis that the difference between the means is zero, an appropriate test statistic is

$$z = \frac{(M_1 - M_2)}{\sqrt{\sigma_D^2/N}} \tag{6.35}$$

which would be compared to the standard normal distribution to make decisions about rejecting the null hypothesis.

As was the case for the other t tests developed in this chapter, this statistic is altered because the population variance σ_D^2 is unknown. As before, we proceed by using the unbiased estimator of σ_D^2:

$$\hat{\sigma}_D^2 = s_D^2 = \frac{\sum_{i=1}^{N} (d_i - M_D)^2}{N - 1} \tag{6.36}$$

Replacing the expression for the population variance in Equation (6.35) by that of the unbiased estimator, we find that the test statistic for the dependent group problem becomes

$$t = \frac{(M_1 - M_2)}{\sqrt{s_D^2/N}} = \frac{(M_1 - M_2)}{\sqrt{S_D^2/(N - 1)}} \tag{6.37}$$

When the population of the difference scores is normally distributed, this t statistic has a t distribution with $(N - 1)$ degrees of freedom, where it is important to remember that N is the number of difference scores (i.e., number of pairs). This test is called the dependent sample t test, paired t test, or correlated t test.

The following example (Example 6.1) will demonstrate the use of the dependent group t test.

EXAMPLE 6.1

Suppose that a researcher is comparing two interventions directed toward improving performance on some learning task. Typically, the intelligence of the subjects is a nuisance factor in such experiments. That is to say, much of the variance of performance is due to varying intelligence and this source of variance may mask the difference between the two interventions. To control for this nuisance factor, subjects were matched for intelligence in the following way. Subjects were randomly selected until two subjects with intelligence test scores in the following ranges were obtained: 75 to 84; 85 to 94; 95 to 104; 105 to 114; and 115 to 124. The two subjects in each range were then randomly assigned to the two interventions and the interventions were administered to the subjects. The null hypothesis was

$$H_0: \mu_1 - \mu_2 = \mu_D = 0$$

versus the alternative

$$H_a: \mu_1 - \mu_2 = \mu_D \neq 0$$

since the researcher had no prior evidence that one intervention was superior to the other. The results of this experiment are presented in Table 6.1 (high scores indicate better performance).

TABLE 6.1

Scores on Learning Task

IQ Range	Treatment 1 (y_{i1})	Treatment 2 (y_{i2})	Difference (d_i)
75–84	5	4	1
85–94	7	7	0
95–104	7	6	1
105–114	11	10	1
115–124	10	8	2
	$M_1 = 8$ $S_1^2 = 4.8$	$M_2 = 7$ $S_2^2 = 4$	$M_D = 1$ $S_D^2 = .4$

The value of the t statistic for the dependent group is easily calculated.

$$t = \frac{M_1 - M_2}{\sqrt{S_D^2/(N-1)}} = \frac{8-7}{\sqrt{.4/4}} = 3.16$$

which, when compared to a t distribution with 4 degrees of freedom, is sufficient to reject the null hypothesis, provided the α level was set a priori at .05.

Confidence intervals for $\mu_D = \mu_1 - \mu_2$ are constructed in the usual manner.

$$P\left(M_1 - M_2 - t_{\alpha/2;N-1} \sqrt{\frac{S_D^2}{N}} \le \mu_1 - \mu_2 \le M_1 - M_2 + t_{\alpha/2;N-1} \sqrt{\frac{S_D^2}{N}} \right)$$
$$= 1 - \alpha \qquad (6.38)$$

or

$$P\left(M_1 - M_2 - t_{\alpha/2;N-1} \sqrt{\frac{S_D^2}{N-1}} \le \mu_1 - \mu_2 \le M_1 - M_2 \right.$$
$$\left. + t_{\alpha/2;N-1} \sqrt{\frac{S_D^2}{(N-1)}} \right) = 1 - \alpha$$

For Example 6.1 the 95% confidence interval is as follows:

$$P\left(8 - 7 - 2.78 \sqrt{\frac{.4}{4}} \le \mu_1 - \mu_2 \le 8 - 7 + 2.78 \sqrt{\frac{.4}{4}} \right) = .95$$

and thus

$$P(.121 \le \mu_1 - \mu_2 \le 1.879) = .95$$

6.12   Dependent Versus Independent Group *t* Test

We now return to the issue of determining under which conditions, which design, independent group or dependent group, is superior. As mentioned previously, the dependent group design can be used to decrease variance due to nuisance factors, thus increasing power. In this section we investigate the conditions under which the variance is decreased. The researcher should beware because there are conditions under which the opposite phenomenon can occur—variance can be increased and the power decreased.

The differences between the two tests can be seen by comparing the test statistics for the dependent group *t* test [Equation (6.37), p. 163] and the independent group *t* test for equal sized groups [Equation (6.24), p. 149] (assume H_0: $\mu_1 - \mu_2 = 0$).

Dependent Group	Independent Group
$t = \dfrac{M_1 - M_2}{\sqrt{\dfrac{S_D^2}{N-1}}}$	$t = \dfrac{M_1 - M_2}{\sqrt{\dfrac{S_1^2 + S_2^2}{N-1}}}$

Disregarding for the moment that the degrees of freedom for the two tests are different, the statistics are equivalent except that the dependent group statistic includes the expression S_D^2, whereas the independent group statistic includes the expression $S_1^2 + S_2^2$. Three cases are possible. In the first case

$$S_D^2 < S_1^2 + S_2^2$$

In this instance the dependent group t statistic is larger than the independent group t statistic. In the second case

$$S_D^2 = S_1^2 + S_2^2$$

and the two statistics are equal. In the third case

$$S_D^2 > S_1^2 + S_2^2$$

In this case the dependent group t statistic is smaller than the independent group t statistic. Given the researcher's goal of rejecting the null hypothesis, the larger t statistic is desired, indicating that in case 1 the dependent group t test is preferred and in case 3 the independent group t test is preferred (again disregarding the degrees of freedom). Thus, it is important to understand the conditions affecting the relative size of the variance expressions.

The relative size of the variance expressions and their effect on the size of the t statistics can be understood by recalling from Section 3.13 [Equation (3.43), p. 70 rewritten with notation related to this context] that

$$S_D^2 = S_1^2 + S_2^2 - 2 \text{ (sample covariance)}$$

Remember that the sample covariance is a measure of the degree to which scores covary: (1) if the scores are positively related (high scores in one sample are paired with high scores in the other sample and vice versa), then the sample covariance is positive, (2) if the scores are unrelated, the sample covariance is zero, and (3) if the scores are negatively related (high scores in one sample are paired with low scores in the other sample and vice versa), then the sample covariance is negative.

We can now consider each of the three cases in turn. With regard to the first case, if the scores are positively related the sample covariance is positive and

$$S_D^2 < S_1^2 + S_2^2$$

Thus, in the first case the dependent group t statistic is larger than the independent group t statistic. Before this result can be stated as a rule for researchers, two issues must be discussed. First, it should be clear that the decision to use an independent or a dependent design must be made prior to conducting an experiment, and thus the sample covariance is unknown to the researcher. However, all is not lost. If the population covariance is positive (i.e., the paired scores are positively related in the population), then most likely the sample covariance will also be positive. Second, the degrees of freedom of the dependent sample t test are less than the degrees of freedom of the independent sample *t* test by a factor of 2 $[(N - 1)$ compared to $2(N - 1)]$. Thus, a price is paid in terms of the degrees of freedom when a dependent sample *t* test is used, although this price becomes relatively small when sample sizes are relatively large. In spite of these two caveats, when there is evidence that a factor is strongly related to the dependent variable, the researcher is well advised to use a dependent sample design in order to reduce the estimate of the variance in the statistical test. The advantage of the dependent sample *t* test when there is a very strong positive relation among the scores is illustrated by reanalyzing the data from Example 6.1 (p. 164). Although the data from this example were obtained from a dependent sample design, suppose that the subjects were not matched (i.e., five subjects were randomly assigned to the first intervention and the remaining five to the second intervention without regard to intelligence). If this were the case, the value of the *t* statistic for the independent sample *t* test would be

$$t = \frac{M_1 - M_2}{\sqrt{\dfrac{S_1^2 + S_2^2}{N - 1}}} = \frac{8 - 7}{\sqrt{\dfrac{4.8 + 4}{4}}} = .67$$

which, when compared to a *t* distribution with 8 degrees of freedom, is not sufficiently large to reject the null hypothesis and importantly is dramatically less than the value of the *t* statistic ($t = 3.16$) obtained when the data were treated as obtained from a dependent sample design.

In the second case the scores are unrelated and

$$S_D^2 = S_1^2 + S_2^2$$

In this case the *t* statistics for the independent and dependent analyses are equal. However, because of the reduced degrees of freedom, one would not want to use the dependent group *t* test when there was insufficient evidence to believe that the factor was related to the dependent variable. That is to say, it makes no sense to match on an irrelevant factor, such as the blood type of subjects in a learning experiment.

In the third case the scores are negatively related and

$$S_D^2 > S_1^2 + S_2^2$$

In this case the independent group test is superior both in terms of the size of the *t* statistic and in terms of the greater number of degrees of freedom. This might be the case where marital partners were used to pair observations, if it were true that introverts tend to select extroverted spouses, and the outcome measure was related to the amount of verbal material produced in a problem-solving task involving both spouses. In this case a spouse who produced large amounts of verbal material would be paired with a spouse who produced small amounts of verbal material and the dependent group *t* test would be an extremely poor choice because the variance would be increased.

NOTES AND SUPPLEMENTARY READINGS

Hays, W. L. (1981). *Statistics* (3rd ed.). New York: Holt, Rinehart and Winston.

See notations regarding Hays' text made in earlier chapters.

PROBLEMS

1. Perform a one-sample t test for the following hypotheses with respect to the given information:
   a. H_0: $\mu = 5$ vs. H_a: $\mu \neq 5$ given $\alpha = .05$ and the set of observations $\{4, 5, 7, 1, 3, 4\}$
   b. H_0: $\mu = 12$ vs. H_a: $\mu \neq 12$ given $\alpha = .01$, $N = 20$, $s^2 = 9$, $M = 14.1$ Also, find the 99% confidence interval for μ.
   c. H_0: $\mu = 100$ vs. H_a: $\mu < 100$ given $\alpha = .05$, $N = 30$, $s^2 = 10$, $M = 98.9$

2. Perform a one-sample t test for the following hypotheses with respect to the given information:
   a. H_0: $\mu = 10$ vs. H_a: $\mu > 10$ given $\alpha = .01$ and the set of observations $\{10, 9, 16, 15, 15, 12, 14\}$
   b. H_0: $\mu = 0$ vs. H_a: $\mu \neq 0$ given $\alpha = .01$, $N = 12$, $s^2 = 6$, $M = 1.5$ Also, find the 99% confidence interval for μ.
   c. H_0: $\mu = 12$ vs. H_a: $\mu < 12$ given $\alpha = .05$, $N = 25$, $s^2 = 14$, $M = 16$

3. Perform a two-independent-sample t test for the following hypotheses with respect to the given information:
   a. H_0: $\mu_1 - \mu_2 = 0$ vs. H_a: $\mu_1 - \mu_2 \neq 0$ given $\alpha = .05$ and the following sets of observations:

   Sample 1: $\{3, 4, 7, 1, 5\}$
   Sample 2: $\{6, 3, 1, 8, 10, 8\}$

   Also, find the 95% confidence interval for $\mu_1 - \mu_2$.
   b. H_0: $\mu_1 - \mu_2 = 0$ vs. H_a: $\mu_1 - \mu_2 \neq 0$ given $\alpha = .01$ and the following summary statistics:

   Sample 1: $N_1 = 11$, $S_1^2 = 16$, $M_1 = 18$
   Sample 2: $N_2 = 11$, $S_2^2 = 11$, $M_2 = 12$

   Also, find the 99% confidence interval for $\mu_1 - \mu_2$.
   c. H_0: $\mu_1 - \mu_2 = 0$ vs. H_a: $\mu_1 - \mu_2 < 0$ given $\alpha = .05$ and the following summary statistics:

   Sample 1: $N_1 = 20$, $s_1^2 = 25.3$, $M_1 = 13.3$
   Sample 2: $N_2 = 22$, $s_2^2 = 17.9$, $M_2 = 15.1$

4. Perform a two-independent-sample t test for the following hypotheses with respect to the given information:

a. $H_0: \mu_1 - \mu_2 = 0$ vs. $H_a: \mu_1 - \mu_2 \neq 0$ given $\alpha = .01$ and the following set of observations:

  Sample 1: {7, 9, 11, 7, 6}
  Sample 2: {3, 6, 2, 5, 9}

Also, find the 99% confidence interval for $\mu_1 - \mu_2$.

b. $H_0: \mu_1 - \mu_2 = 0$ vs. $H_a: \mu_1 - \mu_2 \neq 0$ given $\alpha = .05$ and the following summary statistics:

  Sample 1: $N_1 = 20$, $S_1^2 = 10.2$, $M_1 = 10.13$
  Sample 2: $N_2 = 25$, $S_2^2 = 9.3$, $M_2 = 10.58$

Also, find the 95% confidence interval for $\mu_1 - \mu_2$.

c. $H_0: \mu_1 - \mu_2 = 0$ vs. $H_a: \mu_1 - \mu_2 > 0$ given $\alpha = .01$ and the following summary statistics:

  Sample 1: $N_1 = 12$, $s_1^2 = 23.3$, $M_1 = 105.1$
  Sample 2: $N_2 = 13$, $s_2^2 = 18.9$, $M_2 = 100.7$

5. Test the hypotheses $H_0: \mu_1 - \mu_2 = 0$ vs. $H_a: \mu_1 - \mu_2 \neq 0$ with a two-dependent-sample t test with $\alpha = .05$ given the following observations:

Subject	x_{i1}	x_{i2}
1	3	4
2	5	6
3	4	4
4	7	10
5	6	5
6	9	10
7	7	8
8	4	7
9	3	3
10	2	3

Also find the 95% confidence interval for $\mu_1 - \mu_2$.

6. Test the hypotheses $H_0: \mu_1 - \mu_2 = 0$ vs. $H_a: \mu_1 - \mu_2 \neq 0$ with a two-dependent-sample t test with $\alpha = .01$ given the following observations:

Subject	x_{i1}	x_{i2}
1	11	8
2	16	11
3	8	11
4	17	14
5	11	13
6	17	14
7	13	13
8	20	21

Find the 99% confidence interval for $\mu_1 - \mu_2$.

CHAPTER 7 ANALYSIS OF VARIANCE— FIXED EFFECTS

I n the previous chapter statistical tests were discussed for testing hypotheses about two means. Although a variety of interesting research questions might be answered by a two-sample test, it would be unduly limiting if the researcher were restricted to tests involving only two levels of an independent variable. Clearly there are numerous instances where more than two conditions are needed to adequately answer research questions of interest. For example, a researcher wishing to test the efficacy of a treatment might want to compare it to a placebo condition and to a no-treatment condition. Or a researcher interested in whether the therapist affects the efficacy of psychotherapy might wish to study several therapists. When more than two levels of an independent variable are compared, a data analysis procedure called the analysis of variance can be used. The *analysis of variance*, as we will see, is a flexible procedure that can be adapted to test many variations on the theme. (The *t* test is actually a special case of the analysis of variance.)

One of the most powerful aspects of the analysis of variance is that it can be used to test simultaneously for the effects of two or more independent variables on a dependent variable. For example, a researcher might be interested in the effect of various treatments (a manipulated variable) and in the effect of gender (a status variable). The analysis of variance can be used to analyze the data from a multifactor experiment; tests for the effects for each of the independent variables (main effects) as well as the interaction among the independent variables (interaction effects) are discussed in this chapter and in the following chapter.

Another aspect of the analysis of variance is that it can be adapted to take into account the way in which the levels of the independent variable were selected. Three general models are examined in this chapter and Chapter 8: the *fixed-effects model*, the *random-effects model*, and the *mixed model*. A fixed-effects model is employed when the levels of the experimental treatment are selected arbitrarily and inferences are intended to apply only to the treatments examined and to no others that might have been included. For example, the researcher might be interested in three treatments for depression; if a fixed-effects model is used, the results are specific to those three treatments and to no others. A random-effects model is used when the levels of the independent variable are randomly selected from a much larger set with the intent of examining the effects of the independent variable in general. Such an experiment might involve inferences about the effect of therapists on the outcome of psychotherapy; if the therapists were chosen randomly from a population of therapists, then the conclusions would be applicable to all therapists in the population and not just to the ones included in the experiment. A mixed model is employed when levels of one (or more) of the variables have been selected arbitrarily and the levels of one (or more) of the variables have been selected randomly. As we will see, the choice of the design depends on the inferences that the researcher wishes to make. We discuss fixed-effects analysis of variance in this chapter and random and mixed models in the following chapter.

Before examining the fixed-effects analysis of variance, we need to examine two additional distributions—the chi-square and the F distributions. The test statistics of the t tests discussed in the previous section were compared to a t distribution. The test statistics of the analysis of variance are compared to the F distribution. To understand the F distribution, we should examine the chi-square distribution, which is denoted by χ^2. However, the chi-square distribution is important in its own right and will be an important distribution in many of the other statistical tests discussed later in this book. It turns out that the four continuous distributions studied in this book, the normal, the t, the chi-square, and the F distribution are closely related. In Chapter 13 we will examine the binomial distribution, a discrete distribution that is also related to these four continuous distributions.

The Chi-Square and F Distributions

7.1   The Chi-Square Distribution

As was the case with the t distribution, there is a family of chi-square distributions and each member of the family is determined by the degrees of freedom. All of the chi-square distributions can be derived by sampling from a normal distribution. To be consistent, the degrees of freedom of a chi-square distribution are given as a subscript. A chi-square distribution with ν degrees of freedom is written as χ^2_ν.

We begin the discussion with the chi-square distribution with 1 degree of freedom. Suppose that Y has a normal distribution with mean μ and variance σ^2; that is

$$Y \sim N(\mu, \sigma^2)$$

The first step in forming the chi-square distribution with 1 degree of freedom is to sample one case from this normal distribution. This case is standardized and squared forming a *squared standardized score*

$$z^2 = \frac{(y - \mu)^2}{\sigma^2}$$

The distribution of all such scores is the chi-square distribution with 1 degree of freedom and is denoted by χ^2_1; that is

$$\frac{(y - \mu)^2}{\sigma^2} \sim \chi^2_1 \tag{7.1}$$

The graph of this distribution (as well as the standard normal) is shown in Figure 7.1.

By examining the formula for the chi-square distribution with 1 degree of freedom, we can easily see why it is so positively skewed. First, note that because χ^2_1 is formed by squaring a score, it must always be nonnegative. Further, because most of the cases in a standard normal distribution are near zero, most of the cases of χ^2_1 are just to the right of zero. For the standard

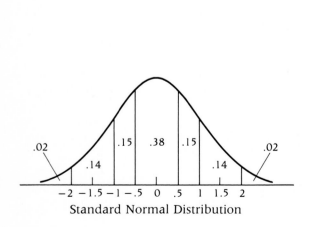

Standard Normal Distribution

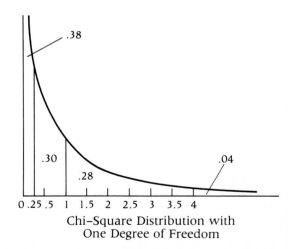

Chi–Square Distribution with
One Degree of Freedom

FIGURE 7.1   Areas Under the Graphs of the Standard Normal Distribution and the Chi-Square Distribution with 1 Degree of Freedom

normal distribution about 38% of the cases falls between $-.50$ and $+.50$; therefore, 38% of the cases of χ_1^2 falls between 0 and .25. Similarly, for the standard normal 15% of the cases falls between -1.00 and $-.50$ and 15% falls between .50 and 1.00; therefore, 30% of the cases of χ_1^2 falls between .25 and 1.00. The relationship between the areas under the standard normal and the χ_1^2 is presented in Figure 7.1.

The chi-square distribution with 2 degrees of freedom is formed in a similar manner. In this instance two cases are independently and randomly selected from a normal distribution Y with mean μ and variance σ^2. Each case is then standardized and squared

$$z_1^2 = \frac{(y_1 - \mu)^2}{\sigma^2}$$

$$z_2^2 = \frac{(y_2 - \mu)^2}{\sigma^2}$$

Finally, the sum of these two squared standardized scores is formed.

$$z_1^2 + z_2^2 = \frac{(y_1 - \mu)^2}{\sigma^2} + \frac{(y_2 - \mu)^2}{\sigma^2}$$

The distribution of all such scores (i.e., obtained from repeated sampling by the process just described) is the chi-square distribution with 2 degrees of freedom, denoted by χ_2^2. The graph of the chi-square distribution with 2 degrees of freedom (as well as several other chi-square distributions) is presented in Figure 7.2. Because the chi-square distribution with 2 degrees of freedom is based on the sum of two squared standardized scores, each time the sampling process is conducted a larger score is obtained than if the score was based on only one of the squared standardized scores (i.e., chi-

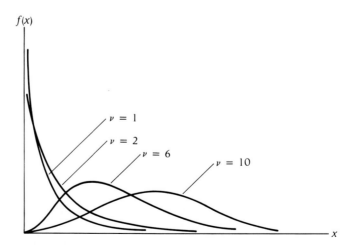

FIGURE 7.2   Graphs of Chi-Square Distributions with Various Degrees of Freedom

square with 1 degree of freedom); therefore, the chi-square with 2 degrees of freedom is less skewed than the chi-square with 1 degree of freedom.

The preceding process can be extended so that chi-square distributions with any desired number of degrees of freedom can be defined. **In general, if ν observations are independently and randomly selected from a normal distribution with mean μ and variance σ^2, then the distribution of the sum of the squared standardized scores is a chi-square distribution with ν degrees of freedom.** Notationally, the distribution of scores

$$\sum_{i=1}^{\nu} \frac{(y_i - \mu)^2}{\sigma^2}$$

is a chi-square with ν degrees of freedom. The graphs of several chi-square distributions are shown in Figure 7.2.

Several properties of the chi-square distribution are interesting and useful in later applications of this distribution. First, as was the case with the t distribution, the chi-square distribution is dependent only on one parameter, the degrees of freedom. That is, specification of ν determines the distribution exactly. Second, as the degrees of freedom increase, the skewness of the chi-square distribution decreases, as illustrated in Figure 7.2. Third, it can easily be shown by using the rules of expectation and variance that the mean and variance of a chi-square distribution with ν degrees of freedom are given by the following expressions:

$$E(\chi_\nu^2) = \nu \tag{7.2}$$

and

$$V(\chi_\nu^2) = 2\nu \tag{7.3}$$

Furthermore, as the degrees of freedom increase, the chi-square distribution approaches a normal distribution. Therefore

$$\frac{\chi_\nu^2 - \nu}{\sqrt{2\nu}} \to N(0, 1) \qquad \text{as } \nu \to \infty$$

Finally, because a chi-square distribution is the sum of squared standardized scores derived from independently and randomly selected samples, the sum of two independent chi-square random variables is easily formed. Suppose that two independent random variables have chi-square distributions with ν_1 and ν_2 degrees of freedom, respectively, then the random variable that is the sum of the two random variables has a chi-square distribution with $(\nu_1 + \nu_2)$ degrees of freedom; that is

$$\chi_{(\nu_1+\nu_2)}^2 = \chi_{\nu_1}^2 + \chi_{\nu_2}^2 \tag{7.4}$$

The table for the chi-square distributions (see Table B.3, p. 452) is arranged in a similar fashion to the table for the t distributions, with the difference that because the chi-square distributions are not symmetric, critical

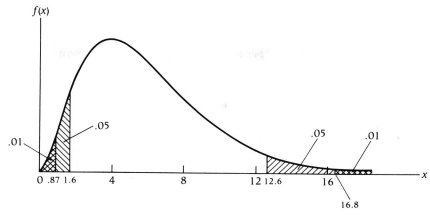

FIGURE 7.3   Some Areas Under the Graph of a Chi-Square Distribution with 6 Degrees of Freedom

values across the entire range of the chi-square distribution are given. The critical value is found by entering the table at the row corresponding to the degrees of freedom and the column corresponding to the area to the right of the critical value. The use of the table is illustrated by Figure 7.3, which presents a chi-square distribution with 6 degrees of freedom and selected critical values from the chi-square tables.

7.2 Sampling Distribution for the Sample Variance and Its Relation to the Chi-Square Distribution

In Chapter 4 we spent a great deal of effort to derive the sampling distribution of the sample mean. From Section 4.9 (p. 99) recall that if

$$Y \sim N(\mu, \sigma^2)$$

then the sampling distribution of the mean obtained from samples of size N is given by

$$M \sim N\left(\mu, \frac{\sigma^2}{N}\right)$$

With knowledge of the chi-square distribution, we can now discuss the sampling distribution of the sample variance S^2 (and also s^2). This discussion is vital to the development of the analysis of variance.

We begin by assuming that the population is normally distributed with mean μ and variance σ^2 [i.e., $Y \sim N(\mu, \sigma^2)$]. Recall the definition of the sample variance S^2 [Equation (3.15), p. 52].

$$S^2 = \frac{\sum_i (y_i - M)^2}{N}$$

Multiplying both sides of this equation by N and dividing both sides by σ^2, we have

$$\frac{NS^2}{\sigma^2} = \frac{\sum_i (y_i - M)^2}{\sigma^2}$$

Except that μ is replaced with M, it appears that the right-hand side of this equation has a chi-square distribution. Replacing μ with M results in a decrease of 1 degree of freedom, and it can be shown that

$$\frac{NS^2}{\sigma^2} \sim \chi^2_{(N-1)} \tag{7.5}$$

Using the relationship between S^2 and s^2, we see that

$$\frac{(N-1)s^2}{\sigma^2} \sim \chi^2_{(N-1)} \tag{7.6}$$

Dividing both sides of Equation (7.6) by $N - 1$, we obtain

$$\frac{s^2}{\sigma^2} \sim \frac{\chi^2_{(N-1)}}{N-1}$$

and it can be shown that, in general, **provided the parent distribution is normal**

$$\frac{\hat{\sigma}^2}{\sigma^2} \sim \frac{\chi^2_\nu}{\nu} \tag{7.7}$$

where ν is the degrees of freedom associated with making the estimate of σ^2. The relation expressed by Statement (7.7) is extremely important in our development of the analysis of variance and we will refer to it often.

7.3   The F distribution

The F distribution, which is named for Sir R. A. Fisher, may be derived from chi-square distributions. First, we form a random variable, called the F ratio, by taking the quotient of two independent chi-square random variables, each divided by its degrees of freedom. That is

$$F = \frac{\chi^2_{\nu_1}/\nu_1}{\chi^2_{\nu_2}/\nu_2} \tag{7.8}$$

The distribution of the *F* ratio is the *F* distribution and depends on two parameters, v_1 and v_2, typically called the degrees of freedom associated with the numerator and denominator, respectively. Again, the degrees of freedom are written as subscripts, with the numerator degrees of freedom and denominator degrees of freedom, separated by a comma, that is, Fv_1,v_2. (Often, an *F* distribution with v_1 degrees of freedom associated with the numerator and v_2 degrees of freedom associated with the denominator is simply referred to as an *F* distribution with v_1 and v_2 degrees of freedom.) It is important to emphasize the difference between the *F* ratio and its distribution, the *F* distribution. The notation for either the ratio or the distribution may contain the subscripted degrees of freedom; however, the context should make distinction clear. To summarize

$$\frac{\chi^2_{v_1}/v_1}{\chi^2_{v_2}/v_2} = F_{v_1,v_2} \sim F_{v_1,v_2} \tag{7.9}$$

As was the case for previous distributions, we have a family of distributions. In this case each pair of parameters v_1 and v_2 defines a unique *F* distribution.

In some respects the *F* distribution is quite removed from the normal distribution. First, the chi-square random variable is formed by summing squared standardized scores derived from independent random sampling from a normal distribution. Then, the *F* ratio is formed by taking the quotient of two independent chi-square random variables, each divided by its degrees of freedom. The *F* ratio thus formed is distributed as an *F* distribution. It would not be surprising if the reader has difficulty guessing the shape of the graph of an *F* distribution; indeed, the shape changes considerably with changes in the degrees of freedom. As well, the probability density function of the *F* distribution is complex. However, experience with *F* distributions will result in some intuitive feeling for the nature of the *F* distribution.

To illustrate some of the properties of the *F* distribution, an *F* distribution with 9 and 15 degrees of freedom (i.e., $F_{9,15}$) is presented in Figure 7.4. This unimodal and positively skewed graph is characteristic of *F* distributions for which v_1 is less than v_2 ($v_2 > 2$), which will be the case for the applications of the *F* distribution presented in this book. The expectation of the *F* ratio is

$$E(F_{v_1,v_2}) = \frac{v_2}{(v_2 - 2)} \tag{7.10}$$

when $v_2 > 2$. It is interesting to note that the expectation of *F* is greater than 1 and approaches 1 as the degrees of freedom associated with the denominator become large.

Tables for *F* distributions are slightly more complicated because of the 2 degrees of freedom parameters. The table for the *F* distribution is found in the Appendix (Table B.4, p. 454). Two panels for this table are given, one for the areas of .05 and .01 to the right of the critical values. For each area the critical value is found by entering the panel corresponding to the desired

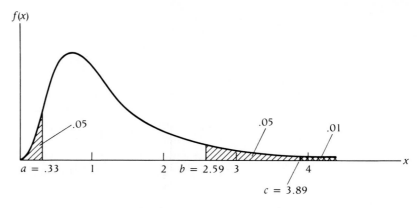

FIGURE 7.4   Some Areas Under the Graph of an F Distribution with 9 and 15 Degrees of Freedom

area at the row corresponding to the degrees of freedom associated with the denominator and the column associated with the degrees of freedom associated with the numerator. For example, we would use panel 1 to find the critical value of an F with 9 and 15 degrees of freedom that cut off an area in the right-hand tail of .05. The critical value found in the table (2.59) is illustrated in Figure 7.4. The critical value that cuts off an area of .01 in the right-hand tail is also illustrated in this figure.

Although critical values for the left-hand tail are not used extensively, it is useful to be able to find them. Table B.4 is set up for the right tail but some algebraic manipulations will allow us to use these tables for finding critical values in the left-hand tail. From the definition of the F ratio [see Equation (7.9)] it should be clear that

$$F_{v_1, v_2} = \frac{1}{F_{v_2, v_1}} \tag{7.11}$$

Using this relationship, we can now find the critical value in the left-hand tail of an F distribution with 9 and 15 degrees of freedom that cuts off an area of .05. Denoting the left-hand critical value by a, we want to find a such that

$$P(F_{9,15} < a) = .05$$

Using the properties of inequalities, we obtain

$$P\left(\frac{1}{F_{9,15}} > \frac{1}{a}\right) = .05$$

Now using the relationship expressed in Equation (7.11), we have

$$P\left(F_{15,9} > \frac{1}{a}\right) = .05$$

Having expressed our problem as a "right-hand tail" problem, Table B.4, p. 454 (Panel 1) can be entered with 15 degrees of freedom associated with the numerator and 9 degrees of freedom associated with the denominator. Thus

$$\frac{1}{a} = 3.01$$

and

$$a = .33$$

which is illustrated in Figure 7.4.

7.4 The *F* Distribution and the Ratio of Estimators for Population Variances

As was the case for the chi-square distribution, the *F* distribution is related to variances. Suppose that two populations are normally distributed with variances σ_1^2 and σ_2^2, respectively. Further, suppose that two independent samples of size N_1 and N_2 are randomly selected from populations 1 and 2, respectively. Let s_1^2 and s_2^2 be the unbiased estimates of the population variances of populations 1 and 2, respectively, based on the two independent samples. We know from Section 7.2 [Statement (7.6), p. 176] that

$$\frac{s_1^2}{\sigma_1^2} \sim \frac{\chi_{N_1-1}^2}{N_1 - 1}$$

and

$$\frac{s_2^2}{\sigma_2^2} \sim \frac{\chi_{N_2-1}^2}{N_2 - 1}$$

Thus

$$F_{N_1-1,N_2-1} = \frac{\chi_{N_1-1}^2/(N_1-1)}{\chi_{N_2-1}^2/(N_2-1)} = \frac{s_1^2/\sigma_1^2}{s_2^2/\sigma_2^2}$$

Furthermore, the *F* ratio formed in this way is distributed as an *F* distribution with $(N_1 - 1)$ and $(N_2 - 1)$ degrees of freedom, provided the parent distributions are normal, as was mentioned previously. To summarize: **The ratio of two independent unbiased estimators of population variances each divided by the respective population variances being estimated is an *F* ratio. Given respective sample sizes of N_1 and N_2 and normal population distributions, then the *F* ratio has an *F* distribution with $(N_1 - 1)$ and $(N_2 - 1)$ degrees of freedom. Notationally**

$$\frac{s_1^2/\sigma_1^2}{s_2^2/\sigma_2^2} = F \sim F_{N_1-1,N_2-1} \tag{7.12}$$

This relationship between unbiased estimates of population variances can be simplified when $\sigma_1^2 = \sigma_2^2$. **Given sample sizes of N_1 and N_2, then, provided the parent distributions are normal and $\sigma_1^2 = \sigma_2^2$, the statistic s_1^2/s_2^2 is an F ratio and is distributed as an F distribution with $(N_1 - 1)$ and $(N_2 - 1)$ degrees of freedom; that is**

$$\frac{s_1^2}{s_2^2} = F \sim F_{N_1-1,N_2-1} \tag{7.13}$$

Different forms of the statistic s_1^2/s_2^2 are used throughout the discussion of the analysis of variance. Each time it is used, recall that two assumptions are necessary in order for this ratio to have an F distribution; namely, normality of parent distributions and homogeneity of variance (these two assumptions should be familiar to the reader!, see Section 6.8, p. 150).

7.5   Relation Between the t and F Distributions

The t and F distributions have an intimate relationship that is crucial to understanding that the t test, which depends on the t distribution, is a special case of the analysis of variance, which depends on the F distribution. We will demonstrate (although not prove) that a particular F distribution is the square of the t distribution. This relationship will be "discovered" by returning to the one-sample t test. Recall that

$$\frac{M - \mu_0}{s/\sqrt{N}} = t \sim t_{N-1}$$

provided the parent distribution is normally distributed (see Section 6.4, p. 138). The strategy is to show that the square of the one-sample t statistic has an F distribution. First, multiply the numerator and denominator of the t statistic by $[1/(\sigma/\sqrt{N})]$:

$$t = \frac{\dfrac{1}{\sigma/\sqrt{N}}(M - \mu_0)}{\left(\dfrac{1}{\sigma/\sqrt{N}}\right)\left(\dfrac{s}{\sqrt{N}}\right)} = \frac{\dfrac{M - \mu_0}{\sigma/\sqrt{N}}}{\dfrac{s}{\sigma}}$$

Now square the expression on the right and make an additional algebraic manipulation:

$$t^2 = \frac{\left(\dfrac{M - \mu_0}{\sigma/\sqrt{N}}\right)^2}{\dfrac{s^2}{\sigma^2}} = \frac{\left(\dfrac{M - \mu_0}{\sigma/\sqrt{N}}\right)^2 \Big/ 1}{\dfrac{(N - 1)s^2}{\sigma^2} \Big/ (N - 1)}$$

Close inspection of the numerator and denominator of the expression on the right reveals that each is a chi-square random variable divided by its degrees of freedom. The numerator is the square of the sample mean standardized, divided by 1, and thus is a chi-square random variable with 1 degree of freedom divided by its degree of freedom. The denominator is a chi-square random variable with $(N - 1)$ degrees of freedom divided by $(N - 1)$ [see Equation (7.6), p. 176]. Although we have not mentioned this previously, M and s are independent when the parent distribution is normal. Thus, we now have the ratio of two independent chi-square variables, each divided by its degrees of freedom, which is of course an F distribution:

$$\frac{\chi_1^2/1}{\chi_{N-1}^2/(N-1)} = t_{N-1}^2 = F_{1,N-1}$$

Because the parent distribution is normal, this F ratio has an F distribution with 1 and $(N - 1)$ degrees of freedom. To summarize: **The square of a t statistic that has a t distribution with $(N - 1)$ degrees of freedom is an F ratio that has an F distribution with 1 and $(N - 1)$ degrees of freedom. Notationally**

$$t_{N-1}^2 = F_{1,N-1} \tag{7.14}$$

Application of the equality between the square of the t distribution and an F distribution with 1 degree of freedom associated with the numerator can be tricky, as we will illustrate using the t and F tables. First, we find the critical value for an F distribution with 1 and 10 degrees of freedom that cuts off an area of .05 in the right-hand tail; that is

$$P(F_{1,10} > c) = .05$$

From Table B.4, p. 454, $c = 4.96$. Now substitute t_{10}^2 for $F_{1,10}$.

$$P(t_{10}^2 > c) = .05$$

Using the rules of inequalities, we have

$$P(t_{10} < -\sqrt{c} \quad \text{or} \quad t_{10} > \sqrt{c}) = .05$$

We are now in a two-tailed context and

$$P(t_{10} < -\sqrt{c}) = .025 \quad \text{and} \quad P(t_{10} > \sqrt{c}) = .025$$

From Table B.2, p. 451

$$\sqrt{c} = 2.228 \quad \text{and} \quad c = 4.96$$

which agrees with the result found using the F tables. The point to remember here is that the relationship between the t and F distributions holds only if we consider a one-tailed rejection region for the F distribution and two-tailed rejection regions for the t distribution.

**One-Way
Analysis
of Variance**

7.6 Introduction to One-Way Analysis of Variance

To understand the strategy used in the analysis of variance, we need to return to the two-sample problem. Consider the two sampling distributions presented in Figure 7.5. In the two-sample problem the null hypothesis typically is

$$H_0: \mu_1 = \mu_2$$

The test statistic used to determine whether or not to reject the null hypothesis was developed in the previous chapter (see Section 6.7, p. 147).

$$t = \frac{M_1 - M_2}{\hat{\sigma}_{(M_1 - M_2)}}$$

Essentially, the t statistic compares the separation of the means to an estimate

Original

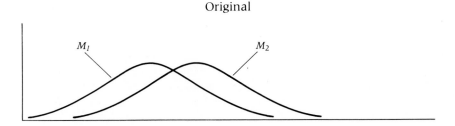

Increase Separation

Decrease Variance

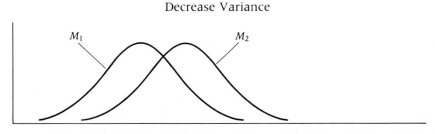

FIGURE 7.5   Increasing t Statistic for Two-Independent-Sample Problem

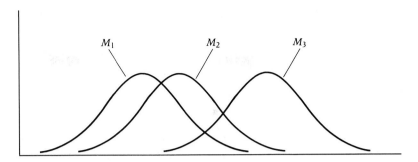

FIGURE 7.6   Three-Independent-Sample Problem

of the standard deviation of the sampling distribution of the difference (i.e., the standard error of the difference). Thus, the value of the t ratio is dependent on the degree to which the means are separated and on an estimate of the standard error of the difference, which is a function of the estimates of the standard deviations (or variances) of each sample. Simply, the t statistic can be increased by increasing the separation of the means or decreasing the variance of the sampling distribution of the difference between the means, as illustrated in Figure 7.5.

Now consider the three-sample problem, which is illustrated in Figure 7.6. The null hypothesis is similar to that for the two-sample problem.

$$H_0: \mu_1 = \mu_2 = \mu_3$$

To test this hypothesis, we need a statistic that is sensitive to both the separation of the sample means and the variability of the sampling distributions. First, let us consider the separation of the means. Separation is essentially a measure of dispersion. The more dispersed the sample means are, the less likely the null hypothesis is true. That is, if the null hypothesis is true, the sample means will be close together (although very unlikely that they will ever be exactly equal); whereas if the null hypothesis is not true (i.e., the population means are not all equal), then the sample means will be more dispersed. A measure of the dispersion of the sample means is the variance of the sample means, which is referred to as the *variance between treatments* (treatments are often referred to as groups and we use both terms interchangeably). The second consideration in forming a test statistic for three treatments is the variability of the sampling distribution for each mean. The *variance within treatments* is an uncontrolled variance because if we knew the source of such error, it could be eliminated (e.g., by selecting homogeneous samples such that all subjects had equivalent levels of the source of the error). Often the variance within treatments is called *error variance*. If the variance between treatments is large in comparison to the variance within treatments, then there is good evidence that the null hypothesis is not true.

The comparison between the variance between treatments and the variance within treatments is made by forming the ratio of the two variances

$$\text{Test statistic} = \frac{\text{variance between treatments}}{\text{error variance}}$$

A few comments about this test statistic are important. First, as we have seen, because the ratio of variances and the F distribution have intimate ties, the test statistic derived for the analysis of variance is called an F ratio and when certain conditions are met the test statistic has an F distribution. Second, although this discussion was framed in terms of three treatments, it is applicable to any number of treatments, including two treatments. Third, it should be clear why the procedure to be derived in this chapter is called the analysis of variance. Finally, because the derivation of the test statistic for the analysis of variance is fairly complex, we must keep in mind that the ultimate goal is to develop a test statistic that compares the variance between treatments to the variance within treatments (error variance).

7.7   The Linear Model and Assumptions of the One-Way Analysis of Variance

We start our development of the analysis of variance by discussing the linear model. In general, a *model* is a mathematical representation of a phenomenon. Typically, the model contains parameters that describe different aspects of the phenomenon. The hypothesis-testing strategy is to assume that the phenomenon fits the model (i.e., the collateral assumptions are true, see Section 5.4, p. 110) and then to test null hypotheses that are related to parameters of the model. The model discussed in the context of the one-way analysis of variance is a model of how J different treatments affect observations; that is, there is one independent variable with J levels (hence, the term *one-way* analysis of variance). It should be noted that there are models other than the linear model that might represent the effects of J treatments on the dependent variable. However, the linear model is parsimonious, is applicable to many behavioral phenomena, and is widely used.

Let J denote the number of treatments involved in the experiment. Here "treatment" is a generic term that includes such terms as conditions, interventions, or groups; in other words, there are J levels of the independent variable (see Section 6.10, p. 155). Because this chapter is concerned with fixed effects, we consider the J treatments as the only conditions of interest to the researcher and whatever conclusions are drawn from the experiment are applicable to those J treatments and to no others. Within each treatment we postulate an infinite population of observations. For example, if a treatment involves a particular intervention, then the population is thought to be the observations of all eligible subjects who could be administered the treatment. Of course, in practice, we randomly assign the subjects, administer the treatment, and then make our observations (see Section 6.10, p. 155). In any event, let the population mean of the jth treatment be denoted by μ_j. Furthermore, let the *grand mean*, denoted by μ, be the mean of the J populations pooled. Assuming that each of the populations contribute equally to the pooled populations (i.e., they are the same size), then the grand mean is the mean of the J population means:

$$\mu = \frac{\sum_{j=1}^{J} \mu_j}{J} \tag{7.15}$$

Any observation in any of the J populations may be thought of as the sum of three components: the grand mean, a treatment effect, and error. To write this sum mathematically, we need the following additional notation. Let Y_{ij} be the ith observation in the jth treatment, α_j be the effect of treatment j, and e_{ij} be the error for the ith observation in the jth treatment. Then

$$Y_{ij} = \mu + \alpha_j + e_{ij} \tag{7.16}$$

Because this equation is so important, it needs some further explanation. Consider for a moment an example in which the three treatments consist of the administration of three drugs. Suppose that the population means of the three treatments are as follows:

Treatment 1	Treatment 2	Treatment 3
$\mu_1 = 3$	$\mu_2 = 5$	$\mu_3 = 10$

Hence $\mu = 6$. The effect of each treatment is simply the difference between the treatment mean and the grand mean; that is

$$\alpha_j = \mu_j - \mu \tag{7.17}$$

So, the effect of treatment for the three treatments are -3, -1, and 4, respectively. Notice that the sum $\mu + \alpha_j$ is equal to the treatment mean. The term e_{ij} is added to the model to reflect the variability of the observations within each treatment. Thus, the linear model can be thought of as

$$Y_{ij} = \mu + \alpha_j + e_{ij}$$

or

$$Y_{ij} = \mu_j + e_{ij} \tag{7.18}$$

Suppose that $Y_{52} = 8$ for the three-drug-treatment example. This can be thought of as the sum

$$Y_{52} = \mu + \alpha_2 + e_{52} = 6 + (-1) + 3 = 8$$

that is, Y_{52} is the sum of the grand mean, the effect for being in the second treatment, and error (viz., $e_{52} = 3$), which reflects variability within treatments.

The null hypothesis in the one-way analysis of variance is that the means associated with the J populations are equal to each other (and therefore equal to the grand mean). That is

$$H_0:\ \mu_1 = \mu_2 = \cdots = \mu_J = \mu$$

The alternative is simply that H_0 is false:

$$H_a:\ H_0 \text{ is false}$$

Note that the alternative does not indicate that all the means are different. The fact that H_0 is false only indicates that at least one mean is different from at least one of the others. For example, when four populations are involved, if

$$\mu_1 = \mu_2 = \mu_3 \neq \mu_4$$

then the null hypothesis is false. The test of a null hypothesis involving the equality of more than two parameters is called an *omnibus* test. Because rejection of the null hypothesis only indicates that the means are not all equal, an omnibus test provides no information about the comparison of individual means. However, there are methods, which are termed post hoc comparisons and planned comparisons (see Chapter 9, p. 260), that are designed to test hypotheses about individual means in the context of the linear model. If the omnibus test results in the rejection of the null hypothesis, post hoc comparisons between individual means are often conducted. When the researcher has specific hypotheses about comparisons between particular means, planned comparisons can be conducted in lieu of an omnibus test.

It is important to recognize that μ, μ_j, and α_j are parameters of the linear model. As such they are fixed quantities that characterize the populations, although typically they are unknown to the researcher. Given the relationship between the treatment mean and the effect of the treatment, $\mu_j = \mu + \alpha_j$, we clearly see that the null hypothesis for the linear model

$$H_0:\ \mu_1 = \mu_2 = \cdots = \mu_J = \mu$$

is equivalent to

$$H_0:\ \alpha_1 = \alpha_2 = \cdots = \alpha_J = 0$$

Furthermore, it is easily shown that the sum of treatment effects always equals zero:

$$\sum_{j=1}^{J} \alpha_j = \sum_{j=1}^{J} (\mu_j - \mu) = \sum_{j=1}^{J} \mu_j - \sum_{j=1}^{J} \mu = \sum_{j=1}^{J} \mu_j - J\mu$$

But by the definition of the grand mean μ, $\sum_{j=1}^{J} \mu_j = J\mu$, and therefore $\sum_{j=1}^{J} \alpha_j = 0$. **However, it should be noted that**

$$\sum_{j} \alpha_j^2 = 0$$

only if the null hypothesis is true. When the null hypothesis is not true

$$\sum_j \alpha_j^2 > 0$$

This characteristic of the sum of the squares of the treatment effects is a vital aspect in the test statistic for the analysis of variance.

The only random variable in the linear model is e_{ij}, the error component associated with the ith observation in the jth group. (Although random variables are typically denoted by upper-case letters, here the lower-case is used to differentiate errors from expectations.) When we discuss the random variable Y_{ij} in the context of the linear model, we can think of it as composed of a random variable e_{ij} plus a constant $\mu + \alpha_j$. The parameters represent a systematic relationship between the independent variable X and the dependent variable Y, whereas the random variable e_{ij} represents unsystematic variation in Y, or error variance.

To develop the analysis of variance, we must place some restrictions on the model. These restrictions, which take the role of collateral assumptions, should be familiar from the discussion of assumptions in the t test context, although here they are expressed in terms of the linear model. The three assumptions that are required for the development of the test statistic in the analysis of variance are as follows. First, for each of the J treatment populations the random variable e_{ij} has a normal distribution. Second, the variance of the random variable e_{ij} is the same for each of the J treatment populations (i.e., homogeneity of variance). This variance is denoted by σ_e^2. Third, the errors associated with any pair of observations are independent.

Several observations about these assumptions are important to understand. We begin the discussion of the assumptions by finding the expectation of the observations in population j. Because μ and α_j are parameters (i.e., constants)

$$E(Y_{ij}) = E(\mu + \alpha_j + e_{ij}) = \mu + \alpha_j + E(e_{ij}) = \mu_j + E(e_{ij})$$

But by definition the mean of population j is μ_j, therefore, to make the preceding equality hold, $E(e_{ij})$ must equal zero. Thus, the first two assumptions can be summarized by the following statement:

$$e_{ij} \sim N(0, \sigma_e^2)$$

Furthermore, by the rules of variance the variance of the jth population is given by

$$V(Y_{ij}) = V(\mu + \alpha_j + e_{ij}) = V(e_{ij}) = \sigma_e^2$$

To summarize, the distribution of Y in treatment j is normal with mean μ_j and variance σ_e^2.

Clearly the first two assumptions considered here are identical to the assumptions discussed in the context of the two-independent-sample t test.

The discussion of the violation of these assumptions made in Section 6.8 (p. 150) applies as well to violations of these assumptions in the context of the analysis of variance.

The assumption that the e_{ij}'s are independent is equivalent to saying that the observations are independent, which we have assumed for all sampling, unless otherwise specified (see Section 4.2, p. 85). It is worth repeating that violation of the independence of observation assumption has very deleterious effects on the decisions to reject the null hypothesis.

7.8   Sampling and the Linear Model

Up to this point the linear model has been discussed in terms of J treatment populations. Although the parameters μ, μ_j, and α_j are fixed constants that characterize the true state of affairs, they are unknown to the researcher. As was the case previously, inferences about the population parameters are made by sampling from the population.

The sampling procedure is to select observations randomly and independently from each of the J treatment populations. The ith observation in the jth sample is denoted by y_{ij}. Further, the size of the sample from the jth population is denoted by N_j. As well, let M_j be the mean based on the sample from the jth population:

$$M_j = \frac{\sum\limits_{i=1}^{N_j} y_{ij}}{N_j}$$

(Because the summation notation becomes fairly complex in the analysis of variance, it is worth emphasizing here that the index i refers to subjects within groups. That is, in the formula for M_j, j can assume any of J values and then the observations are summed within that sample. For example, if $j = 2$, then the observations within the sample selected from the second population are summed and divided by N_2, the size of this sample.) The total sample size N is the sum of the sample sizes of the individual samples

$$N = \sum_{j=1}^{J} N_j$$

(Here the index of summation is j; that is, we are summing over the treatment groups.) The *sample grand mean M* can be found in one of two ways. First, the J samples can be combined into one sample (as was described in Section 6.6, p. 144) and the mean of the combined sample can be determined:

$$M = \frac{\sum\limits_{j=1}^{J} \sum\limits_{i=1}^{N_j} y_{ij}}{N} \qquad (7.19)$$

(This is the first time we have encountered a "double sum." Multiple sums are calculated by starting with the right-hand sum and working to the left. Here the sum with the index i is accomplished first; that is, the observations within each group are summed. Next, the sum over the index j is accomplished; that is, the sums of the observations within groups are summed across groups. In this way $\sum_{j=1}^{J} \sum_{i=1}^{N_j} y_{ij}$ is the sum of all of the observations.) The second way to calculate the sample grand mean is to form the weighted sum of the sample means for each sample:

$$M = \frac{\sum_{j=1}^{J} N_j M_j}{N} \qquad (7.20)$$

which is the extension of the formula for the weighted mean based on two samples [see Equation (6.17), p. 145].

As has continually been the case, sample means are used as estimators for the population means:

$$M_j = \hat{\mu}_j$$

and

$$M = \hat{\mu}$$

Furthermore, the sample means can be used as estimators for the treatment effects. Recall that $\alpha_j = \mu_j - \mu$. Thus, a reasonable estimator for α_j is $M_j - M$. This estimator is unbiased since

$$E(M_j - M) = E(M_j) - E(M) = \mu_j - \mu = \alpha_j$$

and has the other desirable properties of estimators. Notationally

$$M_j - M = \hat{\alpha}_j \qquad (7.21)$$

Given the estimators described previously, the equation for the *sample linear model* is

$$y_{ij} = M + \hat{\alpha}_j + e_{ij} = M_j + e_{ij} \qquad (7.22)$$

where e_{ij} in this context is the obtained difference between the sample treatment mean and the observations within that treatment.

To this point we have examined estimators for means and treatment effects. Estimation of population variances in the context of the linear model is more complex and is the essence of the analysis of variance. The following sections will examine this topic in some detail.

7.9 Partitioning the Sum of Squares

The sum of squares is the central device for deriving the distributional characteristics of the variances involved in the analysis of variance as well as for making the calculations that are necessary to obtain the test statistic. Essentially, sums of squares are deviations squared. Given our knowledge of the sample variance (see Section 3.6, p. 51), it is not surprising that terms involving squared deviations would be used in the analysis of variance. In the one-way analysis of variance context three sums of squares are involved: *sum of squares total (SST)*, *sum of squares between (SSB)* (sometimes called sum of squares between treatments), and *sum of squares error (SSE)* (sometimes called sum of squares within treatments). The sum of squares between is related to the variance between groups and is "good stuff" in the sense that the greater the variance between groups is, the more likely the null hypothesis will be rejected. The sum of squares error is related to the variance within groups and is "bad stuff" in the sense that the greater the variance within groups (i.e., uncontrolled variance) is, the less likely the null hypothesis will be rejected. The sum of squares total is the sum of SSB and SSE.

We present the partitioning of the sum of squares and illustrate its use before proving, as a comment, that indeed, SST = SSB + SSE. The partition of the sum of squares is given by

$$\underbrace{\sum_{j=1}^{J}\sum_{i=1}^{N_j}(y_{ij}-M)^2}_{\textbf{SST}} = \underbrace{\sum_{j=1}^{J}N_j(M_j-M)^2}_{\textbf{SSB}} + \underbrace{\sum_{j=1}^{J}\sum_{i=1}^{N_j}(y_{ij}-M_j)^2}_{\textbf{SSE}} \qquad (7.23)$$

A brief discussion of the expression for each of the sum of squares will help to provide an intuitive idea of their role. The *sum of squares total* is the sum of the squared deviations of each observation from the grand mean (first summed within each group and then summed across groups). The *sum of squares between* is formed from the squared deviations of the treatment mean and the grand mean. Thus, these deviations measure the dispersion among the treatment means—the greater the dispersion of the treatment means is, the greater the sum of squares between will be. The squared deviations are then weighted by the sample size for each group and summed across groups. The *sum of squares error* is the sum of the squared deviations of each observation from the treatment mean for the treatment to which the observation belongs. Thus, the sum of squares error measures the dispersion within groups. In the next section the sum of squares between and the sum of squares error will be used to form estimates of population variances.

Calculation of the sum of squares is illustrated by using the data presented in Table 7.1. First, we calculate the estimated treatment effects. Recall Equation (7.21) (p. 189)

$$M_j - M = \hat{\alpha}_j$$

Thus

$$\hat{\alpha}_1 = 3 - 2 = 1$$
$$\hat{\alpha}_2 = 1 - 2 = -1$$

and

$$\hat{\alpha}_3 = 2 - 2 = 0$$

First, we calculate the sum of squares total:

$$SST = \sum_j \sum_i (y_{ij} - M)^2$$

$$= \underbrace{(2 - 2)^2 + (3 - 2)^2 + (4 - 2)^2}_{j = 1} + \underbrace{(0 - 2)^2 + (0 - 2)^2 + (3 - 2)^2}_{j = 2}$$

$$+ \underbrace{(1 - 2)^2 + (2 - 2)^2 + (3 - 2)^2 + (2 - 2)^2}_{j = 3}$$

$$= 5 + 9 + 2 = 16$$

Next, we calculate the sum of squares between:

$$SSB = \sum_j N_j(M_j - M)^2$$

$$= \underbrace{3(3 - 2)^2}_{j = 1} + \underbrace{3(1 - 2)^2}_{j = 2} + \underbrace{4(2 - 2)^2}_{j = 3} = 3 + 3 + 0 = 6$$

Finally, we calculate the sum of squares error.

$$SSE = \sum_j \sum_i (y_{ij} - M_j)^2$$

$$= \underbrace{(2 - 3)^2 + (3 - 3)^2 + (4 - 3)^2}_{j = 1} + \underbrace{(0 - 1)^2 + (0 - 1)^2 + (3 - 1)^2}_{j = 2}$$

$$+ \underbrace{(1 - 2)^2 + (2 - 2)^2 + (3 - 2)^2 + (2 - 2)^2}_{j = 3}$$

$$= 2 + 6 + 2 = 10$$

Note that for this example SST = SSB + SSE, as expected.

TABLE 7.1

Data for One-Way Design

	Treatment	
1	2	3
2	0	1
3	0	2
4	3	3
		2

$M_1 = 3$, $M_2 = 1$, $M_3 = 2$, $M = 2$.

COMMENT

The proof that SST = SSB + SSE is not difficult or exceedingly complex. However, as the models for the analysis of variance become more involved, the proofs of the corresponding partition of the sum of squares become algebraically "messy." Consequently, in this book only the proof of the partition of the sum of squares for the one-way fixed-effects analysis of variance will be provided.

The strategy used in the proof is to begin with the expression for the sum of squares total and algebraically manipulate it until we can show that it equals the sum of the sum of squares between and the sum of squares error. Accordingly

$$\text{SST} = \sum_j \sum_i (y_{ij} - M)^2$$

$$= \sum_j \sum_i [(y_{ij} - M_j) + (M_j - M)]^2$$

$$= \sum_j \sum_i (y_{ij} - M_j)^2 + \sum_j \sum_i 2(y_{ij} - M_j)(M_j - M) + \sum_j \sum_i (M_j - M)^2$$

We examine this last expression term by term. The first term $\sum_j \sum_i (y_{ij} - M_j)^2$ is the sum of squares error. Now examine the middle term

$$\sum_j \sum_i 2(y_{ij} - M_j)(M_j - M)$$

Because the Σ to the right involves i (and not j), the middle term can be rewritten as

$$2 \sum_j (M_j - M) \sum_i (y_{ij} - M_j)$$

but because $\sum_j (y_{ij} - M_j) = 0$ (i.e., the sum of the deviations equal zero), the middle term is equal to zero. Finally, examine the last term

$$\sum_j \sum_i (M_j - M)^2$$

Because this expression is summed over the index i but i does not appear in the expression, the following algebraic "trick" is used:

$$\sum_j \sum_i (M_j - M)^2 = \sum_j \sum_i (M_j - M)^2(1)$$

$$= \sum_j (M_j - M)^2 \sum_{i=1}^{N_j} 1$$

$$= \sum_j (M_j - M)^2(N_j)$$

$$= \sum_j N_j(M_j - M)^2$$

$$= \text{SSB}$$

Therefore

$$SST = SSE + 0 + SSB = SSB + SSE$$

which completes the proof.

7.10   Distribution Theory

Recall that the goal of our development of the analysis of variance was to develop a test statistic of the form

$$\text{Test statistic} = \frac{\text{variance between treatments}}{\text{error variance}}$$

Using the sum of squares error and the sum of squares between, we can derive expressions that estimate the error variance and the variance between treatments.

We begin our discussion with the error variance. An unbiased estimator for the variance within treatment j is given by

$$\frac{\sum_i (y_{ij} - M_j)^2}{N_j - 1}$$

Assuming that the variances within groups are equal, we form the estimator for the error variance by pooling.

$$\hat{\sigma}_e^2 = \frac{\sum_j \sum_i (y_{ij} - M_j)^2}{(N_1 - 1) + \cdots + (N_j - 1)} = \frac{\sum_j \sum_i (y_{ij} - M_j)^2}{N - J} \tag{7.24}$$

Three important points need to be made about this pooled estimator for the error variance. First, the pooled estimator is simply the sum of squares error (SSE) divided by $N - J$. Because dividing the sum of square error by $N - J$ forms the "average" or mean squared deviation, we call this quantity the *mean square error (MSE)*. To summarize

$$\textbf{MSE} = \frac{\textbf{SSE}}{N - J} = \hat{\sigma}_e^2 \tag{7.25}$$

Second, the formula for the pooled estimate based on J samples is a straightforward generalization of the estimate based on two samples [see Equation (6.19), p. 145]. Finally, because the mean square error is an unbiased estimator for σ_e^2, we write

$$E(\textbf{MSE}) = E\left[\frac{\textbf{SSE}}{N - J}\right] = \sigma_e^2 \tag{7.26}$$

Because this relationship is so important and because the proof is representative of the derivations of the expectations of mean squares in general, the proof of Equation (7.26) is found in a comment at the end of this section.

Recalling that the chi-square distribution was related to the distribution of variances, the distribution of the mean square error can be determined. From Equation (7.7), p. 176, we know that, in general, provided the parent distribution is normal

$$\frac{\hat{\sigma}^2}{\sigma^2} \sim \frac{\chi_\nu^2}{\nu}$$

where ν is the degrees of freedom associated with the estimate of σ^2. Because MSE is an estimator for σ_e^2 with $(N - J)$ degrees of freedom (1 degree of freedom is "lost" for each of the J sample means, see Section 6.3, p. 138)

$$\frac{\hat{\sigma}_e^2}{\sigma_e^2} = \frac{\text{MSE}}{\sigma_e^2} = \frac{\text{SSE}}{(N - J)\sigma_e^2} \sim \frac{\chi_{N-J}^2}{N - J} \tag{7.27}$$

provided the population distributions are normal.

We now turn our attention to the variance between treatments. Recall that the sum of squares between is given by the following expression:

$$\text{SSB} = \sum_j N_j (M_j - M)^2$$

A reasonable way to obtain the variance between groups would be to divide the sum of squares between by J, the number of treatments. However, to obtain an unbiased estimate, we divide by $J - 1$ to account for the estimate of the grand mean. Hence, we define *mean square between* as

$$\text{MSB} = \frac{\text{SSB}}{J - 1} \tag{7.28}$$

To derive the distributional properties of mean square between, we need to determine its expectation. Although the derivation of the expectation of MSB is similar to the derivation of the expectation of MSE (which is found in the comment at the end of this section), it is more complicated and not particularly instructive, and therefore is not presented here. Nevertheless, it can be shown that

$$E(\text{MSB}) = E\left(\frac{\text{SSB}}{J - 1}\right) = \sigma_e^2 + \frac{\sum_j N_j \alpha_j^2}{J - 1} \tag{7.29}$$

provided the assumptions of the analysis of variance discussed earlier are true (see Section 7.7, p. 184).

Equation (7.29) in many ways is the key to the one-way fixed-effects analysis of variance and close inspection of it will yield a wealth of information. **When the null hypothesis that the treatment effects are zero,**

$$E(\text{MSB}) = \sigma_e^2$$

However, when the null hypothesis is false

$$E(\text{MSB}) > \sigma_e^2$$

In this way mean square between treatments is sensitive to the null hypothesis; that is, the larger the treatment effects are, the larger the mean square between will be. Furthermore, when the null hypothesis is true, $E(\text{MSB}) = \sigma_e^2$ (i.e., MSB is an unbiased estimator for σ_e^2). Therefore

$$\frac{\hat{\sigma}_e^2}{\sigma_e^2} = \frac{\text{MSB}}{\sigma_e^2} = \frac{\text{SSB}}{(J-1)\sigma_e^2} \sim \frac{\chi_{J-1}^2}{J-1} \tag{7.30}$$

provided the assumptions of the analysis of variance are true and the null hypothesis is true.

There is an additional property of mean square between and mean square error that is needed to form the test statistic for the analysis of variance. It can be shown that **when J samples of independent observations are each drawn from normally distributed populations, mean square error and mean square between are independent.**

COMMENT

As was mentioned earlier, derivations of the expected values of the mean squares are omitted because they are generally quite involved and not very instructive. Nevertheless, derivation of the simplest case (MSE) will give the reader a flavor of these derivations. Finding the expectation of MSE is straightforward.

$$E(\text{MSE}) = E\left(\frac{\text{SSE}}{N-J}\right)$$

$$= E\left(\frac{\sum_j \sum_i (y_{ij} - M_j)^2}{N-J}\right)$$

$$= \frac{1}{N-J} \sum_j E\left[\sum_i (y_{ij} - M_j)^2\right]$$

$$= \frac{1}{N-J} \sum_j (N_j - 1)E\left(\frac{\sum_i (y_{ij} - M_j)^2}{N_j - 1}\right)$$

Because the variances within each treatment are assumed to be equal to σ_e^2

$$E(\text{MSE}) = \frac{1}{N-J} \sum_j (N_j - 1) E(\sigma_e^2)$$

Because σ_e^2 is a parameter (i.e., a constant)

$$E(\text{MSE}) = \frac{1}{N-J} \sigma_e^2 \sum_{j=1}^{J} (N_j - 1)$$

$$= \frac{1}{N-J} \sigma_e^2 (N-J)$$

$$= \sigma_e^2$$

7.11   The F Test for the One-Way Analysis of Variance

We are now ready to form the test statistic for the one-way fixed-effects analysis of variance. Given the strategy to derive a statistic of the form

$$\text{Test statistic} = \frac{\text{variance between treatments}}{\text{error variance}}$$

we would reasonably think that the test statistic to be used would be

$$\text{Test statistic} = \frac{\text{MSB}}{\text{MSE}}$$

Indeed, this is the statistic used and we have the necessary machinery to derive the distribution of this test statistic.

Recall that

$$\frac{\text{MSE}}{\sigma_e^2} \sim \frac{\chi_{N-J}^2}{N-J}$$

when the assumptions of the analysis of variance are met. Furthermore

$$\frac{\text{MSB}}{\sigma_e^2} \sim \frac{\chi_{J-1}^2}{J-1}$$

when the assumptions of the analysis of variance are met **and when the null hypothesis that there are no treatment effects is true.** Therefore

$$\frac{\text{MSB}/\sigma_e^2}{\text{MSE}/\sigma_e^2} \sim \frac{\chi_{J-1}^2/(J-1)}{\chi_{N-J}^2/(N-J)}$$

and

$$\frac{\text{MSB}}{\text{MSE}} \sim \frac{\chi_{J-1}^2/(J-1)}{\chi_{N-J}^2/(N-J)}$$

Given that MSB and MSE are statistically independent

$$\frac{\text{MSB}}{\text{MSE}} \sim F_{J-1,N-J}$$

To summarize: **If the assumptions of the analysis of variance are met and the null hypothesis**

$$H_0: \alpha_1 = \alpha_2 = \cdots = \alpha_J = 0$$

(or $H_0: \mu_1 = \mu_2 = \cdots = \mu_J = \mu$) is true, then the F ratio

$$\frac{\text{MSB}}{\text{MSE}} = F \tag{7.31}$$

has an F distribution with $(J - 1)$ and $(N - J)$ degrees of freedom. When the null hypothesis is not true, recall that the expectation of MSB is greater than σ_e^2, and hence it would be expected that MSB will be greater than MSE, yielding an *F* ratio greater than 1. That is, an *F* ratio greater than 1 is evidence that the null hypothesis is not true. For this reason the *F* test for the analysis of variance is a one-tailed test. Rejection of the null hypotheses is determined by referring to the *F* tables (B.4, p. 454) for given levels of α set a priori.

Obtaining values of the *F* ratio that are less than 1.00 is a phenomenon worth examining. If the assumptions of the analysis of variance are met, then this result must be due to sampling error. In effect, an *F* of less than 1.00 says that the sample means are closer together (in comparison to the variance) than we would expect by chance. Although an *F* ratio that is much smaller than 1.00 (i.e., one that is statistically significant in the left-hand tail of the *F* distribution), could be due to chance, such small *F* values may give a clue that some violation of the assumptions or some structural aspect of the experiment created a situation where the sample means were closer together than expected. Such structural aspects might include nonrandom selection, floor or ceiling effects, nonindependent observations, and so forth (actually such structural aspects lead to violation of the assumptions). In short, any experiment that yields an *F* value that is quite small should be scrutinized closely.

A useful way to present the results of the analysis of variance is in a *source table*, which is schematically presented in Table 7.2. In practice, the values for the sum of squares, degrees of freedom, mean squares, and *F* are inserted, as will be illustrated in the example that follows.

TABLE 7.2

Source Table for One-Way Analysis of Variance

Source	Sum of Squares	Degrees of Freedom	Mean Squares	*F*
Treatments (between)	SSB	$J - 1$	$\text{MSB} = \text{SSB}/(J - 1)$	$\dfrac{\text{MSB}}{\text{MSE}}$
Error	SSE	$N - J$	$\text{MSE} = \text{SSE}/(N - J)$	
Total	SST	$N - 1$		

Before presenting the example, we should mention that the computational "effort" in the analysis of variance is primarily involved with the sum of squares. Computational formulas that reduce the effort exist, but with the accessibility of computers, they seem superfluous. The reader who is interested in learning about the concepts of the analysis of variance is best advised to use the definitional formulas for the sum of squares given by Equation (7.23) (p. 190). With that in mind, Example 7.1 is presented with the calculation of the sums of squares omitted (an example of the calculation of the sums of squares was presented in Section 7.9, p. 189).

EXAMPLE 7.1

Suppose that a study was conducted to determine the effectiveness of various kinds of devices designed to increase reading comprehension of a passage. Three conditions were of interest. In the control condition subjects read the passage and answered questions about it afterward. In another condition subjects were provided organizers (material that gave the readers a schema of the material). In the third condition subjects were provided organizers as well as advanced questions (questions provided before the passage was read and which were similar to those involved with assessing reading comprehension). At the completion of reading the passage the subjects answered 18 questions that assessed their comprehension. Subjects were randomly selected and assigned to the three conditions. Originally each condition contained 24 subjects, but the data for 12 of the subjects in the organizer condition were lost. The null hypothesis is that the three population means were all equal. A priori the researcher set α at .05. Table 7.3 presents the results of this study.

TABLE 7.3

Data for One-Way Analysis of Variance Example

Treatments				
Control		Organizers	Organizers and Questions	
15	11	15	18	12
14	10	15	17	12
14	10	14	16	12
14	10	14	15	12
13	10	14	14	11
13	10	13	14	11
13	9	12	14	10
13	9	11	14	10
13	9	11	14	10
13	8	10	14	10
13	8	10	13	9
13	8	10	13	8
$M_1 = 11.38$		$M_2 = 12.42$	$M_3 = 12.63$	

The sample means for the three conditions are easily calculated:

$$M_1 = 11.38$$
$$M_2 = 12.42$$

and

$$M_3 = 12.63$$

The grand mean is calculated by using Equation (7.20) (p. 189).

$$M = \frac{\sum_j N_j M_j}{N} = \frac{24(11.38) + 12(12.42) + 24(12.63)}{60} = 12.09$$

Estimates of the treatment effects are calculated by using Equation (7.21) (p. 189).

$$\hat{\alpha}_1 = M_1 - M = 11.38 - 12.09 = -.71$$
$$\hat{\alpha}_2 = M_2 - M = 12.42 - 12.09 = .33$$

and

$$\hat{\alpha}_3 = M_3 - M = 12.63 - 12.09 = .54$$

At this point the researcher is quite pleased that the estimated effects are what would be expected given the nature of the study. The results of the remaining calculations are summarized in a source table.

Source	Sum of Squares	Degrees of Freedom	Mean Squares	F
Treatments	20.416	2	10.208	$1.91 = \dfrac{MSB}{MSE} \sim F_{2,57}$
Error	304.167	57	5.336	
Total	324.583	59		

The *F* ratio of 1.91 is compared to an *F* distribution with 2 and 57 degrees of freedom and is insufficient to reject the null hypothesis that the population means are all equal.

We also note that it is perfectly acceptable to conduct an analysis of variance when only two levels of the independent variable are involved. However, the usual test in this case is the two-independent-sample *t* test. As we have alluded to earlier, the two-independent-sample *t* test is a special case of the analysis of variance (i.e., $J = 2$). The square of the *t* statistic obtained by the independent group *t* test is equal to the *F* ratio obtained by the analysis of variance, which is not surprising given the relationship between the *t* and *F* distributions discussed in Section 7.5 (p. 180). With regard to the independent sample t test and the analysis of variance, as long as a nondirectional alternative is specified, decisions about the null hypothesis will be identical.

7.12   Strength of Association

In the context of two independent samples the population effect size was defined in the following way [see Equation (6.23), p. 148]:

$$\text{Effect size} = \left| \frac{\mu_1 - \mu_2}{\sigma} \right|$$

where σ is the standard deviation of each population (under the assumption of homogeneity of variance). Generalization of this measure to more than two populations is not obvious and we turn to a group of related measures, called measures of strength of association.

Measures of strength of association index the degree to which the dependent variable is associated with the independent variable. When there is no association, knowledge of the independent variable gives no information about the dependent variable. This would be the case when the population means were all equal (and the treatment effects were zero). If one were asked to predict a given observation in the jth treatment, the best guess (in the sense that it minimizes the squared error, see Section 3.17, p. 77) would be the grand mean μ. Uncertainty about the guess would be expressed as the variance of the dependent measure; that is, the uncertainty of Y (given no information is provided by X) is σ_Y^2.

Typically, the independent variable provides some information about the dependent variable. For example, if the population treatment effect for treatment 2 was positive, we know that observations from the second population will be generally larger than the grand mean. Said another way, knowing X (actually the level of X) provides information about Y. When some information is provided by knowing X, the best guess for an observation from the jth population would be the mean of that population μ_j. Uncertainty about this guess is denoted by $\sigma_{Y|X}^2$, the variance in Y given X. When there is perfect association, there is no uncertainty and $\sigma_{Y|X}^2$ is equal to zero, whereas when there is no association, $\sigma_{Y|X}^2$ is equal to σ_Y^2.

A reasonable definition of the strength of association of X and Y would be

$$\sigma_Y^2 - \sigma_{Y|X}^2$$

However, because this measure is dependent on the metric of the variables, **strength of association, denoted by ω^2 (omega squared), is defined as the relative reduction in uncertainty:**

$$\omega^2 = \frac{\sigma_Y^2 - \sigma_{Y|X}^2}{\sigma_Y^2} \tag{7.32}$$

Defined in this way, ω^2 ranges from zero when there is no association ($\sigma_{Y|X}^2 = \sigma_Y^2$) to 1 when there is perfect association ($\sigma_{Y|X}^2 = 0$). For obvious reasons the value of ω^2 is often referred to as the *proportion of variance in Y accounted for by X*. In the analysis of variance context $\sigma_{Y|X}^2 = \sigma_e^2$ and

$$\omega^2 = \frac{\sigma_Y^2 - \sigma_e^2}{\sigma_Y^2} = \frac{\sum_j \alpha_j^2 / J}{\sigma_Y^2} \tag{7.33}$$

which for the two-sample problem reduces to

$$\omega^2 = \frac{(\mu_1 - \mu_2)^2}{4\sigma_Y^2}$$

The value ω^2 is one-fourth the square of the effect size discussed earlier for two independent samples. It should be kept in mind that ω^2 is a population value and we are left with the task of finding an estimator for it.

In the analysis of variance context a sample strength of association measure is given simply by

$$\text{Sample strength of association} = \frac{\text{total variance} - \text{within treatment variance}}{\text{total variance}}$$

This sample value is denoted by η^2 (eta squared; this violates our convention of only denoting population values by Greek letters, but it conforms to the generally accepted notation for this measure). Remembering that η^2 is a sample value (no attempt to estimate population values)

$$\eta^2 = \frac{\text{SST}/N - \text{SSE}/N}{\text{SST}/N} = \frac{\text{SST} - \text{SSE}}{\text{SST}} = \frac{\text{SSB}}{\text{SST}} \tag{7.34}$$

The value η^2 may be thought of as the proportion of variance in Y accounted for by X **in the sample**. Although an extremely easy measure to calculate, η^2 is a dramatically biased estimate of ω^2 in that it leads to an overestimate of the population proportion of variance accounted for by the independent variable. This overestimation can be understood by realizing when the F ratio is equal to 1.00 (the expected value when the population means are equal), η^2 will be greater than zero. For Example 7.1

$$\eta^2 = \frac{\text{SSB}}{\text{SST}} = \frac{20.416}{324.583} = .063$$

which indicates that about 6% of the variance in the Y-scores actually obtained in the experiment is accounted for by X.

A better estimator for ω^2 is given by the following expression:

$$\hat{\omega}^2 = \frac{\text{SSB} - (J - 1)\text{MSE}}{\text{SST} + \text{MSE}} \tag{7.35}$$

This estimate is equal to zero when F is equal to 1.00 and is less than zero when the F ratio is less than 1.00 (when such is the case the estimate is set equal to zero). For Example 7.1

$$\hat{\omega}^2 = \frac{\text{SSB} - (J-1)\text{MSE}}{\text{SST} + \text{MSE}} = \frac{20.416 - 2(5.336)}{324.583 + 5.336} = .030$$

which is considerably less than the value obtained with η^2.

Two-Way Analysis of Variance

7.13 Introduction to the Two-Way Analysis of Variance

In the one-way analysis of variance the influence of one independent variable (with multiple levels) on a dependent variable was examined. However, rarely is it the case that only one independent variable accounts for variability in the dependent variable. Typically, the researcher's goal is to identify and study several important independent variables to determine how they separately as well as jointly influence the dependent variable. For instance, suppose that a researcher is interested in the acquisition of mathematics knowledge in junior high school students. The researcher wishes to compare various teaching strategies (traditional lecture, programmed instruction, and computer-assisted instruction) and to determine whether enrichment is facilitative (enrichment versus no enrichment). The two-way analysis of variance tests the effects of both independent variables as well as the interaction of the independent variables.

Typically, examination of two independent variables involves designing an experiment whereby subjects are studied in each combination of treatments. In the mathematics example six combinations would be studied—traditional lectures with and without enrichment, programmed instruction with and without enrichment, and computer-assisted instruction with and without enrichment. A schematic diagram of these combinations is presented in Figure 7.7. A design of this sort is called a *completely crossed factorial design*.

	Traditional lectures	Programmed instruction	Computer-assisted instruction
No Enrichment			
Enrichment			

FIGURE 7.7 Example of Completely Crossed Factorial Design

Completely crossed refers to the fact that each level of one factor occurs with each level of the other factor.

Factorial experiments have a number of advantages over an approach whereby the researcher might study each experimental variable separately in several one-way designs. First of all, factorial studies are more economical from the standpoint of a researcher's time and money. A second advantage is that it can be determined if the two independent variables produce an interactive effect. Often some particular conditions of one independent variable will interact with certain particular conditions of the other independent variable to generate an effect that would not be evident if the two variables were examined separately. Finally, inclusion of additional variables might well reduce the uncertainty in the dependent variable, which reduces the error variance and produces a more powerful statistical test.

7.14 Linear Model for Two-Way Analysis of Variance

The linear model for the two-way analysis of variance is similar to that for the one-way analysis of variance except that the model contains terms associated with each of the independent variables as well as the interaction. Let R and C be the two independent variables (R and C are used to indicate row and columns, as we will see). Let J denote the number of levels (treatments, conditions, categories, etc.) of R, and K denote the number of levels of C, as illustrated in Figure 7.8. Such a design is called a $J \times K$ factorial design (read "J by K factorial design"). The combination of different levels

Column Variable

Row Variable	Level 1	Level 2	...	Level k	...	Level K	
Level 1	μ_{11}	μ_{12}	...	μ_{1k}	...	μ_{1K}	$\mu_{1\bullet}$
Level 2	μ_{21}	μ_{22}	...	μ_{2k}	...	μ_{2K}	$\mu_{2\bullet}$
⋮	...	...	...	...	...	...	⋮
Level j	μ_{j1}	μ_{j2}	...	μ_{jk}	...	μ_{jK}	$\mu_{j\bullet}$
⋮	...	...	...	...	...	...	⋮
Level J	μ_{J1}	μ_{J2}	...	μ_{Jk}	...	μ_{JK}	$\mu_{J\bullet}$
	$\mu_{\bullet 1}$	$\mu_{\bullet 2}$	...	$\mu_{\bullet k}$	...	$\mu_{\bullet K}$	

FIGURE 7.8 General Two-Way Factorial Design

of R with different levels of C are called cells. Cell jk represents the jth level of R and the kth level of C. Associated with each cell jk is a population of observations with cell mean μ_{jk} (see Figure 7.8). It is useful, as well, to refer to the *marginal means*. The marginal mean for the jth level of R, denoted by $\mu_{j.}$, is the mean for that level of R disregarding (averaged over) the levels of C. More precisely

$$\mu_{j.} = \frac{\sum\limits_{k} \mu_{jk}}{K} \tag{7.36}$$

Similarly

$$\mu_{.k} = \frac{\sum\limits_{j} \mu_{jk}}{J} \tag{7.37}$$

The grand mean μ is given by

$$\mu = \frac{\sum\limits_{k}\sum\limits_{j} \mu_{jk}}{JK} \tag{7.38}$$

The linear model for the fully crossed two-way factorial design is given by

$$Y_{ijk} = \mu_{jk} + e_{ijk} \tag{7.39}$$

where Y_{ijk} is the ith observation in the jth level of R and the kth level of C and e_{ijk} is the error associated with the ith observation in the jth level of R and kth level of C. Furthermore, the cell mean μ_{jk} can be expressed as a composite of the grand mean plus effects due to the independent variables R and C as well as the interaction of R and C. Notationally

$$\mu_{jk} = \mu + \alpha_j + \beta_k + \gamma_{jk} \tag{7.40}$$

where α_j is the treatment effect associated with the jth level of R, β_k is the treatment effect associated with the kth level of C, and γ_{jk} is the treatment effect associated with the jth level of R and the kth level of C. The effects α_j and β_k are called *main effects* and γ_{jk} is called an *interaction effect*. The need to include an interaction effect may not yet be clear, but much of this section is devoted to a discussion of this effect. The linear model can now be expressed as

$$Y_{ijk} = \mu + \alpha_j + \beta_k + \gamma_{jk} + e_{ijk} \tag{7.41}$$

As before, μ, α_j, β_k, and γ_{jk} are parameters and e_{ijk} is a random variable.

The same assumptions that applied to the one-way analysis of variance apply to the two-way analysis of variance. To reiterate, the first assumption is that the random variable e_{ijk} has a normal distribution, the second assumption is that the variance of e_{ijk} is the same for each of the combinations of the row and column variables (denote this variance by σ_e^2), and the third assumption is that the errors associated with any pair of observations are independent.

We now discuss each of the effects in more detail. The *row effect* is the effect due to particular levels of R regardless of the level of C. That is

$$\alpha_j = \mu_{j \cdot} - \mu \qquad (7.42)$$

which is the difference between the marginal mean and the grand mean. With regard to the row effects, the null hypothesis is that all the row effects are zero; that is

$$H_0: \alpha_1 = \alpha_2 = \cdots = \alpha_J = 0$$

As was the case with the one-way linear model

$$\sum_j \alpha_j = 0$$

but

$$\sum_j \alpha_j^2 = 0$$

only if the null hypothesis is true; otherwise

$$\sum_j \alpha_j^2 > 0$$

The *column effects* are defined analogously to the row effects:

$$\beta_k = \mu_{\cdot k} - \mu \qquad (7.43)$$

With regard to the column effects, the null hypothesis is that the column effects are all zero; that is

$$H_0: \beta_1 = \beta_2 = \cdots = \beta_K = 0$$

Similarly

$$\sum_k \beta_k = 0$$

but

$$\sum_k \beta_k^2 = 0$$

only if the null hypothesis is true; otherwise

$$\sum_k \beta_k^2 > 0$$

The *interaction effect* is the effect that is unique to the combination of levels of the row and column variables. Consider the cell jk. The cell mean is due, in part, to the row effect α_j and the column effect β_k. However, the combination of these two treatments might result in an added enhancement (or attenuation) that is due to the *combination* of the treatments. Something may be gained (or lost) over and above the simple addition of the row and column effects. This something extra is called the interaction effect. We will derive an expression for the interaction effect and then illustrate all of the effects in a series of examples. In these examples recall that the cell mean is the sum of the grand mean and all the effects [Equation (7.40) p. 204]:

$$\mu_{jk} = \mu + \alpha_j + \beta_k + \gamma_{jk}$$

Solving for the interaction effect

$$\gamma_{jk} = \mu_{jk} - \alpha_j - \beta_k - \mu \qquad\qquad (7.44)$$

That is, the interaction effect is the cell mean minus the grand mean and the main effects. By noting that $\alpha_j = \mu_{j.} - \mu$ and that $\beta_k = \mu_{.k} - \mu$, we can rewrite the expression for the interaction effect as

$$\gamma_{jk} = \mu_{jk} - \mu_{j.} - \mu_{.k} + \mu \qquad\qquad (7.45)$$

The null hypothesis for the interaction is that the interaction effects are all zero; that is

$$H_0: \gamma_{jk} = 0 \qquad \text{for all } j \text{ and } k$$

As was the case for the main effects

$$\sum_k \sum_j \gamma_{jk} = 0$$

but

$$\sum_k \sum_j \gamma_{jk}^2 = 0$$

only if the null hypothesis is true; otherwise

$$\sum_k \sum_j \gamma_{jk}^2 > 0$$

The concepts of main effects and interaction effects are illustrated by considering six examples of a 2×3 factorial design, which are presented in Figure 7.9. For each example the cell means are given and from them the marginal means, the grand mean, the main effects, and the interaction effects can be calculated. Keep in mind that we are operating in the context of population parameters.

For the first example from Figure 7.9 the row effects are calculated from Equation (7.42)

$$\alpha_1 = \mu_{1.} - \mu = 3 - 4 = -1$$

and

$$\alpha_2 = \mu_{2.} - \mu = 5 - 4 = 1$$

From Equation (7.43) the column effects are calculated

$$\beta_1 = \mu._1 - \mu = 4 - 4 = 0$$
$$\beta_2 = \mu._2 - \mu = 4 - 4 = 0$$

and

$$\beta_3 = \mu._3 - \mu = 4 - 4 = 0$$

Recalling that the interaction effects are the cell means minus the grand mean and the main effects [Equation (7.44)], we get

$$\gamma_{11} = \mu_{11} - \alpha_1 - \beta_1 - \mu = 3 - (-1) - 0 - 4 = 0$$
$$\gamma_{21} = \mu_{21} - \alpha_2 - \beta_1 - \mu = 5 - 1 - 0 - 4 = 0$$

and in the same manner we can see that the other interaction effects are zero. In this example the row effects are the only nonzero effects. Hence, observations are described by a trimmed linear model

$$Y_{ijk} = \mu + \alpha_j + e_{ijk}$$

Typically, it is informative to examine graphs of the effects by plotting the cell means. The dependent measure is found on the vertical axis of such graphs. One of the two independent variables is found on the horizontal axis, whereas the other independent variable is represented by different lines within the graph. Either independent variable can be represented by the horizontal axis although the "gestalt" of the graph is changed, as illustrated for the first example in Figure 7.9.

The second example represents a case in which there are column effects but no row or interaction effects (i.e., the row and interaction effects are zero). Thus, the linear model is

$$Y_{ijk} = \mu + \beta_k + e_{ijk}$$

For the third example there are main effects for both rows and columns but no interaction effect. The linear model is represented by

$$Y_{ijk} = \mu + \alpha_j + \beta_k + e_{ijk}$$

This is called the *additive model* because the observations are due to the addition of the main effects (plus the error).

The fourth example is interesting because it includes an interaction effect. From the graphs we see that the lines are no longer parallel, which indicates the presence of an interaction. In this example clearly there are no main effects. The mean for the **first** level of the row variable averaged over the levels of the column variables (i.e., $\mu_1.$) is equal to the **second** level of the row variable averaged over the levels of the column variables (i.e., $\mu_2.$). The same is true for the column variables. Nevertheless, the combination of levels of the two variables does have an effect. Clearly level 1 of the row variable in conjunction with level 1 of the column variable attenuates the scores, whereas level 1 of the row variable in conjunction with level 3 of the column variable augments the scores. Calculation of the interaction effects

Example 1

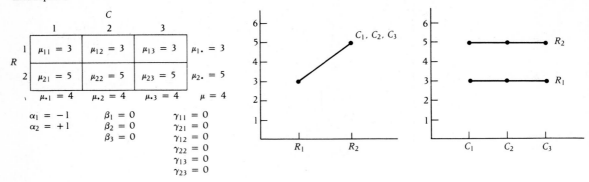

Example 2

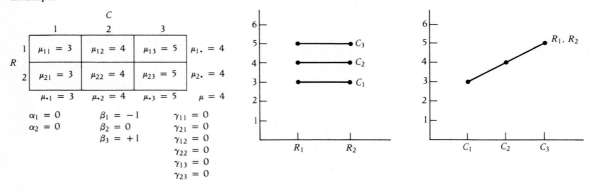

Example 3

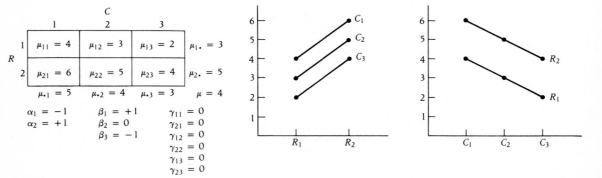

FIGURE 7.9   Six Examples of Main and Interaction Effects

Example 4

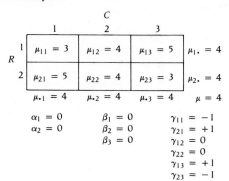

R	C 1	2	3	
1	$\mu_{11} = 3$	$\mu_{12} = 4$	$\mu_{13} = 5$	$\mu_{1.} = 4$
2	$\mu_{21} = 5$	$\mu_{22} = 4$	$\mu_{23} = 3$	$\mu_{2.} = 4$
	$\mu_{.1} = 4$	$\mu_{.2} = 4$	$\mu_{.3} = 4$	$\mu = 4$

$\alpha_1 = 0$ $\beta_1 = 0$ $\gamma_{11} = -1$
$\alpha_2 = 0$ $\beta_2 = 0$ $\gamma_{21} = +1$
       $\beta_3 = 0$ $\gamma_{12} = 0$
               $\gamma_{22} = 0$
               $\gamma_{13} = +1$
               $\gamma_{23} = -1$

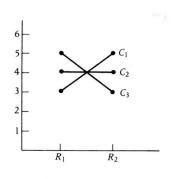
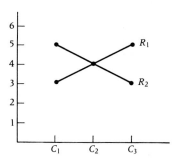

Example 5

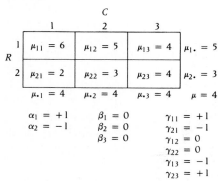

R	C 1	2	3	
1	$\mu_{11} = 6$	$\mu_{12} = 5$	$\mu_{13} = 4$	$\mu_{1.} = 5$
2	$\mu_{21} = 2$	$\mu_{22} = 3$	$\mu_{23} = 4$	$\mu_{2.} = 3$
	$\mu_{.1} = 4$	$\mu_{.2} = 4$	$\mu_{.3} = 4$	$\mu = 4$

$\alpha_1 = +1$ $\beta_1 = 0$ $\gamma_{11} = +1$
$\alpha_2 = -1$ $\beta_2 = 0$ $\gamma_{21} = -1$
        $\beta_3 = 0$ $\gamma_{12} = 0$
               $\gamma_{22} = 0$
               $\gamma_{13} = -1$
               $\gamma_{23} = +1$

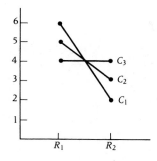
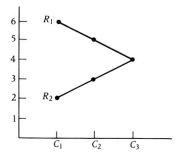

Example 6

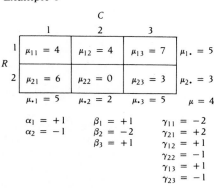

R	C 1	2	3	
1	$\mu_{11} = 4$	$\mu_{12} = 4$	$\mu_{13} = 7$	$\mu_{1.} = 5$
2	$\mu_{21} = 6$	$\mu_{22} = 0$	$\mu_{23} = 3$	$\mu_{2.} = 3$
	$\mu_{.1} = 5$	$\mu_{.2} = 2$	$\mu_{.3} = 5$	$\mu = 4$

$\alpha_1 = +1$ $\beta_1 = +1$ $\gamma_{11} = -2$
$\alpha_2 = -1$ $\beta_2 = -2$ $\gamma_{21} = +2$
        $\beta_3 = +1$ $\gamma_{12} = +1$
               $\gamma_{22} = -1$
               $\gamma_{13} = +1$
               $\gamma_{23} = -1$

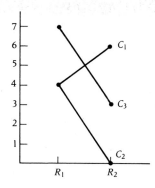
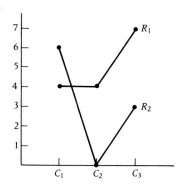

using Equation (7.44) will demonstrate that the interaction effects are not all zero:

$$\gamma_{11} = \mu_{11} - \alpha_1 - \beta_1 - \mu = 3 - 0 - 0 - 4 = -1$$
$$\gamma_{21} = \mu_{21} - \alpha_2 - \beta_1 - \mu = 5 - 0 - 0 - 4 = 1$$
$$\gamma_{12} = \mu_{12} - \alpha_1 - \beta_2 - \mu = 4 - 0 - 0 - 4 = 0$$
$$\gamma_{22} = \mu_{22} - \alpha_2 - \beta_2 - \mu = 4 - 0 - 0 - 4 = 0$$
$$\gamma_{13} = \mu_{13} - \alpha_1 - \beta_3 - \mu = 5 - 0 - 0 - 4 = 1$$

and

$$\gamma_{23} = \mu_{23} - \alpha_2 - \beta_3 - \mu = 3 - 0 - 0 - 4 = -1$$

In this case the linear model reduces to

$$Y_{ijk} = \mu + \gamma_{jk} + e_{ijk}$$

The fifth example illustrates a row effect and an interaction effect but no column effect. Although there is no column effect, it would be a mistake to say that the column variable was not important. The column variable is important because it interacts with the row variable. Although the mean for level 1 of the row variable is greater than the mean for level 2 of the row variable, the highest mean occurs when level 1 of the row variable is combined with level 1 of the column variable. With regard to level 3 of the column variable, the means for levels 1 and 2 of the row variable are equal.

The final example contains main effects and an interaction effect and illustrates the problem in interpreting main effects without considering the interaction effect. Generally, the mean of level 1 of the row variable is larger than the mean of level 2 of the row variable. If the levels represent therapeutic treatments and higher scores indicate better performance, it would be tempting to say that treatment 1 of the row variable was superior to treatment 2 of the row variable. However, that conclusion is mitigated by the interaction because within level 1 of the column variable treatment 2 of the row variable is superior to treatment 1 of the row variable. This illustrates that **when an interaction effect is present, the main effects should always be interpreted in light of the interaction.**

7.15   Sampling and the Two-Way Linear Model

Up to this point we have considered only the population parameters for the two-way model. Sampling from the completely crossed factorial design involves selecting observations from each of the combinations of the row and column variables. It is important to select the subjects in such a way that the main effects are independent of each other and of the interaction effect. If the subjects are selected in such a way that the effects are not independent, then one effect will be confounded with the other effects. To illustrate, suppose that the researcher wishes to study the effect of attitudes toward abortion (coded as either pro, neutral, or con) and religiosity (coded as either religious or not religious) on some outcome measure. Because attitudes toward abortion and religiosity are likely related (e.g., the proportion of religious subjects who are against abortion is likely greater than the proportion of nonreligious subjects who are against abortion), allocation of subjects to cells from a single sample from the general population likely will result in unequal cell size (e.g., more subjects in the con-religious cell than the con-nonreligious cell). When this occurs, the effect of attitudes toward abortion and religiosity are confounded in the sense that their effect on the dependent variable overlap and the effect for one independent variable will *ipso facto* create an effect in the other independent variable. Sampling in this way allows for the structural dependency among the effects (the main effects and the interaction effect).

The effects in a two-way completely crossed factorial design will be independent provided two conditions are met. First, the design must be *orthogonal*. The simplest and most practical means to obtain an orthogonal design is to have an equal number of observations in each cell. Such a sampling strategy is referred to as a *balanced design*. Another less practical means is to have proportional numbers of subjects in each cell. If N_{jk} is the number of observation in cell jk, $N_{j.}$ is the number of subjects in row j, and $N_{.k}$ is the number of subjects in column k, and N is the total number of subjects, then proportionality of subjects occurs provided

$$N_{jk} = \frac{N_{j.}N_{.k}}{N}$$

Our development only considers the case where the cell sizes are equal (i.e., balanced designs). The second requirement for independence is that the subjects are randomly and independently selected from normal populations.

A final requirement of our development of the two-way fixed-effects analysis of variance is that the balanced completely crossed factorial designs must also be *replicated*. A replicated design has at least two observations in each cell. It is possible to analyze nonreplicated designs, but we must assume that there is no interaction effect. Such designs are discussed in the context of the mixed-model analysis of variance presented in the next chapter.

It will be helpful to review the sampling procedure for the two-way analysis of variance. Recall that we are independently and randomly sampling from parent distributions that are normal by using a completely crossed design that is replicated and balanced (i.e., equal cell sizes). Let n denote the sample size within each cell (i.e., $N_{jk} = n$ for all j and k). Then

$$Kn = N_{j.}$$
$$Jn = N_{.k}$$

and

$$JKn = N$$

Furthermore, let M_{jk}, $M_{j\cdot}$, $M_{\cdot k}$, and M denote the sample cell mean, row marginal mean, column marginal mean, and grand mean, respectively. Furthermore, let y_{ijk} denote the ith sampled observation in the jth level of R and the kth level of C.

As was the case in the one-way analysis of variance, the population means and effects are estimated with sample means:

$$\hat{\mu}_{jk} = M_{jk} = \frac{\sum_i y_{ijk}}{n} \tag{7.46}$$

$$\hat{\mu}_{j\cdot} = M_{j\cdot} = \frac{\sum_k \sum_i y_{ijk}}{Kn} = \frac{\sum_k M_{jk}}{K} \tag{7.47}$$

$$\hat{\mu}_{\cdot k} = M_{\cdot k} = \frac{\sum_j \sum_i y_{ijk}}{Jn} = \frac{\sum_j M_{jk}}{J} \tag{7.48}$$

$$\hat{\mu} = M = \frac{\sum_k \sum_j \sum_i y_{ijk}}{JKn} = \frac{\sum_k \sum_j M_{jk}}{JK} = \frac{\sum_j M_{j\cdot}}{J} = \frac{\sum_k M_{\cdot k}}{K} \tag{7.49}$$

$$\hat{\alpha}_j = M_{j\cdot} - M \tag{7.50}$$

$$\hat{\beta}_k = M_{\cdot k} - M \tag{7.51}$$

and

$$\hat{\gamma}_{jk} = M_{jk} - \hat{\alpha}_j - \hat{\beta}_k - M = M_{jk} - M_{j\cdot} - M_{\cdot k} + M \tag{7.52}$$

7.16 Partitioning the Sum of Squares, Distribution Theory, and the F Test

The development of the test statistics for the two-way analysis of variance proceeds in the same sequence as for the one-way analysis of variance. The sum of squares is partitioned, the mean squares are formed, the expectation of the mean squares is determined, and ratios of mean squares are used as the test statistics.

In the two-way analysis of variance the sum of squares total is partitioned into *sum of squares error (SSE)*, *sum of squares for rows (SSR)*, *sum of squares for columns (SSC)*, *and sum of squares for interaction [SS(R × C)]*.

$$\underbrace{\sum_{k=1}^{K}\sum_{j=1}^{J}\sum_{i=1}^{n}(y_{ijk} - M)^2}_{\text{SST}} = \underbrace{\sum_{k=1}^{K}\sum_{j=1}^{J}\sum_{i=1}^{n}(y_{ijk} - M_{jk})^2}_{\text{SSE}}$$

$$+ \underbrace{\sum_{j=1}^{J} N_{j\cdot}(M_{j\cdot} - M)^2}_{\text{SSR}}$$

$$+ \underbrace{\sum_{k=1}^{K} N_{\cdot k}(M_{\cdot k} - M)^2}_{\text{SSC}}$$

$$+ \underbrace{\sum_{k=1}^{K}\sum_{j=1}^{J} n(M_{jk} - M_{j\cdot} - M_{\cdot k} + M)^2}_{\text{SS}(R \times C)} \qquad (7.53)$$

This partition should be examined until it seems intuitive. The sum of squares total is analogous to that for the one-way partition. The sum of squares error is also analogous to the one-way partition in that it is the square of the deviations **within** each cell, summed over the cells. The sum of squares for rows is the square of the deviation of the marginal mean for a level of the row variable from the grand mean weighted by the number of observations in the row and summed over the levels of the row variable. A similar interpretation is found for the sum of squares for the columns. The sum of squares for the interaction is slightly more complicated. The expression within the parentheses for $SS(R \times C)$ is simply the sample interaction effect for cell *jk*. The sum of squares is formed by weighting the square of the sample interaction effects by the sample size within the cell and summing over all cells.

Although algebraically "messy," it is not difficult to prove that this partition is true [i.e., that SST equals the sum of SSE, SSR, SSC, and $SS(R \times C)$]; however, it is more instructive to illustrate the equality. The data for this illustration are found in Table 7.4. The sample means and effect sizes are easily calculated by using Equations (7.46) through (7.52) and are left to the reader. The sum of squares are calculated in the following way:

$$SST = \sum_k \sum_j \sum_i (y_{ijk} - M)^2$$

$$= \underbrace{(2 - 6)^2 + (10 - 6)^2}_{\text{cell 11}} + \underbrace{(6 - 6)^2 + (14 - 6)^2}_{\text{cell 21}}$$

$$+ \underbrace{(6 - 6)^2 + (2 - 6)^2}_{\text{cell 12}} + \underbrace{(2 - 6)^2 + (6 - 6)^2}_{\text{cell 22}}$$

$$= 128$$

$$\text{SSE} = \sum_k \sum_j \sum_i (y_{ijk} - M_{jk})^2$$

$$= \underbrace{(2 - 6)^2 + (10 - 6)^2}_{\text{cell } 11} + \underbrace{(6 - 10)^2 + (14 - 10)^2}_{\text{cell } 21}$$

$$+ \underbrace{(6 - 4)^2 + (2 - 4)^2}_{\text{cell } 12} + \underbrace{(2 - 4)^2 + (6 - 4)^2}_{\text{cell } 22}$$

$$= 80$$

$$\text{SSR} = \sum_j N_{j\cdot}(M_{j\cdot} - M)^2$$

$$= \underbrace{4(5 - 6)^2}_{\text{row } 1} + \underbrace{4(7 - 6)^2}_{\text{row } 2}$$

$$= 8$$

$$\text{SSC} = \sum_k N_{\cdot k}(M_{\cdot k} - M)^2$$

$$= \underbrace{4(8 - 6)^2}_{\text{column } 1} + \underbrace{4(4 - 6)^2}_{\text{column } 2}$$

$$= 32$$

$$\text{SS}(R \times C) = \sum_k \sum_j n(M_{jk} - M_{j\cdot} - M_{\cdot k} + M)^2$$

$$= \underbrace{2(6 - 5 - 8 + 6)^2}_{\text{cell } 11} + \underbrace{2(10 - 7 - 8 + 6)^2}_{\text{cell } 21}$$

$$+ \underbrace{2(4 - 5 - 4 + 6)^2}_{\text{cell } 12} + \underbrace{2(4 - 7 - 4 + 6)^2}_{\text{cell } 22}$$

$$= 8$$

Hence, we have shown for this example that $\text{SST} = \text{SSE} + \text{SSR} + \text{SSC} + \text{SS}(R \times C)$.

We now form the mean square error (MSE), mean square for rows (MSR), mean square for columns (MSC), and mean square for interactions [MS$(R \times C)$] by dividing the sum of squares by the appropriate degrees of freedom:

$$\text{MSE} = \frac{\text{SSE}}{JK(n - 1)} \tag{7.54}$$

$$\text{MSR} = \frac{\text{SSR}}{J - 1} \tag{7.55}$$

TABLE 7.4

Data for Two-Way Factorial Design

		Columns		
		1	2	
Rows	1	2, 10 $M_{11} = 6$	6, 2 $M_{12} = 4$	$M_{\cdot 1} = 5$
	2	6, 14 $M_{21} = 10$	2, 6 $M_{22} = 4$	$M_{\cdot 2} = 7$
		$M_{\cdot 1} = 8$	$M_{\cdot 2} = 4$	$M = 6$

$$\text{MSC} = \frac{\text{SSC}}{K - 1} \tag{7.56}$$

and

$$\text{MS}(R \times C) = \frac{\text{SS}(R \times C)}{(J - 1)(K - 1)} \tag{7.57}$$

Using the rough heuristic of Section 6.3 (p. 138), we can lend some intuition to the number of degrees of freedom. For the mean square error, 1 degree of freedom is "lost" for each cell. Thus, the degrees of freedom are $N - JK = JK(n - 1)$. For the mean squares for rows and columns, 1 degree of freedom is lost for the grand mean. For the interaction, degrees of freedom are lost for the marginal means and the grand mean.

The next step that is necessary to derive the test statistic for the two-way analysis of variance is to determine the expectation of the mean squares. The determination of the expectations is not always straightforward and the proofs are omitted here. However, it should be recognized that the assumptions for the analysis of variance are invoked in several places in these proofs. Although not mentioned at every step, the distributional theory discussed here is true only if the assumptions are met. We begin with the expectation of the mean square error:

$$E(\text{MSE}) = \sigma_e^2 \tag{7.58}$$

Therefore

$$\frac{\text{MSE}}{\sigma_e^2} \sim \frac{\chi_{JK(n-1)}^2}{JK(n - 1)} \tag{7.59}$$

The mean square error forms the denominator of the test statistics for the row, column, and interaction effects.

The expectation of the mean square row contains two terms—one that is the error variance and another that is a function of the row effects squared:

$$E(\text{MSR}) = \sigma_e^2 + \frac{Kn \sum_j \alpha_j^2}{J - 1} \tag{7.60}$$

The second term is zero when the null hypothesis that the row effects are zero is true; otherwise the second term is greater than zero. When the null hypothesis is true

$$\frac{\text{MSR}}{\sigma_e^2} \sim \frac{\chi_{J-1}^2}{J - 1} \tag{7.61}$$

Because mean square rows and mean square error are independent, from Equations (7.59) and (7.61) it is clear that

$$\frac{\text{MSR}}{\text{MSE}} \sim F_{J-1, JK(n-1)} \tag{7.62}$$

when the null hypothesis is true and the assumptions are met. Therefore, to test the null hypothesis that the row effects are all zero, we compare the test statistic MSR/MSE to an F distribution with $(J - 1)$ and $JK(n - 1)$ degrees of freedom.

The test statistic for columns is found similarly. First, the expectation of mean square for columns is given by

$$E(\text{MSC}) = \sigma_e^2 + \frac{Jn \sum_k \beta_k^2}{K - 1} \tag{7.63}$$

The second term of this expression is zero when the null hypothesis of no column effects is true; otherwise it is greater than zero. When the null hypothesis is true

$$\frac{\text{MSC}}{\sigma_e^2} \sim \frac{\chi_{K-1}^2}{K - 1} \tag{7.64}$$

Because mean square columns and mean square error are independent, from Equations (7.59) and (7.64) it is clear that

$$\frac{\text{MSC}}{\text{MSE}} \sim F_{K-1, JK(n-1)} \tag{7.65}$$

when the null hypothesis is true and the assumptions are met. Therefore, to test the null hypothesis that the column effects are all zero, we compare the test statistic MSC/MSE to an F distribution with $(K - 1)$ and $JK(n - 1)$ degrees of freedom.

The test statistic for interactions is found in the same manner as the test statistic for the main effects. First, the expectation of mean square for interaction is given by

$$E[MS(R \times C)] = \sigma_e^2 + \frac{n \sum_k \sum_j \gamma_{jk}^2}{(J-1)(K-1)} \tag{7.66}$$

The second term of this expression is zero when the null hypothesis of no interaction effects is true; otherwise it is greater than zero. When the null hypothesis is true

$$\frac{MS(R \times C)}{\sigma_e^2} \sim \frac{\chi_{(J-1)(K-1)}^2}{(J-1)(K-1)} \tag{7.67}$$

Because mean square interaction and mean square error are independent, from Equations (7.59) and (7.67) it is clear that

$$\frac{MS(R \times C)}{MSE} \sim F_{(J-1)(K-1),JK(n-1)} \tag{7.68}$$

when the null hypothesis is true and the assumptions are met. Therefore, to test the null hypothesis that the interaction effects are all zero, we compare the test statistic MS(R × C)/MSE to an *F* distribution with $(J-1)(K-1)$ and $JK(n-1)$ degrees of freedom.
    The calculations for the two-way fixed effects analysis of variance are often summarized in a source table, which is schematically presented in Table 7.5. The two-way fixed effects analysis of variance is illustrated in Example 7.2.

TABLE 7.5

Source Table for Two-Way Analysis of Variance

Source	Sum of Squares	Degrees of Freedom	Mean Squares	F
Rows (R)	SSR	$J-1$	$MSR = \dfrac{SSR}{J-1}$	$\dfrac{MSR}{MSE}$
Columns (C)	SSC	$K-1$	$MSC = \dfrac{SSC}{K-1}$	$\dfrac{MSC}{MSE}$
Interaction (R × C)	SS(R × C)	$(J-1)(K-1)$	$MS(R \times C) = \dfrac{SS(R \times C)}{(J-1)(K-1)}$	$\dfrac{MS(R \times C)}{MSE}$
Error	SSE	$JK(n-1)$	$MSE = \dfrac{SSE}{JK(n-1)}$	
Total	SST	$JKn - 1 = N - 1$		

EXAMPLE 7.2

Suppose that an educational researcher conducted a study to determine the effect of enrichment (enrichment, no enrichment) and instructional modality (traditional lecture, programmed instruction, and computer-assisted instruction) on the acquisition of mathematics knowledge of junior high school students. The design used was a 2 (enrichment) × 3 (instructional modality) balanced, completely crossed, factorial design with ten subjects in each cell. The dependent variable was the number of problems out of 100 answered correctly. The null hypotheses were that the effects for enrichment, instructional modality, and the interaction of enrichment and modality were all zero. Sixty subjects were randomly and independently selected and assigned to the six combinations of treatments (10 to each cell). The results of the study are summarized tabularly and graphically in Figure 7.10.

It appears from the results that indeed there are main and interaction effects. For example, the lines of the graph are not parallel, suggesting that there is an interaction effect. The question, of course, is whether the observed interaction is sufficiently great that the notion

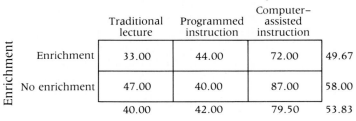

Instructional Modality

	Traditional lecture	Programmed instruction	Computer-assisted instruction	
Enrichment	33.00	44.00	72.00	49.67
No enrichment	47.00	40.00	87.00	58.00
	40.00	42.00	79.50	53.83

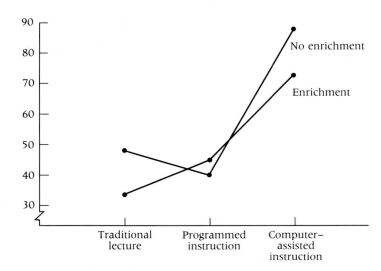

FIGURE 7.10   Cell Means for Two-Way Analysis of Variance Example

218

that there are no **population** interaction effects can be rejected in favor of the notion that there are **population** interaction effects.

First, we calculate the estimates of the population effects [see Equations (7.50)–(7.52), p. 212]:

$$\hat{\alpha}_1 = M_{1\cdot} - M = 49.67 - 53.83 = -4.16$$
$$\hat{\alpha}_2 = M_{2\cdot} - M = 58.00 - 53.83 = 4.17$$
$$\hat{\beta}_1 = M_{\cdot 1} - M = 40.00 - 53.83 = -13.83$$
$$\hat{\beta}_2 = M_{\cdot 2} - M = 42.00 - 53.83 = -11.83$$
$$\hat{\beta}_3 = M_{\cdot 3} - M = 79.50 - 53.83 = 25.67$$
$$\hat{\gamma}_{11} = M_{11} - M_{1\cdot} - M_{\cdot 1} + M = 33.00 - 49.67 - 40.00 + 53.83 = -2.84$$
$$\hat{\gamma}_{21} = M_{21} - M_{2\cdot} - M_{\cdot 1} + M = 47.00 - 58.00 - 40.00 + 53.83 = 2.83$$
$$\hat{\gamma}_{12} = M_{12} - M_{1\cdot} - M_{\cdot 2} + M = 44.00 - 49.67 - 42.00 + 53.83 = 6.16$$
$$\hat{\gamma}_{22} = M_{22} - M_{2\cdot} - M_{\cdot 2} + M = 40.00 - 58.00 - 42.00 + 53.83 = -6.17$$
$$\hat{\gamma}_{13} = M_{13} - M_{1\cdot} - M_{\cdot 3} + M = 72.00 - 49.67 - 79.50 + 53.83 = -3.34$$
$$\hat{\gamma}_{23} = M_{23} - M_{2\cdot} - M_{\cdot 3} + M = 87.00 - 58.00 - 79.50 + 53.83 = 3.33$$

(The fact that the sum of the row, column, and interaction effects differ slightly from zero is due to rounding.) The calculations for the sum of squares, mean squares, and F ratios are summarized in a source table:

Source	Sum of Squares	Degrees of Freedom	Mean Squares	F
Enrichment	1041.7	1	1041.7	$1.70 = \dfrac{\text{MSR}}{\text{MSE}} \sim F_{1,54}$
Instructional modality	19803.3	2	9901.67	$16.13 = \dfrac{\text{MSC}}{\text{MSE}} \sim F_{2,54}$
Interaction	1143.3	2	571.7	$.93 = \dfrac{\text{MS}(R \times C)}{\text{MSE}} \sim F_{2,54}$
Error	33158.0	54	614.0	
Total	55146.3	59		

Comparing the F ratios to the appropriate F distributions, we see that the null hypotheses that the enrichment effects and the interaction effects are zero cannot be rejected, whereas the null hypothesis that the instructional modality effects are zero can be rejected at α equals .01. Although visually there appears to be interaction effects (see Figure 7.10), there is not sufficient evidence to declare that these are not due to sampling error. Given the omnibus nature of the analysis of variance F test, we can only say at this point that we reject the null hypothesis that all instructional modalities are equally effective. To say whether computer-assisted instruction is superior to traditional lecture and programmed instruction will necessitate another test, which will be discussed in Chapter 9.

It should be noted that extension of the analysis of variance beyond the two-way case is straightforward. In the three-way case the sum of squares and mean squares are calculated for error; for factors A, B, and C; and for interactions $A \times B$, $A \times C$, $B \times C$, and $A \times B \times C$. The F ratios formed by dividing the mean square for the effect by the mean square error are the test statistics. It should be mentioned that three-way interactions typically are difficult to interpret and interactions beyond three-way typically are too complicated to have much substantive importance.

7.17   Strength of Association in the Two-Way Analysis of Variance

The sample strength of association measures for each of the effects (η^2) is analogous to that for the one-way case and is easy to calculate:

$$\eta_R^2 = \frac{\text{SSR}}{\text{SST}} \tag{7.69}$$

$$\eta_C^2 = \frac{\text{SSC}}{\text{SST}} \tag{7.70}$$

and

$$\eta_{R \times C}^2 = \frac{\text{SS}(R \times C)}{\text{SST}} \tag{7.71}$$

For Example 7.2 (p. 218)

$$\eta_R^2 = \frac{1041.7}{55,146.3} = .019$$

$$\eta_C^2 = \frac{19,803.3}{55,146.3} = .359$$

and

$$\eta_{R \times C}^2 = \frac{1143.3}{55,146.3} = .021$$

To discuss estimates of the population strength of association, we must note that for the two-way design discussed here it can be shown that

$$\sigma_Y^2 = \sigma_e^2 + \frac{\sum_j \alpha_j^2}{J} + \frac{\sum_k \beta_k^2}{K} + \frac{\sum_k \sum_j \gamma_{jk}^2}{JK} \tag{7.72}$$

Hence

$$\omega_{\text{total}}^2 = \frac{\sigma_Y^2 - \sigma_e^2}{\sigma_Y^2} = \frac{\sum_j \alpha_j^2}{J/\sigma_Y^2} + \frac{\sum_k \beta_k^2}{K/\sigma_Y^2} + \frac{\sum_k \sum_j \gamma_{jk}^2}{JK/\sigma_Y^2} \tag{7.73}$$

Given Equation (7.73), it is reasonable to define ω^2 for rows, columns, and interactions in the following way:

$$\omega_R^2 = \frac{\sum_j \alpha_j^2}{J/\sigma_Y^2} \tag{7.74}$$

$$\omega_C^2 = \frac{\sum_k \beta_k^2}{K/\sigma_Y^2} \tag{7.75}$$

$$\omega_{(R \times C)}^2 = \frac{\sum_k \sum_j \gamma_{jk}^2}{JK/\sigma_Y^2} \tag{7.76}$$

The estimators for these population parameters are given by

$$\hat{\omega}_R^2 = \frac{SSR - (J-1)MSE}{MSE + SST} \tag{7.77}$$

$$\hat{\omega}_C^2 = \frac{SSC - (K-1)MSE}{MSE + SST} \tag{7.78}$$

$$\hat{\omega}_{(R \times C)}^2 = \frac{SS(R \times C) - (J-1)(K-1)MSE}{MSE + SST} \tag{7.79}$$

Estimated values for Example 7.2 are easily calculated:

$$\hat{\omega}_R^2 = \frac{1041.7 - (1)(614.0)}{614.0 + 55,146.3} = .008$$
$$\hat{\omega}_C^2 = \frac{19,803.3 - 2(614.0)}{614.0 + 55,146.3} = .333$$
$$\hat{\omega}_{(R \times C)}^2 = \frac{1143.3 - (1)(2)(614.0)}{614.0 + 55,146.3} = -.002$$

The last estimate being less than zero is set to zero. As discussed previously (Section 7.12, p. 220), the sample values are greater than the estimates of the population proportion of variance accounted for by the independent variables.

NOTES AND SUPPLEMENTARY READINGS

One area where a certain amount of debate has occurred pertains to the effects of violating fundamental assumptions underlying the analysis of variance. Most statistical texts discuss this matter in varying degrees of depth. However, they, and professors using them to teach statistics, tend largely to excuse violations to some degree if certain steps are taken to offset the impact. In some cases the issues are examined superficially to a degree that students begin to dismiss their importance. The Box references are the classics that have been cited for many years in this area. The four references by Bradley suggest that a bit more attention is warranted and provide a flavor of the debate, particularly Bradley (1984b). The review by Glass, Peckham, and Sanders (1972) summarizes assumption violation impact.

Box, G. E. P. (1953). Non-normality and tests on variances. *Biometrika, 40,* 318–335.

Box, G. E. P. (1954a). Some theorems on quadratic forms applied in the study of analysis of variance problems: I. Effect of inequality of variance and of correlation of errors in the one-way classification. *Annals of Mathematical Statistics, 25,* 290–302.

Box, G. E. P. (1954b). Some theorems on quadratic forms applied in the study of analysis of variance problems: II. Effect of inequality of variance and of correlation of errors in the two-way classification. *Annals of Mathematical Statistics, 25,* 484–498.

Bradley, J. V. (1968). Nonrobustness in *Z, t,* and *F* tests at large sample sizes. *Bulletin of the Psychonomic Society, 16,* 333–336.

Bradley, J. V. (1982). The insidious L-shaped distribution. *Bulletin of the Psychonomic Society, 20,* 85–88.

Bradley, J. V. (1984a). The complexity of nonrobustness effects. *Bulletin of the Psychonomic Society, 22,* 250–253.

Bradley, J. V. (1984b). Antinonrobustness: A case study in the sociology of science. *Bulletin of the Psychonomic Society, 22,* 463–466.

Glass, G. V, Peckham, P. D., & Sanders, J. R. (1972). Consequences of failure to meet assumptions underlying the analysis of variance and covariance. *Review of Educational Research, 42,* 237–288.

This article reviews the impact of violating assumptions underlying analysis of variance, a topic of continuing debate and interest to researchers employing this procedure.

Keppel, G. (1982). *Design and analysis: A researcher's handbook* (2nd ed.). Fort Worth, Texas: Harcourt Brace Jovanovich.

Keppel provides a very readable examination of analysis of variance, which is most notable for its close tie to design terminology. His notation varies somewhat from others who adhere to what many consider the conventional or orthodox, but it is translatable and usable by the beginning student. Additionally, this edition and its earlier version include computational procedures for some analyses that are not frequently found in other volumes.

Kirk, R. E. (1968). *Experimental design: Procedures for the behavioral sciences* (1st ed.). Belmont, CA: Brooks/Cole.

Kirk examines analysis of variance in a thorough and readable fashion in this edition. The more recent version may be less readable for some students since it employs matrix notation.

Winer, B. J. (1971). *Statistical principles in experimental design* (2nd ed.). New York: McGraw-Hill.

Winer has long been a standard reference for analysis of variance. On this topic he presents a very comprehensive and in-depth volume. This, along with Hays', is another text to which many statisticians can trace their early training. Occasionally students find the book somewhat difficult to read because of the detail that makes it such an excellent reference. The book is organized in relation to various design configurations.

PROBLEMS

1. Suppose that a sample of size 8 is randomly chosen from a population that has a normal distribution with $\mu = 24$ and $\sigma^2 = 9$. The standardized value corresponding to each observed value is computed by subtracting 24 and dividing the result by 3. Let T denote the sum of the squares of the 8 standardized values. Find the following probabilities:
   a. $P(T \geq 20.09)$
   b. $P(2.73 < T < 5.07)$
   c. $P(T > 2.18)$

2. Suppose that a sample of size 6 is randomly chosen from a population that has a normal distribution with $\mu = 7$ and $\sigma^2 = 16$. The standardized value corresponding to each observed value is computed by subtracting 7 and dividing the result by 4. Let T denote the sum of the squares of the 6 standardized values. Find the following probabilities:
   a. $P(T \geq 12.59)$
   b. $P(0.68 < T < 18.55)$
   c. $P(T > 5.35)$

3. Suppose that the two chi-square random variables referred to in this problem are independent. Find the value of a for the following:

a. $P(F < a) = .95$, given $F = \dfrac{\chi_8^2/8}{\chi_{10}^2/10}$

b. $P(F > a) = .01$, given $F = \dfrac{\chi_1^2/1}{\chi_{12}^2/12}$

c. $P(F < a) = .05$, given $F = \dfrac{\chi_7^2/7}{\chi_{15}^2/15}$

d. $P(F > a) = .05$, given $F = \dfrac{\chi_{20}^2/20}{\chi_{30}^2/30}$

4. Suppose that the two chi-square random variables referred to in this problem are independent. Find the value of a for the following:

a. $P(F < a) = .99$, given $F = \dfrac{\chi_5^2/5}{\chi_{12}^2/12}$

b. $P(F > a) = .05$, given $F = \dfrac{\chi_5^2/5}{\chi_{25}^2/25}$

c. $P(F < a) = .05$, given $F = \dfrac{\chi_9^2/9}{\chi_{30}^2/30}$

d. $P(F > a) = .05$, given $F = \dfrac{\chi_{12}^2/12}{\chi_{40}^2/40}$

5. Suppose that 30 subjects were randomly assigned to three groups and the subjects in each group received a different treatment. The subjects' scores are as follows:

Group I	Group II	Group III
3	12	7
5	13	9
8	11	10
5	14	4
6	10	6
9	8	10
10	15	13
8	17	12
5	10	13
11	20	16

a. State the null hypothesis for the one-way fixed-effects analysis of variance for these data.

b. Construct the source table for the one-way fixed-effects analysis of variance for these data. Would the null hypothesis be rejected ($\alpha = .05$)?

c. Estimate the treatment effects.

d. Calculate η^2 and $\hat{\omega}^2$.

6. Suppose that 22 subjects were randomly assigned to four groups and the subjects in each group received a different treatment. The subjects' scores are as follows:

Group I	Group II	Group III	Group IV
1	1	3	7
9	5	7	9
4	4	5	12
5	7	8	16
7	3	7	
4	4		
	4		

a. State the null hypothesis for the one-way fixed-effects analysis of variance for these data.
b. Construct the source table for the one-way fixed-effects analysis of variance for these data. Would the null hypothesis be rejected ($\alpha = .05$)?
c. Estimate the treatment effects.
d. Calculate η^2 and $\hat{\omega}^2$.

7. Suppose that a researcher wished to study learning of simple and complex skills with two types of media presentations—audio and video. Six subjects were randomly assigned to each of the four conditions formed by crossing material (complex and simple) and media (audio and video). At the end of each presentation subjects were asked 20 questions that were relevant to the material. Suppose that the following scores were observed:

		Media	
		Video	Audio
Material	Complex	12, 12, 10, 16, 11, 11	6, 8, 9, 5, 5, 3
	Simple	12, 13, 15, 16, 13, 15	14, 13, 17, 17, 18, 17

a. Given a fixed-effects model, state the null hypotheses for the rows, columns, and interaction effects.
b. Construct the source table for the two-way analysis of variance and indicate which hypotheses would be rejected ($\alpha = .01$).
c. Estimate the row, column, and interaction effects.
d. Present a graph of the results (see Section 7.14, p. 203).
e. Estimate ω^2 for rows, columns, and interaction.

8. Suppose that one collected data about the relative efficacy of three treatments with men and women and obtained the following results:

		Treatment		
		Treatment I	Treatment II	Treatment III
Gender	Male	5, 7, 7, 10, 6	6, 6, 7, 9, 12	3, 4, 6, 8, 9
	Female	1, 4, 5, 7, 8	4, 5, 7, 8, 6	1, 1, 3, 6, 9

a. Given a fixed-effects model, state the null hypotheses for the rows, columns, and interaction effects.
b. Construct the source table for the two-way analysis of variance and indicate which hypotheses would be rejected ($\alpha = .05$).
c. Estimate the row, column, and interaction effects.
d. Present a graph of the results (see Section 7.14, p. 203).
e. Estimate ω^2 for rows, columns, and interaction.

8 ANALYSIS OF VARIANCE— RANDOM AND MIXED MODELS

I n the fixed-effects analysis of variance the levels of each independent variable were considered to constitute the universe of levels. Suppose that a researcher was interested in determining the effect of classroom teachers on achievement. Further, suppose that the researcher used the first five available teachers, that each teacher taught several two-week summer session courses, and that a fixed-effects analysis of variance revealed a statistically significant difference among the teachers. The results of this study would indicate that **as far as these five teachers were concerned**, the teacher makes a difference in achievement. The utility of the results are restricted because the researcher wants to generalize the results to all potential teachers. To be able to make such a statement, the researcher would need to **randomly select** the teachers to be included in the study from all potential teachers. In this way a statistically significant result would indicate that, **in general**, the teacher makes a difference in achievement.

If it is not feasible to include all levels of an independent variable in a study, randomly selecting a subset of these levels is a valuable, although underused, procedure for making generalized inferences. Such a design is referred to as a *random-effects design* and in this chapter we develop the *random-effects analysis of variance* for analyzing the results derived from such designs. Further, *mixed designs* that include one or more fixed-effects factors and one or more random-effects factors are considered.

Random-Effects Analysis of Variance

8.1   One-Way Random-Effects Analysis of Variance

The one-way random-effects design differs from the one-way fixed-effects design in that the random-effects context the J treatments are randomly selected from a set of possible treatments. The only restriction on the set of possible treatments is that it contains at least $(J + 1)$ treatments. Further, let a sample of N subjects be randomly selected and assigned to each treatment condition. We place the further restriction that an equal number of subjects be assigned to each treatment (the number of subjects in each treatment is denoted by n). Keep in mind that under this scheme there are two sources of sampling variability: treatments and subjects.

Recall that the linear model for the fixed-effects analysis of variance was [see Equation (7.16), p. 185]

$$Y_{ij} = \mu + \alpha_j + e_{ij}$$

where Y_{ij} is the ith observation in the jth group, μ is the grand mean, α_j is the treatment effect for the jth treatment, and e_{ij} is the error associated with the ith observation in the jth treatment. Recall that μ and α_j are parameters and e_{ij} is a random variable. Particular attention needs to be paid to α_j. To each and every treatment in a fixed-effects design there is associated a value of α_j and the sum of the treatment effects is zero. In the random-effects design the mean of the set of all treatments is zero; however, the mean of the effects for those treatments actually involved in the study typically do not equal zero. The effects in a random-effects design vary from one sample of levels of the independent variable to another. Said another way; there is a distribution of effects from which J treatment effects are selected at random to be studied and inferences about the distribution of effects are made from the J effects selected and studied.

In the random-effects design the ith observation in the jth group is given by the linear model

$$Y_{ij} = \mu + A_j + e_{ij} \tag{8.1}$$

where μ is the grand mean, A_j is the treatment effect for the jth treatment selected, and e_{ij} is the error component of Y_{ij}. This model is identical to the fixed-effects model except for the treatment effects. The value of A_j is not fixed, but rather it varies over repetitions of the experiment. Hence, A_j is a random variable and the values a_j that A_j assumes are called *random effects*. Although the sum of the random effects for a particular experiment will not necessarily equal zero (i.e., it is not true that $\sum_j a_j = 0$), the expected value of A_j is zero; that is

$$E(A_j) = 0 \tag{8.2}$$

Let

$$V(A_j) = \sigma_A^2 \tag{8.3}$$

Importantly, note that the variance of A_j is zero if and only if the treatment effects for all possible treatments are zero. Hence, the null hypothesis for the random-effects analysis of variance is

H_0: $\sigma_A^2 = 0$

The additional random variable in the random-effects model requires additional assumptions to develop the statistical test of the null hypothesis of no effects. Particularly, the following assumptions are made for the simple random-effects analysis of variance:

    (*i*) For each treatment *j*, $e_{ij} \sim N(0, \sigma_e^2)$.
    (*ii*) The *J* values of the random variable A_j are independent of each other.
    (*iii*) The values of the random variable e_{ij} are independent of each other.
    (*iv*) The values of A_j and e_{ij} are independent of each other.

Of particular relevance to the random-effects analysis of variance is that the errors in observation must be independent of the treatment. It should be noted that for the application of the random-effects analyses discussed here no restriction is placed on the form of the distribution of the treatment effects (other than the fact that the mean is zero and the variance is finite). More sophisticated analyses can be conducted if it is assumed that the distribution of effects is normal, although these analyses are not discussed here.

Computationally, the development of the test statistic is identical to that of the one-way fixed-effects analysis of variance, except that it must be recalled that **in the random-effects analysis of variance, we restrict ourselves to the case where the number of observations in each treatment is the same.** Given the equal number of observations, the partition of the sum of squares is expressed as

$$\underbrace{\sum_j \sum_i (y_{ij} - M)^2}_{\text{SST}} = \underbrace{\sum_j n(M_j - M)^2}_{\text{SSB}} + \underbrace{\sum_j \sum_i (y_{ij} - M_j)^2}_{\text{SSE}} \tag{8.4}$$

where *n* is the number of observations in each treatment [see Equation (7.23), p. 190]. As well

$$\text{MSE} = \frac{\text{SSE}}{N - J} \tag{8.5}$$

and

$$\text{MSB} = \frac{\text{SSB}}{J - 1} \tag{8.6}$$

As was the case for the fixed-effects analysis of variance

$$E(\text{MSE}) = \sigma_e^2$$

However, the expected value of the mean squares between for the random-effects model is different from that of the fixed-effects model. In the random-effects model

$$E(\text{MSB}) = \sigma_e^2 + n\sigma_A^2 \tag{8.7}$$

That is, the expectation of the mean square between is equal to the sum of the error variance and the variance of the population effects (weighted by n). However, as was the case for the fixed-effects analysis of variance, **when the null hypothesis is true**

$$E(\text{MSB}) = \sigma_e^2$$

When the assumptions discussed earlier are true, it can be shown that the mean square error and the mean square between are independent. Thus, the test statistic for the one-way random-effects analysis of variance is formed in the same way as that for the one-way fixed-effects analysis of variance (see Section 7.11, p. 196). **Specifically, when the assumptions are met and the null hypothesis H_0: $\sigma_A^2 = 0$ is true, then the F ratio**

$$\frac{\text{MSB}}{\text{MSE}} = F$$

has an F distribution with $(J - 1)$ and $(N - J)$ degrees of freedom. If F is sufficiently large, the null hypothesis that there are no treatment effects ($\sigma_A^2 = 0$) is rejected in favor of the alternative that the treatment effects are not all zero ($\sigma_A^2 > 0$). It should be emphasized that although computationally the one-way fixed and random analyses of variance are identical, the conclusions reached are very different. In the fixed-effects case the results are applicable only to those treatments included in the study, whereas in the random-effects case the results apply to the population of effects.

The one-way random-effects analysis of variance is illustrated in the following example (Example 8.1).

In the fixed-effects context the proportion of variance accounted for by the independent variable was given by formulas for ω^2 (see Sections 7.12, p. 200, and 7.17, p. 200). Because the linear model for the random-effects analysis of variance involves specification of the treatment effects as a random variable, the concept of proportion of variance accounted for by the independent variable is slightly different. In the random-effects context the proportion of variance accounted for by the independent variable is denoted by ρ_I ("rho sub I") and is referred to as the *intraclass correlation coefficient*. First, the variance in Y is decomposed into two components:

EXAMPLE 8. 1

Suppose that a researcher has an experimental task with four components to it and he or she is interested in determining whether the order of the components makes a difference in the time that it takes subjects to complete the task. The researcher contemplates a study in which subjects are presented different orders of the components and the time to complete the task is recorded. Because it is desirable to have the results apply to all 24 possible orders, two alternatives are available. Either the study could include all 24 orders and a fixed-effects analysis of variance would be used or the study could include a randomly selected subset of the 24 orders and a random-effects analysis of variance would be used. The economy in the latter alternative should be obvious.

Suppose that 5 of the 24 orders are randomly selected. Forty subjects are randomly selected, of which 8 are randomly assigned to one of the five conditions corresponding to the 5 orders. Each of the subjects completes the task in the specified order and the time of completion is recorded in seconds. The mean time of completion for the 5 orders is as follows:

Order 1	Order 2	Order 3	Order 4	Order 5
30.00	20.00	50.00	25.00	40.00

The sum of squares, mean squares, and F ratios for this hypothetical experiment are presented in Table 8.1. Clearly the null hypothesis that the **variance of the population of effects is zero** is rejected. That is, there is sufficient evidence to give up the belief that order, in general, does not make a difference in the time it takes to complete the task. In any further experiments using this task the researcher is well advised to take into account the order of the presentation of the components of the task.

TABLE 8.1

Source Table for One-Way Random-Effects Analysis of Variance Example

Source	Sum of Squares	df	Mean Squares	F
Between	4640	4	1160.00	$11.79 = \dfrac{\text{MSB}}{\text{MSE}} = F_{4,35}$
Error	3444	35	98.40	
Total	8084	39		

$$V(Y_{ij}) = V(\mu + A_j + e_{ij}) = V(A_j + e_{ij})$$

But because it is assumed that the random variables A_j and e_{ij} are independent

$$V(A_j + e_{ij}) = V(A_j) + V(e_{ij}) = \sigma_A^2 + \sigma_e^2$$

and hence

$$\sigma_Y^2 = \sigma_A^2 + \sigma_e^2 \qquad (8.8)$$

Therefore, we define the intraclass correlation coefficient in the following way:

$$\rho_I = \frac{\sigma_Y^2 - \sigma_e^2}{\sigma_Y^2} = \frac{\sigma_A^2}{\sigma_Y^2} \qquad (8.9)$$

As was the case for ω^2, the range of ρ_I is given by

$$0 \le \rho_I \le 1$$

Because the variance of Y consists of two components, the variance of the treatment effects and the error variance, the random-effects analysis of variance is often referred to as the components of variance analysis.

The intraclass correlation coefficient is easily estimated from the mean squares. Recall that

$$E(\text{MSB}) = \sigma_e^2 + n\sigma_A^2$$

and

$$E(\text{MSE}) = \sigma_e^2$$

Therefore

$$E(\text{MSB}) - E(\text{MSE}) = E(\text{MSB} - \text{MSE}) = n\sigma_A^2 \qquad (8.10)$$

Hence, a good estimator of σ_A^2 is given by

$$\frac{\text{MSB} - \text{MSE}}{n} = \hat{\sigma}_A^2 \qquad (8.11)$$

Because $\text{MSE} = \hat{\sigma}_e^2$,

$$\hat{\sigma}_Y^2 = \hat{\sigma}_A^2 + \hat{\sigma}_e^2 = \frac{\text{MSB} - \text{MSE}}{n} + \text{MSE} = \frac{\text{MSB} + (n-1)\text{MSE}}{n} \qquad (8.12)$$

Therefore, the estimator of ρ_I is given by

$$\hat{\rho}_I = \frac{\hat{\sigma}_A^2}{\hat{\sigma}_Y^2} = \frac{\textbf{MSB} - \textbf{MSE}}{\textbf{MSB} + (n-1)\textbf{MSE}} \qquad \textbf{(8.13)}$$

For Example 8.1 (p. 231)

$$\hat{\rho}_I = \frac{\text{MSB} - \text{MSE}}{\text{MSB} + (n-1)\text{MSE}} = \frac{1160 - 98.4}{1160 + (7)98.4} = .57$$

indicating that it is estimated that 57% of the variability in completing the experimental task is accounted for by the order of the components of the task.

In the derivation of the random-effects analysis of variance no special conditions were set on the distribution of the treatment effects. However, if the distribution of the effects is normal, then confidence intervals for ρ_I can be calculated and statistics exist for testing for particular values of ρ_I (e.g., H_0: $\rho_I = k$, where $0 \leq k \leq 1$). Although not exceedingly complex, these methods are infrequently used and are not discussed here.

8.2   Two-Way Random-Effects Analysis of Variance

The two-way random-effects model involves two experimental factors, each of which is represented by a random effect. That is, the levels of each factor included in the experiment are sampled from a population of levels. As was the case for the two-way fixed-effects model, the two factors are denoted by R and C, designating the row variable and the column variable, respectively. Further, suppose that J levels of the row variable and K levels of the column variable are randomly selected from the population of levels of each factor for inclusion in the study. The linear model for the two-way random-effects analysis of variance is given by

$$Y_{ijk} = \mu + A_j + B_k + G_{jk} + e_{ijk} \tag{8.14}$$

where Y_{ijk} is the ith observation in the jth level of R and the kth level of C, μ is the population grand mean, A_j is the random variable representing the row effects, B_k is the random variable representing the column effects, G_{jk} is the random variable representing the interaction effects, and e_{ijk} is the random variable associated with error. We will be restricted to the case where n observations are randomly assigned to each of the JK combinations of the two factors.

The assumptions for this model are as follows:

(i) $A_j \sim N(0, \sigma_R^2)$
(ii) $B_k \sim N(0, \sigma_C^2)$
(iii) $G_{jk} \sim N(0, \sigma_{(R \times C)}^2)$
(iv) $e_{ijk} \sim N(0, \sigma_e^2)$
(v)   The values of A_j, B_k, G_{jk}, and e_{ijk} are all independent of each other.

As opposed to the one-way random-effects analysis of variance, the two-way model assumes normality of the row, column, and interaction random variables. As well, the row, column, interaction, and error random variables are assumed to be independent of each other. The null hypotheses for this model are given by

$$H_0: \sigma_R^2 = 0$$
$$H_0: \sigma_C^2 = 0$$

and

$$H_0: \sigma_{(R \times C)}^2 = 0$$

Significance testing proceeds in the same fashion as in the two-way fixed-effects analysis of variance. That is, the sums of squares for row, column, interaction, and error are computed. Furthermore, the mean squares are calculated exactly as they were for the two-way fixed-effects analysis of variance. However, this is where the similarity of the hypothesis testing for the two-way fixed and random-effects models ends.

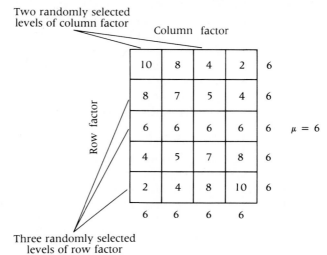

FIGURE 8.1   Cell Means for Population and Randomly Selected Levels for Two-Way Random-Effects Model

The special problem raised in the context of the two-way random model is illustrated in Figure 8.1, which presents a factorial design containing all possible levels of the row and column factors as well as a design containing randomly selected levels of the row and column factors. When interpreting this figure, the reader must realize that the population grand mean remains the same for both designs. As can be seen in the figure, in the design that contains all possible levels there are no row or column effects although there are interaction effects. However, in the design containing randomly selected levels there are main effects for rows and columns as well as interaction effects! Because the interaction effect is present, the sampled row and column effects are produced by the interaction effect. **In the two-way random effects model, the sum of squares for rows (and columns) contains the deviations due to the row (and column) effects *as well as* deviations due to interaction effects.** This phenomenon will have direct implications for the choice of a test statistic. This figure also illustrates another important aspect of the two-way random-effects model. When all the levels of the row and column variables are considered, the effects for rows, columns, and interaction sum to zero. As a consequence

$$E(A_j) = 0$$
$$E(B_k) = 0$$
$$E(G_{jk}) = 0$$

However, it is not true that the sum of the row, column, and interaction effects will sum to zero for the particular levels of the factors chosen to be studied, as illustrated in Figure 8.1.

As indicated earlier, the sums of squares (as well as the partition thereof) and the mean squares for the two-way random-effects model are identical to that for the two-way fixed-effects model. That is

$$\underbrace{\sum_k \sum_j \sum_i (y_{ijk} - M)^2}_{\text{SST}} = \underbrace{\sum_k \sum_j \sum_i (y_{ijk} - M_{jk})^2}_{\text{SSE}}$$

$$+ \underbrace{\sum_j N_{j\cdot}(M_{j\cdot} - M)^2}_{\text{SSR}}$$

$$+ \underbrace{\sum_k N_{\cdot k}(M_{\cdot k} - M)^2}_{\text{SSC}}$$

$$+ \underbrace{\sum_k \sum_j n(M_{jk} - M_{j\cdot} - M_{\cdot k} + M)^2}_{\text{SS}(R \times C)} \qquad (8.15)$$

and

$$\text{MSE} = \frac{\text{SSE}}{JK(n-1)} \qquad (8.16)$$

$$\text{MSR} = \frac{\text{SSR}}{J-1} \qquad (8.17)$$

$$\text{MSC} = \frac{\text{SSC}}{K-1} \qquad (8.18)$$

$$\text{MS}(R \times C) = \frac{\text{SS}(R \times C)}{(J-1)(K-1)} \qquad (8.19)$$

However, with the exception of the mean square error, the expectations of the mean squares for the two-way random-effects model are very different from the expectations for the two-way fixed-effects model. For error

$$E(\text{MSE}) = \sigma_e^2 \qquad (8.20)$$

For rows

$$E(\text{MSR}) = \sigma_e^2 + n\sigma_{(R \times C)}^2 + Kn\sigma_R^2 \qquad (8.21)$$

and when the null hypothesis related to rows is true (viz., $H_0: \sigma_R^2 = 0$)

$$E(\text{MSR}) = \sigma_e^2 + n\sigma_{(R \times C)}^2 \qquad (8.22)$$

The mean square related to row effects is contaminated by the interaction effect, as was expected given the discussion earlier in this section. In the two-way fixed-effects model the test statistic was formed by dividing MSR by MSE; however, because of the presence of the interaction effect in MSR, this is precluded in the random-effects case. As we will see, however, there is another denominator for the test statistic that is appropriate for the random-effects model. Before identifying the appropriate denominator, we turn to the expected values of MSC:

$$E(\text{MSC}) = \sigma_e^2 + n\sigma_{(R \times C)}^2 + Jn\sigma_C^2, \qquad (8.23)$$

and when the null hypothesis related to columns is true (viz., $H_0: \sigma_C^2 = 0$)

$$E(\text{MSC}) = \sigma_e^2 + n\sigma_{(R \times C)}^2 \qquad (8.24)$$

Again, the mean square related to column effects is contaminated by the interaction effect. For interaction

$$E[\text{MS}(R \times C)] = \sigma_e^2 + n\sigma_{(R \times C)}^2, \qquad (8.25)$$

and when the null hypothesis related to the interaction is true (viz., $\sigma_{(R \times C)}^2 = 0$)

$$E[\text{MS}(R \times C)] = \sigma_e^2. \qquad (8.26)$$

Inspection of these expectations [Equations (8.20) to (8.26)] reveals the appropriate test statistics for rows, columns, and interaction. **Notice that when the null hypothesis related to rows is true, the mean square for rows is equal to the mean square for interaction.** Therefore, the test statistic for rows is given by the ratio of MSR to $\text{MS}(R \times C)$. **When the null hypothesis that the population row effects are zero ($H_0: \sigma_R^2 = 0$) is true and the assumptions are met**

$$\frac{\text{MSR}}{\text{MS}(R \times C)} \sim F_{J-1,(J-1)(K-1)} \tag{8.27}$$

and therefore to test the null hypothesis, **we compare the test statistic MSR/MS($R \times C$) to an F distribution with $(J - 1)$ and $(J - 1)(K - 1)$ degrees of freedom.**

The analogous test statistic is used for the column effects. **When the null hypothesis that the population column effects are zero ($H_0: \sigma_C^2 = 0$) is true and the assumptions are met**

$$\frac{\text{MSC}}{\text{MS}(R \times C)} \sim F_{K-1,(J-1)(K-1)} \tag{8.28}$$

and therefore to test the null hypothesis, **we compare the test statistic MSC/MS($R \times C$) to an F distribution with $(J - 1)$ and $(J - 1)(K - 1)$ degrees of freedom.**

From Equations (8.20), (8.25), and (8.26) it is clear that the denominator for the test statistic for the interaction is the mean square error. **When the null hypothesis that the population interaction effects are zero ($H_0: \sigma_{(R \times C)}^2 = 0$) is true and the assumptions are met**

$$\frac{\text{MS}(R \times C)}{\text{MSE}} \sim F_{(J-1)(K-1),JK(n-1)} \tag{8.29}$$

and therefore to test the null hypothesis, **we compare the test statistic MS($R \times C$)/MSE to an F distribution with $(J - 1)(K - 1)$ and $JK(n - 1)$ degrees of freedom.**

The calculations of the test statistics for the two-way random-effects analysis of variance are summarized in the schematic source table presented in Table 8.2.

An illustration of the two-way random-effects analysis of variance is found in Example 8.2.

There is a variation of the tests for the row and column effects that under the proper circumstances leads to increased power. When strong evidence exists that there are no interaction effects (i.e., $\sigma_{(R \times C)}^2 = 0$), the denominator of the test statistic can be formed by pooling mean square error and interaction by using the following formula:

$$\text{Pooled MS} = \frac{\text{SSE} + \text{SS}(R \times C)}{JK(n - 1) + (J - 1)(K - 1)} \tag{8.30}$$

TABLE 8.2

Schematic Source Table for Two-Way Random-Effects Analysis of Variance Example

Source	Sum of Squares	df	Mean Squares	F
Rows	SSR	$J - 1$	$MSR = \dfrac{SSR}{J - 1}$	$\dfrac{MSR}{MS(R \times C)} = F_{J-1,(J-1)(K-1)}$
Columns	SSC	$K - 1$	$MSC = \dfrac{SSC}{K - 1}$	$\dfrac{MSC}{MS(R \times C)} = F_{K-1,(J-1)(K-1)}$
Interaction	$SS(R \times C)$	$(J-1)(K-1)$	$MS(R \times C) = \dfrac{SS(R \times C)}{(J - 1)(K - 1)}$	$\dfrac{MS(R \times C)}{MSE} = F_{(J-1)(K-1),JK(n-1)}$
Error	SSE	$JK(n - 1)$	$MSE = \dfrac{SSE}{JK(n - 1)}$	
Total	SST	$JKn - 1$		

Under the assumption that $\sigma^2_{(R \times C)} = 0$

$$E(\text{pooled MS}) = \sigma^2_e$$

Therefore, the test statistic for rows is given by

$$\frac{MSR}{\text{pooled MS}} = F \qquad (8.31)$$

with $(J - 1)$ and $JK(n - 1) + (J - 1)(K - 1)$ degrees of freedom. Similarly, the test statistic for columns is given by

$$\frac{MSC}{\text{pooled MS}} = F \qquad (8.32)$$

with $(K - 1)$ and $JK(n - 1) + (J - 1)(K - 1)$ degrees of freedom. Of course, there is no test of the interaction effects because they were assumed to be zero. Under the proper conditions pooling results in a decrease in the size of the denominator of the F ratio (as well as an increase in degrees of freedom).

Example 8.2 illustrates the usefulness of pooling in that the denominator of the test statistic is less when the pooled mean square is used in place of the mean square interaction. For Example 8.2

EXAMPLE 8.2

Suppose that a researcher is interested in the effects of classroom size and of teachers on student achievement. Because the researcher is interested in generalizing the findings to classrooms and teachers in general, a two-way random-effects model was used. First, the range of classroom sizes was restricted to the range of 1 to 40, and four sizes from this range were randomly selected (sizes 2, 23, 15, and 31). From the population of teachers at the local school district who were willing to participate in the study, five were randomly selected. During the summer session each teacher taught a two-week unit with three classes of each size. Students were randomly assigned to each class. At the end of the two-week unit a 100-point exam was administered to assess the achievement in each class. In this experiment the class was the unit of analysis, and thus the mean score on the test for each class constituted an observation.

The results of this hypothetical experiment are found in Table 8.3. Given an α level of .05, as presently analyzed, only the null hypothesis related to classroom size is rejected. Thus, it is reasonable to believe that, generally, classroom size affects the average achievement of students. The exact nature of the effects is not known, however. This analysis was not designed to detect the specific relationship between classroom size and achievement. Regression analysis, which will be discussed in Chapter 9, is an alternative that would indicate whether a linear or curvilinear relationship exists between size and achievement.

TABLE 8.3

Source Table for Two-Way Random-Effects
Analysis of Variance Examples

Source	Sum of Squares	df	Mean Square	F
Rows (teachers)	1,992.00	4	498.00	$2.56 = \dfrac{\text{MSR}}{\text{MS}(R \times C)} = F_{4,12}$
Columns (classroom sizes)	4,230.00	3	1,410.00	$7.27 = \dfrac{\text{MSC}}{\text{MS}(R \times C)} = F_{3,12}$
Interaction	2,328.00	12	194.00	$1.46 = \dfrac{\text{MS}(R \times C)}{\text{MSE}} = F_{12,40}$
Error	5,304.00	40	132.60	
Total	13,854.00	59		

$$\text{Pooled MS} = \frac{\text{SSE} + \text{SS}(R \times C)}{JK(n-1) + (J-1)(K-1)}$$

$$= \frac{5304.00 + 2328.00}{(5)(4)(3-1) + (5-1)(4-1)}$$

$$= 146.77$$

which is less than $\text{MS}(R \times C)$. When the mean squares are pooled, the F ratio for teachers is

$$\frac{\text{MSR}}{\text{pooled MS}} = \frac{498.00}{146.77} = 3.39$$

which when compared to an F with 4 and 52 degrees of freedom leads to the decision to reject the null hypothesis of no teacher effects ($\alpha = .05$), a different decision from that reached when the mean square interaction was used in the denominator of the F ratio. The F ratio for classroom size is

$$\frac{\text{MSC}}{\text{pooled MS}} = \frac{1410.00}{146.77} = 9.60$$

which when compared to an F with 3 and 52 degrees of freedom leads to the decision to reject the null hypothesis of no classroom size effects.

It should be understood that the decision to pool is justified only if sufficient reason exists to believe that there are no interaction effects. Although there may be good theoretical or empirical evidence to suggest that the interaction effects are zero, a reasonable way to proceed is to make the inference that the interaction effects are zero from the sample data. A generally applicable rule is Paull's rule for pooling: pool (as described earlier) if the following conditions are true:

(i) $(J-1)(K-1) > 6$

(ii) $JK(n-1) > 6$

(iii) $\dfrac{\text{MS}(R \times C)}{\text{MSE}} < 2.00$

The proportion of variance accounted for by the row and column factors and the interaction is defined in the same manner as that for the one-way random-effects analysis of variance. Specifically

$$\rho_I^R = \frac{\sigma_R^2}{\sigma_Y^2} \tag{8.33}$$

$$\rho_I^C = \frac{\sigma_C^2}{\sigma_Y^2} \tag{8.34}$$

$$\rho_I^{(R \times C)} = \frac{\sigma_{(R \times C)}^2}{\sigma_Y^2} \tag{8.35}$$

where ρ_I^R, ρ_I^C, and $\rho_I^{(R\times C)}$ are the proportions of variance accounted for by the row and column factors and the interaction, respectively. The various components are estimated in the following way:

$$\hat{\sigma}_R^2 = \frac{MSR - MS(R \times C)}{Kn} \tag{8.36}$$

$$\hat{\sigma}_C^2 = \frac{MSC - MS(R \times C)}{Jn} \tag{8.37}$$

$$\hat{\sigma}_{(R\times C)}^2 = \frac{MS(R \times C) - MSE}{n} \tag{8.38}$$

$$\hat{\sigma}_e^2 = MSE \tag{8.39}$$

Estimates less than zero should be set equal to zero. Furthermore, because the row and column factors as well as the interaction are independent

$$\hat{\sigma}_Y^2 = \hat{\sigma}_R^2 + \hat{\sigma}_C^2 + \hat{\sigma}_{(R\times C)}^2 + \hat{\sigma}_e^2 \tag{8.40}$$

Therefore, the following estimates of the intraclass correlation coefficients are used:

$$\hat{\rho}_I^R = \frac{\hat{\sigma}_R^2}{\hat{\sigma}_R^2 + \hat{\sigma}_C^2 + \hat{\sigma}_{(R\times C)}^2 + \hat{\sigma}_e^2} \tag{8.41}$$

$$\hat{\rho}_I^C = \frac{\hat{\sigma}_C^2}{\hat{\sigma}_R^2 + \hat{\sigma}_C^2 + \hat{\sigma}_{(R\times C)}^2 + \hat{\sigma}_e^2} \tag{8.42}$$

$$\hat{\rho}_I^{(R\times C)} = \frac{\hat{\sigma}_{(R\times C)}^2}{\hat{\sigma}_R^2 + \hat{\sigma}_C^2 + \hat{\sigma}_{(R\times C)}^2 + \hat{\sigma}_e^2} \tag{8.43}$$

For Example 8.2

$$\hat{\sigma}_R^2 = \frac{MSR - MS(R \times C)}{Kn} = \frac{498.00 - 194.00}{(4)(3)} = 25.33$$

$$\hat{\sigma}_C^2 = \frac{MSC - MS(R \times C)}{Jn} = \frac{1410.00 - 194.00}{(5)(3)} = 81.07$$

$$\hat{\sigma}_{(R\times C)}^2 = \frac{MS(R \times C) - MSE}{n} = \frac{194.00 - 132.60}{3} = 20.47$$

$$\hat{\sigma}_e^2 = MSE = 132.60$$

Therefore

$$\hat{\rho}_I^R = \frac{\hat{\sigma}_R^2}{\hat{\sigma}_R^2 + \hat{\sigma}_C^2 + \hat{\sigma}_{(R\times C)}^2 + \hat{\sigma}_e^2}$$
$$= \frac{25.33}{25.33 + 81.07 + 20.47 + 132.60}$$
$$= 0.10$$

$$\hat{\rho}_I^C = \frac{\hat{\sigma}_C^2}{\hat{\sigma}_R^2 + \hat{\sigma}_C^2 + \hat{\sigma}_{(R \times C)}^2 + \hat{\sigma}_e^2}$$

$$= \frac{81.07}{25.33 + 81.07 + 20.47 + 132.60}$$

$$= 0.31$$

$$\hat{\rho}_I^{(R \times C)} = \frac{\hat{\sigma}_{(R \times C)}^2}{\hat{\sigma}_R^2 + \hat{\sigma}_C^2 + \hat{\sigma}_{(R \times C)}^2 + \hat{\sigma}_e^2}$$

$$= \frac{20.47}{25.33 + 81.07 + 20.47 + 132.60}$$

$$= 0.08$$

Mixed Models

Mixed models are models that contain one or more fixed factors and one or more random factors, although we will only discuss those instances for which there are one fixed factor and one random factor. Consider again Example 8.2 (p. 239); which described a design with two random factors, class size and teacher. Perhaps a researcher was also interested in the effect of type of classroom (e.g., open classroom versus traditional classroom). This factor, which is fixed because all levels would be included in the experiment, could be added to the previous design, yielding a mixed-model design.

Three types of mixed models are discussed here. The first type is the mixed model that is replicated. That is, more than one subject is assigned to each of the combinations of the factors (we will be restricted to the case where equal numbers of subjects are assigned to each of the combinations). The second and third types are nonreplicated designs. Here only one subject is assigned to each of the combinations. These latter two designs primarily are used to control for nuisance factors (see Sections 8.4, p. 247, and 8.5, p. 250).

8.3   Mixed Model with Replication

In this model we consider two factors, a fixed factor and a random factor. We designate the rows to represent the random factor and the columns the fixed factor. Specifically, let the J levels of the row factor be randomly selected from a population of levels and let the K levels of the column factor represent the entire set of levels that are of interest to the researcher. The linear model is given by the following expression:

$$Y_{ijk} = \mu + A_j + \beta_k + G_{jk} + e_{ijk} \tag{8.44}$$

where μ is the grand mean (a parameter), A_j is the random variable associated with the rows, β_k is the effect for the kth level of the column factor (a parameter), G_{jk} is the random variable associated with the interaction, and e_{ijk} is the random variable associated with error.

The following assumptions are made:

(i) $A_j \sim N(0, \sigma_R^2)$

(ii) $G_{jk} \sim N(0, \sigma_{(R \times C)}^2)$

(iii) $e_{ijk} \sim N(0, \sigma_e^2)$

(iv) A_j and G_{jk} are jointly normal (see Section 3.15, p. 72).

(v)   The values of e_{ijk} are independent of A_j and G_{jk}.

(vi)   The values of e_{ijk} are all independent of each other.

For the random (row) factor the null hypothesis is

$$H_0: \sigma_R^2 = 0$$

For the fixed (column) factor the null hypothesis is

$$H_0: \beta_k = 0 \qquad \text{for} \qquad k = 1 \text{ to } k = K$$

Finally, for the interaction effect, the null hypothesis is

$$H_0: \sigma_{(R \times C)}^2 = 0$$

The sums of squares and mean squares for the mixed model are calculated exactly the way they were for the fixed and random two-way designs. However, the expectation of the mean squares is different and hence the test statistics will be different. The expectation for the random factor (i.e., row factor) is given by

$$E(\text{MSR}) = \sigma_e^2 + Kn\sigma_A^2 \tag{8.45}$$

The expectation for the fixed factor (i.e., column factor) is given by

$$E(\text{MSC}) = \sigma_e^2 + Jn\frac{\sum_k \beta_k^2}{K - 1} + n\sigma_{(R \times C)}^2 \tag{8.46}$$

For the interaction of the fixed and random factors the expectation is given by

$$E[\text{MS}(R \times C)] = \sigma_e^2 + n\sigma_{(R \times C)}^2 \tag{8.47}$$

Finally

$$E(\text{MSE}) = \sigma_e^2 \tag{8.48}$$

Contrary to what one might expect intuitively, the fixed rather than the random factor is contaminated with the interaction effect. Figure 8.2 illustrates this counterintuitive result. In the top panel the cell and marginal means are given for five levels of the row factor and three levels of the column factor. These levels represent the population of all possible levels

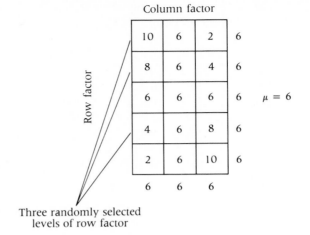

FIGURE 8.2   Cell Means for Population and Randomly Selected Levels of Row Factor for Mixed Model

of these factors. In the bottom panel three of the five levels for the row factor were randomly selected for study, making the rows the random factor. However, all three levels of the column factor were selected; hence, the column factor is a fixed factor. In the population of levels there were no row or column effects but there were interaction effects. When levels of the row factor were randomly selected, the interaction effects caused apparent column effects. That is, the column effects in the bottom panel of Figure 8.2 are due solely to the interaction effect. This explains the presence of the interaction term in the mean square for columns in Equation (8.46). The interaction effects, however, do not affect the random factor, and thus there is no interaction term in the expectation of the mean square for rows [see Equation (8.45)].

Examining the expectations of the mean squares, we can determine the correct test statistics. When the null hypothesis related to the random factor is true (H_0: $\sigma_R^2 = 0$), the mean square for rows is equal to the mean square error [see Equations (8.45) and (8.48)]. Hence, **when the null hypothesis that the population row effects are all zero is true and the assumptions are met**

$$\frac{\text{MSR}}{\text{MSE}} \sim F_{J-1, JK(n-1)} \tag{8.49}$$

and therefore to test the null hypothesis, we compare the test statistic MSR/MSE to an F distribution with $(J - 1)$ and $JK(n - 1)$ degrees of freedom.

When the null hypothesis related to the fixed factor is true (H_0: $\beta_k = 0$), the mean square for columns is equal to the mean square for interaction [see Equations (8.46) and (8.47)]. Hence, **when the null hypothesis that the fixed effects are all zero is true and the assumptions are met**

$$\frac{\text{MSC}}{\text{MS}(R \times C)} \sim F_{K-1, (J-1)(K-1)} \tag{8.50}$$

and therefore to test the null hypothesis, we compare the test statistic MSC/MS($R \times C$) to an F distribution with $(K - 1)$ and $(J - 1)(K - 1)$ degrees of freedom.

When the null hypothesis related to the interaction is true ($\sigma_{(R \times C)}^2 = 0$), the mean square for interaction is equal to the mean square error [see Equations (8.47) and (8.48)]. Hence, **when the null hypothesis that the population of interaction effects are all zero and the assumptions are met**

$$\frac{\text{MS}(R \times C)}{\text{MSE}} \sim F_{(J-1)(K-1), JK(n-1)} \tag{8.51}$$

and therefore to test the null hypothesis, we compare the test statistic MS($R \times C$)/MSE to an F distribution with $(J - 1)(K - 1)$ and $JK(n - 1)$ degrees of freedom.

The calculations for the test statistics for the mixed-model analysis of variance are summarized in the schematic source table found in Table 8.4.

When there is sufficient evidence that no interaction effects are present in the mixed model, it is advisable to use the pooled mean squares as an estimate of error for both the random and fixed factors. Pooling is accomplished in exactly the same way as that used for the random-effects analysis of variance [see Equation (8.30), p. 237], and Paull's rule is also applicable. In the mixed model, calculation of the proportion of variance accounted for by the factors is problematic and analogues to either ω^2 or ρ_I do not exist.

The mixed model is illustrated in Example 8.3, which follows.

TABLE 8.4

Source Table for Mixed Model Analysis of Variance

Source	Sum of Squares	df	Mean Square	F
Rows (random)	SSR	$J - 1$	$\text{MSR} = \dfrac{\text{SSR}}{J - 1}$	$\dfrac{\text{MSR}}{\text{MSE}} = F_{J-1,\,JK(n-1)}$
Columns (fixed)	SSC	$K - 1$	$\text{MSC} = \dfrac{\text{SSC}}{K - 1}$	$\dfrac{\text{MSC}}{\text{MS}(R \times C)} = F_{K-1,\,(J-1)(K-1)}$
Interaction	$\text{SS}(R \times C)$	$(J - 1)(K - 1)$	$\text{MS}(R \times C) = \dfrac{\text{SS}(R \times C)}{(J - 1)(K - 1)}$	$\dfrac{\text{MS}(R \times C)}{\text{MSE}} = F_{(J-1)(K-1),\,JK(n-1)}$
Error	SSE	$JK(n - 1)$	$\text{MSE} = \dfrac{\text{SSE}}{JK(n - 1)}$	
Total	SST	$JKn - 1$		

EXAMPLE 8.3

An example of the mixed model is provided by a researcher who wishes to investigate the efficacy of three psychotherapeutic modalities, but who also suspects that each of the modalities are optimally effective with different therapists (i.e., modality interacts with therapist). To investigate the therapist factor, the researcher randomly selects five therapists from the local register. Sixty clients are randomly assigned so

TABLE 8.5

Cell Means for Two-Way Mixed Model Analysis of Variance Example

		Modality (Fixed)			Therapist Mean
		I	II	III	
Therapist (Random)	1	8	6	4	6
	2	8	2	5	5
	3	4	2	3	3
	4	2	4	6	4
	5	3	1	2	2
Modality Mean		5	3	4	4

that there are four clients in each combination of therapist and modality; that is, each therapist administers each modality to four clients. The outcome is measured by scaling the degree to which each client obtained his or her psychotherapeutic goal (on a 1 to 10 scale). The cell means for this hypothetical experiment are found in Table 8.5.

The source table for these results is found in Table 8.6. The null hypothesis related to therapist is rejected, whereas the null hypotheses related to modality and interaction are not rejected (at least not at the .05 level of α). Thus, it appears that there is good reason to believe that, in general, achievement of psychotherapeutic goals is affected by the therapist who administers the treatment.

TABLE 8.6

Source Table for Two-Way Mixed Model Analysis of Variance Example

Source	Sum of Squares	df	Mean Squares	F
Rows— therapist (random)	120	4	30.00	$4.50 = \dfrac{MSR}{MSE} = F_{4,45}$
Columns— modality (fixed)	40	2	20.00	$1.43 = \dfrac{MSC}{MS(R \times C)} = F_{2,45}$
Interaction	112	8	14.00	$2.10 = \dfrac{MS(R \times C)}{MSE} = F_{8,45}$
Error	300	45	6.67	
Total	572	59		

8.4 Randomized Block Design

In the context of two sample problems we made a distinction between independent and dependent designs (see Section 6.11, p. 160). In the dependent design observations were paired in some way. The paired observations were either associated structurally (e.g., twins, spouses) or matched by the experimenter on the basis of some nuisance factor. The primary objective of the dependent sample design was to reduce the error variance yielding a more powerful test of the statistical hypothesis. The *randomized block design* is an extension of the dependent sample design; the only difference is that the randomized block design is not restricted to two levels of the independent variable. Blocks are formed so that the number of subjects in each block is equal to the number of levels of the independent variable so that one subject in each block can be assigned to each level. As was the case for the dependent sample design, the subjects in each block are matched in some way. Typically, subjects have the same or nearly the same scores on some nuisance factor. The primary purpose of the randomized block design is the same as that of the dependent sample design—to reduce error variance. With that in mind, it is vital to remember that the variable or variables on which the subjects are blocked must be positively related to the dependent variable; otherwise blocking does not reduce error variance and may actually increase the error variance [a review of Section 6.12 (p. 165) should reinforce this concept].

In the randomized block design each of J blocks contain K subjects matched in some manner and the K subjects are randomly assigned to K treatments or levels of the independent variable. The randomized block design is schematically presented in Figure 8.3. The K treatments are fixed (i.e., they constitute the entire population of effects that are of interest). This factor frequently is referred to as *treatments*. The factor that includes the blocks, which is referred to as the *blocking factor*, is considered a random factor given that the subjects are randomly selected from some population. Clearly it is logical that the data derived from a randomized block design could be (and in fact is usually) analyzed with a mixed-model analysis of variance. The only difference is that the randomized block design is non-replicated, which in the present context is not troublesome. The linear model for the mixed analysis of variance expressed by Equation (8.44) (p. 242)

Treatments (fixed)

Blocks (random)	1	2	···	k	···	K
1	$n = 1$	$n = 1$	···	$n = 1$	···	$n = 1$
2	$n = 1$	$n = 1$	···	$n = 1$	···	$n = 1$
⋮	⋮	⋮	⋰	⋮	⋰	⋮
j	$n = 1$	$n = 1$	···	$n = 1$	$n = 1$	$n = 1$
⋮	⋮	⋮	⋰	⋮	⋰	⋮
J	$n = 1$	$n = 1$	···	$n = 1$	···	$n = 1$

FIGURE 8.3 Randomized Block Design

remains valid. However, because we are only interested in testing the effects of the K treatments, the null hypothesis of interest is that the K treatment effects are zero; that is

$$H_0: \beta_k = 0 \quad \text{for} \quad k = 1 \text{ to } k = K$$

Because the randomized block design is nonreplicated, there is no sum of squares error. Therefore, the partition of the sum of squares is given by

$$\underbrace{\sum_k \sum_j \sum_i (y_{ijk} - M)^2}_{\text{SST}} = \underbrace{\sum_j K(M_{j\cdot} - M)^2}_{\text{SS blocks}} + \underbrace{\sum_k J(M_{\cdot k} - M)^2}_{\text{SS treatments}}$$

$$+ \underbrace{\sum_k \sum_j (M_{jk} - M_{j\cdot} - M_{\cdot k} + M)^2}_{\text{SS residuals}} \quad (8.52)$$

In this context we refer to the sum of squares for interaction as the sum of squares residuals, because it is the sum of squares that remains after the sum of squares for blocks and treatments is removed from the total; that is

$$\text{SS residuals} = \text{SST} - \text{SS blocks} - \text{SS treatments}$$

The mean squares are formed by dividing by the appropriate degrees of freedom.

$$\text{MS blocks} = \frac{\text{SS blocks}}{J - 1} \quad (8.53)$$

$$\text{MS treatments} = \frac{\text{SS treatments}}{K - 1} \quad (8.54)$$

$$\text{MS residuals} = \frac{\text{SS residuals}}{(J - 1)(K - 1)} \quad (8.55)$$

As we will see, MS residuals become our estimate of error variance.
     The expectations of the mean squares are as follows:

$$E(\text{MS treatments}) = \sigma_e^2 + \sigma_{(R \times C)}^2 + \frac{J \sum_k \beta_k^2}{K - 1} \quad (8.56)$$

$$E(\text{MS blocks}) = \sigma_e^2 + K\sigma_R^2 \quad (8.57)$$

$$E(\text{MS residuals}) = \sigma_e^2 + \sigma_{(R \times C)}^2 \quad (8.58)$$

When the null hypothesis related to treatments is true ($H_0: \beta_k = 0$), the mean square for treatment is equal to the mean square for residuals. Hence, **when the null hypothesis is true and the assumptions are met**

$$\mathbf{\frac{MS\ treatments}{MS\ residuals}} \sim F_{(K-1),(J-1)(K-1)} \quad \textbf{(8.59)}$$

and therefore to test the null hypothesis, we compare the test statistic MS treatments/MS residuals to an F distribution with $(K - 1)$ and $(J - 1)(K - 1)$ degrees of freedom. Effects for blocks cannot be tested in the nonreplicated design and typically are not of interest anyway. The computations for the nonreplicated mixed-model analysis of variance for the randomized block design is summarized in the schematic source table presented in Table 8.7.

TABLE 8.7

Source Table for Random Block Design (Nonreplicated Mixed Model Analysis of Variance)

Source	Sum of Squares	df	Mean Squares	F
Blocks (rows)	SS blocks	$J - 1$	MS blocks $= \dfrac{\text{SS blocks}}{J - 1}$	
Treatments (columns)	SS treatments	$K - 1$	MS treatments $= \dfrac{\text{SS treatments}}{K - 1}$	$\dfrac{\text{MS treatments}}{\text{MS residuals}} =$ $F_{K-1,(J-1)(K-1)}$
Residual	SS residual	$(J - 1)(K - 1)$	MS residuals $= \dfrac{\text{SS residuals}}{(J - 1)(K - 1)}$	
Total	SS Total	$JK - 1$		

The randomized block design is illustrated in Example 8.4.

EXAMPLE 8.4   The advantages of the randomized block design are illustrated by considering an experiment to determine the time that it takes mice to complete a maze when various neurotransmitters are administered. Specifically, suppose that the researcher is interested in the effects of three transmitters in comparison to each other and to a control group. However, the researcher knows that there is a wide variation in the time that it takes the mice to run the maze and that these individual differences are due to a number of nuisance factors that are difficult to measure. To reduce the error variance due to these nuisance factors, the researcher employs a randomized block design, where each of eight blocks is composed of four litter mates. Each of the litter mates is then randomly assigned to one of four treatments—the three different transmitters and the control group. The time taken to complete the maze, in seconds, for each of the 32 mice is given in Table 8.8.

TABLE 8.8

Time (in Seconds) for Mice to Complete Maze

		Treatment				
		Transmitter 1	Transmitter 2	Transmitter 3	Control Group	Block Mean
	1	6	12	14	16	12
	2	15	16	14	15	15
	3	22	21	20	25	22
Blocks	4	17	20	25	14	19
	5	8	10	8	26	13
	6	10	18	23	21	18
	7	25	24	29	30	27
	8	9	7	11	13	10
Treatment Mean		14	16	18	20	

The mixed-model analysis of variance for these data is displayed in the top of Table 8.9. The null hypothesis that the treatment effects are all zero is rejected (provided α was set at .05). However, if these same data had been obtained from a one-way fixed-effects design (i.e., without blocking), then the null hypothesis would not be rejected, as illustrated in the bottom of Table 8.9. Perusal of the data should make it clear that much of the variance in scores for these mice is attributable to the block. That is, litter mates tend to run the maze at about the same speed, whereas mice from different litters tend to run the maze at different speeds (at least in this hypothetical experiment). The randomized block design eliminates the variance between blocks from the error variance, in this instance greatly reducing the error variance.

TABLE 8.9

Source Table for Randomized Block Example (Mixed Model and Fixed Model)

			Mixed Model	
Source	Sum of Squares	df	Mean Squares	F
Blocks	896	7	128.00	
Treatments	160	3	53.33	$3.20 = \dfrac{\text{MS treatment}}{\text{MS residual}} = F_{3,21}$
Residual	350	21	16.67	
Total	1406	31		

			Fixed Model (Ignoring Blocks)	
Source	Sum of Squares	df	Mean Squares	F
Between	160	3	53.33	$1.20 = \dfrac{\text{MSB}}{\text{MSE}} = F_{3,28}$
Error	1246	28	44.50	
Total	1406	31		

8.5   Repeated Measures Design

In the randomized block design each block contained K subjects who were randomly assigned to each of the K treatments. This can be altered so that each block contains only one subject who receives each of the treatments. The blocks are now called subjects. In other words, each of the J subjects is observed under the K treatments, yielding JK observations, as shown in Figure 8.4. To control for order effects, the order of the treatments should be randomized in some way. The goal of such a design is the same as for the randomized block design, namely, to reduce error variance. In the repeated measures design, nuisance factors, such as intelligence, are accounted for because presumably these factors remain constant across treatments. In this way the repeated measures design is desirable because the design accounts for all person variables that might influence performance. A variant of this design involves observing each subject over time (K time points rather than K treatments). For example, subjects may be observed at the end of some psychotherapeutic treatment and then followed up at the end of 3, 6, 9, and 12 months.

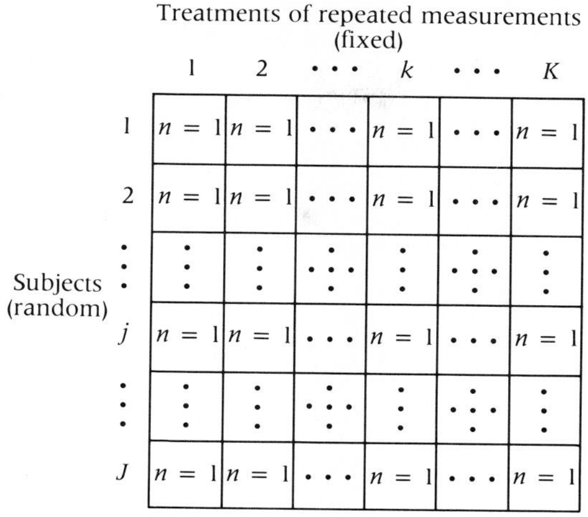

Treatments of repeated measurements
(fixed)

	1	2	$\cdots$	k	$\cdots$	K
1	$n = 1$	$n = 1$	$\cdots$	$n = 1$	$\cdots$	$n = 1$
2	$n = 1$	$n = 1$	$\cdots$	$n = 1$	$\cdots$	$n = 1$
$\vdots$	$\vdots$	$\vdots$	$\ddots$	$\vdots$	$\ddots$	$\vdots$
j	$n = 1$	$n = 1$	$\cdots$	$n = 1$	$\cdots$	$n = 1$
$\vdots$	$\vdots$	$\vdots$	$\ddots$	$\vdots$	$\ddots$	$\vdots$
J	$n = 1$	$n = 1$	$\cdots$	$n = 1$	$\cdots$	$n = 1$

Subjects (random)

FIGURE 8.4   Repeated Measures Design

One of the assumptions of the mixed model is that the error terms are independent of each of the factors. In a repeated measures design it is probably unrealistic to believe that the error terms within subjects are independent across treatments. However, as long as the relationship among the errors that exist across treatments for a particular subject (i.e., within a row) is the same as that for any other subject (row), then the mixed-model analysis of variance is still appropriate to analyze the data from a repeated measures design. However, if this assumption is not valid, alternative analyses, which will be mentioned at the end of this section, should be used.

The sum of squares and mean squares for the repeated measures design are identical to that for the randomized block design [see Equations (8.52) to (8.55), p. 249], although we refer to SS subjects (rather than SS blocks) and if time points are used rather than treatments, we refer to SS repeated factors (rather than SS treatments). However, the expected values of the mean squares for the repeated measures differ from those of the randomized blocks because the expected values contain a parameter that models the dependency of the errors within rows. Nevertheless, this parameter appears in both the numerator and denominator of the F ratio and thus is "canceled out." Therefore, the F ratio for the repeated measures is identical to that of the randomized block design, although it should be remembered that the assumptions for the repeated measures design were less stringent than for the randomized block design. To summarize: **When the null hypothesis is true and the assumptions are met**

$$\frac{\text{MS treatments}}{\text{MS residuals}} \sim F_{(K-1),(J-1)(K-1)} \tag{8.60}$$

and therefore to test the null hypothesis that the treatment effects

TABLE 8.10

**Source Table for Repeated Measures Design
(Nonreplicated Mixed-Model Analysis of Variance)**

Source	Sum of Squares	df	Mean Squares	F
Subjects (rows)	SS subjects	$J-1$	MS subjects = $\dfrac{\text{SS subjects}}{J-1}$	
Treatments or repeated measures	SS treatments	$K-1$	MS treatments = $\dfrac{\text{SS treatments}}{K-1}$	$\dfrac{\text{MS treatments}}{\text{MS residuals}}=$ $F_{K-1,(J-1)(K-1)}$
Residual	SS residual	$(J-1)(K-1)$	MS residuals = $\dfrac{\text{SS residuals}}{(J-1)(K-1)}$	
Total	SS total	$JK-1$		

are zero, we compare the test statistic for MS treatments/MS residuals to an F distribution with $(K-1)$ and $(J-1)(K-1)$ degrees of freedom. The calculations for the repeated measures design is schematically summarized in Table 8.10.

It should be mentioned that the repeated measures analysis presented here is not robust with respect to the assumption that the relationship among the errors is the same for each row. Violation of this assumption may result in rejecting the null hypothesis with a greater probability than realized. One solution is to use the Geisser–Greenhouse conservative test, which protects against the worst possible violations of the assumption. The Geisser–Greenhouse test involves comparing the F ratio obtained to an F distribution with 1 and $(J-1)$ degrees of freedom [rather than to an F with $(K-1)$ and $(J-1)(K-1)$ degrees of freedom]. The reduction in degrees of freedom, of course, makes it more difficult to reject the null hypothesis. The major disadvantage of this test is that it is very conservative and will likely lead to increased Type II errors. Other alternative tests that are not so conservative exist but are beyond the scope of this text. An alternative approach for repeated measures and one that is probably superior to the mixed-model analysis of repeated measures, is the multivariate Hotelling's T^2 test, which is also beyond the scope of this text. The message should be clear: The analysis of repeated measures is difficult and the reader is advised to read more advanced texts or seek assistance for this type of design.

NOTES AND SUPPLEMENTARY READINGS

Readers interested in influences of violating the basic assumptions will find the same references noted for Chapter 7 applicable. Additionally, the other volumes mentioned in Chapter 7 also examine random and mixed analysis of variance models.

PROBLEMS

1. Suppose that an examination consists of six items. The six items can be arranged in 720 different orders. The researcher wishes to determine whether order of presentation of the items affects the scores on the examinations and thus randomly selects five different orders, each of which is then presented to 15 subjects. Based on the examination scores, the following sums of squares are obtained:

   SS between (order) = 100
   SS error = 621

   a. State the appropriate null hypothesis.
   b. Construct the source table and test the null hypothesis (α = .05).
   c. Estimate the proportion of variance in the scores accounted for by the order of the items.

2. To determine whether the particular tutor makes a difference in achievement for a group of probationary students, four tutors were randomly selected from the pool of available tutors at a college tutoring center. Six students were randomly assigned to each of the tutors and at the end of the quarter the grade point average of each student was noted. Based on the grade point averages, the following sums of squares were obtained:

   SS between (tutor) = 150
   SS error = 140

   a. State the appropriate null hypothesis.
   b. Construct the source table and test the null hypothesis (α = .05).
   c. Estimate the proportion of variance in the scores accounted for by the tutor.

3. Suppose that the researcher in Problem 1 now wishes to determine whether or not the proctor of the examination makes a difference as well as determine whether or not the order of the items makes a difference. As before, 5 orders were randomly selected from the 720 possible orders; as well, four proctors were randomly selected from the pool of

available proctors. Each proctor administered each order to 10 subjects (total number of subjects was 200). The following sums of squares were obtained:

    SS rows (order) = 280
    SS columns (proctor) = 60
    SS interaction = 80
    SS error = 1440

   a. State the null hypotheses for order, proctor, and interaction.
   b. Construct the source table and test the null hypotheses ($\alpha = .05$).
   c. Estimate the intraclass correlation coefficients.
   d. Are Paull's criteria for pooling satisfied? If so, perform the analysis of variance by pooling.

4. A researcher is interested in whether different types of diets make a difference in the number of calories consumed by overweight individuals. Furthermore, the researcher suspects that age plays a part in the effectiveness of the diet. To investigate these questions, four diets were randomly selected from the pool of all diets available on the over-the-counter market and six different ages were randomly selected from the range of 21 to 45 years. Twelve subjects were randomly assigned to each of the 24 combinations of diet and age and the following sums of squares were obtained:

    SS rows (diet) = 1950
    SS columns (age) = 4500
    SS interaction = 5400
    SS error = 26,400

   a. State the null hypotheses for diet, age, and interaction.
   b. Construct the source table and test the null hypotheses ($\alpha = .05$).
   c. Estimate the intraclass correlation coefficients.
   d. Are Paull's criteria for pooling satisfied? If so, perform the analysis of variance by pooling.

5. To determine whether door-to-door salespeople make a difference in the amount of sales in various types of geographical areas (rural, urban, and suburban), five salespeople were randomly chosen from the pool of available salespeople and asked to sell various products door-to-door to ten households randomly selected from each of the three types of areas. Based on the dollar amount of the products sold, the following sums of squares were obtained:

    SS rows (salespeople) = 9600
    SS columns (geographical type) = 6720
    SS interaction = 4880
    SS error = 32,400

a. State the null hypotheses for the row, columns, and interaction.
b. Construct the source table and test the null hypotheses ($\alpha = .05$).

6. Suppose that a researcher wishes to investigate whether schools have an effect on the number of hours spent on homework in grades 3, 4, 5, and 6. Four elementary schools were randomly selected from the schools in the district and six students in each grade level and in each school were asked to record the number of minutes that they devoted to homework over a six-day period. Based on the number of minutes devoted to homework each day, the following sums of squares were obtained:

SS rows (schools) = 1263
SS columns (grade level) = 8088
SS interaction = 12,267
SS error = 15,552

a. State the null hypotheses associated with the row, column, and interaction effects.
b. Construct the source table and test the null hypotheses ($\alpha = .05$).

7. Three variations of a program to teach speech skills to handicapped children were compared by using a randomized block design. The children were reinforced for uttering recognizable sounds, where the determination of what is recognizable is determined by (a) a computer-assisted voice recognition system in treatment A, (b) speech pathologists in treatment B, and (c) undergraduate students in treatment C. Each block consisted of three children with similar intelligence. Ten blocks were used and the three children in each block were randomly assigned to the three treatments. Upon completion of the treatments, speech skills were assessed by having each child make 20 requests of strangers (in a simulated situation) and the number of requests understood was the outcome measure. The following means were obtained for the three treatment groups:

Treatment A (computer-assisted)	Treatment B (speech pathologist)	Treatment C (undergraduate)
6.2	9.0	8.8

The following sums of squares were obtained:

SS blocks = 45
SS treatments = 48.8
SS residual = 109

Determine whether or not the treatments are equally effective by calculating a nonreplicated mixed-model analysis of variance (use $\alpha = .05$).

8. A researcher located 12 sets of twins who were taken away from their parents by the state before age 4; one of the twins was adopted shortly thereafter and the other was raised by foster parents. The intelligence of the children was measured at age 12, and the following mean IQ squares were obtained:

Adopted	Foster Parents
98	104

The following sums of squares were obtained:

    SS blocks = 812
    SS treatments = 216
    SS residual = 1133

Determine whether or not the intelligence of children who were adopted and children who were raised by foster parents differed by calculating a nonreplicated mixed-model analysis of variance (use $\alpha = .05$).

9. In an experiment to understand aggressive behavior in hyperactive children, hyperactive children were asked to complete a task with various types of partners and the number of aggressive behaviors of the target child was recorded. Specifically, each of 15 hyperactive children was asked to complete the task three times, once with a hyperactive partner, once with a normal partner, and once with a passive partner (the order of the partners was randomized). The following mean number of aggressive behaviors was obtained:

Hyperactive Partner	Normal Partner	Passive Partner
22	14	17

The following sums of squares were obtained:

    SS subjects = 350
    SS treatments = 390
    SS residuals = 672

   a. Determine whether or not the partner makes a difference in the number of aggressive behaviors emitted by hyperactive children by calculating a nonreplicated mixed-model analysis of variance on the repeated measures (use $\alpha = .01$).
   b. Does the conclusion drawn in (a) remain the same if the Geisser–Greenhouse test is used?

10. Because the effects of psychotherapy are often transitory, 20 clients treated for agoraphobia (fear of open spaces) were assessed immediately after therapy and thereafter every two weeks for a period of ten weeks (a total of six measurements). At each assessment period family members were asked to indicate the hours that the subject spent away from the

house during the previous three days; the following mean numbers of hours away from the house are as follows:

End of Therapy	Two Weeks	Four Weeks	Six Weeks	Eight Weeks	Ten Weeks
5.6	4.0	3.8	2.4	1.0	1.2

The following sums of squares were obtained:

SS subjects = 38
SS repeated measures = 17
SS residuals = 47.5

a. Determine whether or not the time since completion of therapy affects the number of hours away from the house by calculating a non-replicated mixed-model analysis of variance on the repeated measures (use $\alpha = .05$).
b. Does the conclusion drawn in (a) remain the same if the Geisser–Greenhouse test is used?

9 PLANNED AND POST HOC COMPARISONS

In the previous two chapters, which covered several variations of the analysis of variance, the discussion of the test statistics was limited to omnibus tests. That is, the hypotheses tested in those instances were statements about all groups simultaneously and were designed only to identify whether or not there existed some treatment effects. A statistically significant omnibus test does not provide information about where the particular effects are to be found. For example, consider a treatment study with four groups—an innovative treatment, a traditional treatment, a placebo control group, and a no-treatment control group. A sufficiently large F ratio from a one-way fixed-effects analysis of variance leads to rejection of the null hypothesis that the population means for all four groups are not equal (or alternatively, that the treatment effects are not all zero). Yet, we have no information about which particular means are different. It might be that the traditional treatment has a beneficial effect but that the innovative treatment and the two control groups have no effect. Then again, it might be that the two treatments are equally effective and superior to the two control groups. Or it may be that the placebo vis-à-vis no treatment results in augmented scores for the subjects. When a researcher is interested in comparisons between or among specific groups, an omnibus test provides little information.

A further advantage is that tests of specific comparisons are likely to be much more powerful than the test of the omnibus hypothesis. In this chapter methods to compare means of particular groups are discussed.

There are two ways that a researcher may proceed with comparisons involving particular groups. If he or she is clear about the particular comparisons before the experiment is conducted, the method of planned comparisons should be used. In this instance the researcher forgoes an omnibus test and proceeds directly to the planned comparisons of interest. Although there are restrictions related to the use of planned comparisons, as we will see, this strategy provides a very elegant test of the specific hypotheses that are specified in advance. The use of planned comparisons dictates that considerable thought must be given to the experimental questions prior to executing a study. As research in a substantive area progresses, it would be expected that the use of planned comparisons would increase.

When a researcher does not have hypotheses about particular groups, the investigative spirit still leads to an interest in comparisons among particular groups, although the specific groups cannot be specified in advance. In this case the researcher conducts an omnibus test to be assured that there are indeed differences among groups. If the omnibus test is significant, post hoc comparisons can be used to search for the particular groups that contributed to the omnibus effect. However, a price is paid for not specifying in advance the comparisons of interest; the tests of post hoc comparisons are more conservative than the tests of planned comparisons.

Analysis of Planned Comparisons

9.1 Planned Comparisons

Planned comparisons are comparisons between or among means of groups that are specified before the study is conducted. For the sake of illustration, suppose that the researcher has randomly assigned an equal number of subjects to four groups: innovative treatment (Group 1), traditional treatment (Group 2), placebo control (Group 3), and no treatment (Group 4). The omnibus null hypothesis is that the population means for all the groups are equal; that is

$$H_0: \mu_1 = \mu_2 = \mu_3 = \mu_4$$

However, the researcher has three particular comparisons among groups that are of interest:

(*i*) Innovative treatment versus traditional treatment
(*ii*) Placebo control versus no-treatment control
(*iii*) Treatments (innovative and traditional) versus control groups (placebo and no treatment)

(The third comparison involves contrasting the average of the treatment groups to the average of the control groups.) Keep in mind that there are many such comparisons; the three discussed here are the ones that interest this hypothetical researcher. If these comparisons are tested in terms of the means of the groups, then they can be rephrased as

(i) μ_1 versus μ_2

(ii) μ_3 versus μ_4

(iii) $\dfrac{\mu_1 + \mu_2}{2}$ versus $\dfrac{\mu_3 + \mu_4}{2}$

We can now write the null hypotheses related to these comparisons:

(i) H_0: $\mu_1 - \mu_2 = 0$

(ii) H_0: $\mu_3 - \mu_4 = 0$

(iii) H_0: $\dfrac{\mu_1 + \mu_2}{2} - \dfrac{\mu_3 + \mu_4}{2} = 0$

To eliminate the fractions, both sides of the equation contained in the last null hypothesis are multiplied by 2, giving

(iii) H_0: $(\mu_1 + \mu_2) - (\mu_3 + \mu_4) = 0$

or preferably

(iii) H_0: $\mu_1 + \mu_2 - \mu_3 - \mu_4 = 0$

It is customary and convenient to express the null hypotheses in terms of a linear combination of all of the population means, so the three null hypotheses are rewritten again:

(i) H_0: $1\mu_1 + (-1)\mu_2 + 0\mu_3 + 0\mu_4 = 0$

(ii) H_0: $0\mu_1 + 0\mu_2 + 1\mu_3 + (-1)\mu_4 = 0$

(iii) H_0: $1\mu_1 + 1\mu_2 + (-1)\mu_3 + (-1)\mu_4 = 0$

Written this way, the hypotheses are of the general form

H_0: $c_1\mu_1 + c_2\mu_2 + c_3\mu_3 + c_4\mu_4 = 0$

The left-hand side of the equation of this hypothesis is called a *population comparison*. In general, **given J populations with population means μ_1, . . ., μ_J, the population comparison, which is denoted by ψ (psi), is a linear combination of the means:**

$$\psi = c_1\mu_1 + c_2\mu_2 + \cdots + c_J\mu_J = \sum_j c_j\mu_j \tag{9.1}$$

where the coefficients c_j are not all equal to zero. For this model the restriction that the sum of the coefficients must equal zero (i.e., $\sum_j c_j = 0$) is adopted, although the rationale for this will be postponed until Section 9.2 (p. 268). It is important to realize that the methods of this chapter are used to compare means of specific groups; therefore, the methods are applicable only to fixed factors.

Keep in mind that the population comparison is simply a way of expressing a comparison among groups that is of interest to the researcher. For example, if $c_1 = 1$ and $c_2 = -1$ and the remaining coefficients are zero, the population comparison compares the mean of population 1 to the mean of population 2.

The null hypothesis related to any comparison can be stated as

$$H_0\colon \psi = 0$$

Evidence about the null hypothesis is obtained simply by calculating a sample value of the population comparison ψ. The experimental design involves randomly selecting N_j observations from each of the j populations. The *sample comparison* is found by substituting the sample mean M_j based on the N_j observations for the population mean μ_j in Equation (9.1). Thus, **the sample comparison, which is denoted by $\hat{\psi}$, is given by**

$$\hat{\psi} = c_1 M_1 + c_2 M_2 + c_3 M_3 + \cdots + c_J M_J = \sum_j c_j M_j \qquad (9.2)$$

(Again, for now we require that $\sum_j c_j = 0$.) Note that the sample comparison is a statistic. For example, if we are comparing the means of populations 1 and 2 (i.e., $c_1 = 1$ and $c_2 = -1$ and the other coefficients are zero) and $M_1 = 10$ and $M_2 = 8$, then $\hat{\psi} = 2$. Whether the value of 2 is sufficient to reject the null hypothesis that $H_0\colon \psi = 0$ depends on the sampling distribution of ψ.

Recall that when we used the sample mean M to make inferences about the population mean μ, knowledge of the sampling distribution of M was required. To make inferences about ψ, the sampling distribution of $\hat{\psi}$ needs to be investigated, and we now turn to that topic. The sampling distribution of $\hat{\psi}$ is intimately tied to the distribution of the mean of each population; assuming normality and homogeneity of variance

$$M_j \sim N\left(\mu_j, \frac{\sigma_e^2}{N_j}\right)$$

We begin by finding the expectation of $\hat{\psi}$. By the rules of expectation

$$
\begin{aligned}
E(\hat{\psi}) &= E(c_1 M_1 + \cdots + c_J M_J) \\
&= E(c_1 M_1) + \cdots + E(c_J M_J) \\
&= c_1 E(M_1) + \cdots + c_J E(M_J) \\
&= c_1 \mu_1 + \cdots + c_J \mu_J \\
&= \psi
\end{aligned}
\qquad (9.3)
$$

Thus, $\hat{\psi}$ is an unbiased estimator for ψ.

The variance of the sample comparison is now derived. Given the definition of the sample comparison,

$$V(\hat{\psi}) = V(c_1 M_1 + \cdots + c_J M_J)$$

Because the sample means of the various populations are independent

$$V(c_1 M_1 + \cdots + c_J M_J) = V(c_1 M_1) + \cdots + V(c_J M_J)$$

By the rules of variance (and assuming homogeneity of variance)

$$V(c_1 M_1) + \cdots + V(c_J M_J) = c_1^2 V(M_1) + \cdots + c_J^2 V(M_J)$$

$$= \frac{c_1^2 \sigma_e^2}{N_1} + \frac{c_2^2 \sigma_e^2}{N_2} + \cdots + \frac{c_J^2 \sigma_e^2}{N_J}$$

$$= \sigma_e^2 \sum_j \frac{c_j^2}{N_j}$$

So

$$V(\hat{\psi}) = \sigma_e^2 \sum_j \frac{c_j^2}{N_j} \tag{9.4}$$

Because the distribution of the linear combination of normally distributed random variables is normal (see Section 3.15, p. 72)

$$\hat{\psi} \sim N\left(\psi, \sigma_e^2 \sum_j \frac{c_j^2}{N_j} \right) \tag{9.5}$$

Given the mean and variance of $\hat{\psi}$, the sample comparison is standardized:

$$\frac{\hat{\psi} - \psi}{\sqrt{\sigma_e^2 \sum_j (c_j^2/N_j)}} \sim N(0, 1) \tag{9.6}$$

However, because the variance σ_e^2 is unknown, Equation (9.6) cannot serve as a test statistic. Nevertheless, recall that the unbiased estimator for σ_e^2 based on several samples (i.e., two or more groups) is the mean square error (MSE). So, let

$$\hat{\sigma}_e^2 = \text{MSE} = \frac{\text{SSE}}{J - 1} \tag{9.7}$$

It should be noted that MSE is calculated in the usual fashion based on all J samples, even though a particular comparison may involve only two means.

As with the case for the t test, when σ_e^2 is estimated, the standardized statistic has a t distribution:

$$\frac{\hat{\psi} - \psi}{\sqrt{\text{MSE} \sum_{j} (c_j^2/N_j)}} \sim t_{N-J} \tag{9.8}$$

Therefore, **to test the null hypothesis H_0: $\psi = 0$, we compare the value of the test statistic**

$$\frac{\hat{\psi}}{\sqrt{\text{MSE} \sum_{j} (c_j^2/N_j)}} = t \tag{9.9}$$

to a t distribution with $(N - J)$ degrees of freedom provided the assumptions of normality and homogeneity of variance are met. The consequences of violating the assumptions of this test are of the same order as violating the assumptions in the context of the t test or the analysis of variance (see Section 6.8, p. 150). The t test for a planned comparison can be used to test either one- or two-tailed alternative hypotheses.

Often it is convenient to multiply the coefficients of a planned comparison by a constant, thereby avoiding the use of fractions. Suppose that the coefficients c_j are multiplied by the constant a, resulting in j coefficients ac_j. Clearly the size of the sample comparison is altered by a factor of a. However, the expectation of the sample comparison is also altered by a factor of a and the variance is altered by a factor of a^2, leaving the size of the test statistic unchanged. To reiterate: **Multiplying the coefficients of a planned comparison by a constant does not change the size of the test statistic nor does it affect the decisions about rejection/acceptance of the null hypothesis.**

The test of a planned comparison is illustrated in Example 9.1, which follows.

An important issue raised by the example is related to whether there are restrictions placed on the planned comparisons that may be tested, both as regards to the number of comparisons and the particular characteristics of the comparisons. As we will see in the next section, there are restrictions. However, before turning to that topic, let us mention the similarity of the test for planned comparisons to other tests as well as confidence intervals.

When there are only two groups in the experiment (i.e., $J = 2$), the test for the only planned comparison possible (i.e., $c_1 = 1$, $c_2 = -1$ or an equivalent comparison formed by multiplying by a constant) is equivalent to the independent sample t test, because MSE is equivalent to the pooled estimate used in the context of the t test (see Section 6.6, p. 144). However, when more than two groups are involved, the test of a planned comparison that includes only two samples (e.g., $c_1 = 1$, $c_2 = -1$, and all the other coefficients are zero) is not equivalent to the two-independent-sample t test because the estimate of σ_e^2 (i.e., MSE) is based on all of the groups rather than on just the two groups involved in the comparison.

EXAMPLE 9.1

Suppose that a researcher is interested in comparing the means of four groups: an innovative treatment, the traditional treatment, a placebo control group, and a no-treatment control group. The researcher randomly selects 60 subjects and randomly assigns 15 subjects to each group. Suppose that higher scores indicate better performance and the following means and sample sizes (after attrition) are obtained:

$$\text{Innovative treatment:} \quad M_1 = 12 \quad N_1 = 14$$
$$\text{Traditional treatment:} \quad M_2 = 10 \quad N_2 = 13$$
$$\text{Placebo control:} \quad M_3 = 6 \quad N_3 = 15$$
$$\text{No-treatment control:} \quad M_4 = 4 \quad N_4 = 14$$

Based on the four groups, we find that MSE = SSE/$N - J$) = 416/52 = 8.00.

The first question of interest to the researcher is whether one of the two treatments is more effective. The null hypothesis related to this question is

$$H_0: \mu_1 - \mu_2 = 0$$

In terms of planned comparisons, this null hypothesis is written as a linear combination of means.

$$H_0: \psi = 1\mu_1 + (-1)\mu_2 + 0\mu_3 + 0\mu_4 = 0$$

Therefore, the coefficients of the planned comparison are $c_1 = 1$, $c_2 = -1$, $c_3 = 0$, $c_4 = 0$. The alternative hypothesis is stated nondirectionally:

$$H_a: \psi \neq 0$$

The value of the sample comparison is

$$\hat{\psi} = 1M_1 + (-1)M_2 + 0M_3 + 0M_4$$
$$= 1(12) + (-1)(10) + 0(6) + 0(4) = 2$$

To calculate the value of the test statistic, we need to determine the term $\sum_j (c_j^2 / N_j)$:

$$\sum_j \frac{c_j^2}{N_j} = \frac{1^2}{14} + \frac{(-1)^2}{13} + \frac{0^2}{15} + \frac{0^2}{14} = .148$$

The value of the test statistic is

$$\frac{\hat{\psi}}{\sqrt{\text{MSE}\sum_j (c_j^2/N_j)}} = \frac{2}{\sqrt{8(.148)}} = 1.84$$

which, when compared to a t distribution with $(N - J) = 52$ degrees of freedom, is insufficient to reject the null hypothesis and the researcher cannot conclude that one treatment is more effective than the other treatment.

A second hypothesis that interests the researcher relates to the two treatment groups vis-à-vis the two control groups. That is, a contrast of the average of the two treatment groups to the average of the two control groups is desired.

$$H_0: \frac{\mu_1 + \mu_2}{2} - \frac{\mu_3 + \mu_4}{2} = \left(\frac{1}{2}\right)\mu_1 + \left(\frac{1}{2}\right)\mu_2 + \left(-\frac{1}{2}\right)\mu_3$$
$$+ \left(-\frac{1}{2}\right)\mu_4 = 0$$

After multiplying by the constant 2, we obtain the desired planned comparison:

$$\psi = 1\mu_1 + 1\mu_2 + (-1)\mu_3 + (-1)\mu_4$$

To test the null hypothesis

$$H_0: \psi = 0$$

versus the alternative

$$H_a: \psi > 0$$

(i.e., the treatments are more effective than the controls), we obtain the value of the sample comparison:

$$\hat{\psi} = 1M_1 + 1M_2 + (-1)M_3 + (-1)M_4$$
$$= 1(12) + 1(10) + (-1)(6) + (-1)(4) = 12$$

The value of the test statistic is

$$\frac{\hat{\psi}}{\sqrt{\text{MSE}\sum_j (c_j^2/N_j)}} = \frac{12}{\sqrt{8(.286)}} = 7.93$$

which compared to a t distribution with 52 degrees of freedom, leads to rejection of the null hypothesis in favor of the alternative that the treatments are more effective than the control groups.

The confidence interval for ψ is formed in the typical manner. Specifically, the $(1 - \alpha)100\%$ confidence interval is given by

$$P\left[\hat{\psi} - t_{\alpha/2;N-J}\sqrt{\text{MSE}\sum_j(c_j^2/N_j)} \leq \psi \leq \hat{\psi} + t_{\alpha/2;N-J}\sqrt{\text{MSE}\sum_j(c_j^2/N_j)}\right]$$

$$= 1 - \alpha \quad (9.10)$$

The 95% confidence interval for the first planned comparison test in Example 9.1 ranges from

$$2 - 2.01\sqrt{8(.148)} = 2 - 2.19 = -.19$$

to

$$2 + 2.01\sqrt{8(.148)} = 2 + 2.19 = 4.19$$

which does include the value zero (this is expected given the two-tailed test of the null hypothesis that $\psi = 0$ was not rejected at the α level of .05).

9.2   Multiple Planned Comparisons

In the context of the analysis of variance it was emphasized that an omnibus test was conducted because conducting a series of t tests can lead to spurious conclusions. Specifically, α is the probability that a single test leads to the decision to reject the null hypothesis when the null is true. However, when more than one test is conducted, the probability of rejecting at least one null hypothesis when that null is true is much greater than α. This is especially troublesome when the tests are not independent. Therefore, it is inappropriate to arbitrarily select a set of planned comparisons and to test each at a selected level of α. However, it would be unduly restrictive to limit the number of planned comparisons that can be tested to one. Fortunately, there are methods that allow for more than one planned comparison to be tested.

The first method involves the use of the *Bonferroni* inequality. The Bonferroni inequality states that given K statistical tests, the probability that at least one Type I error occurs is equal to or less than $K\alpha$, where the probability of a Type I error for each test is set to α. It should be emphasized that this property applies to dependent tests. According to the Bonferroni inequality, to ensure that the probability of a Type I error overall is less than a given α, each test is conducted at the level of α/K. In the context of planned comparisons, if there are five planned comparisons of interest and the researcher wishes to have an overall α of .05, each planned comparison is tested at the .01 level. Although this is a straightforward method, for even a moderately large number of comparisons, the α level for an individual comparison is stringent.

A less conservative and generally preferred method is to construct independent comparisons and test each at the α level desired. For two independent comparisons the results of one sample comparison will have no bearing on the results of another sample comparison. (Independence of planned comparisons is the same as the concept of independence of factors in a balanced factorial design discussed in Section 7.15, p. 210.) Because the sample value obtained for one of two independent planned comparisons has no effect whatsoever on the sample value of the other, each can be tested at the given level of α.

Several conditions must be met in order for any two sample comparisons to be independent. Let $\hat{\psi}_1$ and $\hat{\psi}_2$ be two sample comparisons, such that

$$\hat{\psi}_1 = c_{11}M_1 + c_{12}M_2 + \cdots + c_{1J}M_J$$

and

$$\hat{\psi}_2 = c_{21}M_1 + c_{22}M_2 + \cdots + c_{2J}M_J$$

(here the first subscript on the coefficient c refers to the comparison). **Two comparisons $\hat{\psi}_1$ and $\hat{\psi}_2$ are independent provided the following conditions are met:**

(i) The J populations are normally distributed and have equal variances (i.e., $Y_j \sim N(\mu_j, \sigma_e^2)$ for $j = 1$ to $j = J$)

(ii) The sample sizes for the J samples are equal (i.e., $N_1 = N_2 = \cdots = N_J$)

(iii) $c_{11}c_{21} + c_{12}c_{22} + \cdots + c_{1J}c_{2J} = \sum_j c_{1j}c_{2j} = 0$

Although infrequently used, the second condition can be relaxed to include the instance when

$$\sum_j \frac{c_{1j}c_{2j}}{N_j} = 0$$

Often there is confusion around the term *orthogonal planned comparisons*. One frequently hears this term applied when comparisons satisfy the third condition. Actually two comparisons are orthogonal only if the third condition **and** the second condition (including the infrequently used relaxation just mentioned) are satisfied. However, it is vital to remember that orthogonal comparisons lead to statistical independence only if the first condition is also met. A set of mutually orthogonal comparisons is a set of comparisons for which every two comparisons are orthogonal to each other.

The idea of orthogonal comparisons is illustrated by examining the following coefficients for comparisons involving four independent samples (assume that the sample sizes are equal):

Samples

	I	II	III	IV
$\hat{\psi}_1$	1	−1	0	0
$\hat{\psi}_2$	0	0	1	−1
$\hat{\psi}_3$	1	1	−1	−1
$\hat{\psi}_4$	1	1	0	−2

It should be clear from this example that orthogonality is not related to the values of the means for the J samples nor to the value of the sample comparisons; orthogonality is completely dependent on the coefficients and the sample sizes. Clearly comparisons $\hat{\psi}_1$ and $\hat{\psi}_2$ are orthogonal because

$$\sum_j c_{1j}c_{2j} = 1(0) + (-1)(0) + 0(1) + 0(-1) = 0$$

The reader should verify the following:

$\hat{\psi}_1$ and $\hat{\psi}_2$: orthogonal

$\hat{\psi}_1$ and $\hat{\psi}_3$: orthogonal

$\hat{\psi}_1$ and $\hat{\psi}_4$: orthogonal

$\hat{\psi}_2$ and $\hat{\psi}_3$: orthogonal

$\hat{\psi}_2$ and $\hat{\psi}_4$: not orthogonal

$\hat{\psi}_3$ and $\hat{\psi}_4$: not orthogonal

It is not difficult to see why $\hat{\psi}_2$ and $\hat{\psi}_4$ are not orthogonal; comparison $\hat{\psi}_2$ compares the average of samples I and II to the average of III and IV, whereas comparison $\hat{\psi}_4$ compares the average of I and II to IV (which is part of the average of III and IV). Because comparisons $\hat{\psi}_1$, $\hat{\psi}_2$, and $\hat{\psi}_3$ are each orthogonal from the other, they form a set of mutually orthogonal comparisons.

Previously the condition that the sum of the coefficients for any comparison is equal to zero (i.e., $\Sigma_j c_j = 0$) had been imposed on planned comparisons. This condition was applied so that the sample value of the comparison is orthogonal to the sample grand mean. This means that provided the normality and homogeneity of variance assumptions are met, the value of the sample comparison is statistically independent of the value of the sample grand mean. These two values do not affect each other in any way. Nevertheless, there may be instances for which comparisons that are not orthogonal to the grand mean are of interest. Suppose that a researcher wishes to test the claim that a mental patient with one type of diagnosis receives twice as much nursing care (minutes of care per shift) as a patient with another type of diagnosis. The null hypothesis would be

$$H_0: \hat{\psi} = \mu_1 - 2\mu_2 = 0$$

Such a comparison is legitimate provided that the dependent variable is measured on a ratio scale.

There is a limit to the number of comparisons that are mutually orthogonal and orthogonal to the grand mean. **Given J independent samples, a set of mutually orthogonal comparisons that are also orthogonal from the grand mean contains at most $(J - 1)$ comparisons.** For example, in the four-sample problem discussed earlier, after three mutually orthogonal comparisons have been found, it is not possible to find a fourth that does not contain information in one of the previous three. Of course, the three comparisons found are not unique; there are other sets of three comparisons that are mutually orthogonal. The researcher determines which set of comparisons best answers the research questions. We should also note that it is not necessary to test all $(J - 1)$ comparisons; only those that are substantively interesting should be examined (more about this in the next section). Furthermore, if the researcher wishes to test more than $(J - 1)$ comparisons or to test comparisons that are not orthogonal, then the solution based on the Bonferroni inequality, which was discussed at the beginning of this section, can be used.

9.3   Relation Between Planned Comparisons and the Analysis of Variance

It is not a coincidence that for a J-independent-sample problem there are $(J - 1)$ degrees of freedom associated with treatments in the analysis of variance as well as $(J - 1)$ possible orthogonal planned comparisons. Essentially, the F test for treatments in the analysis of variance can be decomposed into $(J - 1)$ tests of $(J - 1)$ mutually independent planned comparisons.

To demonstrate the relation between the F test for the analysis of variance and the tests for planned comparisons, we need to transform the t test of a planned comparison into an F test. Recall that under the normality and homogeneity assumptions, when the null hypothesis that $\psi = 0$ is true

$$\frac{\hat{\psi}}{\sqrt{\text{MSE} \sum_j (c_j^2/N_j)}} \sim t_{N-J}$$

Using the relation between the t and F distributions (see Section 7.5, p. 180)

$$\frac{(\hat{\psi})^2}{\text{MSE} \sum_j (c_j^2/N_j)} \sim F_{1, N-J}$$

resulting in an F test of the planned comparison. We now rewrite this ratio as

$$\frac{(\hat{\psi})^2 \Big/ \sum_j (c_j^2/N_j)}{\text{MSE}}$$

If the numerator is defined as the mean square of the comparison, denoted by $\text{MS}(\hat{\psi})$, the test statistic has a familiar form:

$$\frac{\text{MS}(\hat{\psi})}{\text{MSE}} = F_{1,N-J}$$

Because there is 1 degree of freedom

$$\mathbf{MS}(\hat{\psi}) = \frac{\mathbf{SS}(\hat{\psi})}{1} = \mathbf{SS}(\hat{\psi}) = \frac{(\hat{\psi})^2}{\sum_j (c_j^2/N_j)} \tag{9.11}$$

To summarize: **When the null hypothesis that $\psi = 0$ is true and the assumptions of the analysis of variance are met (viz., normality and homogeneity of variance), the test statistic**

$$\frac{\mathbf{MS}(\hat{\psi})}{\mathbf{MSE}} = \frac{\mathbf{SS}(\hat{\psi})}{\mathbf{MSE}} \tag{9.12}$$

has an F distribution with 1 and $(N - J)$ degrees of freedom.

Now consider the case of a set of $(J - 1)$ mutually independent comparisons $\hat{\psi}_1, \hat{\psi}_2, \ldots, \hat{\psi}_{J-1}$. It can be shown that

$$\text{SS}(\hat{\psi}_1) + \text{SS}(\hat{\psi}_2) + \cdots + \text{SS}(\hat{\psi}_{(J-1)}) = \text{SS between} \tag{9.13}$$

That is, the sum of squares between can be partitioned into $(J - 1)$ sum of squares each corresponding to one of the $(J - 1)$ independent comparisons. This relation between the analysis of variance and planned comparisons is summarized in Table 9.1. The sum of squares between is placed in parentheses in the source table to indicate that the omnibus F test is not conducted if one chooses to test planned comparisons. It is interesting to note that the value of the omnibus F ratio for the analysis of variance is equal to the mean of the $(J - 1)$ F values for the $(J - 1)$ independent planned comparisons. This fact has implications for understanding the power of planned comparisons: It is entirely possible to have a statistically nonsignificant omnibus F and have one or more statistically significant planned comparisons. Failure to ask specific questions can cost the researcher dearly. Another way of looking at the omnibus F test is that it is a test of the null hypothesis

$$H_0\colon \psi_1 = 0,\ \psi_2 = 0,\ \ldots,\ \psi_{(J-1)} = 0$$

for any $(J - 1)$ independent comparisons. That is, the omnibus F test contains

TABLE 9.1

Source Table for $(J - 1)$ Independent Planned Comparisons

Source	Sum of Squares	df	Mean Squares	F
(Between	SSB	$J - 1$)		
$\hat{\psi}_1$	$SS(\hat{\psi}_1)$	1	$MS(\hat{\psi}_1) = SS\,(\hat{\psi}_1)$	$\dfrac{SS(\hat{\psi}_1)}{MSE} = F_{1,N-J}$
$\hat{\psi}_2$	$SS(\hat{\psi}_2)$	1	$MS(\hat{\psi}_2) = SS\,(\hat{\psi}_2)$	$\dfrac{SS(\hat{\psi}_2)}{MSE} = F_{1,N-J}$
.	.	.	.	.
.	.	.	.	.
.	.	.	.	.
$\hat{\psi}_{J-1}$	$SS(\hat{\psi}_{J-1})$	1	$MS(\hat{\psi}_{J-1}) = SS\,(\hat{\psi}_{J-1})$	$\dfrac{SS(\hat{\psi}_{J-1})}{MSE} = F_{1,N-J}$
Error	SSE	$N - J$	$MSE = \dfrac{SSE}{N - J}$	
Total	SST	$N - 1$		

$(J - 1)$ specific independent questions that might have been asked. Of course, one must choose between the omnibus test and tests of planned comparisons.

Often research questions lead to less than $(J - 1)$ planned comparisons. Suppose that the researcher is interested in testing g independent planned comparisons, where $g < (J - 1)$. This presents no special problems and the tests of the g comparisons are conducted in the usual manner. However, there may be some interesting effects remaining in the $(J - 1 - g)$ remaining comparisons, although they were not of interest to the researcher a priori. It is acceptable to test the $(J - 1 - g)$ remaining comparisons as a group. The sum of squares for the remaining comparisons, denoted by SS remainder, is given by

$$SS \text{ remainder} = SS \text{ between} - SS(\hat{\psi}_1) - SS(\hat{\psi}_2) - \cdots - SS(\hat{\psi}_g) \quad (9.14)$$

Thus, the test statistic for the remainder is given by

$$\frac{SS \text{ remainder}/(J - 1 - g)}{MSE} = F \quad (9.15)$$

which under the assumptions of the analysis of variance has an F distribution with $(J - 1 - g)$ and $(N - J)$ degrees of freedom. The relation of the test for the remaining comparison to the analysis of variance and the other

TABLE 9.2

Source Table for g Independent Planned Comparisons and Remainder

Source	Sum of Squares	df	Mean Squares	F
(Between	SSB	$J - 1$)		
$\hat{\psi}_1$	$SS(\hat{\psi}_1)$	$\hat{\psi}_1$	$MS(\hat{\psi}_1) = SS(\hat{\psi}_1)$	$\dfrac{SS(\hat{\psi}_1)}{MSE} = F_{1,N-J}$
.	.	.	.	.
.	.	.	.	.
.	.	.	.	.
$\hat{\psi}_g$	$SS(\hat{\psi}_g)$	$\hat{\psi}_1$	$MS(\hat{\psi}_g) = SS(\hat{\psi}_g)$	$\dfrac{SS(\hat{\psi}_g)}{MSE} = F_{1,N-J}$
Remainder	SS remainder	$J - 1 - g$	$\begin{aligned}\text{MS remainder} = \\ \dfrac{\text{SS remainder}}{(J-1-g)}\end{aligned}$	$\dfrac{\text{MS remainder}}{MSE} = F_{J-1-g,N-J}$
Error	SSE	$N - J$	$MSE = \dfrac{SSE}{N-J}$	
Total	SST	$N - 1$		

comparisons is presented in Table 9.2. If the F test for the remaining planned comparisons is statistically significant, there is reason to believe that some comparisons contained in the remainder are of interest; the post hoc methods discussed in Section 9.4 (p. 279) are available to investigate comparisons contained in the remainder.

Example (9.2), which follows, illustrates the major points made in this section so far.

To this point the discussion of planned comparisons has been limited to one-way designs. Extensions to factorial designs present no special problems and illustrate some useful concepts. Consider the following 2×2 design:

Factor B

	B_1	B_2	
A_1	μ_{11}	μ_{12}	$\mu_{1\cdot}$
A_2	μ_{21}	μ_{22}	$\mu_{2\cdot}$
	$\mu_{\cdot 1}$	$\mu_{\cdot 2}$	

Factor A

EXAMPLE 9.2   Suppose that a researcher is interested in the effects of reading instruction delivered during summer school on the reading achievement level of students who have just finished the fourth grade. Two instructional methods are available and the researcher wants to know whether it is advantageous to use trained teachers as opposed to paraprofessionals with each of the instructional methods. To determine whether the instructional methods are superior to no instruction, two control groups are used: a no-instruction control group and a group that participates in physical activities at the same time the other students are receiving reading instruction. Because the trained teacher versus paraprofessional cannot be crossed with the control groups, the design is conceptualized as a one-way design with six groups.

Group I	Group II	Group III	Group IV	Group V	Group VI
Method *A* Trained Teacher	Method *B* Trained Teacher	Method *A* Para-professional	Method *B* Para-professional	No Treatment	Physical Activities

The three specific questions that the researcher wishes to ask are

(*i*)   Is the achievement of students instructed with Method *A* superior to those instructed with Method *B*?

(*ii*) Is the achievement of students instructed by trained teachers superior to those instructed by paraprofessionals?

(*iii*) Is the achievement of students who receive reading instruction superior to those who do not receive reading instruction?

Each question will be tested with a directional test at the .05 level. Sixty subjects are randomly selected from a population of fourth graders with ten randomly assigned to each group and the treatments are implemented. At the end of the treatment period the reading level of each subject was assessed and expressed as a grade level equivalent.

The sample means for each group as well as the coefficients for the three planned comparisons corresponding to the three research questions are as follows:

	Group I	Group II	Group III	Group IV	Group V	Group VI
	Method *A* Trained Teacher	Method *B* Trained Teacher	Method *A* Para-professional	Method *B* Para-professional	No Treatment	Physical Activities
M_j	7.2	6.5	6.0	6.2	5.4	5.3
$\hat{\psi}_1$	1	-1	1	-1	0	0
$\hat{\psi}_2$	1	1	-1	-1	0	0
$\hat{\psi}_3$	1	1	1	1	-2	-2

Clearly these three comparisons are mutually orthogonal; assuming normality and homogeneity of variance, they are mutually independent as well. Because the researcher does not have other specific research questions, the remaining two possible independent comparisons will be expressed as a remainder.

First, the value of SS between is determined:

$$SS\ between = \sum_j N_j(M_j - M)^2 = 25.2$$

Next, the values of the planned comparisons are calculated:

$$\hat{\psi}_1 = \sum_j c_j M_j = (1)(7.2) + (-1)(6.5) + (1)(6.0)$$

$$+ (-1)(6.2) + 0(5.4) + 0(5.3)$$

$$= .5$$

$$\hat{\psi}_2 = 1(7.2) + 1(6.5) + (-1)(6.0) + (-1)(6.2)$$

$$+ 0(5.4) + 0(5.3)$$

$$= 1.5$$

$$\hat{\psi}_3 = 1(7.2) + 1(6.5) + 1(6.0) + 1(6.2)$$

$$+ (-2)(5.4) + (-2)(5.3)$$

$$= 4.5$$

The values of the sums of squares for the three planned comparisons are calculated using Equation (9.11).

$$SS(\hat{\psi}_1) = \frac{(\hat{\psi}_1)^2}{\sum(c_j^2/N_j)} = \frac{(.5)^2}{4/10} = .625$$

$$SS(\hat{\psi}_2) = \frac{(1.5)^2}{4/10} = \quad 5.625$$

$$SS(\hat{\psi}_3) = \frac{(4.5)^2}{12/10} = \quad 16.875$$

The value of sum of squares remainder is

$$SS\ remainder = SS\ between - SS(\hat{\psi}_1) - SS(\hat{\psi}_2) - SS(\hat{\psi}_3)$$

$$= 25.2 - .625 - 5.625 - 16.875$$

$$= 2.075$$

The only other value needed is that for the sum of squares error, which is given to be 73.980. The statistical tests of the planned comparisons and the remainder are arranged in a source table.

Source	Sum of Squares	Degrees of Freedom	Mean Squares	F
(Between	25.200	5)		
$\hat{\psi}_1$	.625	1	.625	.46
$\hat{\psi}_2$	5.625	1	5.625	4.11
$\hat{\psi}_3$	16.875	1	16.875	12.32
Remainder	2.075	2	1.038	.76
Error	73.980	54	1.370	
Total	99.180	59		

When compared to an F distribution with 1 and 54 degrees of freedom, the values of F for $\hat{\psi}_2$ and $\hat{\psi}_3$ are sufficiently large to reject the null hypothesis that the population comparisons are zero, indicating that the training of the instructor (trained teacher versus paraprofessional) as well as the instructional methods (vis-à-vis the control groups) contribute to reading achievement. The F values for $\hat{\psi}_1$ and the remainder, being less than 1, clearly do not lead to rejection of the null hypotheses. It does not appear that Method A is superior to Method B. The nonsignificant F for the remainder indicates that there does not seem to be any additional comparisons that are meaningful and post hoc tests would not be conducted on the remainder.

This factorial design can be laid as a one-way design:

A_1B_1	A_1B_2	A_2B_1	A_2B_2
μ_{11}	μ_{12}	μ_{21}	μ_{22}

So arranged, it is clear that there are 3 degrees of freedom that can be expressed as planned comparisons. The following set of mutually orthogonal comparisons is chosen:

	μ_{11}	μ_{12}	μ_{21}	μ_{22}
$\hat{\psi}_1$	1	1	-1	-1
$\hat{\psi}_2$	1	-1	1	-1
$\hat{\psi}_3$	1	-1	-1	1

This particular set of comparisons was chosen because $\hat{\psi}_1$ represents the main effect for factor A (i.e., row 1 versus row 2), $\hat{\psi}_2$ represents the main effect for factor B (column 1 versus column 2), and $\hat{\psi}_3$ represents the main effect for the interaction factor (μ_{11} and μ_{22} versus μ_{12} and μ_{21}). A test of these planned comparisons will yield identical results to the tests of the main effects and the interaction effect in the analysis of variance. Further, notice that the coefficients of $\hat{\psi}_3$ are the products of the respective coefficients of $\hat{\psi}_1$ and $\hat{\psi}_2$. This property will be discussed further in Section 12.7 (p. 398).

In general, for a two-way factorial design there are $(J - 1)$ mutually independent comparisons that comprise the main effect for rows, $(K - 1)$ mutually independent comparisons that comprise the main effect for columns, and $(J - 1)(K - 1)$ mutually independent comparisons that comprise the interaction effects. Furthermore, the comparisons for rows are independent of the comparisons for columns and interactions, the comparisons for columns are independent of the comparisons for rows and interactions, and so forth. Not surprisingly, any $(J - 1)$ independent planned comparisons that relate to row effects, when taken together (i.e., the sum of squares for the comparisons is summed), result in an equivalent test to the omnibus test for row effects.

Analysis of Post Hoc Comparisons

To this point we have discussed planned comparisons, which are related to specific questions that the researcher has about various means and that are conducted in lieu of the omnibus F test. However, often the researcher has no specific questions about means and thus chooses to conduct the omnibus test. When an omnibus null hypothesis that all the group means are equal is rejected, the naturally curious investigator will want to know which particular comparisons led to the statistically significant omnibus F. At this point in the research, *post hoc comparisons* are used to detect differences between or among means.

A post hoc comparison is defined exactly in the same way as a planned comparison.

$$\psi = c_1\mu_1 + c_2\mu_2 + \cdots + c_J\mu_J = \sum_j c_j\mu_j \tag{9.16}$$

The difference between a planned comparison and a post hoc comparison is the manner in which they are used. Planned comparisons are formulated a priori and are tested in lieu of an omnibus test. Post hoc comparisons are formulated and tested only after a statistically significant omnibus test is obtained. The price to be paid for not formulating the comparison a priori is that the post hoc test of the comparison is **more conservative** than the test for the planned comparison. However, because of the conservative nature of post hoc tests, **there is no requirement that post hoc comparisons be independent.**

9.4   Post Hoc Comparisons Using the Scheffé Method

Many different tests of post hoc comparisons have been developed; each has advantages and disadvantages. The method discussed in this section is the Scheffé method. The advantage of the Scheffé method is that it is simple, accommodates unequal sample sizes, is applicable to any comparison (as opposed to methods that are applicable only to pairwise comparisons, that is, comparison of two means), and is robust with regard to normality and homogeneity of variance. Nevertheless, it is conservative vis-à-vis other methods for post hoc tests. Once the general method of post hoc comparisons is learned, it is relatively easy to use any of the methods.

According to the Scheffé method, the value of the F ratio

$$\frac{SS(\hat{\psi})/(J-1)}{MSE} = F \tag{9.17}$$

is calculated; under the assumption of normality and homogeneity of variance, this F ratio is distributed as an F distribution with $(J-1)$ and $(N-J)$ degrees of freedom.

To illustrate the use of the Scheffé method, we return to Example 9.2, (p. 275). Suppose that an omnibus test was conducted on the data yielding the following source table:

Source	Sum of Squares	Degrees of Freedom	Mean Squares	F
Between	25.200	5	5.04	3.68
Error	73.980	54	1.37	
Total	99.180	59		

When compared to an F distribution with 5 and 54 degrees for freedom, the value of F is sufficiently large to reject the omnibus null hypothesis. Because the omnibus null hypothesis was rejected, it is permissible to test any or all of the possible comparisons post hoc. For illustration, only the comparison related to training of the instructor (ψ_2 in Example 9.2) is tested. The value of the F ratio is

$$\frac{SS(\hat{\psi})/(J-1)}{MSE} = \frac{5.625/5}{1.370} = .82$$

which is compared to an F distribution with 5 and 54 degrees of freedom, and clearly is insufficient to reject the null hypothesis for this comparison. Recall that this comparison led to rejection of the null hypothesis when planned comparisons were used, Illustrating the conservative nature of the post hoc method.

The difference between the methods for planned comparisons and post hoc comparisons can be accentuated by examining the confidence intervals for the two methods. Previously, the confidence intervals for planned comparisons had been written in terms of values of the t distribution [see Equation (9.10), p. 268]:

$$P\left[\hat{\psi} - t_{\alpha/2;N-J} \sqrt{\text{MSE} \sum_j (c_j^2/N_j)} \leq \psi \leq \hat{\psi} + t_{\alpha/2;N-J} \sqrt{\text{MSE} \sum_j (c_j^2/N_j)} \right]$$

$$= 1 - \alpha$$

In terms of values of the F distribution (using the $\pm$ notation), the confidence interval for the planned comparisons is

$$\hat{\psi} \pm \sqrt{F_{\alpha;1,N-J}} \sqrt{\text{MSE} \sum_j (c_j^2/N_j)} \tag{9.18}$$

The corresponding Scheffé confidence interval is

$$\hat{\psi} \pm \sqrt{(J - 1)F_{\alpha;J-1,N-J}} \sqrt{\text{MSE} \sum_j (c_j^2/N_j)} \tag{9.19}$$

As expected, the confidence interval for the planned comparisons is shorter than the confidence interval for the post hoc comparison. However, there is another important distinction between the confidence intervals. The confidence interval for the planned comparison gives the probability that the confidence interval contains the true value of ψ. For 95% confidence intervals it is expected that of 100 samples, 95 will yield confidence intervals constructed by Equation (9.18) which contain the true value of ψ for that particular comparison. The interpretation of the confidence intervals for post hoc comparisons is different. The confidence intervals for post hoc comparisons are true simultaneously for all possible comparisons. Suppose that the post hoc confidence intervals for all possible comparisons are computed for 100 samples; it is expected that 95 samples will produce post hoc confidence intervals that contain every one of the true values of the comparisons.

It is worth reiterating the relative advantages of planned versus post hoc comparisons. When a few specific questions are vital to the investigation, planned comparisons provide a powerful means to test these questions. The researcher, however, places emphasis on these questions to the exclusion of others. Post hoc comparisons allow an investigation of any question involving comparisons of means after the researcher established that there are indeed differences among means (i.e., a statistically significant omnibus test). However, the post hoc tests are conservative. It is possible that had the researcher opted for planned comparisons, the specific comparisons that were chosen would not have been statistically significant whereas comparisons not chosen might have been. Had the post hoc comparison approach been adopted, though, it is also possible that a comparison would fail to be statistically significant when if tested a priori it would have been statistically significant. It is also possible that a planned comparison would be statistically significant and the omnibus F test would not be statistically significant, precluding even the possibility of testing this comparison post hoc. However, it should be noted that if the omnibus F test is statistically significant at some level of α, then there is at least one comparison that when tested with the Scheffé method would yield a statistically significant result at the same level of α. Of course, the comparison may not be one that is substantively interesting.

9.5   Post Hoc Pairwise Comparisons Using the Newman–Keuls Method

A pairwise comparisons is used to contrast two means from the set of means obtained in the context of the analysis of variance. For example

$$\hat{\psi}_1 = 1\mu_1 - 1\mu_2$$

is a pairwise comparison. There are a variety of methods used to test pairwise comparisons post hoc that are much less conservative than the Scheffé method (see Winer, 1971, for a discussion and comparison of these methods). The basic pairwise methods available to the researcher are quite similar so that if one method is understood, application of the other methods is straightforward. The method presented here is the Newman–Keuls method, which is appropriate when we wish to examine all pairwise comparisons.

The Newman–Keuls method is explained by examining all of the pairwise comparisons from the six groups discussed in Example (9.2)(p. 275). First, the groups are arranged in ascending order based on the size of the mean.

	Group VI	Group V	Group III	Group IV	Group II	Group I
Means	5.3	5.4	6.0	6.2	6.5	7.2
Order	1	2	3	4	5	6

To test the difference between any two means, we use the statistic

$$q = \frac{\text{larger mean} - \text{smaller mean}}{\sqrt{\text{MSE}/n}}$$

The distribution of q is called the *studentized range distribution* and depends on two parameters. The first parameter, denoted by r, is the number of means between the two means being compared plus 2. For a comparison of Group V with Group IV, $r = 3$. The second parameter is the degrees of freedom associated with MSE. The upper .05 and .01 points of the studentized range distribution are found in Table B.5 (p. 456).

To facilitate testing all pairwise comparisons, we find that the critical value for the Newman–Keuls method for the difference between the larger and the smaller mean is:

$$q_\alpha \sqrt{\frac{MSE}{n}} \qquad (9.20)$$

where q_α is the value of the studentized range distribution that cuts off an area of α. With one important consideration to be discussed shortly, if the difference between the larger and smaller mean exceeds $q_\alpha \sqrt{MSE/n}$, then the null hypothesis that the two population means are equal is rejected. To discuss this important consideration and for ease in calculations, we present all the pairwise comparisons for Example 9.2 (p. 275) tabularly, as illustrated in Table 9.3. Each entry in the body of the table is the difference between the mean of the group represented by the column and the mean of the group represented by the row. The critical values ($\alpha = .05$) given by Equation (9.20) are presented in the far right-hand column. The broken lines connect comparisons with a common value of r.

TABLE 9.3

Newman–Keuls Pairwise Post Hoc Tests for Example 9.2

	VI	V	III	IV	II	I	r	$q(r, 54) \sqrt{\dfrac{MS\ error}{n}}$
	5.3	5.4	6.0	6.2	6.5	7.2		
VI		.1	.7	.9	1.2	1.9* –	6	$1.55 = 4.20 \sqrt{\dfrac{1.37}{10}}$
V			.6	.8	1.1	1.8* –	5	$1.48 = 4.01 \sqrt{\dfrac{1.37}{10}}$
III				.2	.5	1.2 –	4	$1.39 = 3.76 \sqrt{\dfrac{1.37}{10}}$
IV					.3	1.0 –	3	$1.27 = 3.42 \sqrt{\dfrac{1.37}{10}}$
II						.7 –	2	$1.05 = 2.84 \sqrt{\dfrac{1.37}{10}}$
I								

*Comparisons significant at $\alpha = .05$.

Decisions to reject the null hypothesis of no difference for each comparison are made in the following way. First, the comparison for the largest r is made (in this example, for $r = 6$). If the difference between means for this comparison does not exceed the critical value, the null hypothesis is not rejected and no further tests are conducted because it would be illogical to state that other pairwise comparisons would be significant when the comparison for the greatest difference was not (although with the Newman–Keuls method, some of the other differences may exceed the critical values). If the first comparison leads to rejection of the null hypothesis (i.e., the difference exceeds the critical value, as is the case in this example because the difference 1.9 exceeds the critical value 1.55), then the comparisons for the next lowest value of r are tested (in this example, $r = 5$). As before, restrictions need to be placed on subsequent comparisons to avoid the problem that a comparison with a smaller difference may lead to rejection when larger differences do not. This is done by eliminating for consideration all differences in the table that are in the triangular region formed by the column and row of an entry that did not lead to rejection. In this example, at $r = 5$, the difference between the means for Group VI and Group II is equal to 1.2, which does not exceed the critical value of 1.48; therefore, all entries to the left and below 1.2 are not tested. This process is continued until all comparisons are tested or are eliminated from consideration. As illustrated in Table 9.3, two comparisons lead to rejection in this example.

An informative way to summarize the results of a Newman–Keuls analysis is to list the groups and underline the groups so that groups that do not differ share a common underline, whereas groups that do differ do not have a common underline. Thus, for our example

$$\underline{\text{VI} \quad \text{V} \quad \text{III} \quad \text{IV} \quad \text{II}} \quad \text{I}$$

To this point the Newman–Keuls method has been presented with equal sample sizes. When sample sizes are unequal, a good approximation is made by letting n equal $N/\sum_i(1/x_i)$.

NOTES AND SUPPLEMENTARY READINGS

Winer, B. J. (1971). *Statistical principles in experimental design* (2nd ed.). New York: McGraw-Hill.

This text was mentioned earlier, but what is worth noting here is that it contains a thorough and understandable discussion of comparisons.

PROBLEMS

1. To determine the effectiveness of various methods to teach the oxidation-reduction method of balancing chemical equations, an educational researcher randomly assigned 80 students to groups (20 in each group). The methods for the four groups were traditional lecture, traditional lecture with visual aids (e.g., filmstrips), programmed instruction presented with text materials, and computer-assisted programmed instruction, respectively. The outcome measure was the number of correct responses for a 20-item test, which was given at the end of the instructional period. The mean number of correct responses for the four groups was as follows:

Lecture	Lecture + Visual Aids	Programmed Instruction with Text Materials	Computer-Assisted Programmed Instruction
6.0	8.0	10.0	12.0

Further

$$SS\ error\ =\ 760$$

The researcher wanted to test three comparisons: (a) lecture versus lecture plus visual aids, (b) programmed instruction with text materials versus computer-assisted programmed instruction, and (c) the lecture methods versus the programmed instruction methods.
   a. Show that the three comparisons are orthogonal.
   b. Test the three comparisons ($\alpha = .05$).
   c. Construct the 95% confidence intervals for the three comparisons.

2. A researcher interested in the obsessionality of women with eating disorders administered a scale that measured obsessionality ($10 = $ low obsessionality, $30 = $ high obsessionality) of college women undergraduates. Specifically, 15 anorexics, 15 bulimic/anorexics, 15 normal weight bulimics, 15 noneating disordered (clinical controls), and 15 nondisordered (normal controls) women completed the scale. The mean obsessionality scores were as follows:

Anorexics	Bulimic/ Anorexics	Normal Weight Bulimics	Clinical Controls	Normal Controls
28	26	23	24	19

Further

$$SS\ error\ =\ 1400$$

The researcher wanted to test three comparisons: (a) eating disordered women versus control groups, (b) anorexics and bulimic/anorexics versus normal weight bulimics, and (c) clinical controls versus normal controls.

    a. Show that the three comparisons are orthogonal.

    b. Test the three comparisons ($\alpha = .05$).

    c. Construct the 95% confidence intervals for the three comparisons.

3. Because the omnibus test for an analysis of variance for the data in Problem 1 was sufficiently large to reject the null hypothesis ($\alpha = .05$), it is permissible to conduct post hoc tests.

    a. Calculate the Scheffé tests for the three comparisons described in Problem 1 ($\alpha = .05$).

    b. Construct the 95% confidence intervals for the three comparisons in the context of Scheffé tests.

    c. Conduct all pairwise comparisons post hoc by using the Neuman–Keuls method ($\alpha = .05$).

4. Because the omnibus test for an analysis of variance for the data in Problem 2 was sufficiently large to reject the null hypothesis ($\alpha = .05$), it is permissible to conduct post hoc tests.

    a. Calculate the Scheffé tests for the three comparisons described in Problem 2 ($\alpha = .05$).

    b. Construct the 95% confidence intervals for the three comparisons in the context of Scheffé tests.

    c. Conduct all pairwise comparisons post hoc by using the Neuman–Keuls method ($\alpha = .05$).

10 BIVARIATE RELATIONS— CORRELATION AND ASSOCIATION

U p to this point statistical methods have been presented that primarily are applicable to experimental designs. Typically, in these designs the researcher randomly assigns subjects to various levels of the independent variable(s) and the goal is to identify the way in which manipulation of the independent variable affects the dependent variable. However, there is another frequently used research paradigm that investigates the relation of two variables, neither of which was "controlled" by the researcher. In this paradigm the researcher's goal is to determine the nature of the bivariate relation between the two variables. Although typically the two variables are referred to as X and Y, they have equal status (i.e., there is no independent or dependent variable). For example, we may want to know whether intelligence and achievement are related. It is not difficult to imagine some study that measures both the intelligence and achievement of subjects—in this instance neither variable would be manipulated by the researcher.

As different as experimental and relational paradigms are, the underlying inferential statistical strategy remains the same. The two random variables X and Y have some true but unknown relation determined by their bivariate distribution. That is to say, there is a population parameter that indexes the relation between X and Y; the index measures the degree to which the two variables are dependent. Inferences about the population index are made by examining a random sample from the joint distribution of X and Y. The typical method is to select N subjects randomly and to make observations on both X and Y, yielding a set of N bivariate observations if possible $\{(x_1, y_1), (x_2, y_2), \ldots, (x_N, y_N)\}$. A sample value of the index of relation is calculated and the value of this sample statistic is used to test a null hypothesis about the population index. The null hypothesis typically is that X and Y are independent.

Different indices are used depending on the nature of the variables. If the variables are measured on interval or ratio scales, the Pearson correlation coefficient is used. If the variables are measured on the ordinal scale, the Spearman rank correlation coefficient is used. Because the Pearson coefficient is ubiquitous, we will refer to it as the correlation coefficient; when reference is made to another correlation coefficient (e.g., the intraclass or the Spearman coefficients), the appropriate adjective will be used to differentiate it from the Pearson coefficient. Finally, if the variables are measured on the categorical scale, the phi coefficient or Cramer's statistic is used. In this chapter these coefficients as well as the corresponding tests of independence are discussed.

Correlation Coefficient

10.1   Population Correlation Coefficient

It should be remembered (and probably reviewed) that the covariance of X and Y is a measure of the degree to which the random variables X and Y are linearly related (see Section 3.13, p. 65). Remember that

$$\mathrm{Cov}(X, Y) = E(X - \mu_X)(Y - \mu_Y) \tag{10.1}$$

$\mathrm{Cov}(X, Y)$ is a good measure of dependence in that if X and Y are independent, the covariance of X and Y is zero and if the covariance of X and Y is not equal to zero, X and Y are dependent.

There are two disadvantages of the covariance as a measure of dependence. The first problem is that covariance only indexes one type of dependence. Recall from Section 3.13 (p. 65) that $\mathrm{Cov}(X, Y) = 0$ does not imply that X and Y are independent. This is an issue that will be discussed later in this section and in the following sections.

The second problem is that the size of $\mathrm{Cov}(X, Y)$ is dependent on the scale of the random variables. That is, a linear transformation of either of the random variables will change the size of the covariance although the degree to which the variables are dependent is left unchanged. This is not a problem because the covariance is easily standardized by dividing by the product of the standard deviations of X and Y. The coefficient so formed is called the *Pearson product-moment correlation coefficient* (or simply the correlation coefficient) and is denoted by ρ_{XY} (or simply ρ when the context is clear). Formally, **the population correlation coefficient ρ_{XY} is defined as**

$$\rho_{XY} = \frac{\mathbf{Cov}(X, Y)}{\sigma_X \sigma_Y} \tag{10.2}$$

Several properties of the population correlation coefficient can be seen by algebraically manipulating Equation (10.2). By the definition of covariance

$$
\begin{aligned}
\rho_{XY} &= \frac{\mathrm{Cov}(X, Y)}{\sigma_X \sigma_Y} \\[2mm]
&= \frac{E(X - \mu_X)(Y - \mu_Y)}{\sigma_X \sigma_Y} \\[2mm]
&= E\left(\frac{X - \mu_X}{\sigma_X}\right)\left(\frac{Y - \mu_Y}{\sigma_Y}\right) \\[2mm]
&= E(Z_X Z_Y)
\end{aligned}
\tag{10.3}
$$

where Z_X and Z_Y are the standardized random variables associated with X and Y. In other words, ρ_{XY} is the expected value of the product of the standardized random variables. Furthermore, because the mean of each standardized random variable is zero

$$\rho_{XY} = E[(Z_X - \mu_{Z_X})(Z_Y - \mu_{Z_Y})] = \mathrm{Cov}(Z_X, Z_Y) \tag{10.4}$$

Hence, the population correlation coefficient is the covariance of the standardized random variables (and explains why the correlation coefficient was described as the standardized covariance).

Another property of the population correlation coefficient is that it ranges from -1.00 to $+1.00$; that is

$$-1.00 \leq \rho_{XY} \leq +1.00 \tag{10.5}$$

From the definition of the population correlation coefficient it is clear that when X and Y are independent, ρ_{XY} is equal to zero. However, as was discussed in some detail when the topic of covariance was explained (see Section 3.13, p. 65), the converse is not true; that is, $\rho_{XY} = 0$ does not imply that X and Y are independent. Provided a special condition is true, however, the converse will also be true: **If X and Y are bivariate normal, then X and Y are independent** *if and only if* $\rho_{XY} = 0$. In fact the parameter ρ_{XY} appearing in the probability density function of the bivariate normal distribution [see Equation (3.48), p. 75] is the correlation coefficient. In other words, the bivariate normal distribution of X and Y has parameters related to X (μ_X and σ_X^2), related to Y (μ_Y and σ_Y^2), and related to the correlation between X and Y (ρ_{XY}). When X and Y are bivariate normal, the only relation between X and Y that is possible is the one indexed by the correlation coefficient. As will become clearer in the next sections, the correlation coefficient indexes the linear relation between X and Y and is not sensitive to other types of relations. The role of bivariate normality in the context of correlation will be discussed more fully in the following sections.

Because the correlation coefficient is standardized, its size is not affected by linear transformations of the random variables X and Y. That is, the magnitude of the correlation of X and Y is equal to the magnitude of the correlation of $BX + A$ and $DY + C$, where A, B, C, and D are constants (the signs of the correlations will be different if B is negative and D is positive, or vice versa).

10.2 Sample Correlation Coefficient

A random sample of size N from a joint distribution of X and Y yields a set of N pairs of observations: $\{(x_1, y_2), (x_2, y_2), \ldots, (x_N, y_N)\}$. Typically, this occurs when there are two observations for each of N subjects. For instance, the x_i's might refer to intelligence test scores and the y_i's might refer to achievement scores. The *sample Pearson product-moment correlation coefficient* (or simply the sample correlation coefficient), which is referred to as r_{xy} (or simply r when the context is clear) is defined analogously to the population correlation coefficient:

$$r_{xy} = \frac{\textbf{sample covariance}}{S_x S_y} = \frac{S_{xy}}{S_x S_y} \tag{10.6}$$

where S_{xy} was defined as follows (see Section 3.13, p. 65)

$$S_{xy} = \frac{1}{N} \sum_i (x_i - M_x)(y_i - M_y)$$

The remainder of this section discusses several aspects of the sample correlation coefficient and the following section will discuss the relation between the sample correlation coefficient r and the population correlation coefficient ρ.

Given the definition of the sample covariance, the definition of r can be modified to yield an informative variation:

$$r = \frac{(1/N)\sum_i (x_i - M_x)(y_i - M_y)}{S_x S_y} = \frac{1}{N}\sum_i \left(\frac{x_i - M_x}{S_x}\right)\left(\frac{y_i - M_y}{S_y}\right)$$

Using the definition of standardized scores, we obtain

$$r_{xy} = \frac{\sum_i z_{x_i} z_{y_i}}{N} \tag{10.7}$$

TABLE 10.1

Calculation of the Sample Correlation Coefficient

Observation	x	z_x	y	z_y	$z_x z_y$
1	5	.000	9	−1.316	0
2	3	−.707	12	.000	0
3	7	.707	15	1.316	.930
4	9	1.414	14	.877	1.240
5	1	−1.414	10	−.877	1.240

$$M_x = 5 \qquad\qquad M_y = 12 \qquad\qquad \Sigma z_{x_i} z_{y_i} = 3.41$$
$$S_x = 2.83 \qquad\qquad S_y = 2.28$$

Definitional formula:

$$S_{xy} = \frac{1}{N}\sum (x_i - M_x)(y_i - M_y) = \frac{1}{5}[(0)(-3) + (-2)(0) + (2)(3) + (4)(2)$$
$$+ (-4)(-2)]$$
$$= 4.4$$

$$r_{xy} = \frac{S_{xy}}{S_x S_y} = \frac{4.4}{(2.83)(2.28)} = .682$$

z-score formula:

$$r_{xy} = \frac{1}{N}\sum z_{x_i} z_{y_i} = \frac{1}{5}(3.41) = .682$$

That is, the sample correlation coefficient is the mean of the product of the standardized scores [and is analogous to Equation (10.3) for the population correlation coefficient]. Equation (10.7) often is referred to as the "z-score" formula for the sample correlation coefficient. There are several other variations of the formula for r including "computational" and "raw score" formulas, but given the availability of modern computers and calculators, these formulas are infrequently used.

Table 10.1 illustrates the calculation of the sample correlation coefficient using the definitional and the z-score formulas.

As was the case for the population correlation coefficient, r ranges from −1.00 to +1.00; that is

$$-1.00 \leq r_{xy} \leq +1.00 \qquad\qquad (10.8)$$

In the case of perfect correlation $z_{x_i} = z_{y_i} = z_i$ and

$$r = \frac{\sum_i z_{x_i} z_{y_i}}{N} = \frac{\sum_i z_i^2}{N} = 1$$

Similarly, when $z_{x_i} = -z_{y_i}$, $r = -1.00$.

To this point the presentation of the sample correlation coefficient has been algebraic. However, graphical presentation of various relations is illustrative of several characteristics of the correlation coefficient. Some of these relations, which are best understood in terms of standardized scores, are presented in Figure 10.1, which is a bivariate "scatter plot" of the scores. When the standardized scores for x and the standardized scores for y are unrelated, the sample correlation is zero [see panel (a)]. When positive standardized scores for x are associated with positive standardized scores for y (and vice versa), the sample correlation is positive; when the sample correlation coefficient is equal to 1, the points fall along a straight line that goes from lower left to upper right at a 45° angle [see panel (b)]. When positive standardized scores for x are associated with negative standardized scores for y (and vice versa), the sample correlation is negative; when the sample correlation coefficient is equal to -1.00, the points fall along a straight line that goes from upper left to lower right at a 45° angle [see panel (c)]. However, it should be noted that some definite relations produce a correlation of zero [see panel (d)]. From these graphs one fact should now be clear: **The correlation coefficient (as well as the covariance) indexes the degree of *linear* relation between a set of scores.** Methods to detect a curvilinear relation, such as that depicted in panel (d), will be discussed in the following chapter.

10.3   Inferences About ρ_{XY} Based on r_{xy}

Although r is often used as a descriptive statistic, the most potent use of r is to make inferences about ρ, an index of the linear relation between X and Y in the population. The most basic inference is related to the size of ρ; typically, the null hypothesis tested is that the population correlation coefficient is zero, although any value of ρ in the interval -1.00 to $+1.00$ may be hypothesized, as we will see. A method to form confidence intervals for ρ will also be discussed.

For all the procedures discussed in this section the bivariate normality of X and Y is assumed. Recall when X and Y have a bivariate normal distribution, X and Y are independent if and only if $\rho_{XY} = 0$ (see Section 10.1, p. 287). Under this condition the hypothesis that $\rho_{XY} = 0$ is equivalent to the hypothesis that X and Y are independent. Another implication of the bivariate normality assumption is that the only relation between X and Y that is possible is a linear relation. If one assumes bivariate normality, it makes no sense to search for nonlinear relations.

Implications of departures from the bivariate normal assumption need to be discussed. First, when the sample correlation coefficient r is used simply to **describe** the relation between scores for a set of observations, no special conditions need to be placed on distributions or the measurement scales used to obtain the observations. In this instance no inferences are made about the

(a)

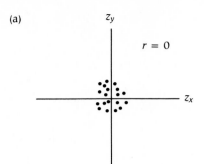

(b)

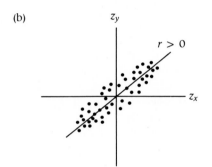
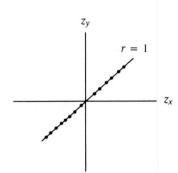

(c)

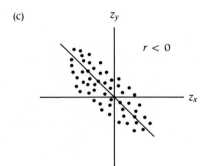
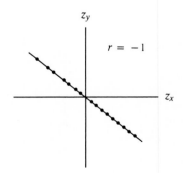

(d)

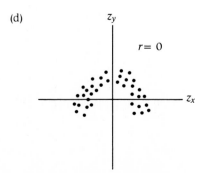

FIGURE 10.1   Different Patterns for Sample Correlations

population correlation coefficient, but it is perfectly acceptable and often useful to say that for a set of scores, generally, the scores are associated in a certain way (e.g., high scores for one variable tend to be associated with high scores for the other variable and vice versa). Second, when X and Y are not bivariate normally distributed, the range of ρ_{XY} may very well be restricted. For example, for a particular bivariate distribution the maximum value of ρ that is possible may be .40. Many researchers have been frustrated trying to obtain large sample values of the correlation coefficient when the population correlation coefficient is restricted to values that are considerably less than 1 because of the nature of the population distributions. A third and related point is that when one of the random variables is restricted in some particular manner, specialized correlation coefficients exist. For example, if one random variable is dichotomous (i.e., restricted to two values) and the other is continuous, the point-biserial correlation coefficient is appropriate. Later in this chapter two of these other coefficients will be discussed, the rank order correlation coefficient for ordinal data and the phi coefficient for categorical data. Fourth, it should be noted that to make appropriate inferences not only must the variables X and Y have a bivariate normal distribution, but the observations must be randomly selected from the population. For example, assume that intelligence and achievement have a bivariate normal distribution; however, if the relation between intelligence and achievement is studied for an elite group of subjects, this bias affects the correlation coefficient. Generally, restriction of the range of either or both of the variables attenuates the size of the correlation coefficient, as illustrated

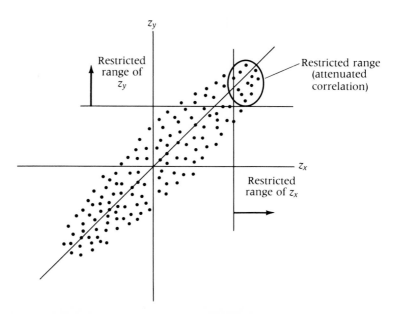

FIGURE 10.2   Example of Attenuated Correlation Resulting from Restricted Range

in Figure 10.2. Finally, it should be emphasized that violation of the bivariate assumption has implications for decisions to reject the null hypothesis. Again, the degree to which the decisions are affected depends on the degree to which the assumption is violated. For correlation problems the larger the sample size is, the smaller will be the impact of violation of the bivariate normal assumption on decisions to reject the null hypothesis.

The first null hypothesis to be tested is that the population correlation coefficient is equal to zero; that is

$$\rho_{XY} = 0$$

The alternative may be one-tailed

$$\rho_{XY} < 0 \qquad \text{or} \qquad \rho_{XY} > 0$$

or two-tailed

$$\rho_{XY} \neq 0$$

A test statistic for this null hypothesis is

$$t = \frac{r_{xy}\sqrt{N - 2}}{\sqrt{1 - r_{xy}^2}} \tag{10.9}$$

which under the bivariate normal assumption has a _t_ distribution with $(N - 2)$ degrees of freedom. Although no basis for this formula is given here, it takes on a general form of many of the tests of correlation and regression coefficients, which will be discussed in the next chapter. To illustrate the use of this formula, suppose that 20 married couples are randomly selected to determine whether intelligence of husbands is linearly related to intelligence of wives (i.e., do intelligent people tend to marry intelligent spouses?). If the sample correlation coefficient is equal to .70, the value of the test statistic is given by

$$t = \frac{(.70)\sqrt{20 - 2}}{\sqrt{1 - (.70)^2}} = 4.16$$

which, when compared to a _t_ distribution with 18 degrees of freedom, leads to the rejection of the null hypothesis at conventional levels of α for either a one- or two-tailed test (although in this case a one-tailed test seems appropriate).

It should be noted that typically the sample correlation coefficient is used as an estimator for the population correlation coefficient ρ. Nevertheless, _r_ is a biased estimator of ρ, although the degree of bias is small, especially for samples of the size usually used in correlation studies. Because of this, the correction factor, which is complicated, is used seldomly.

There are instances for which it is useful to specify in the null hypothesis a value other than zero. For instance, suppose that a large number of studies indicate that the correlation between two variables is about .40. A researcher might want to demonstrate that for the population from which he or she sampled the correlation between these two variables is greater than .40. Consider the null hypothesis

$$\rho_{XY} = c$$

where $-1.00 \leq c \leq +1.00$. When ρ_{XY} is unequal to zero, the sampling distribution of r_{xy} is skewed. This is reasonable if we consider a population correlation coefficient ρ_{XY}, say, of .80; the sampling distribution of r_{xy} will have a long tail to the left (i.e., a negatively skewed distribution). There is a particular transformation that reduces the skewness of the sampling distribution (see Section 6.9, p. 152, for a discussion of transformations). **This transformation, which is called the Fisher's r to Z transformation, is given by**

$$Z = \frac{1}{2}\log_e\left(\frac{1 + r_{xy}}{1 - r_{xy}}\right) \tag{10.10}$$

Although this formula may appear foreboding, tables exist for making the transformation (see Table B.6, p. 458). For moderate sized samples the sampling distribution of Z is approximately normal. To form a test statistic the obtained value of Z (derived by transforming the obtained value of r) is standardized. (The notation Z is unfortunate because we have used it to indicate standardization, whereas here Z is unstandardized.) The expectation and variance of the sampling distribution are approximately

$$E(Z) = \frac{1}{2}\log_e\left(\frac{1 + \rho_{XY}}{1 - \rho_{XY}}\right) \tag{10.11}$$

and

$$V(Z) = \frac{1}{N - 3} \tag{10.12}$$

respectively, where N is the size of the sample. **Therefore, the test statistic for the hypothesis $\rho_{XY} = c$ is given by**

$$\frac{Z - E(Z)}{\sqrt{V(Z)}} \tag{10.13}$$

where Z is the transformed value of the obtained sample correlation coefficient, $V(Z)$ is given by Equation (10.12) and $E(Z)$ is the expectation of Z when the null hypothesis is true; that is

$$E(Z) = \frac{1}{2}\log_e\left(\frac{1 + c}{1 - c}\right) \tag{10.14}$$

The test statistic is compared to a standard normal distribution.

Suppose that the researcher mentioned at the beginning of the previous paragraph obtains a correlation of .55 from a random sample of 50 drawn from a bivariate normal population. To test the null hypothesis that $\rho_{XY} = .40$, we need to find Z, $E(Z)$, and $V(Z)$. The value of Z is found by transforming the obtained correlation coefficient:

$$r_{xy} = .55 \qquad \text{corresponds to} \qquad Z = .6184$$

$E(Z)$ is found by transforming the hypothesized value of .40:

$$\rho_{XY} = .40 \qquad \text{corresponds to} \qquad E(Z) = .4236$$

Finally

$$V(Z) = \frac{1}{50 - 3} = .0213$$

Therefore, the value of the test statistic is given by

$$\frac{.6184 - .4236}{\sqrt{.0213}} = 1.33$$

which when compared to the standard normal distribution does not lead to rejection of the null hypothesis at conventional levels of α.

Although null hypotheses involving values other than zero for ρ_{XY} are rare, the Fisher's r to Z transformation provides a method to find confidence intervals for ρ_{XY}. Essentially, the method involves finding the confidence interval for $E(Z)$ and then using the inverse of the r to Z transformation (i.e., Z to r transformation) to find the confidence intervals for ρ_{XY}. Recall that when X and Y have a bivariate normal distribution, the sampling distribution of Z is approximately normal; consequently

$$P\left(Z - z_{\alpha/2} \sqrt{\frac{1}{N - 3}} \leq E(Z) \leq Z + z_{\alpha/2} \sqrt{\frac{1}{N - 3}} \right) = 1 - \alpha \qquad (10.15)$$

(It is necessary to keep track of the Z's: Z is the transformed value corresponding to the obtained sample correlation coefficient and $z_{\alpha/2}$ is the critical point for the standard normal distribution). The limits for the confidence interval for ρ_{XY} are found by performing the inverse r to Z transformation (i.e., Z to r) on the limits of the confidence interval for $E(Z)$. Given a sample correlation coefficient of .55 for a sample of 50, the 95% confidence interval for ρ_{XY} is found in the following way:

$$r_{xy} = .55 \qquad \text{corresponds to} \qquad Z = .6184$$

So

$$P(.6184 - 1.96\sqrt{1/47} \leq E(Z) \leq .6184 + 1.96\sqrt{1/47}) = .95$$

which simplifies to

$$P(.3325 \leq E(Z) \leq .9043) = .95$$

To find the corresponding confidence interval for ρ_{XY}, we transform the limits for the confidence interval for $E(Z)$ to correlation coefficients:

$$Z = .3325 \qquad \text{corresponds to} \qquad r_{xy} = .321$$

and

$$Z = .9043 \qquad \text{corresponds to} \qquad r_{xy} = .719$$

Therefore

$$P(.321 \leq \rho_{XY} \leq .719) = .95$$

Before leaving the topic of the correlation coefficient, we should mention that there exists a very convenient measure of the strength of association in correlation problems. However, because the sample value and an estimate of the population value of the strength of association can be discussed more appropriately in the context of regression, this topic is deferred to the next chapter.

Bivariate Relations for Ordinal and Categorical Data

10.4 Rank Order Correlation Coefficient

When the observations for the two variables whose relation is being investigated are expressed ordinally, the *Spearman rank order correlation coefficient*, which is denoted by r_s is appropriate. Suppose that N subjects are randomly selected from a population and each subject is "measured" on two variables. The two observations for each subject consists of the ranks for that subject in the sample. Specifically, ranks 1 to N are assigned to the subjects for one variable and similarly ranks 1 to N are assigned for the second variable. If the variables X and Y are positively related, then the ranks assigned to one subject will tend to be the same; that is, low ranks will be associated with low ranks and high ranks will be associated with high ranks. If X and Y are negatively related, the opposite will tend to be true; low ranks on one variable will be associated with high ranks on the other variable, and vice versa.

The rank correlation coefficient is simply the Pearson product-moment correlation coefficient calculated on the ranks. However, when there are no tied ranks, an alternate form of r_s is given by

$$r_s = 1 - \frac{6 \sum_i d_i^2}{N(N^2 - 1)} \qquad (10.16)$$

where d_i is the difference in the ranks for subject i. As was the case for r_{xy}

$$-1.00 \leq r_s \leq 1.00$$

For relatively large samples ($N > 10$) the null hypothesis that the population correlation of X and Y, which are represented by rankings, is zero is tested with the following test statistic:

$$t = \frac{r_s\sqrt{N-2}}{\sqrt{1-r_s^2}} \tag{10.17}$$

The value of the statistic t is compared to a t distribution with $(N-2)$ degrees of freedom. Clearly the test for r_s is similar to that for r_{xy} [see Equation (10.9), p. 294)].

As an example of the use of r_s, suppose that a teacher is asked to rank order ten students in a reading group based on his or her judgment of their intelligence and then these rankings are compared to the students' scores on a standard measure of intelligence. To calculate r_s, we also rank the intelligence scores, as shown in Table 10.2. Difference scores are obtained by subtracting the ranks for the intelligence test scores from the teacher ranks (the opposite difference yields the same value of r_s) and r_s is obtained:

$$r_s = 1 - \frac{6\sum_i d_i^2}{N(N^2-1)} = 1 - \frac{6(56)}{10(100-1)} = .66$$

To test the null hypothesis that the population correlation is zero, we use the value of the test statistic given in Equation (10.17)

$$t = \frac{r_s\sqrt{N-2}}{\sqrt{1-r_s^2}} = \frac{.66\sqrt{10-2}}{\sqrt{1-(.66)^2}} = 2.48$$

TABLE 10.2

Rank Order of Teacher Judgments and Intelligence Test Scores

Student	Teacher's Rank Ordering	Intelligence Test Score	Rank Order of Intelligence Scores	d_i	d_i^2
1	3	108	5	-2	4
2	10	98	8	2	4
3	4	115	2	2	4
4	2	103	7	-5	25
5	9	96	10	-1	1
6	1	132	1	0	0
7	7	97	9	-2	4
8	5	109	4	1	1
9	6	112	3	3	9
10	8	106	6	2	4

$$\sum d_i^2 = 56$$

which when compared to a t distribution with 8 degrees of freedom would lead to the decision to reject the null hypothesis provided a one-tailed test and an α level of .05 were stipulated before the study was conducted.

The relation between the rank order correlation coefficient and the product-moment correlation coefficient needs to be explored further. It was mentioned that r_s is simply the product-moment correlation coefficient calculated on the ranks—in the case of ranks for which there are no ties, an alternative formula is given by Equation (10.16). Regardless of how r_s is calculated, the question arises about whether or not r_s is a good estimator for the population correlation coefficient. Consider the case where X and Y are continuous random variables and N observations are sampled from the bivariate distribution of X and Y, yielding N pairs $(x_1, y_1), \ldots, (x_N, y_N)$. Now suppose that the observations on X are ranked from 1 to N and the observations on Y are ranked from 1 to N. We know that r_{xy} calculated on the initial observations is a good estimator for the population correlation ρ_{XY}. The question remains whether or not r_s is a good estimator for ρ_{XY}. When X and Y are bivariate normal, r_s can be used as an estimator for ρ_{XY}. However, under violations of the bivariate normality assumption r_s may not be an adequate estimator for the population correlation coefficient ρ_{XY}. Of course, under such circumstances we would want to exercise caution anyway. There are many instances in which the ranks are not based on an underlying continuous distribution, such as the case where judges rank order various objects (e.g., the teacher who ranked ordered the students based on his or her judgment of intelligence). In such instances it makes little sense to speak of a population correlation coefficient. However, r_s remains a useful descriptive statistic of the association between the ranks. The statistical test of r_s takes on a slightly different meaning, though. These issues will be discussed in more detail in Chapter 13, which examines methods of this kind in more depth.

10.5 The χ^2 Test of Association

It is often informative to determine whether or not two variables measured on categorical scales are independent. Typically, N subjects are randomly and independently selected and each subject is categorized in two ways. For example, subjects could be classified as to their political party and as to their religion. The question of independence can be phrased as "Are religion and political affiliation independent?"

Recall from Section 2.6 (p. 29) that an attribute is a set of mutually exclusive and exhaustive categories. In this section we discuss the independence of two attributes A and B, where attribute A contains the categories $A_1, A_2, \ldots, A_j, \ldots, A_R$ (R refers to rows) and where attribute B contains the categories $B_1, B_2, \ldots, B_k, \ldots, B_C$ (C refers to columns). The probability that a subject drawn at random from the population will be classified as A_j and B_k is

$$P(A_j \cap B_k)$$

From Section 2.6 (p. 29) we know that the two events A_J and B_k are independent provided

$$P(A_j \cap B_k) = P(A_j)P(B_k)$$

Thus, it follows that **the attributes A and B are independent if and only if**

$$P(A_j \cap B_k) = P(A_j)P(B_k)$$

for all j and k. Of course, the researcher is not able to assess independence by direct access to the probabilities and must make that judgment based on a sample from a population characterized by the joint probability distribution for the two attributes.

To illustrate the relation of the sample to the population joint distribution, we consider the case where each attribute has only two categories: A contains A_1 and A_2 and B contains B_1 and B_2. The joint distribution of the attributes can be described tabularly:

	B_1	B_2	
A_1	$P(A_1 \cap B_1)$	$P(A_1 \cap B_2)$	$P(A_1)$
A_2	$P(A_2 \cap B_1)$	$P(A_2 \cap B_2)$	$P(A_2)$
	$P(B_1)$	$P(B_2)$	

Each of the subjects selected from this joint distribution is classified as A_1 or A_2 and is classified as B_1 or B_2. Tests of the independence of A and B are based on the number of subjects who are classified in each of the possible combinations of the categories of A and B. Let f_{ojk} be the number of subjects who are classified as A_j and B_k (f refers to frequency and o refers to observed). Typically, the results of such a cross classification are displayed in a $R \times C$ *contingency table*—the 2×2 table would be

	B_1	B_2	
A_1	f_{o11}	f_{o12}	$f_{o1\cdot}$
A_2	f_{o21}	f_{o22}	$f_{o2\cdot}$
	$f_{o\cdot1}$	$f_{o\cdot2}$	N

The notation f_{o1}. refers to the marginal frequency for the first row and, in general, is equal to $\Sigma_k f_{o1k}$. The other marginal frequencies are calculated in a similar fashion.

Suppose that 100 women are randomly selected from a population and classified as either married or unmarried and classified as either of the opinion that children enhance the quality of life or of the opinion that children do not enhance the quality of life (subjects were induced to select one and only one opinion). Let

A_1 = married
A_2 = unmarried

and

B_1 = opinion that children enhance the quality of life
B_2 = opinion that children do not enhance the quality of life

Suppose that the following results were obtained:

	Children Enhance Life	Children Do Not Enhance Life	
Married	42	20	62
Unmarried	22	16	38
	64	36	100

These sample data are then used to make a judgment about the independence of marital status and opinion about children and the quality of life.

The null hypothesis in the context of attributes is

H_0: A and B are independent

or, more formally

H_0: $P(A_j \cap B_k) = P(A_j)P(B_k)$ for all j and k

The alternative hypothesis is that the attributes A and B are not independent, or, in other words, the attributes A and B are *associated*. Under the null hypothesis the expected probability for cell jk in the contingency table is $P(A_j)P(B_k)$, and therefore the expected number of subjects from a sample of size N classified as A_j and B_k is

$$(N)P(A_j)P(B_k) \tag{10.18}$$

Moreover, $P(A_j)$ and $P(B_k)$ are estimated from the relative frequencies obtained from the sample data.

$$\widehat{P(A_j)} = \frac{f_{oj\cdot}}{N} \tag{10.19}$$

$$\widehat{P(B_k)} = \frac{f_{o\cdot k}}{N} \tag{10.20}$$

Using these estimates and Expression (10.18), we obtain the expected number of subjects classified as A_j and B_k, denoted by f_{ejk}, which is given by

$$f_{ejk} = (N)\widehat{P(A_j)}\widehat{P(B_k)} = N\left(\frac{f_{oj\cdot}}{N}\right)\left(\frac{f_{o\cdot k}}{N}\right) = \frac{(f_{oj\cdot})(f_{o\cdot k})}{N} \tag{10.21}$$

That is, under independence the expected number of subjects classified as A_j and B_k is the product of the respective marginals divided by the sample size. For the marital example the expected number of married women who expressed the opinion that children enhance the quality of life is

$$f_{e11} = \frac{(62)(64)}{100} = 39.68$$

which is less than the observed frequency of 42. The remaining three expected frequencies (under the null hypothesis of independence) are given in the following table along with the observed frequencies:

	Children Enhance Life	Children Do Not Enhance Life	
Married	$f_{o11} = 42$ $f_{e11} = 39.68$	$f_{o12} = 20$ $f_{e12} = 22.32$	$f_{o1\cdot} = 62$
Unmarried	$f_{o21} = 22$ $f_{e21} = 24.32$	$f_{o22} = 16$ $f_{e22} = 13.68$	$f_{o2\cdot} = 38$
	$f_{o\cdot 1} = 64$	$f_{o\cdot 2} = 36$	$N = 100$

Clearly the expected frequencies are not equal to the observed frequencies nor would we expect them to be. The question to be asked is whether the observed frequencies are so out of line with what we would expect under independence that the hypothesis of independence should be rejected.

The test statistic for making decisions is based on the difference of the observed frequencies and the expected frequencies (i.e., $f_{ojk} - f_{ejk}$). However, to make positive contributions this difference is squared. Further, each squared difference is divided by the expected frequencies so that each cell makes a comparable contribution to the statistic. **The test statistic for the null hypothesis that A and B are independent is referred to as the Pearson chi-square statistic (χ^2) and is given by**

$$\chi^2 = \sum_k \sum_j \frac{(f_{ojk} - f_{ejk})^2}{f_{ejk}} \tag{10.22}$$

For relatively large sample sizes, the χ^2 statistic is approximately distributed as a χ^2 distribution with $(R - 1)(C - 1)$ degrees of freedom. Clearly only large values of the test statistic lead to rejection of the null hypothesis, which is tested with a one-tailed test. For the continuing example of opinions about children and the quality of life

$$\chi^2 = \frac{(42 - 39.68)^2}{39.68} + \frac{(22 - 24.32)^2}{24.32} + \frac{(20 - 22.32)^2}{22.32} + \frac{(16 - 13.68)^2}{13.68}$$

$$= .9916$$

which when compared to a χ^2 distribution with 1 degree of freedom is not sufficiently large to reject the null hypothesis that marital status and opinion about the quality of life and children are independent (a value of 3.8415 is needed to reject H_0 at an α level of .05).

Three cautionary notes about the use of the χ^2 test of association need to be mentioned. First, the observations must be independent of each other. That is, a pair of observations should not have a structural influence on any other pairs of observations, such as might be the case in repeated observations. Second, each observation must be classified in one and only one cell. As an example of a violation of this assumption, consider the instance when a subject classified on gender selects his or her two favorite colors from a set of 20 colors and a contingency table is formed where each subject contributes to two cells. Violation of either of these assumptions may have dramatic and deleterious effects on decisions about the null hypothesis. The third note is concerned with sample size. The degree to which the χ^2 statistic approximates the χ^2 distribution depends on the sample size. Although there is considerable disagreement about what constitutes sufficient sample size, the following conservative rule offers guidance: For 2×2 tables the expected value of each cell should be greater than or equal to 10 and for other tables the expected value of each cell should be greater than or equal to 5. However, because it is frequently easy to collect data for studies utilizing contingency tables (e.g., data based on questionnaires), such studies often include tables containing a very large number of subjects. Here there is the distinct possibility that the researcher may conclude that the attributes are associated when the degree to which they are associated is very small. The next section will discuss measures of the degree of association as well as illustrate the χ^2 test of association for larger tables.

10.6   Measures of Association for Categorical Data

To illustrate measures of association, consider the data, presented in Table 10.3, of a hypothetical study of the preference of three candidates by 1000 voters randomly selected from a county containing three communities. Because the value of the χ^2 statistic (16.2434) for this table is sufficiently large when compared to a χ^2 distribution with 6 degrees of freedom (α set at .05), the null hypothesis that the place of residence and preference for candidates are independent in the population is rejected in favor of the alternative that these two attributes are associated. However, because the sample size for this study is large, it might well be that the size of the association is quite small and, practically speaking, uninteresting to the researcher.

TABLE 10.3

Cross Classification of Community and Preference for Candidate (expected values in parentheses)

	Candidate A	Candidate B	Candidate C	No Preference	
Community 1	24 (32.39)	105 (115.26)	84 (78.03)	42 (29.32)	255
Community 2	40 (38.99)	132 (138.77)	101 (93.94)	34 (35.30)	307
Community 3	63 (55.62)	215 (197.97)	121 (134.03)	39 (50.38)	438
	127	452	306	115	$N = 1000$

An intuitively appealing measure of association simply is the χ^2 statistic divided by the sample size. This statistic, which is denoted by ϕ^2 (phi squared), is referred to as *Pearson's index of mean square contingency*:

$$\phi^2 = \frac{\chi^2}{N} \qquad (10.23)$$

Often, to obtain units that are not squared, the square root of this index is discussed:

$$\phi = \sqrt{\frac{\chi^2}{N}} \qquad (10.24)$$

For our example of the association of place of residence and preference for candidate

$$\phi^2 = \frac{16.2434}{1000} = .001624$$

and

$$\phi = \sqrt{\phi^2} = \sqrt{.001624} = .0403$$

Although the index of mean square contingency is denoted by a Greek letter, it should be emphasized that the index is a sample statistic.

The index of mean square contingency has a major disadvantage as a measure of association because its range is different from the other measures of dependence discussed in this chapter, such as r_{xy}. The lower limits of ϕ^2 and ϕ are zero. This is understandable given that we are considering unordered categories so that a negative association makes no sense. Unfortunately, the upper limit of ϕ^2 is not 1.00. The limits of ϕ^2 are

$$0 \leq \phi^2 \leq \text{minimum of } (R - 1) \text{ and } (C - 1)$$

where R and C are the number of rows and columns, respectively, in the contingency table. If we let

$$L = \text{minimum of } (R - 1) \text{ and } (C - 1) \tag{10.25}$$

ϕ can be adjusted to yield another statistic, which is referred to as *Cramer's statistic*, and is denoted by ϕ':

$$\phi' = \sqrt{\frac{\phi^2}{L}} = \sqrt{\frac{\chi^2}{NL}} \tag{10.26}$$

The statistic ϕ' now ranges from zero, indicating independence to 1, indicating perfect association. For the continuing example for candidate preference

$$\phi' = \sqrt{\frac{\chi^2}{NL}} = \sqrt{\frac{16.2434}{1000(2)}} = .09$$

The measures of association discussed in this section have special properties in a 2 × 2 contingency table. First, because in this instance $L = 1$, $\phi = \phi'$. For this reason ϕ is often used as a measure of association for 2 × 2 tables. Second, if we give particular numerical labels to the unordered categories for the two attributes, a relation between ϕ and r_{xy} exists. Particularly, define two variables in the following way:

$$x_i = 0 \quad \text{if subject } i \text{ is classified as } A_1$$
$$x_i = 1 \quad \text{if subject } i \text{ is classified as } A_2$$
$$y_i = 0 \quad \text{if subject } i \text{ is classified as } B_1$$
$$y_i = 1 \quad \text{if subject } i \text{ is classified as } B_2$$

Then

$$|r_{xy}| = \phi$$

Numerically coding categorically measured variables will be discussed more fully in Chapter 12.

NOTES AND SUPPLEMENTARY READINGS

Guilford, J. P., & Fruchter, B. (1977). *Fundamental statistics in psychology and education* (6th ed.). New York: McGraw-Hill.

Guilford and Fruchter present a clear examination of topics pertaining to bivariate relations, correlation, and association. This volume is very readable.

PROBLEMS

1. Calculate the correlation coefficient for the following data:

Observation	x	y
1	3	8
2	5	12
3	6	10
4	9	11
5	12	14

2. Calculate the correlation coefficient for the following data:

Observation	x	y
1	5	3
2	7	5
3	8	4
4	10	6
5	12	4
6	13	8
7	15	5

3. Test the following hypotheses:
   a. H_0: $\rho = 0$ versus H_a: $\rho > 0$ when $r = .30$, $N = 17$, $\alpha = .05$.
   b. H_0: $\rho = 0$ versus H_a: $\rho \neq 0$ when $r = -.50$, $N = 42$, $\alpha = .01$. Also calculate the 99% confidence interval for ρ.
   c. H_0: $\rho = 0$ versus H_a: $\rho \neq 0$ when $r = .26$, $N = 29$, $\alpha = .05$. Also calculate the 95% confidence interval for ρ.
   d. H_0: $\rho = .50$ versus H_a: $\rho > .50$ when $r = .80$, $N = 35$, $\alpha = .05$.

4. Test the following hypotheses:
   a. $H_0: \rho = 0$ versus $H_a: \rho < 0$ when $r = -.20$, $N = 50$, $\alpha = .05$.
   b. $H_0: \rho = 0$ versus $H_a: \rho \neq 0$ when $r = .35$, $N = 32$, $\alpha = .01$. Also calculate the 99% confidence interval for ρ.
   c. $H_0: \rho = 0$ versus $H_a: \rho \neq 0$ when $r = -.32$, $N = 15$, $\alpha = .05$. Also calculate the 95% confidence interval for ρ.
   d. $H_0: \rho = .50$ versus $H_a: \rho > .50$ when $r = .59$, $N = 70$, $\alpha = .05$.

5. Suppose that two judges ranked ten contestants in a dancing competition.

Contestant	Judge 1	Judge 2
1	1	3
2	4	2
3	2	5
4	6	1
5	3	4
6	10	7
7	5	8
8	9	6
9	7	10
10	8	9

Calculate the Spearman rank order correlation coefficient for these rankings and test the null hypothesis that the population correlation coefficient is zero (two-tailed test, $\alpha = .05$).

6. Consider the following 12 students who are ranked for their academic and athletic performance.

Student	Academic Rank	Athletic Rank
1	1	6
2	3	2
3	5	4
4	6	1
5	2	8
6	9	9
7	8	3
8	10	5
9	4	10
10	11	12
11	12	7
12	7	11

Calculate the Spearman rank order correlation coefficient for these rankings and test the null hypothesis that the population correlation coefficient is zero (one-tailed test, $\alpha = .01$).

7. Of 100 employees who applied to take a management training course, 40 were accepted. After the acceptances were announced, the employees were asked if they thought that the process was (a) fair, (b) unfair, or (c) had no opinion. The following contingency table summarizes the results:

| | | Opinion of Process | | |
		Fair	Not Fair	No Opinion
Acceptance Decision	Rejected	25	20	15
	Accepted	25	10	5

a. Test the null hypothesis that the acceptance decision and opinions of the process are independent versus the alternative that they are associated by using the χ^2 test of association ($\alpha = .05$).
b. Calculate the sample value of the strength of association between acceptance decision and opinion of the process using Cramer's statistic.

8. In a study of the relation between dominance and marital satisfaction a sample of 200 newly wed couples were classified as either (a) wife dominant, (b) husband dominant, and (c) egalitarian. Five years later the couples were classified as (a) married and satisfied, (b) married and dissatisfied, (c) divorced, or (d) other (e.g., widowed). The following contingency table summarizes the data:

	Wife Dominant	Husband Dominant	Egalitarian
Married, Satisfied	20	30	30
Married, Dissatisfied	25	25	10
Divorced	20	15	5
Other	5	10	5

a. Test the null hypothesis that dominance and marital satisfaction are independent versus the alternative that they are associated by using the χ^2 test of association ($\alpha = .05$).
b. Calculate the sample value of the strength of association between dominance and marital satisfaction using Cramer's statistic.

CHAPTER 11 REGRESSION ANALYSIS WITH ONE INDEPENDENT VARIABLE

U
p to this point two major types of analysis have been discussed. In one design the independent variable is categorical and the dependent variable is quantitative. When the independent variable consists of two levels, the *t* test is an appropriate statistical test; when the independent variable consists of more than two levels, the analysis of variance is an appropriate statistical test. The second type of analysis is related to sampling from a bivariate distribution to determine the relation between two variables, each of which has equal status (there is no independent or dependent variable). Statistical tests for this type of design involve correlation coefficients or tests of association.

A third type of analysis, referred to as *regression analysis*, involves a quantitative dependent variable and a quantitative independent variable. Regression analysis, in its classical application, differs from a correlational analysis in that in regression analysis the researcher determines the quantitative nature of the independent variable. A few examples will illustrate how this is accomplished. A researcher interested in the relation between drug dosage and some outcome measure might design an experiment whereby subjects receive various dosages set by the researcher and the relation between dosage (a quantitative independent variable) and some quantitative dependent variable is studied. Regression analysis could indicate the degree to which the outcome changed for every unit change in dosage. A second example is provided by Example 8.2 (p. 239), which involved classroom size and achievement. In that example classrooms of various sizes were randomly selected and, using a random-effects analysis of variance, it was found that, in general, classroom size influences achievement. However, the nature of the relation between classroom size and achievement remained unclear. Regression analysis could be used to determine whether the relation between classroom size and achievement was linear or curvilinear. In a regression analysis the researcher would likely determine the classroom sizes to be included in the study rather than randomly selecting them, as was the case in Example 8.2.

Although classical regression analysis involves determination of the independent variable by the researcher, connections to correlational analysis seem apparent in that both rely on determination of the bivariate relation between two variables. The statistical theory of regression analysis relies heavily on the correlation coefficient, which indexes the linear relation between two variables.

We have made a distinction between classical regression analysis whereby a researcher determines the nature of the independent variable and correlational analysis whereby the researcher samples from a bivariate distribution. However, current applications of regression analysis often relax the requirement that the levels of an independent variable be chosen by the researcher and the distinctions between regression analysis and correlation analysis become blurred. The dangers of this practice are not great and the gain is immense in that regression analysis can be applied to many types of data resulting from experimental and nonexperimental designs. We make our presentation of regression analysis in the classical way and then indicate how the restrictions can be relaxed.

The presentation of regression analysis differs from that of previous topics in that sample statistics will be discussed prior to the population parameters and a population model. The first portion of this chapter considers regression models with one independent variable and for which there is a linear relation between the independent and dependent variables and the second portion deals with applications of regression analysis with one independent variable. The next chapter will discuss models with more than one independent variable. Because our discussion focuses primarily on linear relations, unless otherwise specified the term "regression analysis" refers to linear regression.

Basic Concepts of Regression Analysis with One Independent Variable

11.1 Example to Be Used for Regression Analysis with One Independent Variable

The following example (Example 11.1) from a classical regression problem illustrates the concepts of this portion of the chapter.

EXAMPLE 11.1   Suppose that a researcher is interested in attempting to improve the memory of brain impaired subjects by using intravenous administration of a neurotransmitter. The researcher wants to determine the relation between memory and the quantity (dosage) of the transmitter. The following experiment is designed. Sixteen subjects are randomly selected from a population of specified brain damaged persons; four subjects are randomly selected to receive 2, 4, 6, or 8 milliliters (ml) of the transmitter. In this instance the independent variable consists of four groups, each of which is associated with a numerical value indicating the dosage of the drug. After administration of the drug, a memory task is given to the subjects and the dependent variable is the number of items recalled from a list of 50 items. The results of this hypothetical example are found in Table 11.1 and presented graphically in Figure 11.1.

TABLE 11.1

Data for Neurotransmitter Example

Subject i	Milliliters of Neurotransmitter x	Number of Items Recalled y	Mean of y for Group	y'	$y - y'$
1	2	22		23.1	−1.1
2	2	17	23	23.1	−6.1
3	2	27		23.1	3.9
4	2	26		23.1	2.9
5	4	21		27.7	−6.7
6	4	25	27	27.7	−2.7
7	4	30		27.7	2.3
8	4	32		27.7	4.3
9	6	36		32.3	3.7
10	6	31	34	32.3	−1.3
11	6	40		32.3	7.7
12	6	29		32.3	−3.3
13	8	36		36.9	−0.9
14	8	45	36	36.9	8.1
15	8	30		36.9	−6.9
16	8	33		36.9	−3.9

$M_x = 5$ $M_y = 30$
$S_x = 2.236$ $S_y = 6.964$
$S_x^2 = 5.00$ $S_y^2 = 48.50$

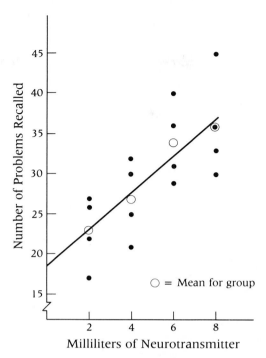

FIGURE 11.1   Relation Between Dosage of Neurotransmitter and Number of Items Recalled

It is informative to compare the goal of regression analysis to that of the basic analysis of variance. If this were an analysis of variance, the goal would be to determine whether or not the means for the four groups were all the same. In regression analysis the goal is to determine the relation between the **value** for each group (i.e., the value of the independent variable: 2, 4, 6, and 8 milliliters) and the mean of the dependent variable in each group. In Figure 11.1 the mean of the dependent variable for each group is indicated by an open circle. It appears that the four open circles roughly fall along a line, indicating a linear relation between the dosage of the drug and the mean number of items recalled. From the figure it is clear that as the dosage increases, the number of items recalled increases.

The line that represents the linear relation between the independent and dependent variables is called the *regression line*. In general, the equation of the regression line is written as

$$y'_i = Bx_i + A \tag{11.1}$$

where x_i refers to the score for the ith subject on the independent variable, y'_i refers to the predicted value for the ith subject for the dependent variable given information about the independent variable for the ith subject, B is called the *regression coefficient*, and *A* is called the *regression constant*. The notation y' is used rather than y because it is clear that the regression line does not pass through each of the data points; the regression equation characterizes the line that "best" predicts the scores on the dependent variable from values of the independent variable. The criterion for what constitutes the best line will be discussed in the next section. The values of *A* and *B* are chosen in such a way that the line is the best line. Determining the equation of the best line, assessing how well it "fits" the data, and performing the related statistical tests comprise regression analysis and will be the subject of the following sections.

TABLE 11.2

Regression Analysis for Neurotransmitter Example with Standardized Scores

Subject i	Milliliters of Transmitter x	z_x	Items Recalled y	z_y	$z_x z_y$	z_y'	$z_y - z_y'$	$(z_y - z_y')^2$
1	2	−1.342	22	−1.152	1.546	−.993	−.159	.0253
2	2	−1.342	17	−1.873	2.514	−.993	−.880	.7744
3	2	−1.342	27	−.432	.580	−.993	.561	.3147
4	2	−1.342	26	−.576	.773	−.993	.417	.1739
5	4	−.447	21	−1.296	.579	−.331	−.965	.9312
6	4	−.447	25	−.720	.322	−.331	−.389	.1513
7	4	−.447	30	.000	.000	−.331	.331	.1096
8	4	−.447	32	.288	−.129	−.331	.619	.3832
9	6	.447	36	.864	.386	.331	.533	.2841
10	6	.447	31	.144	.064	.331	−.187	.0350
11	6	.447	40	1.441	.644	.331	1.110	1.2321
12	6	.447	29	−.144	−.064	.331	−.475	.2256
13	8	1.342	36	.864	1.159	.993	−.129	.0166
14	8	1.342	45	2.161	2.900	.993	1.168	1.3642
15	8	1.342	30	.000	.000	.993	−.993	.9860
16	8	1.342	33	.432	.580	.993	−.561	.3147
					11.854			7.3219

$M_x = 5$, $S_x = 2.24$, $M_y = 30$, $S_y = 6.96$.

11.2   Regression Equation with Standardized Scores

Characterizing the line that best fits the data from a regression problem is most easily understood by finding the equation of the line when the scores for the independent and dependent variables are standardized. For the continuing neurotransmitter example the standardized scores for the independent and dependent variables are found in Table 11.2 and are graphically presented in Figure 11.2.

Because the goal at this point is to explain or predict the standardized scores for Y in terms of a linear function of the standardized scores for X, we write

$$z'_{y_i} = b_{y \cdot x} z_{x_i} + a \cdot \tag{11.2}$$

Because we are in the context of standardized scores, $b_{y \cdot x}$ is called the *standardized regression coefficient for predicting y from x*. Often this coefficient is simply denoted by b and is referred to as the standardized regression coefficient. (It should be mentioned that the standardized regression coefficient is

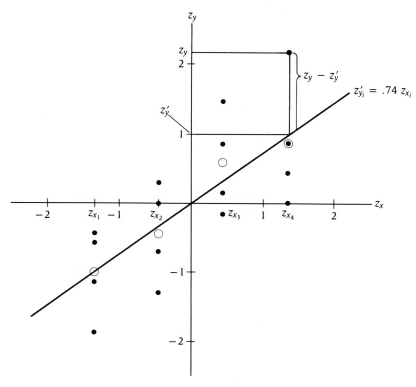

FIGURE 11.2   Relation Between Dosage of Neurotransmitter and Items Recalled with Standardized Scores

often referred to as the *beta coefficient* and denoted by β; however, this notation is not used in this text and β will be used to denote the population standardized coefficient. Further, the constant *a* is not given a special name because in the case of standardized scores *a* is equal to zero, as will be shown later.) The task is to find *a* and *b* so that the line for this equation is the best line. Essentially, the desired regression line should be that line that is "closest" to the actual data points. Therefore, we want the error

$$z'_{yi} - z_{yi}$$

to be small for all *i*. However, because the negative errors will "cancel out" the positive errors, the average of the squared errors are used. **Therefore, the criterion used to define the regression line for standardized scores is that the constants *a* and *b* must minimize the quantity**

$$\frac{\sum_i (z'_{yi} - z_{yi})^2}{N} = \frac{\sum_i (z_{yi} - z'_{yi})^2}{N} \tag{11.3}$$

where *N* is the total number of subjects. Often this criterion is called the "least squared errors" or "least squares" criterion and the regression line, which is based on this criterion, thus is often referred to as the "least squared errors line" or the "least squares line." Subsequently it will be shown that the average of the squared errors is minimized for a set of standardized scores when

$$a = 0$$

and

$$b_{y \cdot x} = r_{xy}$$

That is, the standardized regression coefficient is equal to the correlation coefficient of the independent and dependent variables, which shows the expected close relation between regression and correlation. To summarize: **The regression equation for standardized scores is**

$$z'_{yi} = r_{xy} z_{xi} \tag{11.4}$$

It will not be shown that for $z'_{yi} = b_{y \cdot x} z_{xi} + a$,

$$\frac{\sum_i (z'_{yi} - z_{yi})^2}{N}$$

is minimized when $a = 0$ and $b_{y \cdot x} = r_{xy}$. Before presenting the proof, however, we should mention that the following proof is not the "standard" proof of this result as knowledge of calculus greatly simplifies it; however, the version to be presented gives an extremely important supplementary result. To simplify the notation, the subscript *i* is omitted and $b_{y \cdot x}$ is referred to as *b*. We begin by examining the average of the squared error:

$$\frac{\sum (z_y' - z_y)^2}{N} = \frac{1}{N} \sum (bz_x + a - z_y)^2$$

because the regression equation is $z_y' = bz_x + a$ [see Equation (11.2)]. The proof continues with algebraic manipulations:

$$\frac{1}{N} \sum (bz_x + a - z_y)^2 = \frac{1}{N} \sum [(bz_x - z_y) + a]^2$$

$$= \frac{1}{N} \sum [(bz_x - z_y)^2 + 2a(bz_x - z_y) + a^2]$$

$$= \frac{1}{N} \sum (bz_x - z_y)^2 + \frac{1}{N} \sum 2a(bz_x - z_y) + \frac{1}{N} \sum a^2$$

This last expression contains three terms. First, notice that the middle term $(1/N)\sum 2a(bz_x - z_y)$ equals zero because $\sum bz_x = 0$ and $\sum z_y = 0$ (i.e., the sum of the standardized scores for a set of observations is equal to zero). The third term $(1/N)\sum a^2$ simplifies easily by the rules of summation:

$$\frac{1}{N} \sum a^2 = \frac{1}{N} Na^2 = a^2$$

Therefore, the average of the squared errors simplifies to the following expression:

$$\frac{\sum (z_y' - z_y)^2}{N} = \frac{1}{N} \sum (bz_x - z_y)^2 + a^2$$

Because the constant a is squared, it should be clear that this quantity is minimized when

$$a = 0$$

as we sought to show. Therefore, when a equals zero

$$\frac{\sum (z_y' - z_y)^2}{N} = \frac{1}{N} \sum (bz_x - z_y)^2$$

Hence, the remainder of the proof involves demonstrating that the quantity $(1/N)\sum (bz_x - z_y)^2$ is minimized when $b = r_{xy}$. Again, we algebraically manipulate this quantity:

$$\frac{1}{N} \sum (bz_x - z_y)^2 = \frac{1}{N} \sum (b^2 z_x^2 - 2bz_x z_y + z_y^2)$$

$$= \frac{1}{N} b^2 \sum z_x^2 - \frac{2b}{N} \sum z_x z_y + \frac{1}{N} \sum z_y^2$$

But $(1/N)\sum z_x^2 = 1$ and $(1/N)\sum z_y^2 = 1$ because the sample variance of a set of standardized scores is equal to 1. Furthermore, by the definition of the sample correlation coefficient $(1/N)\sum z_x z_y = r_{xy}$ [see Equation (10.7), p. 289]. So, the average of the squared errors (the quantity to be minimized) is equal to

$$b^2 - 2br_{xy} + 1$$

Those who have a knowledge of calculus will see that the obvious way to minimize this quantity is to set the derivative equal to zero and solve for b. As an alternative, we take the awkward tack of letting

$$b = r_{xy} + c$$

and showing that the quantity is minimized when c is equal to zero (i.e., $b = r_{xy}$, the desired result). Recall that we have reduced the average of the squared error to

$$b^2 - 2br_{xy} + 1$$

and if $b = r_{xy} + c$, then

$$\begin{aligned} b^2 - 2br_{xy} + 1 &= (r_{xy} + c)^2 - 2(r_{xy} + c)r_{xy} + 1 \\ &= r_{xy}^2 + 2cr_{xy} + c^2 - 2r_{xy}^2 - 2cr_{xy} + 1 \\ &= c^2 - r_{xy}^2 + 1 \end{aligned}$$

which is clearly minimized when c equals zero. Because we let $b = r_{xy} + c$, when $c = 0$, $b = r_{xy}$. That is, the average of the squared errors is minimized when $b = r_{xy}$, which completes the proof.

In the last steps of the proof the average of the squared errors was reduced to $c^2 - r_{xy}^2 + 1$, but when c equals zero (i.e., $b = r_{xy}$), the average of the squared error simplifies to

$$1 - r_{xy}^2 \tag{11.5}$$

This quantity, which is vital to an understanding of regression analysis, is called the *sample variance of estimate for standardized scores* and is denoted by $S_{z_y \cdot z_x}^2$. Clearly it is desirable that the variance of estimate be small (i.e., small errors), and this is accomplished when the absolute value of the correlation coefficient is large. That is to say, the larger the correlation coefficient (in terms of the absolute value) is, the better the regression line fits the data.

It is time to summarize the important points of the development of regression analysis with standardized scores. **The equation for the line that minimizes the average squared errors**

$$\frac{\sum (z_y' - z_y)^2}{N}$$

is the regression equation

$$z_y' = b_{y \cdot x} z_x = r_{xy} z_x$$

Furthermore, for this regression equation

$$\frac{\sum (z_y' - z_y)^2}{N} = 1 - r_{xy}^2 = S_{z_y \cdot z_x}^2 \tag{11.6}$$

that is, the average of the squared errors is equal to 1 minus the square of the correlation coefficient and this quantity is called the sample variance of estimate of standardized scores.

The sample variance of estimate for standardized scores is useful in deriving two other important properties of correlation and regression. First, because the average of the squared error is always greater than or equal to zero, the variance of estimate must be greater than or equal to zero. That is

$$\frac{1}{N}\sum(z_y' - z_y)^2 \geq 0$$

implies that

$$1 - r_{xy}^2 \geq 0$$

Therefore

$$1 \geq r_{xy}^2$$

and

$$-1 \leq r_{xy} \leq +1 \tag{11.7}$$

which is a result that we took for granted in Chapter 10 [see Equation (10.8), p. 290].

The second property related to the variance of estimate is finding a proportion of variance in the standardized scores for the dependent variable explained by the standardized scores for the independent variable. Remember that the variance of estimate is the average of the squared errors; that is

$$S_{z_y \cdot z_x}^2 = \frac{\sum(z_y' - z_y)^2}{N}$$

Therefore, the variance of the estimate for standardized scores may be thought of as the variance of the standardized scores given the estimates z_y' based on the regression equation. Recall from Section 7.12 (p. 200) that the proportion of variance explained by the independent variable was defined as a relative reduction in uncertainty:

$$\frac{\sigma_Y^2 - \sigma_{Y|X}^2}{\sigma_Y^2}$$

In the context of the regression equation for standardized scores, the *sample value* of the relative reduction in uncertainty is

$$\frac{S_{z_y}^2 - S_{z_y \cdot z_x}^2}{S_{z_y}^2} = \frac{1 - (1 - r_{xy}^2)}{1} = r_{xy}^2 \tag{11.8}$$

The quantity r_{xy}^2 is called the *coefficient of determination* and is interpreted as the proportion of variance in the dependent variable explained by the independent variable in the sample. Here we did not mention that we are in the context of standardized scores because, as we will see, the coefficient of determination is unchanged when we move to raw scores. If r_{xy}^2 is the proportion of variance in the dependent variable accounted for by the independent variable, then

$$1 - r_{xy}^2$$

is the proportion of variance in the dependent variable *unaccounted for* or *unexplained* by the independent variable. Often r_{xy}^2 is used in the context of correlation and in that case it may be interpreted as the proportion of variance *shared* by the two variables since neither is designated as independent or dependent. It should also be stressed that r_{xy}^2 is a **sample** value of the proportion of variance explained; its relation to the population value will be explored later.

The concepts of this section are illustrated by referring to Example 11.1 (p. 312), for which the standardized scores are displayed in Table 11.2 (p. 314) and Figure 11.2 (p. 315). The correlation coefficient is found in the following manner:

$$r_{xy} = \frac{1}{N} \sum z_{y_i} z_{x_i} = \frac{1}{16}(11.85) = .74$$

Therefore, the regression equation is equal to

$$z'_{y_i} = r_{xy} z_{x_i} = .74 z_{x_i}$$

This equation can be used to predict the standardized score on the dependent variable from the standardized score on the independent variable. For instance, for the sixth subject the prediction is

$$z'_{y_6} = .74 z_{x_6} = .74(-.447) = -.331$$

Clearly there is an error here because the actual standardized score on the dependent variable for the sixth subject was $-.720$. The error is thus

$$z_{y_6} - z'_{y_6} = -.720 - (-.331) = -.389$$

The predicted scores and the errors for each of the 16 subjects are presented in Table 11.2. As well, the squared errors are found in Table 11.2. From these data we can find the sample variance of estimate for the standardized scores in two ways. First, it is equal to the average of the squared errors:

$$S_{z_y \cdot z_x}^2 = \frac{\sum (z_{y_i} - z'_{y_i})^2}{N} = \frac{7.3219}{16} = .46$$

Second, by Equation (11.6)

$$S_{z_y \cdot z_x}^2 = 1 - r_{xy}^2 = 1 - (.74)^2 = .45$$

(the difference in the second decimal place is due to rounding error). Finally, the proportion of variance in the dependent variable accounted for by the independent variable (i.e., the coefficient of determination) is

$$r_{xy}^2 = (.74)^2 = .55$$

That is, 55% of the variance in performance is due to the linear relation between the dosage of the transmitter and performance.

Although for simplicity the mechanics of regression analysis have been developed with standardized scores, it is desirable to have analogous formulas and equations based on the raw scores. With that in mind, we now turn to the development of regression analysis with raw scores.

11.3   Regression Equation with Raw Scores

In general, the equation of a line for predicting y from x is

$$y_i' = B_{y \cdot x}(x_i) + A$$

where $B_{y \cdot x}$ is the *unstandardized regression coefficient for predicting y from x* and A is the *unstandardized regression constant*. Often the term "unstandardized" is dropped and $B_{y \cdot x}$ is referred to as the *regression coefficient* (whereas $b_{y \cdot x}$ is referred to as the *standardized regression coefficient*) and A is referred to as the *regression constant*. Frequently it is said that this equation reflects the *regression of y onto x*. The goal is to find $B_{y \cdot x}$ and A so that the average of the squared errors is minimized. This is easily accomplished by using the regression equation for standardized scores

$$z_{y_i}' = r_{xy} z_{x_i}$$

Substituting

$$z_{y_i}' = \frac{y_i' - M_y}{S_y}$$

and

$$z_{x_i} = \frac{x_i - M_x}{S_x}$$

into the regression equation for standardized scores, we have

$$\frac{y_i' - M_y}{S_y} = r_{xy}\left(\frac{x_i - M_x}{S_x}\right)$$

Solving for y' yields

$$y_i' = r_{xy}\frac{S_y}{S_x}(x_i - M_x) + M_y = r_{xy}\frac{S_y}{S_x}x_i + \left(M_y - M_x r_{xy}\frac{S_y}{S_x}\right) \tag{11.9}$$

Hence, **the regression equation for raw scores is**

$$y_i' = B_{y \cdot x} x_i + A \tag{11.10}$$

where

$$B_{y \cdot x} = r_{xy} \frac{S_y}{S_x} \tag{11.11}$$

and

$$A = M_y - M_x r_{xy} \frac{S_y}{S_x} \tag{11.12}$$

A useful alternative form of the regression equation is given by

$$y_i' = B_{y \cdot x}(x_i - M_x) + M_y \tag{11.13}$$

where $B_{y \cdot x}$ is defined in the same way.

Before discussing some characteristics and implications of the regression equation for raw scores, we will illustrate the regression equation with Example 11.1 (p. 312), which relates the dosage of a neurotransmitter with memory performance. Table 11.1 (p. 312) summarizes the raw scores for the 16 subjects. First the correlation coefficient is found by dividing the sample covariance [which can be calculated by using Equation (3.36), p. 65, or Equation (3.37), p. 67] by the product of the sample standard deviations:

$$r_{xy} = \frac{S_{xy}}{S_x S_y} = \frac{11.5}{(2.236)(6.964)} = .74$$

(This is the same value as that obtained in the previous section using standardized scores.) The regression coefficient is simply

$$B_{y \cdot x} = r_{xy} \frac{S_y}{S_x} = (.74) \frac{6.96}{2.24} = 2.30$$

Using Equation (11.13) we find that the regression equation is

$$y_i' = B_{y \cdot x}(x_i - M_x) + M_y = 2.30(x_i - 5) + 30$$
$$= 2.30 x_i + 18.5$$

For the sixth subject the predicted value y' is

$$y_6' = 2.30 x_6 + 18.5 = 2.30(4) + 18.5 = 27.7$$

The actual value for the sixth subject was 25, yielding an error of

$$y - y' = 25 - 27.7 = -2.7$$

The errors for each of the 16 subjects are also presented in Table 11.1.

We now return to a discussion of the regression equation for raw scores

$$y_i' = B_{y \cdot x}(x_i - M_x) + M_y = r_{xy} \frac{S_y}{S_x} (x_i - M_x) + M_y$$

Notice that when the correlation is zero (i.e., $r_{xy} = 0$)

$$y_i' = M_y$$

As expected, when there is no relation between the independent and the dependent variables, the best prediction for any subject is simply the mean of the dependent variable. When the correlation is nonzero, the regression equation may be thought of in this way: the difference between the score on the independent variable and the mean of the scores on the independent variable is multiplied by the regression constant and is then added to the mean of the scores on the dependent variable. The quantity that is added to the M_y can be conceptualized as a gain in prediction by knowing the linear relation between the independent variable and the dependent variable.

Consider now the form of the regression equation given by

$$y_i' = B_{y \cdot x} x_i + A = r_{xy} \frac{S_y}{S_x}(x_i) + \left(M_y - M_x r_{xy} \frac{S_y}{S_x}\right)$$

First, it should be recognized that the unstandardized regression coefficient $B_{y \cdot x}$ is the correlation coefficient adjusted to reflect the relative variation of the dependent and independent variables. Hence, linear transformations of either the independent or dependent variable that affect the variation change the size of the unstandardized regression coefficient, although the standardized regression coefficient (i.e., the correlation coefficient) remains unchanged. Second, the regression equation written in this way is the "slope-intercept" equation for a line; the coefficient on x is the slope and the constant is the intercept. Thus, the slope of the raw score regression line is the regression coefficient. In the continuing example the slope of the regression line, which is depicted in Figure 11.1 (p. 313) is 2.3. It should be recalled that the slope of a line is equal to the change in y due to a unit change in x. That is, a unit change in the independent variable will result in a predicted change in the dependent variable equal to the regression coefficient. Therefore, in the neurotransmitter example it is predicted that an additional milliliter of the transmitter will result in the recall of an additional 2.3 items. The regression constant A is that value where the line intersects the vertical axis, which in the continuing example is 18.5. (See Figure 11.1, p. 313.)

The average of the squared error resulting from a regression equation is called the *sample variance of estimate*, is denoted by $S_{y \cdot x}^2$, and is given by

$$S_{y \cdot x}^2 = \frac{1}{N}\sum(y_i - y_i')^2 = S_y^2(1 - r_{xy}^2) \tag{11.14}$$

This is similar to the sample variance of estimate for standardized scores with the exception that the variance of estimate for raw scores is the variance in the dependent variable multiplied by the proportion of variance unaccounted for by the dependent variable. An important quantity is the square root of the variance of estimate, which is referred to as the *sample standard error of estimate*, which is denoted by $S_{y \cdot x}$ and is equal to

$$S_{y \cdot x} = S_y \sqrt{1 - r_{xy}^2} \tag{11.15}$$

For the continuing neurotransmitter example

$$S_{y \cdot x}^2 = S_y^2(1 - r_{xy}^2) = 48.50[1 - (.74)^2] = 21.94$$

and

$$S_{y \cdot x} = \sqrt{21.94} = 4.68$$

The proportion of variance explained by the independent variable is given by

$$\frac{S_y^2 - S_{y \cdot x}^2}{S_y^2} = \frac{S_y^2 - S_y^2(1 - r_{xy}^2)}{S_y^2} = r_{xy}^2$$

As was the case for standardized scores, the coefficient of determination is equal to r_{xy}^2 and is interpreted as the proportion of variance in the dependent variable accounted for by the independent variable. Thus, in either case the proportion of variance unaccounted for by the independent variable is equal to

$$1 - r_{xy}^2$$

and is referred to as the *coefficient of alienation*. It is important to understand the distinction between the sample variance of estimate $S_{y \cdot x}^2$ and the coefficient of alienation $1 - r_{xy}^2$. $S_{y \cdot x}^2$ is the **amount** of variance in the dependent variable that remains after the predictions are made and $1 - r_{xy}^2$ is the **proportion** of variance that remains.

11.4   Linear Model for Regression, Significance Tests, and Confidence Intervals

To this point regression analysis has been discussed only in terms of sample equations and statistics. Clearly, to be of much use, the results of a sample must be applicable to a population so that the results of the regression analysis can be generalized. In the neurotransmitter example the predicted scores for any of the 16 subjects are of limited use because the actual scores for these individuals are known. If it turns out that the sample regression equation is a good estimate of the population relation between dosage and memory, then a number of useful applications can be made. For example, it is extremely useful to be able to make predictions about the performance of individuals who were in the population but who were not in the sample. However, as the concepts of regression analysis are developed, it will be clear that the applications of regression analysis extend well beyond simple pre-diction.

The goal of classical regression analysis is to relate the numerical values of various categories determined by the researcher to the dependent variable. Let Y_{ij} be the ith observation in the jth group, X_j be the numerical value associaton with the jth group, and μ_{y_j} be the mean of Y for the jth group. The mean of Y for the jth group μ_{Y_j} is the sum of the grand mean μ and a term that reflects the linear relation between X and Y. The linear relation between X and Y is written in terms of the *population regression coefficient*, which is denoted by $\beta_{Y \cdot X}$ (remember that many sources refer to the stan-dardized regression coefficient with β). Thus

$$\mu_{Y_j} = \mu_Y + \beta_{Y \cdot X}(X_j - \mu_X)$$

However, because the goal of regression analysis is to predict the mean of Y for each group, the population regression equation can be rewritten as

$$\textbf{True } Y'_j = \mu_Y + \beta_{Y \cdot X}(X_j - \mu_X) \tag{11.16}$$

or

$$\textbf{True } Y'_j = \beta_{y \cdot x}X_j + \textbf{(population regression constant)} \tag{11.17}$$

where the population regression constant is equal to $\mu_Y - \beta_{Y \cdot X}\mu_X$. A number of important points about these equations need to be made. First, these population regression equations are analogous to the sample regression equations [(11.13), p. 322 and (11.10), p. 322], except that the subscript i has been replaced by j. However, this is not a problem because the value of X for the ith individual is X_j, the value for the group to which the ith subject belongs. The second point is that the population regression coefficient $\beta_{Y \cdot X}$ can be written in terms of the population correlation coefficient in the following way:

$$\beta_{Y \cdot X} = \rho_{XY}\frac{\sigma_X}{\sigma_Y} \tag{11.18}$$

which is also analogous to the sample statistic. Third, the sample statistics are used as estimators for the population parameters:

$$B_{y \cdot x} = \hat{\beta}_{Y \cdot X} \tag{11.19}$$
$$M_y = \hat{\mu}_Y \tag{11.20}$$
$$y'_j = \text{true } Y'_j = \hat{\mu}_{Y_j} \tag{11.21}$$

A related and subtle point to be made here is that μ_X is not estimated, because technically speaking X is not a random variable inasmuch as it was fixed by the researcher and thus

$$M_x = \mu_X$$

Population analogues also exist for sample statistics such as the variance of estimate, standard error of estimate, and the proportion of variance accounted for by linear regression. The *population variance of estimate* for predicting Y from X is given by

$$\sigma^2_{Y \cdot X} = \sigma^2_Y (1 - \rho^2_{XY}) \qquad (11.22)$$

Similarly, the *population standard error of estimate* is given by

$$\sigma_{Y \cdot X} = \sigma_Y \sqrt{1 - \rho^2_{XY}} \qquad (11.23)$$

Finally, the *population proportion of variance accounted for by linear regression* is

$$\frac{\sigma^2_Y - \sigma^2_{Y \cdot X}}{\sigma^2_Y} \qquad (11.24)$$

To model the ith observation in the jth group (i.e., Y_{ij}), we must introduce a term that reflects the deviations from the group means (i.e., error). Therefore, Equation (11.16) is altered to yield the *linear model for regression*

$$\mathbf{Y_{ij} = \mu_Y + \beta_{Y \cdot X}(X_j - \mu_X) + e_{ij}} \qquad \mathbf{(11.25)}$$

This model is similar to the linear model for the one-way fixed-effects analysis of variance (see Section 7.7, p. 184)

$$Y_{ij} = \mu_Y + \alpha_j + e_{ij}$$

the difference being that the term α_j, which reflects the treatment effect for group j, is replaced by the term $\beta_{Y \cdot X}(X_j - \mu_X)$, which reflects the linear relation between the numerical value attached to group j and the dependent variable.

Three assumptions about the linear model expressed by Equation (11.25) are needed to derive the distributions of the test statistics to be discussed later in this section. The first assumption is that the errors are normally distributed with a mean of zero and the second assumption is that the variance of the errors for each group is the same. That is

$$e_{ij} \sim N(0, \sigma^2_e) \qquad \text{for all } j$$

The assumption that the variance of the errors for each group is the same is often referred to as the *homoscedasticity* assumption. The third assumption is that the observations are independent.

The customary statistical test for linear regression centers around hypotheses related to the population regression coefficient $\beta_{Y \cdot X}$. The typical null hypothesis is the population regression coefficient is equal to zero; that is

$$H_0: \beta_{Y \cdot X} = 0$$

We approach the test of this hypothesis by taking a tack similar to that used in the analysis of variance. Although the resultant test statistic for $\beta_{Y \cdot X}$ will be familiar, the derivation of the test statistic will reveal numerous interesting and useful properties of regression as well as emphasize alternative methods for deriving the sample regression equation. Furthermore, a close connection between regression analysis and the analysis of variance will be demonstrated. At each point in the derivations the data from Example 11.1 (p. 312), which regressed items recalled onto the dosage of a neurotransmitter, is used to illustrate the derivations.

Development of the statistical test for $\beta_{Y \cdot X}$ is begun by partitioning the sum of squares for the sample data. Reference to Figure 11.3 illustrates that the following is true:

$$(y_{ij} - M_y) = (y_{ij} - M_{y_j}) + (M_{y_j} - y'_j) + (y'_j - M_y) \tag{11.26}$$

Discussion of each of these terms will clarify many of the concepts of regression. First, recall that if there is no relation between the independent variable and the dependent variable, the best estimate for Y is the mean of Y. The term on the left of the equality

$$y_{ij} - M_y$$

is a deviation from the grand mean. The last term to the right

$$y'_j - M_y$$

is the advantage that is gained by using regression since it is the difference between the predicted score using regression and the prediction that would be made in the absence of regression. The two remaining terms represent

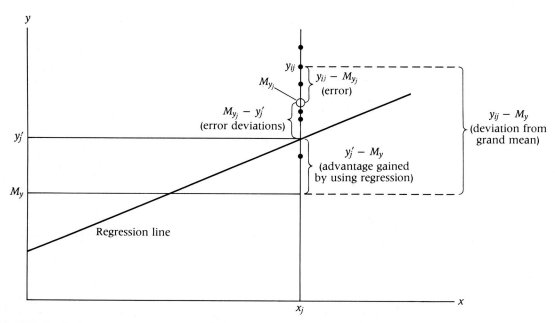

FIGURE 11.3   Errors and Advantage Gained in Linear Regression

two types of error. The first of these

$$y_{ij} - M_{y_j}$$

is the difference between the score and the group mean; the second of these

$$M_{y_j} - y_j'$$

is the difference between the group mean and the predicted value (via the regression equation) for that group. These errors are illustrated in Figure 11.3.

Squaring both sides of Equation (11.26)(p. 327) summing over all individuals i, and simplifying, we obtain the following partition of the sum of squares (below the sums of squares are their commonly used names):

$$\underbrace{\sum_j \sum_i (y_{ij} - M_y)^2}_{\text{SS total}} = \underbrace{\sum_j \sum_i (y_{ij} - M_{y_j})^2}_{\text{SS error}} + \underbrace{\sum_j N_j (M_{y_j} - y_j')^2}_{\text{SS deviations}} + \underbrace{\sum_j N_j (y_j' - M_y)^2}_{\text{SS linear regression}}$$

$$(11.27)$$

where N_j is the number of observations in group j. The reader can verify that for Example 11.1

$$\begin{aligned}
\text{SS total} &= 776 \\
\text{SS error} &= 336 \\
\text{SS deviations} &= 16.8 \\
\text{SS linear regression} &= 423.2
\end{aligned}$$

Two mean squares are needed in the following derivations. First, there is 1 degree of freedom associated with linear regression, so that

$$\begin{aligned}
\text{MS(linear regression)} &= \frac{\text{SS(linear regression)}}{1} \\
&= \text{SS(linear regression)}
\end{aligned} \qquad (11.28)$$

Second, the mean square error used for the statistical test associated with $\beta_{Y \cdot X}$ is found by pooling errors and deviations. Thus

$$\begin{aligned}
\text{MS(error + deviations)} &= \frac{\text{SS error + SS deviations}}{N - J + J - 2} \\
&= \frac{\text{SS error + SS deviations}}{N - 2}
\end{aligned} \qquad (11.29)$$

The importance of the role of MS(error + deviations) is illustrated by noting that it is the unbiased estimator of the population variance of estimate $\sigma_{Y \cdot X}^2$, whereas the sample variance of estimate $S_{y \cdot x}^2$ is biased:

$$\hat{\sigma}_{Y \cdot X}^2 = \text{MS(error + deviations)} = \frac{N}{N - 2} S_{Y \cdot X}^2 \qquad (11.30)$$

Because the sum of squares for linear regression is a key player in the development of the statistical test of $\beta_{Y \cdot X}$, it is useful to examine alternative forms to that given in the partition of the sum of squares. Although the alternatives are not obvious, derivation of them relies only on algebraic manipulations. The alternatives are as follows:

$$\text{SS linear regression} = \sum_{j=1}^{J} N_j (y_j' - M_y)^2$$

$$= N B_{y \cdot x}^2 S_x^2 \tag{11.31}$$

$$= N r_{xy}^2 S_y^2 \tag{11.32}$$

$$= \frac{N(\text{sample covariance of } x \text{ and } y)^2}{S_x^2} = \frac{N S_{xy}^2}{S_x^2} \tag{11.33}$$

where recall that S_{xy} is the notation for the sample covariance of x and y (see Section 3.13, p. 65). These equations are useful because solving for $B_{y \cdot x}$ gives an alternative method to finding $B_{y \cdot x}$:

$$B_{y \cdot x} = \sqrt{\frac{\text{SS linear regression}}{N S_x^2}} = \frac{S_{xy}}{S_x^2} \tag{11.34}$$

For the continuing neurotransmitter example

$$B_{y \cdot x} = \sqrt{\frac{423.20}{16(5)}} = \sqrt{5.29} = 2.3$$

which is the result obtained earlier.

Equation (11.33), which relates SS linear regression to the sample covariance, leads to the statistical test for the regression coefficient. Remember that the regression coefficient $\beta_{Y \cdot X}$ is the correlation coefficient ρ_{XY} multiplied by the quotient of the standard deviations; furthermore, the correlation coefficient ρ_{XY} is the covariance divided by the product of the standard deviations. That is

$$\beta_{Y \cdot X} = \frac{\sigma_Y}{\sigma_X} \rho_{XY} = \frac{\sigma_Y}{\sigma_X} \left(\frac{\text{Cov}(X, Y)}{\sigma_X \sigma_Y} \right) = \frac{\text{Cov}(X, Y)}{\sigma_X^2} \tag{11.35}$$

The covariance of X and Y and the regression coefficient differ by a factor of σ_X^2. Therefore, $\text{Cov}(X, Y)$ equals zero if and only if $\beta_{Y \cdot x}$ equals zero and therefore the hypothesis

$$H_0: \beta_{Y \cdot X} = 0$$

is equivalent to

$$H_0: \text{Cov}(X, Y) = 0$$

It will be shown that the sample covariance is a sample comparison (see Section 9.1, p. 261) and that this sample comparison can be used to test the null hypothesis that the population covariance is zero.

Given the model for regression, the sample covariance can be written as

$$S_{xy} = \sum_{j=1}^{J} \frac{N_j(x_j - M_x)}{N} M_{yj} \tag{11.36}$$

We let

$$\sum_j c_j = \sum_j \frac{N_j(x_j - M_x)}{N} \tag{11.37}$$

so that

$$S_{xy} = \sum_j c_j M_{yj} \tag{11.38}$$

which is a sample comparison provided $\sum_j c_j = 0$. The fact that $\sum_j c_j = 0$ is shown algebraically:

$$\sum_j c_j = \sum_j \frac{N_j(x_j - M_x)}{N} = \frac{1}{N} \sum_j (N_j x_j - N_j M_x)$$

$$= \frac{1}{N} \left(\sum_j N_j x_j - M_x \sum_j N_j \right)$$

but both $\sum_j N_j x_j$ and $M_x \sum_j N_j$ are the sum of the x values for all subjects. Therefore

$$\sum_j c_j = 0$$

and the sample covariance is a sample comparison. That is

$$\text{Sample covariance of } X \text{ and } Y = S_{xy} = \hat{\psi} = \sum_j c_j M_{yj}$$

$$= \sum_j \frac{N_j(x_j - M_x)}{N} M_{yj} \tag{11.39}$$

This relationship is easily illustrated with the continuing neurotransmitter example. First, the four values of c_j are calculated.

$$c_1 = \frac{N_1(x_1 - M_x)}{N} = \frac{4(2 - 5)}{16} = -\frac{3}{4}$$

$$c_2 = \frac{4(4 - 5)}{16} = -\frac{1}{4}$$

$$c_3 = \frac{4(6 - 5)}{16} = \frac{1}{4}$$

$$c_4 = \frac{4(8 - 5)}{16} = \frac{3}{4}$$

Thus

$$S_{xy} = \hat{\psi} = \sum_j c_j M_{y_j}$$

$$= \left(-\frac{3}{4}\right)(23) + \left(-\frac{1}{4}\right)(27) + \left(\frac{1}{4}\right)(34) + \left(\frac{3}{4}\right)(36)$$

$$= 11.5$$

which agrees with the value of the sample covariance obtained earlier (p. 322).

Because the sample covariance is a sample comparison, the hypothesis

$$H_0: \psi = 0$$

is equivalent to

$$H_0: \text{Cov}(X, Y) = 0$$

which is also equivalent to

$$H_0: \beta_{Y \cdot X} = 0$$

The statistic used to test $H_0: \psi = 0$ is

$$\frac{SS(\hat{\psi})}{MSE}$$

which, under the appropriate assumptions, has an F distribution with 1 and $(N - J)$ degrees of freedom (see Section 9.3, p. 271). We first examine the numerator of the test statistic and show that $SS(\hat{\psi})$ is equal to SS linear regression. From Section 9.3 (p. 271) we know that

$$SS(\hat{\psi}) = \frac{(\hat{\psi})^2}{\sum (c_j^2/N_j)}$$

The denominator of $SS(\hat{\psi})$ is simplified algebraically:

$$\sum_j \frac{c_j^2}{N_j} = \sum_j \frac{N_j^2 (x_j - M_x)^2}{N^2} \Big/ N_j$$

$$= \frac{1}{N}\left[\frac{1}{N}\sum_j N_j (x_j - M_x)^2\right]$$

$$= \frac{1}{N}\left[\frac{1}{N}\sum_{j=1}^{J} \sum_{i=1}^{N_j} (x_j - M_x)^2\right]$$

$$= \frac{S_x^2}{N}$$

So

$$SS(\hat{\psi}) = \frac{(\hat{\psi})^2}{\sum (c_j^2/N_j)} = \frac{NS_{xy}^2}{S_x^2}$$

$$= \frac{N(\text{sample covariance of } x \text{ and } y)^2}{S_x^2} \tag{11.40}$$

By Equation (11.33)

$$SS \text{ linear regression} = \frac{NS_{xy}^2}{S_x^2}$$

and therefore

$$SS(\hat{\psi}) = SS \text{ linear regression} \tag{11.41}$$

The denominator of the test statistic

$$\frac{SS(\hat{\psi})}{MSE}$$

is found by pooling SS error and SS deviations to increase the power of the test. By Equation (11.29)

$$MS(\text{deviations} + \text{error}) = \frac{SS \text{ error} + SS \text{ deviations}}{N - 2}$$

Therefore, **when the null hypothesis that $\beta_{Y \cdot x}$ equals zero is true and the assumptions of linear regression are met**

$$\frac{\textbf{SS linear regression}}{\textbf{MS(deviations + error)}} = \frac{\textbf{SS linear regression}}{\textbf{SS(deviations + error)}/(N-2)} \sim F_{1,N-2}$$

$$\tag{11.42}$$

and therefore to test the null hypothesis, we compare the value of the test statistic

$$\frac{\textbf{SS linear regression}}{\textbf{MS(deviations + error)}}$$

to an F distribution with 1 and $(N - 2)$ degrees of freedom.

From the sum of squares calculated earlier for the neurotransmitter example (p. 328) the null hypothesis that $\beta_{Y \cdot x}$ equals zero can be tested:

$$\frac{SS \text{ linear regression}}{SS(\text{deviations} + \text{error})/(N - 2)} = \frac{423.2}{(336 + 16.8)/(16 - 2)} = 16.79$$

which, when compared to an F distribution with 1 and 14 degrees of freedom, leads to the rejection of the null hypothesis that $\beta_{Y \cdot x}$ equals zero at the conventional levels of α.

The F ratio used to test the null hypothesis that the population regression coefficient is zero has an F distribution with 1 degree of freedom in the numerator; consequently, the square root of the F ratio yields a statistic that has a t distribution (see Section 7.5, p. 180). Derivation of this test statistic will further consolidate the connection between regression and correlation. From the relations discussed in this section it can be shown algebraically that

$$\text{SS linear regression} = N r_{xy}^2 S_y^2 \tag{11.43}$$

and

$$\text{SS(deviations + error)} = N S_{y \cdot x}^2 \tag{11.44}$$

Therefore

$$F = \frac{\text{SS linear regression}}{\text{SS(deviations + error)}/(N-2)} = \frac{N r_{xy}^2 S_y^2}{N S_{y \cdot x}^2/(N-2)} = \frac{r_{xy}^2(N-2)}{1 - r_{xy}^2}$$

and

$$t = \sqrt{\frac{r_{xy}^2(N-2)}{1 - r_{xy}^2}} = \frac{r_{xy}\sqrt{N-2}}{\sqrt{1 - r_{xy}^2}}$$

which is distributed as a t distribution with $(N-2)$ degrees of freedom. This statistic can be used as an alternative and equivalent test of the null hypothesis $\beta_{Y \cdot X}$ equals zero. **When the assumptions of linear regression are met and the null hypothesis that $\beta_{Y \cdot X}$ equals zero is true**

$$\frac{r_{xy}\sqrt{N-2}}{\sqrt{1 - r_{xy}^2}} \sim t_{N-2} \tag{11.45}$$

and therefore to test the null hypothesis, we compare the value of the test statistic

$$\frac{r_{xy}\sqrt{N-2}}{\sqrt{1 - r_{xy}^2}}$$

to a t distribution with $(N-2)$ degrees of freedom. Notice that the statistic used to test $\beta_{Y \cdot X}$ is identical to that used to test ρ_{XY} (see Section 10.3, p. 291). This is not surprising given the relation between $\beta_{Y \cdot X}$ and ρ_{XY}.

$$\beta_{Y \cdot X} = \rho_{XY}\frac{\sigma_X}{\sigma_Y}$$

However, the differences between the correlation and regression models should be kept in mind; the correlation model involves sampling from a bivariate distribution; whereas the regression model involves fixed levels of the independent variable. Accordingly, the assumptions underlying the statistical tests are different: the correlation model relies on bivariate normality, whereas the regression model relies on normality of the dependent variable within levels of the independent variable as well as homoscedasticity. The next section discusses the importance of assumptions in the regression model.

The final topic of this section involves finding confidence intervals for two population parameters, $\beta_{Y \cdot X}$ and the true score Y_{ij} for individual i in the population. We begin by giving the standard error of the sample regression coefficient:

$$\sigma_B = \frac{\sigma_{Y \cdot X}}{\sigma_X \sqrt{N-1}} \tag{11.46}$$

As expected, the standard error of the regression coefficient is a function of the standard error of estimate. The estimator for the standard error of the sample regression coefficient is thus

$$\hat{\sigma}_B = \frac{\hat{\sigma}_{Y \cdot X}}{\hat{\sigma}_X \sqrt{N-1}} = \frac{\sqrt{\text{MS(error + deviations)}}}{s_x \sqrt{N-1}}$$

$$= \frac{\sqrt{\text{MS(error + deviations)}}}{S_x \sqrt{N}}$$

$$= \frac{S_{y \cdot x}}{S_x \sqrt{N-2}}$$

$$= \frac{S_y \sqrt{1 - r_{xy}^2}}{S_x \sqrt{N-2}} \tag{11.47}$$

Therefore, when the assumptions of the regression model are true, the limits for the $(1 - \alpha)100\%$ confidence interval for $\beta_{Y \cdot X}$ are

$$B_{y \cdot x} \pm t_{\alpha/2; N-2} \hat{\sigma}_B$$

or

$$B_{y \cdot x} \pm t_{\alpha/2; N-2} \frac{\sqrt{\text{MS(error + deviations)}}}{S_x \sqrt{N}} \tag{11.48}$$

For the continuing neurotransmitter example the limits of the 95% confidence interval are

$$2.30 - \frac{2.145\sqrt{(25.2)}}{2.236\sqrt{16}} = 2.30 - 1.20 = 1.10$$

and

$$2.30 + \frac{2.145\sqrt{(25.2)}}{2.236\sqrt{16}} = 2.30 + 1.20 = 3.50$$

The confidence interval related to the true value of the dependent variable for a given individual in the population is very useful. Suppose that a person is drawn from the population of brain damaged individuals who are discussed in the neurotransmitter example and is administered 4 milliliters (ml) of the drug. Using the regression equation for this example, we can predict that this person would answer 27.7 items correctly. However, of interest is an interval estimate of the true rate of responding on the memory task for this individual. When the assumptions of the regression model are true, the limits of the $(1 - \alpha)100\%$ confidence interval for the true score are

$$y'_j \pm t_{\alpha/2;N-2}\sqrt{\text{MS}(\text{error} + \text{deviations})}\sqrt{1 + \frac{1}{N} + \frac{(x_j - M_x)^2}{NS_x^2}} \quad (11.49)$$

where the subscript j refers to the level of the independent variable to which the ith subject belongs. For the continuing example the limits of the 95% confidence interval of the true value of y for an individual who receives 4 ml of the transmitter is

$$27.7 - 2.145\sqrt{25.2}\sqrt{1 + \frac{1}{16} + \frac{(4-5)^2}{16(5)}} = 27.7 - 11.16 = 16.54$$

and

$$27.7 + 2.145\sqrt{25.2}\sqrt{1 + \frac{1}{16} + \frac{(4-5)^2}{16(5)}} = 27.7 + 11.16 = 38.86$$

That is, the probability is approximately .95 that the interval from 16 items recalled to 39 items recalled covers the true rate of recall for a person in the population who receives 4 ml of the transmitter.

Examining the confidence interval for the true value of y [Equation (11.49)], we can see that the interval is smallest at the mean of x (i.e., when $x_j = M_x$) and diverges as the values of the independent variable depart from the mean. This "diverging confidence interval" phenomenon is illustrated in Figure 11.4. A second observation about this confidence interval is that, as expected, the interval becomes smaller as N increases, and as well, the interval becomes smaller as S_x^2 increases. Therefore, when the numerical values associated with the categories of an independent variable are spread out, the confidence interval is smaller. This is why in the neurotransmitter example, although the correlation was substantial ($r_{xy} = .74$), the confidence interval formed earlier was quite large. It would have been preferable to have a larger range in dosages (provided additional dosages were of interest to the researcher).

11.5   Relaxation of the Restrictions Placed on the Regression Model

To this point only the classical regression model in which the levels of the independent variable are fixed by the researcher has been discussed.

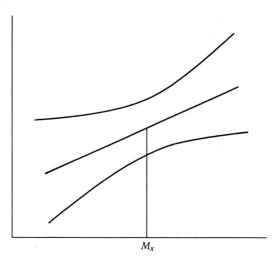

M_x

FIGURE 11.4   Diverging Confidence Intervals for True Y_j

Needless to say, this is a rather restrictive model. A variety of situations exists where the researcher samples from a bivariate distribution but still wishes to predict scores on one variable from scores on the other variable. For example, it would be of interest to investigate the relation between the size of a city and the crime rate. Although the correlation coefficient for the size of city and the crime rate is informative, a regression equation that predicts the crime rate from the size of the city provides important additional information. There is nothing in the development of the regression equation and the related sample statistics that prevents their application to data that have been sampled from a bivariate distribution. Further, when regression analysis is used in the context of sampling from a bivariate distribution, the statistical tests (i.e., the t or F test for $\beta_{Y \cdot X}$) and related confidence intervals are valid **as long as the two variables are bivariate normally distributed.** As was the case for the correlation model, however, violation of this assumption can lead to erroneous conclusions about the relation between the two variables. For example, when the two variables are not bivariate normally distributed, it would be incorrect to say that failure to reject the null hypothesis that $\beta_{Y \cdot X}$ equals zero indicates that there is no relation between the two variables, since the relation may be curvilinear. As stated before, statistical tests should be used cautiously when there is some evidence that violation of the assumptions has occurred.

Although a regression analysis may be conducted on data obtained from studies that used either the classical regression model or the bivariate distribution model, the inferences made are quite different. In the classical regression model the levels of the independent variable are fixed by the researcher, so, provided the study is otherwise well conducted (i.e., free of confounds), an attribution of causality can be made. That is, rejection of the null hypothesis that $\beta_{Y \cdot X}$ equals zero indicates that the independent variable is causally related to the dependent variable. In the neurotransmitter example control of the independent variable permits the researcher to say that "the number of items recalled is caused, in part, by the dosage of the neurotransmitter administered." Further, the nature of the causality is explicated by the various statistics related to regression analysis (the regression equation, confidence intervals, etc.).

When the design of an experiment involves sampling from a bivariate distribution, attributions of causality cannot be made on the basis of the experiment. Envisage a study in which elementary aged children are randomly sampled and their hand size and strength are measured. Further, suppose that the researcher performs a regression analysis in which strength is predicted from hand size and the null hypothesis that the regression coefficient is zero is rejected. It would be incorrect to infer that hand size causes strength when clearly age is responsible for both (i.e., older children have larger hands and are stronger). In this example it would make as much sense to conduct a regression analysis in which hand size was predicted from strength. Because there exists no basis related to the design of the study to assign dependent and independent status to the two variables, it is permissible to predict Y from X or X from Y. Thus, two regression coefficients can be calculated:

$$B_{y \cdot x} = \frac{S_y}{S_x} r_{xy}$$

or

$$B_{x \cdot y} = \frac{S_x}{S_y} r_{xy}$$

Similarly, two regression equations can be determined.

$$y_i' = B_{y \cdot x}(x_i - M_x) + M_y$$

or

$$x_i' = B_{x \cdot y}(y_i - M_y) + M_x$$

Of course, one of these two options may be more desirable to the researcher. Certainly, a researcher investigating college grades and high school SAT scores would want to predict college grades from high school SAT rather than vice versa. Nevertheless, use of regression analysis does not raise the inferential status of the data anymore than does using a t test with data derived from a nonexperimental design (see Section 6.10, p. 155). **It is the nature of the design, and not the statistical procedures, that permit statements of causality.** Generally speaking, experimental designs that involve random assignment of subjects to groups to which various treatments are administered permit statements of causality. If the treatments are qualitative (i.e., the independent variable is categorical), then the analysis of variance typically is used to analyze the data; if the treatment is quantitative (i.e., the classical regression model), then regression analysis typically is used to analyze the data. However, it should be noted that regression analysis can be used to analyze data from designs whereby the independent variable is categorical, and in fact regression analysis so applied is equivalent to an analysis of variance (see Section 12.7, p. 398). Nevertheless, data from nonexperimental designs, whether analyzed using correlational, regression, or analysis of variance models, do not have the same capability to infer causality. At this juncture it should be recognized that there are modern (and complex) methods that attempt to utilize correlation to make inferences about causality; however, the logic leading to statements about causality for these analyses is much different from that of an experimental design in which a researcher manipulates the independent variable.

Extensions of Regression Analysis with One Independent Variable

Linear regression with one independent variable can be used to develop several related statistical procedures. Besides being useful in their own right, several of these procedures are helpful in understanding regression with two or more independent variables, the subject of the next chapter.

11.6   Partial Correlation

Often the correlation between two variables is due to a third variable. For example, the correlation between performance on two tasks for a group of children will likely be due to a large extent to the age of the children. Older children tend to do well on both tasks, whereas younger children tend to do poorly on both tasks. It is desirable to determine the correlation of performance on the two tasks with age held constant. One way to do this is to correlate the performance on the tasks for a specified age. However, this is restrictive in that the results apply only to that age. Fortunately, there exists methods to remove the effects of age statistically from the correlation of performance on the two tasks over a range of ages.

Let X_1, X_2, and X_3 be three random variables. The goal is to make inferences about the correlation of two of these variables, say, X_1 and X_2, holding the third constant. Let $\rho_{12\cdot3}$ be the population correlation of X_1 and X_2 holding X_3 constant. **The coefficient $\rho_{12\cdot3}$ is called the *population partial correlation coefficient of X_1 and X_2 (holding X_3 constant)* and indexes the linear relation between X_1 and X_2 after the contribution of X_3 has been removed from X_1 and X_2.**

To make inferences about the population partial correlation coefficient, we sample from the trivariate distribution of X_1, X_2, and X_3 to develop a sample statistic $r_{12\cdot3}$, which is called the *(sample) partial correlation coefficient* and indexes the degree to which scores for variables X_1 and X_2 are linearly related after the effect of X_3 has been removed. Suppose that a random sample of N subjects is selected from the population and each subject has a score on each variable. To remove the contribution of X_3 from X_1, X_1 is regressed onto X_3:

$$x'_{1_i} = B_{x_1 \cdot x_3} x_{3_i} + A$$

The error in regression

$$x_{1_i} - x'_{1_i}$$

is that part of X_1 that remains after the influence of X_3 is removed. Similarly

$$x'_{2_i} = B_{x_2 \cdot x_3} x_{3_i} + A$$

and the error in regression

$$x_{2_i} - x'_{2_i}$$

is that part of X_2 that remains after the influence of X_3 is removed. Therefore, **the correlation coefficient of the errors in predicting X_1 from X_3 with the errors in predicting X_2 from X_3 is a measure of the linear relation between X_1 and X_2 after the influences of X_3 have been removed.** It can be shown algebraically, that the correlation of the errors in regression reduces to

$$r_{12\cdot3} = \frac{r_{12} - r_{13}r_{23}}{\sqrt{(1 - r_{13}^2)(1 - r_{23}^2)}} \tag{11.50}$$

where r_{12} is the sample correlation coefficient for the scores for variables X_1 and X_2, r_{13} is the sample correlation coefficient for the scores for variables X_1 and X_3, and r_{23} is the sample correlation coefficient for the scores for the variables X_2 and X_3.

The fact that the partial correlation coefficient is based on two regression equations should help to justify the use of the "dot" notation used in this context. Recall that $B_{y \cdot x}$ was the regression equation for predicting Y from X. To form the partial correlation coefficient, we predicted both X_1 and X_2 from X_3, thus the subscript "12·3."

To illustrate the use of the partial correlation coefficient, suppose that the performance on tasks 1 and 2 as well as age was recorded for 50 randomly chosen elementary schoolchildren. Let X_1 represent task 1, X_2 represent task 2, and X_3 represent age, and suppose that

$$r_{12} = .40$$
$$r_{13} = .50$$

and

$$r_{23} = .60$$

Schematically

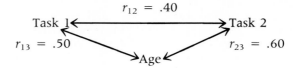

$$r_{12} = .40$$

Task 1 ⟷ Task 2

$$r_{13} = .50 \qquad r_{23} = .60$$

Age

Then the correlation of task 1 with task 2, holding age constant, is given by

$$r_{12 \cdot 3} = \frac{r_{12} - r_{13}r_{23}}{\sqrt{(1 - r_{13}^2)(1 - r_{23}^2)}} = \frac{.40 - (.50)(.60)}{\sqrt{(1 - .25)(1 - .36)}} = .144$$

which is considerably less than the correlation between performance on the two tasks when age is ignored.

The obtained partial correlation coefficient may be used to test the null hypothesis that the population partial correlation coefficient is equal to zero; that is

$$H_0: \rho_{12 \cdot 3} = 0$$

The test involves using Fisher's r to Z transformation in much the same way as it was used to test hypotheses about the regular population correlation coefficient (although in that context it was used only for hypotheses involving nonzero values; see Section 10.3, p. 291). The test statistic is

$$\frac{Z}{\sqrt{V(Z)}} = \frac{Z}{\sqrt{1/(N - 4)}} \qquad (11.51)$$

where Z is the transformed value corresponding to $r_{12 \cdot 3}$. Under the assumption of trivariate normality the value of the test statistic is compared to the standard normal distribution to make a decision with regard to the null hypothesis. For the continuing example the obtained partial correlation coefficient is transformed:

$$r_{12 \cdot 3} = .144 \qquad \text{corresponds to} \qquad Z = .145$$

Thus, the value of the test statistic is equal to

$$\frac{.145}{\sqrt{1/(50 - 4)}} = .98$$

which, when compared to the standard normal, is insufficient to reject the null hypothesis that $\rho_{12 \cdot 3} = 0$ at conventional levels of α. That is to say, after the influence of age is removed, there is little reason to believe that a true linear relation exists between performances on the two tasks; if age had not been considered, a very different conclusion would have been reached (viz., there is a true linear relation between performance on the tasks because the obtained correlation of .40 between the tasks leads to rejection of the hypothesis that $\rho_{12} = 0$).

When one variable is held constant, the partial correlation coefficient is often referred to as the *first-order partial correlation coefficient*. Formulas for calculating and testing *second-order partial correlation coefficients*, in which two variables are held constant, and higher-order partial correlations, in which more than two variables are held constant, also exist. According to this naming convention, the ordinary correlation coefficient is often referred to as the *zero-order correlation coefficient*.

COMMENT   Failure to consider an important third variable can drastically affect the conclusions made about the relation between two variables. This can be seen by examining the ranges of the various coefficients. As was the case with zero-order correlation coefficient, the first-order partial correlation coefficient ranges from -1.00 to 1.00.

$$-1.00 \leq \frac{r_{12} - r_{13}r_{23}}{\sqrt{(1 - r_{13}^2)(1 - r_{23}^2)}} \leq 1.00$$

Solving this inequality for r_{12}, we get

$$r_{13}r_{23} - \sqrt{(1 - r_{13}^2)(1 - r_{23}^2)} \leq r_{12}$$
$$\leq r_{13}r_{23} + \sqrt{(1 - r_{13}^2)(1 - r_{23}^2)} \quad (11.52)$$

which is a range that is more restricted than the typical -1.00 to 1.00. For the example of this section the correlations between age and the tasks were .50 and .60 for task 1 and task 2, respectively. Consequently, the range of the correlation between the tasks (i.e., r_{12}) is

$$(.50)(.60) - \sqrt{(1 - .25)(1 - .36)} \leq r_{12}$$
$$\leq (.50)(.60) + \sqrt{(1 - .25)(1 - .36)}$$

or

$$-.393 \leq r_{12} \leq .993$$

The lower limit of the range of r_{12} is dramatically restricted.

11.7   Semipartial Correlation

Semipartial correlation is similar to the partial correlation in that it indexes the linear relation between two variables when a third variable is taken into account. However, in the context of the semipartial correlation the influence of the third variable is removed from only one of the two other variables. Suppose that a researcher is interested in the correlation between achievement in history (e.g., grades in history class) with a test of history knowledge. Unfortunately, the knowledge test is timed and thus relies on reading speed. The researcher would be interested in the correlation between history grades and test scores after the influence of reading speed has been removed from the test scores.

Let X_1, X_2, and X_3 be three random variables. **We define $\rho_{1(2 \cdot 3)}$ to be the population semipartial correlation coefficient, which indexes the linear relation between X_1 and X_2 when the influence of X_3 has been removed from X_2.** The sample semipartial correlation coefficient, which is denoted by $r_{1(2 \cdot 3)}$, is found by correlating the scores on X_1 with the errors in predicting X_2 from X_3. That is, X_2 is regressed onto X_3:

$$x'_{2i} = B_{x2 \cdot x3} x_{3i} + A$$

and the errors that remain

$$x_{2i} - x'_{2i}$$

are correlated with the scores x_{1i}. Algebraically, it can be shown that this correlation reduces to

$$r_{1(2 \cdot 3)} = \frac{r_{12} - r_{13} r_{23}}{\sqrt{1 - r_{23}^2}} \tag{11.53}$$

The "dot" notation, as previously, follows the general form of that used in regression. The portion of the subscript in parentheses [viz., "$(2 \cdot 3)$"] refers to the regression of X_2 onto X_3. Thus, the entire subscript "$1(2 \cdot 3)$" indicates that the semipartial correlation coefficient is the correlation of X_1 with the errors in predicting X_2 from X_3. Of course, the labeling of the variables was arbitrary, and hence Equation 11.53 can be adapted to find $r_{2(1 \cdot 3)}$. As well, the idea of semipartial correlations can be extended to the case where the influence of more than one variable is removed.

To illustrate the use of the semipartial correlation coefficient, again consider the case of grade point average in history courses and a timed test of history knowledge. Let

X_1 = grade point average
X_2 = timed test of history
X_3 = reading speed

and suppose that

$r_{12} = .30$
$r_{13} = .05$
$r_{23} = .80$

Schematically

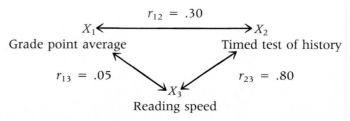

The semipartial correlation of grade point average with the timed test of history when the effects of reading speed are removed from the timed test of history is given by

$$r_{1(2\cdot3)} = \frac{.30 - (.05)(.80)}{\sqrt{1 - .64}} = .43$$

which in this example is considerably greater than the zero-order correlation between grade point average and the timed test.

Several observations about the semipartial correlation coefficient need to be mentioned. First, the semipartial correlation coefficient (and to a lesser extent, the partial correlation coefficient) has an intimate connection with regression with multiple independent variables. Second, because of that connection, statistical tests of the population semipartial correlation coefficient will be discussed in the context of regression with multiple independent variables (see Chapter 12). Third, the semipartial correlation coefficient is not necessarily greater than the zero-order correlation coefficient, as was the case in the example in this section. The relative size of the zero-order and semipartial correlation (and the partial correlation for that matter) is a function of the zero-order correlations among the variables. This important issue will be discussed further in the next chapter as well.

11.8 Analysis of Covariance

Previously we discussed the randomized block design (see Section 8.4, p. 247). The goal of the randomized block design was to reduce the error variance and thereby yield a more powerful test of the statistical hypothesis. An alternative means to reduce error variance in the context of the analysis of variance is the *analysis of covariance*. The analysis of covariance is a statistical means to extract variance due to some nuisance factor(s), whereas the randomized block method approaches the problem from a design standpoint.

Consider the following example, which will be developed further at the end of this section. Suppose that a researcher is investigating the effectiveness of two psychotherapies for the treatment of depression. Subjects, identified as depressed on a pretest, are randomly assigned to the two treatment groups as well as to a control group. Upon completion of the treatment, a posttest is given. Although a one-way fixed-effects analysis of variance is an appropriate statistical test, the researcher realizes that much of the variance in the scores on the posttest are due to initial differences among the subjects before the study began. Had the researcher been concerned with this problem when designing the study, he or she might have blocked on the basis of the pretest scores. The analysis of covariance is a statistical method that removes the variance in the scores due to some other factor (such as, but certainly not limited to, pretest scores).

The linear model for the analysis of covariance is a combination of the model for the one-way analysis of variance with the model for linear regression. As was the case for the one-way analysis of variance, suppose that there are J levels (groups, treatments) of the independent variable and a dependent variable Y. Further, let C be a *covariable* such that each observation on Y is paired with an observation on C. For subject i in group j there are two observations. One observation y_{ij} is the score on the dependent variable and the other observation c_{ij} is the score on a covariable. Again the goal is to extract the variance in the dependent variable that is due to a covariable (i.e., a nuisance factor). Observations on the dependent variable are due to three factors, the treatment effect for treatment j, the linear relation with the covariable, and error. The fact that the observations are due partially to the effect for group j is expressed as

$$Y_{ij} = \mu + \alpha_j + \text{other factors} + \text{error}$$

The fact that the observations are due partially to the linear relation between the dependent variable and the covariable is expressed as

$$Y_{ij} = \beta_{Y \cdot C}(C_{ij} - \mu_C) + \text{other factors} + \text{error}$$

Thus, **the linear model for the analysis of covariance takes into account the treatment effect *and* the linear relation of the dependent variable and the covariable:**

$$Y_{ij} = \mu + \alpha_j + \beta_{Y \cdot C}(C_{ij} - \mu_C) + e_{ij} \tag{11.54}$$

The null hypothesis is the same as for the analysis of variance; namely,

$$H_0: \alpha_j = 0 \quad \text{for } j = 1 \text{ to } j = J$$

At this point it should be recognized that the linear relation between the dependent variable and the covariable has not been expressed in terms of the classical regression model; that is, we have considered C_{ij} to be a random variable and not fixed by the researcher. Technically speaking, the linear model for the analysis of covariance does treat C_{ij} as fixed by the researcher. However, as was the case in the regression model, the consequences of treating C_{ij} as a random variable are not great. Nevertheless, adding the covariable to the linear model for the analysis of variance requires an additional assumption. For the analysis of covariance we assume that the linear relation between X and C is the same for each level of the independent variable. That is, $\beta_{Y \cdot C}$ is not dependent on j. Often this assumption is referred to as *homogeneity of regression*.

Sampling in the context of the analysis of covariance is similar to that for the analysis of variance with the exception that two scores for each subject are recorded, one on the dependent variable y_{ij} and one on the covariable c_{ij}. It is important to realize that the subjects are measured on the covariable **before** the treatment is administered, so as to avoid what is called treatment/covariable confounding. Let N_j be the number of subjects in treatment j, N be the total number of observations, M_y and M_c be, respectively, the sample grand means for the dependent variable and the covariable, and M_{y_j} and M_{c_j} be, respectively, sample means for the dependent variable and the covariable for treatment j.

It is clear from the linear model for the analysis of covariance that the test of the null hypothesis requires that the parameter $\beta_{Y \cdot C}$ be estimated. When the null hypothesis is true [i.e., $Y_{ij} = \mu + \beta_{Y \cdot C}(C_{ij} - \mu_c)$], the best estimator for $\beta_{Y \cdot C}$ is the sample regression coefficient on all of the observations combined. Thus, this estimator, which is denoted by B_t, refers to the estimator based on the total sample, and is given by

$$B_t = \frac{S_{yc}}{S_c^2} = \frac{\sum_j \sum_i (y_{ij} - M_y)(c_{ij} - M_c)}{\sum_j \sum_i (c_{ij} - M_c)^2} \tag{11.55}$$

[see Equation (11.34), p. 329]. However, when the null hypothesis is not true, this estimator does not give optimal predictions for observations within groups (i.e., when group membership is known). In this instance, because it is assumed that the regression coefficients are the same for each group (homogeneity of regression), the best estimator for $\beta_{y \cdot c}$ is obtained by pooling the sample regression coefficients found for each group. This estimator, which is denoted by B_w (w refers to within group estimator), is given by

$$B_w = \frac{\sum_j \sum_i (y_{ij} - M_{y_j})(c_{ij} - M_{c_j})}{\sum_j \sum_i (C_{ij} - M_{c_j})^2} \tag{11.56}$$

which is similar to the equation for B_t, except that here group means are used in the sample covariance and sample variance of the covariable C. The differences between B_t and B_w for a two-group problem are illustrated in Figure 11.5.

We are now ready to develop the statistical test for the null hypothesis that the treatment effects are all zero. Recall that

$$Y_{ij} = \mu + \alpha_j + \beta_{Y \cdot C}(C_{ij} - \mu_c) + e_{ij}$$

The observations in the population Y_{ij} are adjusted by using the linear relation of Y with C to form residuals, denoted by Y_{ij} (adj):

$$Y_{ij}(\text{adj}) = Y_{ij} - \beta_{Y \cdot C}(C_{ij} - \mu_c) = \mu + \alpha_j + e_{ij} \tag{11.57}$$

That is, the residuals (or adjusted observations) are that portion of the observations Y_{ij} that remain after the linear relation between Y and C is accounted for. Conceptually, the analysis of covariance can be thought of as an analysis of variance of these residuals. However, care must be taken when discussing residuals for the sampled scores because it is important to recognize that both B_t and B_w are used in the analysis of covariance.

The sums of squares and mean squares associated with the analysis of covariance are adjusted to reflect the linear association between the depen-

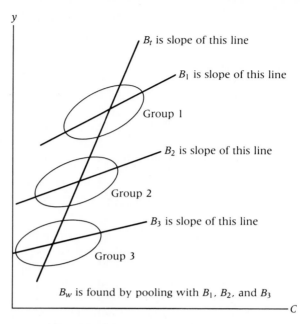

FIGURE 11.5   Schematic of Analysis of Covariance with B_t and B_w Illustrated

dent variable and the covariable. Because the adjustment involves the linear relation between Y and C, cross-product terms are used and defined as follows:

$$\text{Cross product total} = \text{CP total} = \sum_j \sum_i (y_{ij} - M_y)(c_{ij} - M_c) \qquad (11.58)$$

$$\text{Cross product between} = \text{CP between} = \sum_j N_j(M_{y_j} - M_y)(M_{c_j} - M_c) \qquad (11.59)$$

$$\text{Cross product error} = \text{CP error} = \sum_j \sum_i (y_{ij} - M_{y_j})(c_{ij} - M_{c_j}) \qquad (11.60)$$

As expected

$$\text{CP total} = \text{CP between} + \text{CP error}$$

Let SS_y total, SS_y between, and SS_y error denote the sum of squares associated with the dependent variable and SS_c total, SS_c between, and SS_c error denote the sum of squares associated with the covariable. It should be noted that both B_t and B_w can be expressed in terms of sum of squares:

$$B_t = \frac{S_{yc}}{S_c^2} = \frac{\sum_j \sum_i (y_{ij} - M_y)(c_{ij} - M_c)}{\sum_j \sum_i (c_{ij} - M_c)^2} = \frac{\text{CP total}}{SS_c \text{ total}} \qquad (11.61)$$

and

$$B_w = \frac{\sum_j \sum_i (y_{ij} - M_{y_j})(c_{ij} - M_{c_j})}{\sum_j \sum_i (c_{ij} - M_{c_j})^2} = \frac{\text{CP error}}{\text{SS}_c \text{ error}} \qquad (11.62)$$

Returning to the adjusted quantities, we determine the sum of squares total for the analysis of covariance by adjusting the scores on Y using B_t:

$$y_{ij}(\text{adj}) = y_{ij} - B_t(c_{ij} - M_c)$$

It can be shown that the sum of squares total associated with these adjusted scores, which is denoted by *SS(adj) total*, is given by

$$\text{SS(adj) total} = \text{SS}_y \text{ total} - \frac{(\text{CP total})^2}{\text{SS}_c \text{ total}} \qquad (11.63)$$

The adjusted sum of squares error is found from the residuals that are found by taking into account the treatment group j; in this instance

$$y_{ij}(\text{adj for group } j) = y_{ij} - B_w(c_{ij} - M_{c_j})$$

It can be shown that the sum of squares error associated with these adjusted scores, which is denoted by *SS(adj) error*, is given by

$$\text{SS(adj) error} = \text{SS}_y \text{ error} - \frac{(\text{CP error})^2}{\text{SS}_c \text{ error}} \qquad (11.64)$$

The sum of squares between for the adjusted scores, *SS(adj) between*, is found by noting that

$$\text{SS(adj) between} = \text{SS(adj) total} - \text{SS(adj) error}$$

$$= \text{SS}_y \text{ between} + \frac{(\text{CP error})^2}{\text{SS}_c \text{ error}} - \frac{(\text{CP total})^2}{\text{SS}_c \text{ total}} \qquad (11.65)$$

Clearly calculation of SS(adj) between depends on both B_t and B_w.

The adjusted mean squares are found by dividing the sum of squares by the appropriate degrees of freedom:

$$\text{MS(adj) between} = \frac{\text{SS(adj) between}}{J - 1} \qquad (11.66)$$

$$\text{MS(adj) error} = \frac{\text{SS(adj) error}}{N - J - 1} \qquad (11.67)$$

The degrees of freedom for error are 1 less than that of the analysis of variance; this is due to the 1 degree of freedom associated with the regression part of the analysis of covariance.

With regard to the null hypothesis that the treatment effects are all zero, when the mean squares are formed in the way described earlier and the assumptions of the analysis of covariance are met, the test statistic

TABLE 11.3

Analysis of Covariance Source Table

Source	Sum of Squares	df	Mean Squares	F
Between (adj)	SS between(adj)	$J - 1$	MS between(adj) $= \dfrac{\text{SS between(adj)}}{J - 1}$	$\dfrac{\text{MS between(adj)}}{\text{MS error(adj)}} = F \sim F_{J-1, N-J-1}$
Error (adj)	SS error(adj)	$N - J - 1$	MS Error(adj) $= \dfrac{\text{SS error(adj)}}{N - J - 1}$	
Total	SS Total(adj)	$N - 2$		

$$\frac{\textbf{MS(adj) between}}{\textbf{MS(adj) error}} = F \qquad\qquad \textbf{(11.68)}$$

has an F distribution with $(J - 1)$ and $(N - J - 1)$ degrees of freedom. Computationally, the analysis of covariance can be summarized in the source table presented in Table 11.3.

Principles of and calculations involved with the analysis of covariance will be illustrated by referring to Example 11.2, which follows.

EXAMPLE 11.2   At the beginning of this section we mentioned a study involving the two psychotherapeutic treatments for depressed clients. Suppose that 15 moderately to severely depressed subjects are randomly selected from a population of depressed individuals where degree of depression is measured with the Beck Depression Inventory. Scores on this inventory range from 0 to 63 and, typically, the label "severely depressed" is attached to subjects who score above 26, the label "moderately depressed" is attached to subjects who score in the range of 21 to 26, and "mildly depressed" is attached to subjects who score in the range of 14 to 20. The 15 moderately to severely depressed subjects (pretest scores equal to or greater than 21) are randomly assigned to three groups: control, treatment A, and treatment B. The two treatments are administered and posttest scores on the Beck Depression Inventory are obtained for all subjects. The results as well as some summary statistics for this hypothetical study are found in Table 11.4.

TABLE 11.4

**Sample Data for Analysis of Covariance:
Psychotherapeutic Treatment of Depression**

			GROUP		
Control		Treatment A		Treatment B	
c_{i1}	y_{i1}	c_{i2}	y_{i2}	c_{i3}	y_{i3}
21	25	29	21	25	13
27	24	35	23	31	19
24	19	21	16	23	14
30	29	28	20	27	23
28	23	27	30	29	16
$M_{c_1} = 26$	$M_{y_1} = 24$	$M_{c_2} = 28$	$M_{y_2} = 22$	$M_{c_3} = 27$	$M_{y_3} = 17$

$M_c = 27$, $M_y = 21$; correlations for control: $r_{cy} = .45$; treatment A: $r_{cy} = .39$; treatment B: $r_{cy} = .51$.

The null hypothesis is that the treatment effects are all zero; that is,

$$H_0: \alpha_j = 0, \qquad \text{for } j = 1, 2, \text{ and } 3$$

Before beginning the calculations for this example, we should note that there is indeed a strong linear relation between the covariable and the dependent variable for this example; the correlations between pretest and posttest for the three groups are .45, .39, and .51. We begin by calculating the sum of squares for the dependent variable (posttest scores), the sum of squares for the covariable (pretest scores), and the cross products:

$$\text{SS}_y \text{ between} = \sum_j N_j (M_{y_j} - M_y)^2 = 130$$

$$\text{SS}_y \text{ error} = \sum_j \sum_i (y_{ij} - M_{y_j})^2 = 224$$

$$\text{SS}_y \text{ total} = \sum_j \sum_i (y_{ij} - M_y)^2 = 354$$

$$\text{SS}_c \text{ between} = \sum_j N_j (M_{c_j} - M_c)^2 = 10$$

$$\text{SS}_c \text{ error} = \sum_j \sum_i (c_{ij} - M_{c_j})^2 = 190$$

$$\text{SS}_c \text{ total} = \sum_j \sum_i (c_{ij} - M_c)^2 = 200$$

$$\text{CP between} = \sum_j N_j (M_{y_j} - M_y)(M_{c_j} - M_c) = -10$$

$$\text{CP error} = \sum_j \sum_i (y_{ij} - M_{y_j})(c_{ij} - M_{c_j}) = 89$$

$$\text{CP total} = \sum_j \sum_i (Y_{ij} - M_y)(c_{ij} - M_c) = 79$$

From these quantities the estimates of $\beta_{Y \cdot X}$ are calculated. When the null hypothesis that all the treatment effects are zero is true, the best estimate of $\beta_{Y \cdot X}$ (viz., B_t) is found by finding $B_{y \cdot x}$ based on all observations grouped together [see Equation (11.61), p. 346]:

$$B_t = \frac{\text{CP total}}{\text{SS}_c \text{ total}} = \frac{79}{200} = .395$$

However, when the null hypothesis is not true (i.e., there are treatment effects), the best estimate of $\beta_{Y \cdot X}$ (viz., B_w) is found by pooling the estimates for the three groups [see Equation (11.62), p. 347]:

$$B_w = \frac{\text{CP error}}{\text{SS}_c \text{ error}} = \frac{89}{190} = .47$$

The adjusted sum of squares necessary to conduct the F test for the analysis of covariance are

$$\text{SS(adj) between} = \text{SS}_y \text{ between} + \frac{(\text{CP error})^2}{\text{SS}_c \text{ error}} - \frac{(\text{CP total})^2}{\text{SS}_c \text{ total}}$$

$$= 130 + \frac{(89)^2}{190} - \frac{(79)^2}{200} = 140.48$$

$$\text{SS(adj) error} = \text{SS}_y \text{ error} - \frac{(\text{CP error})^2}{\text{SS}_c \text{ error}} = 224 - \frac{(89)^2}{190} = 182.31$$

$$\text{SS(adj) total} = \text{SS}_y \text{ total} - \frac{(\text{CP total})^2}{\text{SS}_c \text{ total}} = 354 - \frac{(79)^2}{200} = 322.79$$

Calculations for the F test are summarized in the following source table:

Source	Sum of Squares	Degrees of Freedom	Mean Squares	F
Between (adjusted)	140.48	2	70.24	4.23
Error (adjusted)	182.31	11	16.58	
Total	322.79	13		

If α was set at .05 a priori, the value of the F ratio, when compared to an F distribution with 2 and 11 degrees of freedom, is sufficiently large to reject the null hypothesis that the treatment effects are all zero.

The power of the analysis of covariance is demonstrated by examining the following source table for the analysis of variance for these data (i.e., ignoring the covariable):

Source	Sum of Squares	Degrees of Freedom	Mean Squares	F
Between	130.00	2	65.00	3.48
Error	224.00	12	18.67	
Total	354.00	14		

In this case the F ratio is insufficiently large to reject the null hypothesis at the same level of α, even though the degrees of freedom for the numerator is 1 greater than for the analysis of covariance.

There are a number of important points to make about the analysis of covariance. Often when the results of an analysis of covariance are reported, quantities referred to as the *adjusted means* are reported rather than the means of the dependent variable. The adjusted means for each group, denoted by $M_{y_j}(\text{adj})$, are the means on the dependent variable adjusted for the linear relation between the dependent variable and the covariable:

$$M_{yj}(\text{adj}) = M_{yj} - B_w(M_{cj} - M_c) \qquad\qquad (11.69)$$

Using this formula, we find the following adjusted means for the depression study:

Group Means

	Control ($j = 1$)	Treatment A ($j = 2$)	Treatment B ($j = 3$)
M_{yj}	24.00	22.00	17.00
$M_{yj}(\text{adj})$	24.47	21.53	17.00

Clearly in this example the adjusted means are not very different from the unadjusted means. This is not surprising because the subjects were randomly assigned to groups, and so it is expected that the means on the covariable for each group would not be too different from each other and any difference would not be systematic. Therefore, when there has been random assignment, the difference between the adjusted and unadjusted means is due to sampling error and is not very interesting. The important adjustment that is made in the analysis of covariance is to the variance. In the analysis of variance, the variance in the dependent variable is separated into error and between group components, whereas in the analysis of covariance, the portion of the error variance that is due to the linear relation between the dependent variable and the covariable is removed, as illustrated in Figure 11.6. Thus, the error term in the F test is reduced. It should be kept in mind that the advantage gained in the analysis of covariance occurs only if there is a linear relation between the dependent variable and the covariable. If

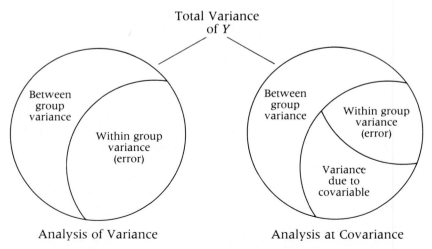

FIGURE 11.6   Heuristic Depiction of Error Variance in Analysis of Variance and Analysis of Covariance

there is no relation or a curvilinear relation (see following section), the analysis of covariance will likely be **less** powerful than the analysis of variance conducted on the same data.

Occasionally we see the analysis of covariance used to "equate" groups to which subjects have not been randomly assigned. The argument is that the regression portion of the analysis of covariance removes differences between groups that are due to nuisance factors and thus equalizes the groups. Unfortunately, this procedure is justifiable only under strict and unrealistic conditions. Although the mathematics of this issue are complicated, as a general rule the analysis of covariance should be thought of as a procedure to reduce error variance and thereby to increase the power of the F test for differences in means rather than as a procedure to equalize groups (see Porter & Raudenbush, 1987).

Even though the analysis of covariance has been discussed only in the context of a one-way analysis of variance with one covariable, the procedure can be generalized to instances with more than one covariable, factorial designs, and planned and post hoc tests, among others.

One of the important questions in any analysis of covariance is whether or not the assumption of homogeneity of regression is reasonable. The answer to this problem will be deferred until the analysis of covariance is revisited in the following chapter.

11.9 Curvilinear Regression

To this point in the regression context we have only considered the linear relation between the independent variable and the dependent variable. Clearly there are instances for which the relation between the independent variable and the dependent variable will not be linear. Consider the relation between anxiety and performance in some tasks: extremely low levels of anxiety can be insufficient to motivate performance, moderate amounts of anxiety can motivate but not interfere with performance, and high levels of anxiety can interfere with performance.

In this section the methods of classical regression are generalized to investigate various forms of curvilinear relations. The term *curvilinear* in this context refers to a nonlinear relation between the independent variable and the dependent variable that can be expressed as a polynomial equation. In general, a polynomial equation is of the following form:

$$y = A + B_1x + B_2x^2 + B_3x^3 + \cdots + B_p x^p$$

where A and B_i are constants. When $B_i = 0$ for $i > 1$, the polynomial equation reduces to

$$y = A + B_1x \tag{11.70}$$

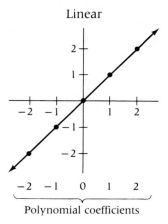

Linear

Polynomial coefficients

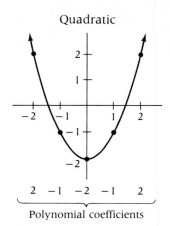

Quadratic

Polynomial coefficients

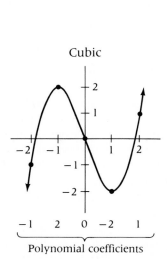

Cubic

Polynomial coefficients

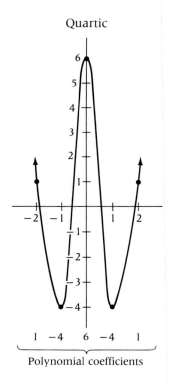

Quartic

Polynomial coefficients

FIGURE 11.7   Graphs of Linear and Curvilinear Relations

which, as we have discussed previously, is a linear equation. Because the exponent on x is 1, the linear equation is often referred to as a first-degree polynomial equation. When $B_i = 0$ for $i > 2$, the general polynomial equation reduces to

$$y = A + B_1 x + B_2 x^2 \qquad (11.71)$$

which is referred to as a quadratic or second-degree polynomial equation. The graph of a quadratic equation is a parabola, as shown in Figure 11.7 (here the parabola is pictured as "concave up" but can also be "concave down"). When $B_i = 0$ for $i > 3$, the general polynomial equation reduces to

$$y = A + B_1 x + B_2 x^2 + B_3 x^3 \qquad (11.72)$$

which is referred to as a cubic or third-degree polynomial equation. The graph of a cubic as well as a quartic or fourth-degree polynomial are also pictured in Figure 11.7.

The goal of a curvilinear regression is to determine whether any of the polynomial equations represent the relation between the independent variable and dependent variable. As was the case with linear regression, curvilinear regression will be developed from a planned comparison approach. This approach is described by considering Example 11.3 which follows.

From the graphical presentation of the data for Example 11.3 it appears that there is a curvilinear relation between the rate of reinforcement and the mean number of problems attempted. The goal of curvilinear regression is

---

EXAMPLE 11.3 Suppose that a researcher is interested in studying the effect of reinforcing recalcitrant children's attempts to work on arithmetic problems. Specifically, the researcher is interested in the rate of reinforcement, believing that a moderate rate is optimal; a nominal rate is insufficiently reinforcing, whereas a saturated rate leads to satiation. The following experimental arrangement is used. Twenty-five children who are resistant to working on arithmetic problems are randomly assigned to five treatment groups so that there are five children in each group. Each group is assigned 20 problems daily and the mean number of problems attempted over a two-week period represents the score on the dependent variable. Each treatment group is reinforced at a different rate: Group I receives one gold star for each 5 problems attempted, Group II receives two gold stars for each five problems attempted, Group III receives three gold stars for each five problems attempted, Group IV receives four gold stars for each 5 problems attempted and Group V receives five gold stars for each 5 problems attempted. The mean number of problems attempted (whole numbers are used for convenience) for each subject is presented tabularly as well as graphically in Figure 11.8.

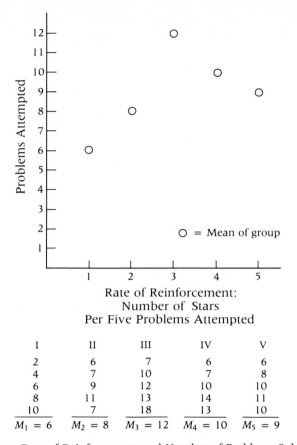

I	II	III	IV	V
2	6	7	6	6
4	7	10	7	8
6	9	12	10	10
8	11	13	14	11
10	7	18	13	10
$M_1 = 6$	$M_2 = 8$	$M_3 = 12$	$M_4 = 10$	$M_5 = 9$

FIGURE 11.8   Relation Between Rate of Reinforcement and Number of Problems Solved

to identify the type of the relation (linear, quadratic, cubic, etc.). The term "identify" is used loosely here; specifically, we want to know if there is sufficient evidence in the sample data to accept the hypothesis that a true (i.e., population) curvilinear relationship exists between the independent variable (here, rate of reinforcement) and the dependent variable (here, mean number of problems attempted).

We begin the presentation by placing further restrictions on the classical regression model. To simplify the computations, we limit our development of curvilinear regression to instances in which the number of subjects in each treatment group is equal (i.e., $N_1 = N_2 = \cdots = N_J$) and the treatment groups are equally spaced (i.e., $x_{j+1} - x_j = $ constant). Clearly the reinforcement example meets these criteria because there are five subjects in each group and the difference in the independent variable for adjacent groups is equal to one gold star for each five problems attempted. In the following chapter we will again consider curvilinear regression and these restrictions will be relaxed.

In this section curvilinear regression is developed by creating a set of orthogonal planned comparisons that correspond to the various curvilinear (and linear) relations of interest. If we were using planned comparisons, the omnibus analysis of variance would not be used. However, to demonstrate the connection between curvilinear regression and previous approaches, we first conduct a simple one-way fixed-effects analysis of variance on the reinforcement example. The sums of squares are as follows:

$$SS \text{ between} = \sum_j N_j (M_j - M)^2 = 100$$

$$SS \text{ error} = \sum_j \sum_i (y_{ij} - M_j)^2 = 188$$

$$SS \text{ total} = \sum_j \sum_i (y_{ij} - M)^2 = 288$$

Thus, the following analysis of the variance results:

Source	Sum of Squares	Degrees of Freedom	Mean Squares	F
Between	100	4	25.00	2.66
Error	188	20	9.40	
Total	288	24		

Although the size of F (when compared to an F distribution with 4 and 20 degrees of freedom) is not sufficient to reject the null hypothesis that all the treatment means are equal (at conventional levels of α), it must be remembered that there remains the possibility that one or more planned comparisons exist that are statistically significant. As we will see, this is the case.

We now turn to linear relations in the data. The reader should recall that linear regression was approached from a planned comparison standpoint and the first planned comparison that will be examined for the reinforcement example is this one. It will turn out that the curvilinear relations can also be expressed as planned comparisons that are orthogonal from each other and from the linear comparison. We continue by examining the linear comparison in some detail before progressing to the curvilinear comparisons.

The planned comparison for the linear relation is denoted by ψ_1. Recall that the sample planned comparison for the linear relation can be expressed in terms of sample statistics as follows [see Equation (11.39), p. 330]:

$$\hat{\psi}_1 = \sum_j c_j M_{y_j}$$

where

$$c_j = \frac{N_j (x_j - M_x)}{N}$$

The coefficients c_j for the planned comparison for the linear relation are easily calculated for the reinforcement example:

$$c_1 = \frac{5}{25}(1 - 3) = -\frac{2}{5}$$

$$c_2 = \frac{5}{25}(2 - 3) = -\frac{1}{5}$$

$$c_3 = \frac{5}{25}(3 - 3) = 0$$

$$c_4 = \frac{5}{25}(4 - 3) = \frac{1}{5}$$

$$c_5 = \frac{5}{25}(5 - 3) = \frac{2}{5}$$

Because the fractions are cumbersome, these coefficients are multiplied by 5, and the following coefficients are used to test for a linear relation:

$$c_1 = -2, \quad c_2 = -1, \quad c_3 = 0, \quad c_4 = 1, \quad c_5 = 2$$

Hence

$$\hat{\psi}_1 = \sum_j c_j M_{yj} = -2(6) - 1(8) + 0(12) + 1(10) + 2(9) = 8$$

To calculate $SS(\hat{\psi}_1)$, recall that [Equation (9.11), p. 272]

$$SS(\hat{\psi}_1) = \frac{(\hat{\psi}_1)^2}{\sum_j (c_j^2/N_j)}$$

Therefore

$$SS(\hat{\psi}_1) = \frac{64}{2} = 32$$

and the statistical test for the planned comparison for the linear relation is summarized in the following source table:

Source	Sum of Squares	Degrees of Freedom	Mean Squares	F
(Between	100	4)		
$\hat{\psi}_1$: Linear	32	1	32.00	3.40
Remainder	68	3	22.67	2.41
Error	188	20	9.40	
Total	288	24		

The size of F is not sufficient (when compared to an F distribution with 1 and 20 degrees of freedom) to reject the null hypothesis related to a linear relation in the population; that is, there is not sufficient evidence in the sample data to accept that a true linear relation exists between rate of reinforcement and performance. The size of F for the remainder is not sufficient either (when compared to an F distribution with 3 and 20 degrees of freedom) to reject the null hypothesis. Nevertheless, as we will see there is a planned comparison contained in the remainder that is significant (although again, in actual practice, the planned comparisons would be stipulated a priori and the analysis would not be conducted in the sequential fashion presented here).

Before discussing planned comparisons for curvilinear relations, we need to compare the test for the linear relation conducted here to the test conducted previously in which the SS remainder was handled differently. Recall from Section 11.4 (p. 324) that the test for linear regression involved pooling SS deviations and SS error, whereas the test for linear regression conducted in this section (i.e., in the context of curvilinear regression), SS remainder (which is equal to SS deviations) was kept separate. Although it is natural to inquire about which strategy is correct, the answer is that they both are, although in different contexts. In Section 11.4 we were considering only linear regression and the model did not allow for other types of relations (i.e., curvilinear relations), and therefore any deviations from linear regression were considered error. In the context of curvilinear regression, deviations from linear regression (i.e., SS remainder) may be due to curvilinear relations and are investigated in their own right. So, it depends on the model the researcher wishes to use (linear relations only vs. linear and curvilinear relations). Of course, as we will see, there are advantages and disadvantages of choosing any particular model (see especially Section 12.5, p. 387).

As we have discussed in this section and in a previous section (see Section 11.4, p. 324), the coefficients for the planned comparison for a linear relation are easily calculated from the sample data. Although determination of the coefficients for curvilinear relations is more complex (and beyond the scope of this text), the idea behind the coefficients is straightforward. The goal is to develop a set of coefficients for each curvilinear relation (quadratic, cubic, etc.) that detects the particular curvilinear relation and is orthogonal to the set of coefficients for the other curvilinear relations and to the set for the linear relation. When there are five treatment groups, the coefficients for the linear, quadratic, cubic, and quartic relations are as follows:

	c_1	c_2	c_3	c_4	c_5
Linear	-2	-1	0	1	2
Quadratic	2	-1	-2	-1	2
Cubic	-1	2	0	-2	1
Quartic	1	-4	6	-4	1

Clearly these coefficients form orthogonal sets. Their connection to the linear and curvilinear relations is also clear if these coefficients are graphed at equally spaced intervals; the coefficients for the linear comparison fall on a straight line, the coefficients for the quadratic comparison fall on a parabola (graph of a second-degree polynomial), and so on (see Figure 11.7, p. 354).

As was indicated earlier, the number of possible mutually orthogonal planned comparisons is always 1 less than the number of groups (i.e., number of levels of the independent variable—see Section 9.2, p. 268). In the present example there are five groups, and thus there are four orthogonal comparisons—linear, quadratic, cubic, and quartic. The tables in Appendix B.7 (p. 460) give the $(J - 1)$ polynomial coefficients that are used to conduct planned comparisons for the linear and $(J - 2)$ curvilinear relations.

Using the polynomial coefficients, we can easily calculate the planned comparisons for curvilinear relations for the rate of reinforcement example. Previously, $\hat{\psi}_1$ was the sample value for the linear planned comparison; let $\hat{\psi}_2$, $\hat{\psi}_3$, and $\hat{\psi}_4$ represent the sample values for the planned comparisons related to the quadratic, cubic, and quartic relations, respectively. Using the coefficients mentioned earlier, we obtain

$$\hat{\psi}_2 = \sum_j c_j M_j = 2(6) - 1(8) - 2(12) - 1(10) + 2(9) = -12$$

$$\hat{\psi}_3 = -1(6) + 2(8) + 0(12) - 2(10) + 1(9) = -1$$

$$\hat{\psi}_4 = 1(6) - 4(8) + 6(12) - 4(10) + 1(9) = 15$$

The sums of squares for these planned comparisons are calculated in the usual manner:

$$SS(\hat{\psi}_2) = \frac{(\hat{\psi}_2)^2}{\sum_j (c_j^2/N_j)} = \frac{144}{14/5} = 51.43$$

$$SS(\hat{\psi}_3) = \frac{1}{10/5} = .50$$

$$SS(\hat{\psi}_4) = \frac{225}{70/5} = 16.07$$

Recalling that $SS(\hat{\psi}_1) = 32$, we construct the source table for the linear and curvilinear trends as follows:

Source	Sum of Squares	Degrees of Freedom	Mean Squares	F
(Between	100.00	4)		
$\hat{\psi}_1$: Linear	32.00	1	32.00	3.40
$\hat{\psi}_2$: Quadratic	51.43	1	51.43	5.47
$\hat{\psi}_3$: Cubic	.50	1	.50	.05
$\hat{\psi}_4$: Quartic	16.07	1	16.07	1.71
Error	188.00	20	9.40	
Total	288.00	24		

When modeled in this way, only the quadratic comparison is statistically significant ($\alpha = .05$), indicating that there is sufficient evidence in the sample data to decide to reject the null hypothesis that no true (i.e., population) quadratic relation exists between rate of reinforcement and problems attempted in favor of the alternative hypothesis that a quadratic relation exists.

In summary, the approach to investigating curvilinear relations taken in this section was to create planned comparisons for the linear and curvilinear relations. To be able to do this, restrictions were necessary, namely, the classical regression paradigm with equal sample sizes in equally spaced intervals of the independent variable. In the next chapter an alternative approach will be used that relaxes these restrictions (although a price will be paid for doing so!). It should be noted that in the present approach no attempt was made to identify the curvilinear regression equation; in the reinforcement example we did not attempt to write the particular quadratic equation that best describes the quadratic relation between the rate of reinforcement and problems attempted. This will be discussed in the next chapter as well. Another point needs to be emphasized. In the reinforcement example the four possible relations were tested (i.e., linear, quadratic, cubic, and quartic). It is quite likely that a researcher has hypotheses about a limited number of these relations, say, linear and quadratic. In this case it makes little sense to test for other relations and the relations that are not tested would collapse into a remainder term. Whether or not the remainder should be pooled with the error terms is a question that needs to be answered by the researcher. A complete discussion of this topic will wait until the following chapter, too. A final point is related to the use of post hoc comparisons. Although the discussion in this section focused on planned comparisons, it is appropriate to conduct an omnibus F test and, if a significant F was obtained, use Scheffé's method to test for particular curvilinear relations.

NOTES AND SUPPLEMENTARY READINGS

Cohen, J., & Cohen, P. (1983). *Applied multiple regression/correlation analysis for the behavioral sciences* (2nd ed.). Hillsdale, NJ: Lawrence Erlbaum Associates.

This volume is primarily a book on multiple regression as indicated by its title. The authors also examine the regression/correlation connection to analysis of covariance.

There are many sources that specifically discuss analysis of covariance. We present the following as a sample of some that relate to various aspects of this procedure.

Cochran, W. G. (1957). Analysis of covariance: Its nature and uses. *Biometrics, 13,* 261–281.

   A classic early summary of the analysis of covariance model and its application.

The following represent discussions pertaining to the interpretation of results from analysis of covariance:

Huitema, B. E. (1980). *The analysis of covariance and alternatives.* New York: Wiley.

   A comprehensive discussion of the many variations on the theme of the analysis of covariance.

Lord, F. M. (1967). A paradox in the interpretation of group comparisons. *Psychological Bulletin, 68,* 304–305.

Lord, F. M. (1969). Statistical adjustments when comparing preexisting groups. *Psychological Bulletin, 72,* 336–337.

Maxwell, S., & Cramer, E. M. (1975). A note on analysis of covariance. *Psychological Bulletin, 82,* 187–190.

Porter, A. C., & Raudenbush, S. W. (1987). Analysis of covariance: Its model and use in psychological research. *Journal of Counseling Psychology, 34,* 383–392.

   This article presents an understandable and concise, yet comprehensible, introduction to analysis of covariance. Important for its discussion of using this method for research with nonrandomly assigned subjects.

PROBLEMS

1. Suppose that a researcher was interested in the conjecture that deprivation from sunlight is related to depression. Sixteen subjects were selected and exposed to sunlight for a given amount of time each day for

a period of two weeks. Groups I, II, III, and IV received 2, 4, 6, and 8 hours, respectively, of sunlight each day. At the end of the two weeks, the Beck Depression Inventory was administered (see Example 11.2, p. 349; recall that higher scores indicate greater levels of depression). The results of this study follow:

Subject	Hours of Daylight x	Beck Depression Inventory Scores y
1	2	34
2	2	20
3	2	18
4	2	32
5	4	28
6	4	24
7	4	23
8	4	21
9	6	20
10	6	11
11	6	14
12	6	15
13	8	17
14	8	7
15	8	8
16	8	12

The correlation between hours of sunlight and depression was equal to $-.785$. Further, $M_x = 5$, $S_x^2 = 5$, $M_y = 19$, $S_y^2 = 59.12$.

a. Calculate the regression coefficient $B_{y \cdot x}$ and write the regression equation for predicting scores on the Beck Depression Inventory from hours of sunlight.

b. How much of the variance in depression scores is due to hours of sunlight?

c. Calculate SS error, SS deviations, and SS linear regression.

d. Test the null hypothesis that the population regression coefficient is zero [use Statement (11.42), p. 332, $\alpha = .05$].

e. Find the 95% confidence interval for $\beta_{Y \cdot X}$ and for the true value of the dependent variable (viz., depression scores) given that a person was exposed to sunlight for 6 hours per day.

2. To determine the effect of the number of clients on the production of case notes of counselor-trainees, the trainees were assigned from one to five clients and after ten weeks the number of pages of case notes per client was noted.

Trainee	Number of Clients	Pages of Notes per Client
1	1	15
2	1	13
3	1	8
4	2	9
5	2	11
6	2	4
7	3	14
8	3	6
9	3	7
10	4	7
11	4	11
12	4	3
13	5	3
14	5	6
15	5	3

The correlation between the number of clients and the number of pages per client of case notes was equal to $-.614$. Further, $M_x = 3$, $S_x^2 = 2$, $M_y = 8$, $S_y^2 = 15.33$.

a. Calculate the regression coefficeint $B_{y \cdot x}$ and write the regression equation for predicting the number of case notes per client from the number of clients.

b. How much of the variance in the number of pages per client is due to the number of clients?

c. Calculate SS error, SS deviations, and SS linear regression.

d. Test the null hypothesis that the population regression coefficient is zero [use Statement (11.42), p. 332, $\alpha = .01$].

e. Find the 99% confidence interval for $\beta_{Y \cdot X}$ and for the true value of the dependent variable (viz., number of pages per client) given that a trainee had one client.

3. An instructor asked each of eight students to record the number of hours that they spent studying for the final examination. The following table presents the number of hours spent studying as well as the number of problems solved correctly on the final (ten problems total).

Student	Hours of Study x	Number of Problems Solved y
1	1	4
2	3	6
3	6	10
4	5	5
5	2	2
6	4	6
7	6	8
8	5	7

The correlation between the number of hours studied and problems solved was .819. Further, $M_x = 4$, $S_x^2 = 3$, $M_y = 6$, $S_y^2 = 5.25$. In this problem we are concerned with predicting problems solved from the number of hours studied.

a. Calculate the regression coefficient $B_{y \cdot x}$ and write the regression equation for predicting problems solved from hours studied.

b. For each student calculate the predicted number of problems solved on the final based on the number of hours studied.

c. For each student calculate the error in prediction (i.e., $y - y'$). As well, calculate the average of the squared errors; that is

$$\frac{1}{N} \sum (y_i - y_i')^2$$

d. How much variance in the number of problems solved is accounted for by the number of hours studied?

e. Calculate the sample standard error of estimate and indicate how this value is related to the average of the errors that was calculated in (c).

f. Test the null hypothesis that the population regression coefficient is zero (one-tailed test, $\alpha = .01$).

4. A researcher interested in short-term memory in older persons administered a memory task to ten randomly selected subjects over the age of 55. The following table gives the ages of the subjects as well as the number of words recalled from a list of 25 words:

Subject	Age x	Number of Words Recalled y
1	55	18
2	60	15
3	62	12
4	58	14
5	73	15
6	79	10
7	58	13
8	61	12
9	64	15
10	60	16

The correlation between age and the number of words recalled is $-.539$. Further, $M_x = 63$, $S_x^2 = 49.4$, $M_y = 14$, $S_y^2 = 4.8$. In this problem we are concerned with predicting words recalled from age.

a. Calculate the regression coefficient $B_{y \cdot x}$ and write the regression equation for predicting words recalled from age.

b. For each subject calculate the predicted number of words recalled based on the subject's age.

c. For each subject calculate the error in prediction (i.e., $y - y'$). As well, calculate the average of the squared errors; that is

$$\frac{1}{N}\sum (y_i - y_i')^2$$

d. How much variance in the number of words recalled is accounted for by age?
e. Calculate the sample standard error of estimate and indicate how this value is related to the average of the errors that was calculated in (c).
f. Test the null hypothesis that the population regression coefficient is zero (one-tailed test, $\alpha = .05$).

5. Let X_1 represent yearly income, X_2 represent years of education, and X_3 represent father's years of education. Suppose that for a sample size of 25

$r_{12} = .40$
$r_{13} = .30$

and

$r_{23} = .50$

a. Calculate the sample partial correlation between income and education holding father's education constant (i.e., $r_{12 \cdot 3}$).
b. Test the null hypothesis that the population partial correlation coefficient is zero versus the alternative that it is greater than zero ($\alpha = .05$).

6. In the context of professional basketball, let X_1 represent scoring average, X_2 represent salary, and X_3 represent height. Suppose that for a sample size of 75

$r_{12} = .50$
$r_{13} = .60$

and

$r_{23} = .40$

a. Calculate the sample partial correlation between scoring average and salary holding height constant (i.e., $r_{12 \cdot 3}$).
b. Test the null hypothesis that the population partial correlation coefficient is zero versus the alternative that it is greater than zero ($\alpha = .05$).

7. An instructor was interested in determining whether or not three different means of teaching students computer skills were equally effective. Fifteen students were randomly assigned to the three treatments and after the instruction was completed, the students attempted 20 content relevant problems. The following table presents the number of problems successfully completed on the computer as well as the students' scores on the quantitative section of the Scholastic Aptitude Test (SAT):

Treatment I		Treatment II		Treatment III	
SAT	Problems Solved	SAT	Problems Solved	SAT	Problems Solved
500	10	550	11	600	12
550	14	500	8	450	5
600	15	400	5	550	14
450	13	650	10	400	9
500	8	400	6	400	10

The correlation between the SAT scores and the number of problems solved is .604.

a. Test the null hypothesis that the three treatments are equally effective taking into account the covariable SAT scores by constructing the appropriate source table ($\alpha = .05$).

b. Calculate the adjusted means for each of the three treatment groups.

8. A researcher wanted to test the potency of four reinforcers for rats. After ten trials using the reinforcer, the time in seconds that it took a rat to complete the maze was recorded. To control for individual differences, the previous month, each rat was tested on a similar maze using the standard reinforcer (which also was the first reinforcer used in the experiment, viz., reinforcer I). The following table lists the times it took the rats to complete a similar maze the previous month (pretest) and to complete the maze in the experiment (posttest)

Reinforcer I		Reinforcer II		Reinforcer III		Reinforcer IV	
Pretest	Posttest	Pretest	Posttest	Pretest	Posttest	Pretest	Posttest
19	22	27	25	20	20	23	31
24	29	22	14	27	21	29	37
26	25	27	20	32	24	20	27
27	32	24	21	29	27	24	25

The correlation between the pretest and posttest scores was .324.

a. Test the null hypothesis that the four reinforcers are equally potent taking into account the pretest by constructing the appropriate source table for the analysis of covariance ($\alpha = .01$).

b. Calculate the adjusted means for each of the three treatment groups.

9. To determine the relation between the number of problems assigned as homework that are related to a topical area in arithmetic and the students' acquisition of knowledge in that area, four students were assigned to each of six groups: After instruction in the area the students in the first group were assigned one problem, the students in the second group were assigned two problems, the students in the third group were assigned three problems, and so forth. A knowledge test containing ten

problems similar to the homework problems was administered and the scores for the 24 students were as follows:

Problems Assigned

1	2	3	4	5	6
0	4	5	3	3	3
2	5	6	6	5	5
4	7	7	6	6	6
2	8	10	9	2	6

Test for a linear and quadratic relation between the number of problems assigned and scores on the knowledge test ($\alpha = .01$)

10. A researcher interested in the effect of providing respite care to elderly people caring for spouses with Alzheimer's disease on the desire to institutionalize the patient conducted the following experiment. Twenty-five couples were randomly assigned to five groups and the couples in each group received a varying amount of respite care. For the couples in the first group no respite care was provided. For the other groups 1, 2, 3, or 4 hours of care were provided per day so that the caretaker spouse could be out of the house without the Alzheimer's patient. The respite care was provided for three months and the dependent measure was the number of times during the three-month period that the care-taker makes attempts to institutionalize the patient. The following table presents the number of attempts to institutionalize for the 25 couples:

No Respite Care	1 Hour of Respite	2 Hours of Respite	3 Hours of Respite	4 Hours of Respite
3	3	0	3	5
5	2	2	1	6
7	4	5	1	5
11	6	4	2	3
9	5	4	3	1

Test the linear and curvilinear relations possible between hours of respite care and attempts to institutionalize ($\alpha = .05$).

12 MULTIPLE REGRESSION

The previous chapter discussed regression analysis for the case in which there was one independent variable. In the classical regression model the levels of a quantitative independent variable are fixed by the researcher; relaxation of this model allowed for bivariate observations. In the latter instance each subject was observed on the independent variable as well as on the dependent variable. Although regression analysis with one independent variable is extremely useful, complex research questions often require more than one independent variable. In this chapter the discussion of regression analysis is extended to the case where there are observations on more than one independent variable.

Regression analysis with more than one independent variable and one dependent variable typically is referred to as *multiple regression*. (Extensions to more than one dependent variable are covered in multivariate texts.) For example, consider the case where the independent variables, mathematics achievement, the number of high school mathematics credits earned, grades in mathematics high school courses, mathematics anxiety, and mathematics self-efficacy are used to predict the dependent variable, the number of college mathematics courses taken. Multiple regression will examine the influence of each of these independent variables on the dependent variable, taking into account all the other independent variables. It should be mentioned that the classical regression model has not been completely abandoned. It is possible to fix the levels of two quantitative independent variables in a factorial arrangement. Nevertheless, in this chapter the paradigm related to observations from a multivariate distribution is emphasized. Three major topics are examined in this chapter: multiple regression with two independent variables, multiple regression with more than two independent variables, and applications of multiple regression.

Multiple Regression with Two Independent Variables

12.1 Linear Model and Sample Equation for Multiple Regression with Two Independent Variables

The concepts and notation for multiple regression are similar to those of the univariate regression discussion in Chapter 11. We begin by giving the population linear model for multiple regression with two independent variables and then discussing the regression equation for a sample selected from the population.

Let X_1, X_2, and Y be three quantitative random variables, where X_1 and X_2 are designated as the independent variables and Y is designated as the dependent variable. The population linear equation, which describes the linear relation of the two independent variables with the dependent variable is given by

$$\text{True } Y'_i = \beta_{Y1 \cdot 2} X_{i1} + \beta_{Y2 \cdot 1} X_{i2} + \text{regression constant} \qquad (12.1)$$

where X_{i1} is the ith observation on X_1, X_{i2} is the ith observation on X_2, $\beta_{Y1 \cdot 2}$ is the *population partial regression coefficient* for X_1, $\beta_{Y2 \cdot 1}$ is the population partial regression coefficient for X_2, and Y'_i is the predicted value of the dependent variable. Clearly in the instance of two independent variables the True Y'_i is a function of the two independent variables. The coefficient for X_1 (i.e., $\beta_{Y1 \cdot 2}$) is called a "partial" coefficient because it accounts for the inclusion of X_2 in the equation, in much the same way as partial correlation coefficients account for another variable. Similarly, $\beta_{Y2 \cdot 1}$ is a partial coefficient. The notation for these coefficients reflects the partial aspect of the coefficient: $\beta_{Y1 \cdot 2}$ is the regression coefficient of Y with X_1 when the effects of X_2 are accounted for. When there is no ambiguity, $\beta_{Y1 \cdot 2}$ simply is written as β_1; thus, the population multiple regression equation with two independent variables simplifies to

$$\text{True } Y'_i = \beta_1 X_{i1} + \beta_2 X_{i2} + \text{regression constant} \qquad (12.2)$$

Understanding the partial nature of the coefficients is important to understanding the intricacies and will be discussed in depth in Section 12.3 (p.

376). Before turning to the sample regression equation, note that, as was the case for univariate regression, the population regression equation can be rewritten in terms of deviations from the means of the independent variables:

$$\text{True } Y'_i = \mu_Y + \beta_1(X_{i1} - \mu_{X_1}) + \beta_2(X_{i2} - \mu_{X_2}) \tag{12.3}$$

We now turn to sampling from the trivariate distribution of X_1, X_2, and Y. A total of N subjects is randomly selected from the population and the scores on the three variables (i.e., x_{i1}, x_{i2}, and y_i) are noted. The sample multiple regression equation with two independent variables is written as

$$y'_i = B_{y1 \cdot 2}X_{i1} + B_{y2 \cdot 1}X_{i2} + A \tag{12.4}$$

which is analogous to the population equation. The coefficients $B_{y1 \cdot 2}$ and $B_{y2 \cdot 1}$ are the *sample partial regression coefficients* and are frequently written as B_1 and B_2, respectively. The sample regression equation may also be written in terms of deviations from the means of the two independent variables M_1 and M_2:

$$y'_i = M_y + B_1(x_{i1} - M_1) + B_2(x_{i2} - M_2) \tag{12.5}$$

The first problem is to find the regression coefficients B_1 and B_2 and the regression constant of Equation (12.4) so that the sample regression equation results in the best possible predictions. As was the case for univariate regression, these quantities are chosen so that the average squared error in prediction is as small as possible. That is, B_1, B_2, and A are chosen so that

$$\frac{1}{N}\sum(y'_i - y_i)^2$$

is minimized.

The problem of finding B_1, B_2, and A is approached by examining the standardized score multiple regression equation wth two independent variables

$$z'_{yi} = b_1 z_{i1} + b_2 z_{i2} + a \tag{12.6}$$

where z_{i1} is the ith observation on X_1 standardized, z_{i2} is the ith observation on X_2 standardized, and b_1 and b_2 are the *standardized partial regression coefficients*. Unfortunately, the least squares criterion in this instance results in a set of equations that must be solved simultaneously. Solving these equations is best approached by using matrix algebra (especially when more than two independent variables are involved) and is beyond the scope of this text. Nevertheless, it can be shown that the average squared error for z-scores [i.e., $(1/N)\sum(z'_{yi} - z_{yi})^2$] is minimized when $a = 0$,

$$b_1 = \frac{r_{y1} - r_{y2}r_{12}}{1 - r_{12}^2} \tag{12.7}$$

and

$$b_2 = \frac{r_{y2} - r_{y1}r_{12}}{1 - r_{12}^2} \tag{12.8}$$

where the numeric subscripts on the Pearson correlation coefficients refer to the independent variables (e.g., r_{y1} is the correlation of scores on X_1 with the scores on Y). Notice that the standardized partial regression coefficients are functions solely of the correlation coefficients among the two independent variables and the dependent variable, which is not surprising given that b_1 and b_2 are standardized coefficients and thus do not contain terms indicating the scale of the variables.

The partial regression coefficients are easily found from the standardized partial regression coefficients by noting that

$$B_1 = b_1 \frac{S_y}{S_1} \tag{12.9}$$

and

$$B_2 = b_2 \frac{S_y}{S_2} \tag{12.10}$$

The partial regression coefficients B_1 and B_2 are used to construct the raw score multiple regression equation and are, as well, estimators for the population partial regression coefficients (i.e., $B_1 = \hat{\beta}_1$, and $B_2 = \hat{\beta}_2$). One interpretation of the partial regression coefficient is straightforward: for a unit change in X_i, the predicted value of the dependent variable (i.e., y') will change B_i units when the other independent variable is unchanged. It should be noted, however, as will be discussed throughout this chapter, the standardized partial regression coefficients are the quantities that often attract our attention because they are uncontaminated with the scale of the variables. Discussion of the test of significance for partial regression coefficients will be postponed until multiple regression is extended to more than two independent variables.

The following example (Example 12.1) illustrates multiple regression with two independent variables.

EXAMPLE 12.1   Consider the problem of predicting monthly income (in dollars) on the basis of years of education and intelligence test scores. Suppose that 23 subjects are randomly selected from a population of employed adults and each subject is measured on the three variables. Further, suppose that the means and standard deviations of the three variables are as follows:

Monthly income Y: $M_y = 1200$ $S_y = 300$
Education X_1: $M_1 = 13$ $S_1 = 3$
Intelligence X_2: $M_2 = 100$ $S_2 = 15$

The correlations of multiple variables are often presented as a correlation matrix, such as the following for this example:

	Y	X_1	X_2
Y = Income		.40	.50
X_1 = Education			.60
X_2 = Intelligence			

We now regress monthly income onto years of education and intelligence test scores by first finding the standardized partial regression coefficients b_1 and b_2:

$$b_1 = \frac{r_{y1} - r_{y2}r_{12}}{1 - r_{12}^2} = \frac{.40 - (.50)(.60)}{1 - (.60)^2} = .156$$

$$b_2 = \frac{r_{y2} - r_{y1}r_{12}}{1 - r_{12}^2} = \frac{.50 - (.40)(.60)}{1 - (.60)^2} = .406$$

The standardized partial regression coefficients are used to find the partial regression coefficients:

$$B_1 = b_1 \frac{S_y}{S_1} = .156\frac{300}{3} = 15.60$$

$$B_2 = b_2 \frac{S_y}{S_2} = .406\frac{300}{15} = 8.12$$

Hence, the sample regression equation for these data is [see Equation (12.5), p. 371]

$$y_i' = M_y + B_1(x_{i1} - M_1) + B_2(x_{i2} - M_2)$$
$$= 1200 + 15.60(x_{i1} - 13) + 8.12(x_{i2} - 100)$$
$$= 15.60x_{i1} + 8.12_{i2} + 185.20$$

This equation is chosen such that the average squared error

$$\left(\frac{1}{N}\right) \sum (y_i' - y_i)^2$$

is minimized across the 23 subjects in the study. The regression equation is used as an estimate of the population regression equation and thus can be used to predict income from education and intelligence. If an individual i from the population has 11 years of education (i.e., $x_{i1} = 11$) and an intelligence score of 108 (i.e., $x_{i2} = 108$), then the best prediction of monthly income for this individual is

$$y' = 15.60(11) + 8.12(108) + 185.20 = 1233.76$$

It is instructive to compare the regression equation using two independent variables to that for one independent variable, say, education. When only education is considered, the regression equation is

$$y'_i = 40x_{i1} + 680$$

In the case where intelligence test scores are ignored an increase of one year of education translates into a predicted increase of $40.00, whereas when intelligence test scores are part of the regression analysis, the increase is $15.60. Similarly, the standardized regression coefficients are dependent on the variables considered. Considering only education, the standardized regression coefficient is

$$b = r_{y1} = .40$$

whereas when education and intelligence are considered

$$b_{y1 \cdot 2} = .156$$

The effect of adding additional variables to an equation is discussed further throughout this chapter.

12.2   Multiple Correlation Coefficient

The *multiple correlation coefficient* for two independent variables, which is denoted by $R_{y \cdot 12}$ (or when the context is clear, simply R), is a measure of how well the predicted scores y' correspond to the actual scores y. If the regression equation is able to predict the dependent variable well, then the correspondence of y and y' will be high. However, if the independent variables are generally unrelated to the dependent variable, there will be little correspondence between y and y'. The correspondence of y and y' that is used is simply the Pearson correlation coefficient. That is, **the multiple correlation coefficient $R_{y \cdot 12}$ is defined to be equal to the correlation between the actual values of y with the predicted values y' obtained by the regression equation [see Equations (12.4) or (12.5), p. 371]:**

$$R_{y \cdot 12} = r_{yy'} \qquad\qquad (12.11)$$

Because y and y' are either positively related or unrelated

$$0 \le R \le 1$$

Actually the definition of the multiple correlation coefficient is general in that it applies to regression analysis with any number of independent variables. When applied to one independent variable

$$R_{y \cdot 1} = r_{yy'} = \text{correlation of } y \text{ with } (Bx_1 + A)$$

but because the size of the correlation coefficient is unaffected by a linear transformation and because R is nonnegative

$$R_{y \cdot 1} = |r_{y1}|$$

That is, when there is only one independent variable, the multiple correlation coefficient R is equal to the absolute value of the zero-order correlation coefficient.

The multiple correlation coefficient can be written in terms of the standardized partial regression coefficients and the zero-order correlations

$$R_{y \cdot 12} = \sqrt{b_1 r_{y1} + b_2 r_{y2}} \tag{12.12}$$

which is computationally easy to use. Applied to Example 12.1 (p. 372), the multiple correlation coefficient for predicting monthly income from years of education and intelligence test scores is

$$R_{y \cdot 12} = \sqrt{.156(.40) + .406(.50)} = .515$$

The square of the multiple correlation coefficient has an extremely important interpretation:

$$R_{y \cdot 12}^2 = \frac{\text{variance of } Y \text{ accounted for by two independent variables}}{\text{total variance of } Y} \tag{12.13}$$

that is, $R_{y \cdot 12}^2$ is the proportion of variance of the dependent variable accounted for when two independent variables are included in the regression equation. This should be intuitively appealing because r^2 is the proportion of variance accounted for by a single independent variable. Keep in mind, this is a sample statistic and refers to the proportion of variance of the particular scores in the sample. For Example 12.1

$$R_{y \cdot 12}^2 = (.515)^2 = .265$$

indicating that 26.5% of the variance in the 23 monthly income scores is accounted for by years of education and intelligence test scores. It is interesting to note (and will be discussed further in the next section) that the proportion of the variance accounted for by both independent variables is greater than the proportion of variance accounted for by either of the independent variables considered by themselves [for the continuing example, $r_{y1}^2 = (.40)^2 = .16$; $r_{y2}^2 = (.50)^2 = .25$], although clearly the proportion of variance accounted for by both independent variables is not the sum of the proportions of the individual variables.

Not surprisingly, the proportion of the variance in the dependent variable that is unaccounted for by the two independent variables is

$$1 - R_{y \cdot 12}^2 =$$

$$\frac{\text{Variance of } Y \text{ unaccounted for by two independent variables}}{\text{total variance of } Y}$$

For the continuing monthly income example

$$1 - R_{y\cdot12}^2 = 1 - .265 = .735$$

73.5% of the variance in income scores is unaccounted for by the two independent variables.

12.3   Interpreting Multiple Regression

When two or more independent variables are involved in a regression analysis, a major issue is related to the interpretation of the analysis. In the previous section it became clear that the proportion of variance accounted for by two independent variables is not necessarily the sum of the proportions of the two independent variables considered separately. It is vital to realize that the regression analysis involving two independent variables **is dependent on the correlation between the two independent variables as well as the correlations of the independent variables with the dependent variable.** This section discusses the influence of the relations among the independent and dependent variables on the regression analysis.

We approach this subject diagrammatically. Consider Figure 12.1 where the circles represent variances of the standardized variables (i.e., unit variances) and the overlap of the circles represent the variance that variables have in common. Before using these diagrams extensively, we should note that they are visual aids only. It is not possible to represent all relationships realistically with these diagrams, although they are particularly useful to understand some basic concepts.

Consider the prototypic relation among two independent variables and a dependent variable on panel (a) of Figure 12.1. In this diagram all variables are correlated with all other variables; furthermore, it is assumed that the correlations are positive. That is, $r_{y1} > 0$, $r_{y2} > 0$, and $r_{12} > 0$. The proportion of variance of Y accounted for by the first independent variable is given by r_{y1}^2 (or $R_{y\cdot1}^2$), and according to the diagram is equal to the sum of the areas denoted by a and b. That is

$$R_{y\cdot1}^2 = r_{y1}^2 = a + b$$

The proportion of variance in Y accounted for by both independent variables is the square of the multiple correlation coefficient for the two independent variables, $R_{y\cdot12}^2$. Diagrammatically, this proportion is equal to the sums of the areas denoted by a, b, and c:

$$R_{y\cdot12}^2 = a + b + c$$

By adding X_2, the proportion of variance accounted for by the independent variables was increased by the area denoted by c:

$$R_{y\cdot12}^2 = r_{y1}^2 + c = R_{y\cdot1}^2 + c$$

(a)

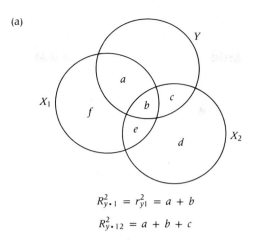

$$R_{y \cdot 1}^2 = r_{y1}^2 = a + b$$

$$R_{y \cdot 12}^2 = a + b + c$$

(b)

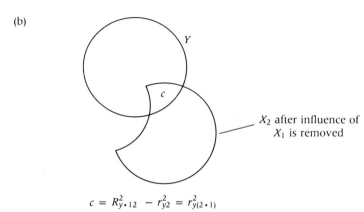

X_2 after influence of X_1 is removed

$$c = R_{y \cdot 12}^2 - r_{y2}^2 = r_{y(2 \cdot 1)}^2$$

(c)

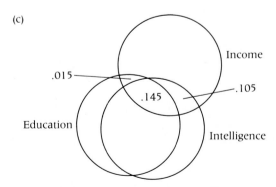

Example 12.1 Illustrated

FIGURE 12.1   Diagrammatic Presentation of the Relation Between Two Independent Variables and a Dependent Variable

That is

$$R^2_{y\cdot12} - R^2_{y\cdot1} = c$$

From the diagram it can be seen that the area denoted by c represents the proportion of variance of Y accounted for by X_2 after the influence of X_1 has been removed from X_2, as shown in panel (b) of Figure 12.1. Hence, the area c is related to the semipartial correlation coefficient $r_{y(2\cdot1)}$ (see Section 11.7, p. 341) in that it is equal to the square of the semipartial correlation coefficient (not surprisingly, because the squares of correlation coefficients are proportions of the variance accounted for). The following result, which summarizes the conclusions reached by examining Figure 12.1, is applicable regardless of the interrelations of the three variables:

$$R^2_{y\cdot12} = r^2_{y1} + r^2_{y(2\cdot1)} \qquad\qquad \textbf{(12.14)}$$

To reiterate, the proportion of variance of Y accounted for by two independent variables is equal to the proportion accounted for by the first plus the proportion accounted for by the second, after the influence of the first variable has been removed. The notion that the square of the semipartial correlation coefficient is the proportion of variance accounted for by the second variable over and above that contributed by the first is demonstrated by rewriting Equation (12.14):

$$r^2_{y(2\cdot1)} = R^2_{y\cdot12} - r^2_{y1} \qquad\qquad (12.15)$$

The quantity $r^2_{y(2\cdot1)}$ is often referred to as an *increment in predictive ability* when the second independent variable is added to the regression equation after the first independent variable. Of course, the notation of the two independent variables is arbitrary; thus, Equations (12.14) and (12.15) can be rewritten so that the scores on X_2 are considered first.

$$R^2_{y\cdot12} = r^2_{y2} + r^2_{y(1\cdot2)}$$

and

$$r^2_{y(1\cdot2)} = R^2_{y\cdot12} - r^2_{y2}$$

These concepts are illustrated for the monthly income example (Example 12.1, p. 372) in panel (c) of Figure 12.1.

Two other concepts related to multiple regression can be illustrated with Figure 12.1. The first concept is related to the various coefficients that index the relation of an independent variable to the dependent variable. One such coefficient is the zero-order correlation coefficient r_{y1} (or its square), which indexes the linear relation of the independent variable X_1 and the dependent variable when the other independent variables are not considered. The coefficient r^2_{y1} is represented by the area denoted by $a + b$. The square of the multiple correlation coefficient $R^2_{y\cdot12}$ is the proportion of the variance of Y accounted for by both independent variables together and is represented by the area denoted by $a + b + c$. The square of the semipartial correlation coefficient $r^2_{y(1\cdot2)}$ is the proportion of the variance in Y accounted for by the first independent variable after the influence of the second is removed and is represented by the area denoted by a, as illustrated in panel (a). Another coefficient discussed in the context of multiple regression is the partial regression coefficient. The partial regression coefficient can be illustrated by Figure 12.1 in that the square of the standardized partial regression coefficient can be represented as a quotient of areas. In particular

$$b^2_1 = \frac{a}{a+f} = \frac{r^2_{y(1\cdot2)}}{1 - r^2_{12}} \tag{12.16}$$

indicating that the square of the standardized regression coefficient is equal to the square of the semipartial correlation coefficient (removing the influence of the second independent variable), divided by the proportion of variance in the first independent variable unaccounted for by the second independent variable. For the continuing monthly income example

$$b^2_1 = \frac{.015}{1 - (.6)^2} = .023$$

and thus $b_1 = .152$, which is equal to the value of b_1 obtained in the previous section (with a minor rounding error).

The second concept that is illustrated by Figure 12.1 is *multicollinearity*. Multicollinearity refers to a linear dependency among the independent variables. One way that such a dependency occurs is when two independent variables are measuring the same thing, resulting in high sample correlations between the two independent variables (i.e., the absolute value of r_{12} is equal to or close to 1.00). Multicollinearity is undesirable for several reasons, which can be understood by examining the instance for which $r_{12} = 1.00$. In this instance the semipartial correlations are equal to zero, indicating that the proportion of variance accounted for by each independent variable over and above the other is equal to zero, even if each independent variable is substantially correlated with the dependent variable. However, it would be incorrect to conclude that the independent variables were unimportant since each alone is related to the dependent variable. Furthermore, when $|r_{12}| = 1.00$, the partial regression coefficient is indeterminate [as can be seen by noting that the denominator in Equation (12.16) is zero in this instance]. Even when $|r_{12}|$ is not equal to 1, if the absolute value of the correlation coefficient of the two independent variables is large, the obtained partial coefficients (e.g., b_1, B_1, $r_{y(1\cdot2)}$) are unstable estimates of the corresponding population values. Finally, computer algorithms for making computations in a regression analysis may produce (unknown) errors when the absolute values of the sample correlations among independent variables are extremely high. If the absolute value of the correlation between two independent variables is high, the researcher should consider (1) obtaining a new sample, because the high obtained correlation may be due to sampling error, (2) use one or the other, but not both, of the independent variables in the analysis, or (3) combine the two independent variables into one, forming a more reliable measure of the psychological construct.

It is important to note that there is another way in which a linear dependency among the independent variables can exist. A linear dependency results when one of the independent variables is equal to a linear combination of other variables. Suppose that X_1 represents a wife's expressed marital satisfaction and X_2 represents the husband's expressed marital satisfaction. It might be tempting to include a variable $X_3 = X_1 - X_2$, which measures the difference in marital satisfaction in the regression analysis. Such a dependency will not necessarily result in high obtained sample correlations; for example, if $r_{12} = .5$ and $X_3 = X_1 - X_2$, then $r_{13} = .5$ and $r_{23} = -.5$. Nevertheless, the partial regression coefficients that result from an analysis including linear combination of variables will be indeterminate (i.e., algorithms for calculating the coefficients will not give unique solutions). One should never include in a regression analysis independent variables that are a linear combination of other independent variables that are also in the analysis.

The complexities of multiple regression are illustrated further by considering two additional possibilities of intercorrelations among the two independent variables and the dependent variable. First, consider the case when the two independent variables are uncorrelated (i.e., $r_{12} = 0$). In this instance

$$r_{y(1\cdot2)}^2 = r_{y1}^2 = b_1^2$$

$$r_{y(2\cdot1)}^2 = r_{y2}^2 = b_2^2$$

and

$$R_{y\cdot12}^2 = r_{y1}^2 + r_{y2}^2$$

That is, because the independent variables are uncorrelated, the semipartial coefficients are equal to the zero-order correlation coefficients and the proportion of variance accounted for by the two independent variables is equal to the sum of the proportions accounted for by each of the independent variables.

More interesting is the interrelationship summarized in the following correlation matrix (which is the same interrelationship as that illustrated in Section 11.7, p. 341, with grade point average as the dependent variable):

	Y	X_1	X_2
Y		.30	.05
X_1			.80
X_2			

Now examine the values of the coefficients generated by regression analysis for each of the independent variables considered by themselves as well as regression analysis with both independent variables:

Considering only X_1	Considering only X_2
$r_{y1}^2 = .0900$	$r_{y2}^2 = .0025$
$b = r_{y1} = .3000$	$b = r_{y2} = .0500$

Considering X_1 and X_2

$$b_1 = \frac{r_{y1} - r_{y2}r_{12}}{1 - r_{12}^2} = \frac{.30 - (.05)(.80)}{1 - (.80)^2} = .7222$$

$$b_2 = \frac{r_{y2} - r_{y1}r_{12}}{1 - r_{12}^2} = \frac{.05 - (.30)(.80)}{1 - (.80)^2} = -.5277$$

$$R_{y \cdot 12}^2 = b_1 r_{y1} + b_2 r_{y2} = .7222(.30) + (-.5277)(.05) = .1903$$

There are several aspects of this example that might seem counterintuitive. First, note that the proportion of variance accounted for by both independent variables (viz., .1903) is greater than the sum of the proportion of each considered separately (viz., .0900 + .0025 = .0925)! That is, it is possible that $R_{y \cdot 12}^2 > r_{y1}^2 + r_{y2}^2$. In this example the addition of X_2 to the regression analysis raised the proportion of variance accounted for by the independent variables from about 9% to 19% even though X_2 was essentially uncorrelated with the dependent variable. Second, note the relation of the values of the standardized partial regression coefficients to the values of the standardized regression coefficients for each independent variable considered alone. The value of the standardized partial regression coefficient for X_1 (viz., .7222) is larger than the value of the standardized regression coefficient for that variable when it is considered by itself. In fact it is possible that when more than one independent variable is included in a regression analysis, the value of the standardized partial regression may be greater than 1.00 (or less than -1.00), unlike the standardized regression coefficient with one independent variable. Further, the value of the standardized partial regression coefficient for X_2 is negative (viz., $-.5277$), although the standardized regression for this variable considered by itself is positive (viz., .0500). When X_1 is considered, scores on X_2 and Y are negatively related, although when X_1 was not considered, the scores were positively related! It is vital to understand that the addition of an independent variable to a regression analysis can dramatically alter the size and direction of partial coefficients for variables. Interpretation of the regression analysis in these instances depends on the substantive meaning of the variables and merits especially close scrutiny by the researcher.

Regression Analysis with More Than Two Independent Variables

12.4 Model, Sample Equation, and Coefficients for Regression Analysis with More Than Two Independent Variables

The development of regression analysis with more than two independent variables parallels the development of the regression analysis with two independent variables. In fact, for the most part, the concepts related to two independent variables apply to more than two independent variables and the sections pertaining to this topic should be read carefully and understood (Sections 12.1, 12.2, and 12.3). Later in this section the concept of sets of independent variables will be introduced.

Let X_1, X_2, ..., X_K be the K independent variables involved in the regression analysis. The population linear regression analysis is given by

$$\text{True } Y_i' = \beta_1 X_{i1} + \cdots + \beta_K X_{iK} + \text{regression constant} \qquad (12.17)$$

or

$$\text{True } Y_i' = \mu_Y + \beta_1(X_{i1} - \mu_{X_1}) + \cdots + \beta_K(X_{iK} - \mu_{X_K}) \qquad (12.18)$$

where β_k is the population partial regression coefficient corresponding to the random variable X_K and X_{ik} is the ith observation of X_k. The partial regression coefficient β_k indexes the linear relation of X_k to the dependent variable Y partialing out the effects of the other $(K - 1)$ independent variables.

The sample regression equation is found by randomly sampling N observations from the multivariate distribution of Y, X_1, X_2, ..., X_K. Typically, N subjects are selected and measured on each of the $(K + 1)$ variables, resulting in $(K + 1)$ scores for each subject. Based on the sample of N subjects, the sample regression equation takes one of two forms:

$$y_i' = B_1 x_{i1} + \cdots + B_K x_{iK} + A \qquad (12.19)$$

or

$$y_i' = M_y + B_1(x_{i1} - M_1) + \cdots + B_K(x_{iK} - M_K) \qquad (12.20)$$

The coefficient B_k is the sample partial regression coefficient. The relation between the sample partial regression coefficient and the standardized sample partial regression coefficient is the same as for univariate and bivariate regression:

$$B_k = b_k \frac{S_y}{S_k} \qquad (12.21)$$

As has been the case throughout regression, the sample regression coefficients and constants are chosen such that the squared error averaged over the N subjects, $(1/N)\Sigma(y' - y)^2$, is minimized. The process of actually calculating partial regression coefficients when K is greater than 1 involves solving simultaneous equations and is most efficiently accomplished with matrix algebra, and thus is beyond the scope of this book. Previously formulas were given for calculating the standardized partial regression coefficients when two independent variables were involved [Equations (12.7) and (12.8), p. 371]; when K is greater than 2, the analogous formulas are tediously complex. Typically, the calculations of multiple regression are accomplished with the use of computers.

The multiple correlation coefficient for K independent variables, denoted by $R_{y \cdot 12 \ldots K}$ is defined in the same way as for two independent variables:

$$R_{y \cdot 12 \ldots K} = r_{yy'} = \sqrt{b_1 r_{y1} + \cdots + b_K r_{yK}} \tag{12.22}$$

Similarly, the square of the multiple correlation coefficient is used to calculate the proportion of variance in Y accounted for and unaccounted for by the K independent variables:

$$R_{y \cdot 12 \ldots K}^2 = \frac{\text{variance of } Y \text{ accounted for by } K \text{ independent variables}}{\text{total variance of } Y} \tag{12.23}$$

and

$$1 - R_{y \cdot 12 \ldots K}^2 = $$
$$\frac{\text{variance of } Y \text{ unaccounted for by } K \text{ independent variables}}{\text{total variance of } Y} \tag{12.24}$$

In the context of multiple regression the sample variance of estimate is referred to as the *sample variance of multiple estimate* and is denoted by $S_{y \cdot 12 \ldots K}^2$. As was the case for regression with one independent variable, the sample variance of multiple estimate is defined as the average squared error:

$$S_{y \cdot 12 \ldots K}^2 = S_y^2 (1 - R_{y \cdot 12 \ldots K}^2) = \frac{1}{N} \sum (y' - y)^2 \tag{12.25}$$

Logically, the *sample standard error of multiple estimate, $S_{y \cdot 12 \ldots K}$,* is given by

$$S_{y \cdot 12 \ldots K} = S_y \sqrt{1 - R_{y \cdot 12 \ldots K}^2} \tag{12.26}$$

We now turn to increments in predictive ability that are gained by adding independent variables to the regression equation. The proportion of variance in Y gained by adding the Kth independent variable to the regression equation after the other $(K - 1)$ independent variables is the square of the semipartial correlation coefficient of X_K with Y after the influences of the other $(K - 1)$ independent variables have been removed from X_K:

$$r_{y[K \cdot 12 \ldots (K-1)]}^2 = R_{y \cdot 12 \ldots K}^2 - R_{y \cdot 12 \ldots (K-1)}^2 \tag{12.27}$$

Increments in predictive ability need not be limited to adding a single independent variable to the equation. It is often substantively interesting, and as we will see often necessary, to determine the increase in the proportion of variance accounted for by a **set** of independent variables over and above another set of independent variables. For now sets of independent variables are denoted by the letters A, B, and T. Let A be a set of G independent variables and B be a set of H independent variables, such that A and B are mutually exclusive. Furthermore, let T be the union of sets A and B such that T contains $G + H = K$ independent variables. Thus, A and B form a partition of T. To summarize

A is a set containing G independent variables
B is a set containing H independent variables
$A \cap B = \phi$
$A \cup B = T$, a set containing $G + H = K$ independent variables

To simplify the notation, we use A and B in place of individual independent variables. For example

$R^2_{y \cdot A}$ = proportion of variance of Y accounted for by the G independent variables contained in set A

$R^2_{y \cdot AB}$ = proportion of variance of Y accounted for by the $G + H = K$ independent variables contained in sets A and B

$r^2_{y(j \cdot A)}$ = the square of the semipartial correlation of Y with X_j after the influences of the G independent variables contained in A have been removed from X_j

For the moment return to the problem of the increment in predictive ability of one independent variable, say, X_j, over and above a set of other independent variables, say, set A, discussed previously. Let B be the set that contains the variable X_j (i.e., $B = \{X_j\}$). Equation (12.27) written in terms of set notation simplifies to

$$r^2_{y(j \cdot A)} = R^2_{y \cdot AB} - R^2_{y \cdot A} \qquad (12.28)$$

The idea of increments in predictive ability now is extended to the case where the set added contains more than one independent variable. As before, the increase in the proportion of variance accounted for by a set, say, set B, over and above another set, say, set A, is given by

$$R^2_{y \cdot AB} - R^2_{y \cdot A}$$

However, this difference is not equal to the square of a semipartial correlation coefficient because the semipartial correlation coefficient was defined in terms of the correlation between two variables (removing the influence of other variables) but not in terms of sets of variables. Nevertheless, the difference between $R^2_{y \cdot AB}$ and $R^2_{y \cdot A}$ is an increment in predictive ability and is denoted by $R^2_{y(B \cdot A)}$. That is to say, we define the *increment in predictive ability for set B over and above set A*, $R^2_{y(B \cdot A)}$, as

$$R^2_{y(B \cdot A)} = R^2_{y \cdot AB} - R^2_{y \cdot A} \qquad (12.29)$$

When the set B contains only one independent variable X_j

$$R^2_{y(B \cdot A)} = r^2_{y(j \cdot A)}$$

The concepts related to multiple regression with more than two independent variables are illustrated in Example 12.2, which follows. This example will also be used in the following section on the relation of sample statistics to the population parameters.

EXAMPLE 12.2

Suppose that a university administrator is interested in the factors that affect faculty salaries. The administration has selected three variables of interest: years since highest degree (X_1), the number of publications (X_2), and the quality of graduate program, rated on a 9-point scale (X_3). A sample of 50 faculty members is selected at random and the following sample correlations are computed:

	Y	X_1	X_2	X_3
Y: Yearly Salary		.40	.50	.40
X_1: Years since degree			.60	.10
X_2: Number of publications				.40
X_3: Quality of graduate program				

The sample means and standard deviations are also found for this hypothetical example.

$$Y: \quad M_y = 30{,}000 \qquad S_y = 10{,}000$$
$$X_1: \quad M_1 = 11 \qquad S_1 = 5$$
$$X_2: \quad M_2 = 7 \qquad S_2 = 3$$
$$X_3: \quad M_3 = 5 \qquad S_3 = 2$$

Although the method for determining the standardized partial regression coefficients when more than two independent variables are included in the regression analysis has not been discussed, the values of the standardized regression coefficients are equal to

$$b_1 = .22$$
$$b_2 = .26$$
$$b_3 = .27$$

The values of the partial regression coefficients are easily found from the standardized coefficients:

$$B_1 = b_1 \frac{S_y}{S_1} = .22 \frac{10000}{5} = 440.0$$

$$B_2 = b_2 \frac{S_y}{S_2} = .26 \frac{10000}{3} = 866.7$$

$$B_3 = b_3 \frac{S_y}{S_3} = .27 \frac{10000}{2} = 1350.0$$

Therefore, the sample regression equation is given by [Equations (12.19) and (12.20), p. 382]

$$
\begin{aligned}
y_i' &= M_y + B_1(x_{i1} - M_1) + B_2(x_{i2} - M_2) + B_3(x_{i3} - M_3) \\
&= 30{,}000 + 440(x_{i1} - 11) + 866.7(x_{i2} - 7) + 1350(x_{i3} - 5) \\
&= 440x_{i1} + 866.7x_{i2} + 1350x_{i3} + 12{,}343.10
\end{aligned}
$$

A partial regression coefficient can be interpreted as the predicted change per unit change of the independent variable with the remaining independent variables unchanged. For example, an additional publication is equal to a predicted increase of $866.70 when years since graduation and the quality of the graduate program are considered in the regression analysis and remain unchanged.

The proportion of the variance in salary accounted for by the three independent variables in this sample is equal to [see Equation (12.22), p. 383]

$$
\begin{aligned}
R_{y \cdot 123}^2 &= b_1 r_{y1} + b_2 r_{y2} + b_3 r_{y3} \\
&= .22(.40) + .26(.50) + .27(.40) \\
&= .33
\end{aligned}
$$

That is, 33% of the variance in the scores on the dependent variable is accounted for by the scores on the three independent variables, and thus 67% is unaccounted for. The value of the standard error of multiple estimate for this sample is equal to [Equation (12.26), p. 383]

$$S_{y \cdot 123} = S_y \sqrt{1 - R_{y \cdot 123}^2} = 10{,}000 \sqrt{1 - .33} = 8185.4$$

It is informative to find out the proportion of variance that each independent variable adds over and above the other independent variables. For X_1 the increment in predictive ability is given by [Equation (12.27), p. 383]

$$r_{y(1 \cdot 23)}^2 = R_{y \cdot 123}^2 - R_{y \cdot 23}^2$$

By the procedures for two independent variables (Section 12.2, p. 374) it is easily shown that

$$R_{y \cdot 23}^2 = .30$$

Therefore, the increment in predictive ability for X_1 is

$$r^2_{y(1\cdot23)} = R^2_{y\cdot123} - R^2_{y\cdot23} = .33 - .30 = .03$$

that is, the first independent variable accounts for 3% of the variance of the dependent variable over and above that accounted for by the other two independent variables. Similarly

$$r^2_{y(2\cdot13)} = R^2_{y\cdot123} - R^2_{y\cdot13} = .33 - .29 = .04$$

and

$$r^2_{y(3\cdot12)} = R^2_{y\cdot123} - R^2_{y\cdot12} = .33 - .27 = .06$$

The use of set notation can be illustrated with this example by letting

$$A = \{X_1\} \qquad \text{and} \qquad B = \{X_2, X_3\}$$

The proportion of variance accounted for by set B over and above that accounted for by set A is given by

$$R^2_{y(B\cdot A)} = R^2_{y\cdot AB} - R^2_{y\cdot A} = .33 - .16 = .17$$

12.5   Estimation and Significance Tests for Multiple Regression

Although the sample regression equation and related statistics are interesting in their own right, sample values typically are used to make inferences about the population regression model. The sample regression coefficients (e.g., partial regression coefficients, semipartial correlation coefficients, multiple correlation coefficients, and increments in predictive ability) are important statistics to help draw some conclusions about the true relation among random variables in the population.

One of the most useful sample statistics for a regression analysis is the square of the multiple correlation coefficient, which was denoted by either $R^2_{y\cdot12\ldots K}$ or $R^2_{y\cdot T}$, where T is the set containing the K independent variables. This statistic is interpretable as the proportion of variance in the sampled y-scores that is accounted for by the independent variables. Of primary interest, however, is the population proportion of variance accounted for by the independent variables. As expected, the sample value is used to make inferences about the population parameter.

Let $\mathrm{P}_{Y\cdot12\ldots K}$ (capital rho) indicate the population multiple correlation coefficient; $\mathrm{P}^2_{Y\cdot12\ldots K}$ is then the population proportion of variance in Y accounted for by the K independent random variables $X_1, \ldots, X_K$. The first question to be considered is whether $R^2_{y\cdot12\ldots K}$ is a good estimator for $\mathrm{P}^2_{Y\cdot12\ldots K}$.

To answer this question, consider two samples of size N randomly selected from the multivariate distribution of $Y, X_1, X_2, ..., X_K$. For the moment ignore the second sample. The proportion of variance accounted for by the independent variables in the first sample (i.e., $R^2_{y \cdot 12...K}$) is based on the regression equation (i.e., $y' = B_1 x_1 + \cdots + B_K x_K + A$) calculated from the first sample. Keep in mind, this regression equation is the "best" regression equation for this sample in the sense that the average squared error across the N multivariate observations is minimized. However, if the regression equation based on the first sample was applied to the second sample, the proportion of variance accounted for by the independent variables would most likely be less than that obtained from applying it to the sample on which it was based. The reason for this is that the regression equation based on the first sample is most probably not the best equation for the second sample. Every sample of size N has associated with it a sample regression equation, which is the best equation for that sample. However, when the regression equation for a particular sample is used with another sample, it is less than optimal and thus the sample proportion of variance accounted for overestimates the population proportion of variance accounted for. That is

$$E(R^2_{y \cdot 12...K}) > P^2_{Y \cdot 12...K}$$

The sample value of the proportion of variance accounted for, $R^2_{y \cdot 12...K}$, can be adjusted so that the population value of $P^2_{y \cdot 12...K}$ can be estimated appropriately:

$$\hat{P}^2 = 1 - \frac{(1 - R^2_{y \cdot 12...K})(N - 1)}{N - K - 1} \tag{12.30}$$

Clearly the estimate of $P^2_{y \cdot 12...K}$ found by adjusting $R^2_{y \cdot 12...K}$ is less than the original value (the estimate is often referred to as "shrunken $R^2_{y \cdot 12...K}$").

The estimate of the population proportion of variance accounted for can be illustrated by referring to Example 12.2, (p. 385) where faculty salary was predicted from three independent variables. Recall that the sample consisted of 50 subjects and the sample value $R^2_{y \cdot 12...K}$ was equal to .33. Thus

$$\hat{P}^2 = 1 - \frac{(1 - R^2_{y \cdot 12...K})(N - 1)}{N - K - 1} = 1 - \frac{(1 - .33)(50 - 1)}{50 - 3 - 1} = .29$$

Interestingly, when the population correlations among the dependent and independent variables are all zero

$$E(R^2_{y \cdot 12...K}) = \frac{K}{N - 1} \tag{12.31}$$

The implication of this formula should be recognized: If the number of independent variables is relatively large (vis-à-vis the number of subjects), the expected value of $R^2_{y \cdot 12...K}$ will be sizable even when there are no linear relations among the variables.

The sample variance of multiple estimate also needs to be adjusted to form the estimator for the population value. Let $\sigma^2_{y \cdot 12 \ldots K}$ be the population variance of multiple estimate. Then

$$\hat{\sigma}^2_{y \cdot 12 \ldots K} = \left(\frac{N}{N - K - 1} \right) S^2_{y \cdot 12 \ldots K} \tag{12.32}$$

Although this estimator will not be used explicitly in our discussion, it is intimately involved in significance testing and confidence intervals for multiple regression.

We now turn to significance tests for various aspects of multiple regression. Because the development of these tests is similar to that of previous statistical tests and because it is rather complex, the details of the development are forsaken in favor of a conceptual discussion. We begin with the test of the square of the multiple correlation coefficient. Explicitly

$$H_0: P^2_{Y \cdot 12 \ldots K} = 0$$

is tested against the alternative that

$$H_a: P^2_{Y \cdot 12 \ldots K} > 0$$

(Of course, this is equivalent to the test of the multiple correlation coefficient, i.e., $H_0: P_{Y \cdot 12 \ldots K} = 0$.) We begin by partitioning the sum of squares for regression in much the same way as we did in regression with one independent variable (see Section 11.4, p. 324, but note that the model has been relaxed to include sampling from a multivariate distribution as opposed to levels of the independent variables fixed by the researcher). The difference between a value of the dependent variable y_i and the sample mean of the dependent variable M_y can be written as

$$y_i - M_y = (y_i - y'_i) + (y'_i - M_y) \tag{12.33}$$

Algebraically, it can be shown that

$$\text{SS total} = \text{SS error} + \text{SS regression} \tag{12.34}$$

where

$$\text{SS total} = \sum (y_i - M_y)^2$$

$$\text{SS error} = \sum (y_i - y'_i)^2$$

$$\text{SS regression} = \sum (y'_i - M_y)^2$$

To obtain unbiased estimators for the variances, we divide the sum of squares by the appropriate degrees of freedom, so that

$$\text{MS error} = \frac{\text{SS error}}{N - K - 1} \tag{12.35}$$

and

$$\text{MS regression} = \frac{\text{SS regression}}{K} \tag{12.36}$$

When the null hypothesis is true, the ratio

$$F = \frac{(\text{SS regression})/K}{(\text{SS error})/(N - K - 1)} = \frac{\text{MS regression}}{\text{MS error}} \tag{12.37}$$

has an F distribution with K and $(N - K - 1)$ degrees of freedom. But note that

$$R^2_{y \cdot 12 \ldots K} = \frac{\text{sample variance accounted for}}{\text{total sample variance}} = \frac{(\text{SS regression})/N}{(\text{SS total})/N}$$

and

$$1 - R^2_{y \cdot 12 \ldots K} = \frac{\text{sample variance unaccounted for}}{\text{total sample variance}} = \frac{(\text{SS error})/N}{(\text{SS total})/N}$$

Therefore

$$F = \frac{(\text{SS regression})/K}{(\text{SS error})/(N - K - 1)} = \frac{R^2_{y \cdot 12 \ldots K}/K}{(1 - R^2_{y \cdot 12 \ldots K})/(N - K - 1)}$$

To summarize: **Under the null hypothesis, the ratio**

$$F = \frac{R^2_{y \cdot 12 \ldots K}/K}{(1 - R^2_{y \cdot 12 \ldots K})/(N - K - 1)} \tag{12.38}$$

has an F distribution with K and $(N - K - 1)$ degrees of freedom and the null hypothesis that $P^2_{Y \cdot 12 \ldots K} = 0$ is rejected provided the F ratio is sufficiently large. Because this statistic is similar to many test statistics in the context of multiple regression, its form is worth noting. Basically, it is the quotient of the variance accounted for and the variance unaccounted for, each divided by the appropriate degrees of freedom. The numerator degrees of freedom is the number of independent variables associated with the variance accounted for and the degrees of freedom associated with the denominator is the sample size minus the number of independent variables minus 1. Several variations of this statistic are presented in this chapter.

As an example, consider again the faculty salary example (Example 12.2, p. 385) where $R^2_{y \cdot 123} = .33$, $N = 50$, and $K = 3$. To test the null hypothesis that $P^2_{Y \cdot 123} = 0$, the value of the F ratio is

$$F = \frac{R^2_{y \cdot 12 \ldots K}/K}{(1 - R^2_{y \cdot 12 \ldots K})/(N - K - 1)} = \frac{(.33)/(3)}{(1 - .33)/(50 - 3 - 1)} = 7.55$$

which when compared to an F distribution with 3 and 46 degrees of freedom results in rejection of the null hypothesis that the true proportion of the variance accounted for is zero.

We now turn to the test of the increment in predictive ability obtained by adding an independent variable to a set of previously entered independent variables. For simplicity, we use the set notation introduced in the previous section. Let set A contain G independent variables and let set B contain the single independent variable X_j, and let there be a total of K independent variables (i.e., in this instance $G + 1 = K$). As discussed in the previous section, the increment in predictive ability for X_j over and above the G independent variables contained in set A is given by

$$r^2_{y(j\cdot A)} = R^2_{y\cdot AB} - R^2_{y\cdot A}$$

To test the null hypothesis that $\rho^2_{Y\cdot(j\cdot A)} = 0$, the F ratio

$$F = \frac{r^2_{y(j\cdot A)}/1}{(1 - R^2_{y\cdot AB})/(N - K - 1)} \tag{12.39}$$

is compared to an F distribution with 1 and $(N - K - 1)$ degrees of freedom. Of course, the division by 1 in the numerator can be omitted but is a remainder that one independent variable contributed to the increase in the proportion of the variance accounted for that is being tested, and thus this test statistic follows the general form discussed previously. It should be recognized that this test of the increment in predictive ability is also a test that the population semipartial correlation coefficient is zero, a test that was alluded to when semipartial correlations were discussed (Section 11.7, p. 341).

To illustrate the use of the test for the increment in predictive ability for one independent variable, again refer to Example 12.2 (p. 385), where the increment in predictive ability for X_1 was .03 ($N = 50$, $K = 3$, $R^2_{y\cdot AB} = .33$):

$$F = \frac{r^2_{y(j\cdot A)}/1}{(1 - R^2_{y\cdot AB})/(N - K - 1)} = \frac{(.03)}{(1 - .33)/(50 - 3 - 1)} = 2.06$$

which when compared to an F distribution with 1 and 46 degrees of freedom does not lead to rejection of the null hypothesis at conventional levels of α. That is, years since the degree does not account for a significant proportion of the variance in faculty salaries after the proportion of variance accounted for by the other two independent variables has been considered.

In the previous section the increment in predictive ability due to a set of variables was discussed. Let the set A contain G independent variables and the set B contain H different independent variables such that there are a total of $G + H = K$ independent variables. Recall that the increment in predictive ability for set B over and above set A was given by

$$R^2_{y(B\cdot A)} = R^2_{y\cdot AB} - R^2_{y\cdot A}$$

Let $P^2_{y(B\cdot A)}$ be the population parameter corresponding to $R^2_{y(B\cdot A)}$. **To test the null hypothesis $P^2_{Y(B\cdot A)} = 0$, we compare the ratio**

$$F = \frac{R^2_{y(B \cdot A)}/H}{(1 - R^2_{y \cdot AB})/(N - K - 1)} \qquad (12.40)$$

to an F distribution with H and $(N - K - 1)$ degrees of freedom. Because H independent variables contributed to the increment in predictive ability $R^2_{y(B \cdot A)}$, the increment in the numerator is divided by H. This test can also be illustrated with Example 12.2 (p. 385), where $A = \{X_1\}$ and $B = \{X_2, X_3\}$. From the example it was shown that $R^2_{y(B \cdot A)} = .17$, $R^2_{y \cdot AB} = .33$, and $K = 50$. Thus

$$F = \frac{R^2_{y(B \cdot A)}/H}{(1 - R^2_{y \cdot AB})/(N - K - 1)} = \frac{.17/2}{(1 - .33)/(50 - 3 - 1)} = 5.84$$

which when compared to an F distribution with 2 and 46 degrees of freedom leads to rejection of the null hypothesis that $P^2_{Y(B \cdot A)} = 0$ at the .01 level of α.

The tests for increments in predictive ability represent a general type of test that bears some further examination. Often a researcher wishes to test the relative ability of two models to explain a psychological phenomenon where one model is a restricted version of the other. We refer to the two models as the full model and the restricted model. In the multiple regression context the model that contains both sets of independent variables A and B is the full model, whereas the model that contains only the set A is the restricted model. The null hypothesis in this framework is that the restricted model explains the phenomena as well as the full model versus the alternative that the full model is superior. With regard to regression, the null hypothesis is that the proportion of variance in the dependent variable accounted for by the independent variables in set A is equal to the proportion of variance accounted for by the independent variables in sets A and B. If the criterion for how well a model fits the data is the least squares criterion, then the general form of the test of the full model versus the restricted model is

$$\frac{[\text{SSE (restricted)} - \text{SSE (full)}]/(\text{degrees of freedom for difference})}{[\text{SSE (full)}]/(\text{degrees of freedom for full model})} \qquad (12.41)$$

where SSE (restricted) is the sum of squares error associated with the restricted model and SSE (full) is the sum of squares error associated with the full model. Given the appropriate assumptions, this ratio frequently has an F distribution under the null hypothesis of no differences between the full and the restricted models.

The test expressed in Equation (12.41) for comparing the restricted model with the full model is easily adapted to increments in predictive ability. Let the restricted model contain the G independent variables in set A and the full model contain the $G + H = K$ independent variables in sets A and B. Given the relationship between the variance unaccounted for, $1 - R^2_y$, and the sum of squares error, it can be shown that when the appropriate degrees of freedom are used, Equation (12.41) reduces to

$$\frac{[(1 - R_{y \cdot A}^2) - (1 - R_{y \cdot AB}^2)]/H}{(1 - R_{y \cdot AB}^2)/(N - K - 1)} = \frac{(R_{y \cdot AB}^2 - R_{y \cdot A}^2)/H}{(1 - R_{y \cdot AB}^2)/(N - K - 1)}$$

$$= \frac{R_{y(B \cdot A)}^2/H}{(1 - R_{y \cdot AB}^2)/(N - K - 1)}$$

which is the ratio discussed earlier [see Equation (12.40), p. 392] for the increment in predictive ability for set B over and above set A. Comparing a restricted model to a full model will be applied to the analysis of covariance when the analysis of covariance is revisited in Section 12.9, (p. 405).

The final test to be discussed here is the one related to the standardized partial regression coefficient b_j. Let B be the set containing the independent variable X_j and let A be the set containing the remaining $(K - 1)$ independent variables. To test the null hypothesis that the population standardized regression coefficient for the jth independent variable is equal to zero, we compare the ratio

$$t = \frac{b_j}{\sqrt{\dfrac{1 - R_{y \cdot AB}^2}{(1 - R_{j \cdot A}^2)(N - K - 1)}}} \tag{12.42}$$

to a t distribution with $(N - K - 1)$ degrees of freedom. In this instance a one- or two-tailed test can be conducted. Notice that the formula for the t statistic contains the term $R_{j \cdot A}^2$, which is the proportion of the variance in X_j accounted for by the remaining $(K - 1)$ independent variables. Clearly hand calculating values for this statistic is cumbersome when $K > 2$; statistical software routinely calculates the value of the t statistic for the user.

12.6 Analytic Strategies

One of the most troublesome areas for researchers is related to the manner in which independent variables are entered into the regression equation and the interpretations made from the regression analysis. There are three basic strategies—simultaneous, stepwise, and hierarchical. Each of these strategies has a number of variations on the theme as well as relative advantages and disadvantages. In this section the strategies are discussed briefly and the reader should seek additional information for complex situations (see Cohen & Cohen, 1983).

In *simultaneous regression* the K independent variables of interest to the researcher are considered simultaneously, resulting in a regression equation with K partial regression coefficients B_1, B_2, ..., B_K [see Equations (12.19) and 12.20, p. 382]. The next step would be to determine whether the proportion of variance accounted for by the K independent variables (viz., $R_{y \cdot 12 \dots K}^2$) was statistically significant; if not, there is not much point in going further because the notion that the proportion of variance accounted for in the sample was due to chance cannot be rejected. If the value of $R^2_{y.12...K}$ was statistically significant, then the researcher would examine the contribution of the individual independent variables in one of two ways. First, the standardized partial regression coefficients could be tested [see Equation (12.42), p. 393]. Second, the increment in predictive ability for each independent variable over and above *all* the remaining independent variables could be tested [see Equation (12.39), p. 391]. Actually the tests of the partial regression coefficient and the increment in predictive ability (over and above all the remaining independent variables) lead to the same conclusion with regard to the null hypothesis, so choice of a statistical test for the individual independent variables is cosmetic. It is possible that the test of $R^2_{y.12...K}$ will be statistically significant but that the tests of all of the individual independent variables (either the partial regression coefficient or the increment in predictive ability) will not be statistically significant. This is one of the consequences of multicollinearity, which was discussed in Section 12.3 (p. 376).

A second analytic strategy is *stepwise regression*. In stepwise regression some algorithm is used to select variables to be entered into the regression equation. Although there are many stepwise regression algorithms from which to select, the following is the simplest. The first independent variable entered into the equation is the one with the highest zero-order correlation with the dependent variables. The second independent variable entered is the one of the remaining variables that results in the largest increase in the increment in predictive ability over and above the first entered variable. The third independent variable entered is the one of the remaining variables that results in the largest increase in the increment in predictive ability over and above the first two entered variables. This process is continued until the increment in predictive ability fails to meet some criterion, such as statistical significance. There are many variations on this theme, some of which include backward stepping (all variables are entered and then one by one eliminated) and procedures that allow for previously entered variables to be removed from the equation.

It should be realized that the increment in predictive ability discussed in the context of stepwise regression is different from that discussed previously. In the context of simultaneous regression the increment in predictive ability for an independent variable was determined by noting the increase in the proportion of variance accounted for by that independent variable over and above that accounted for by *all* the other $(K - 1)$ independent variables. In stepwise regression the increment in predictive ability refers to the increase in the proportion of variance accounted for by that independent variable over and above that accounted for by those independent variables **entered previously**. In a stepwise regression those variables that are not entered into the equation are treated as if they did not exist. For example, the increment in predictive ability for the third entered variable in a stepwise regression would be tested with the following formula:

$$F = \frac{(R^2_{y \cdot 123} - R^2_{y \cdot 12})/1}{(1 - R^2_{y \cdot 123})/(N - 3 - 1)}$$

which would be compared to an F distribution with 1 and $(N - 4)$ degrees of freedom.

There are a number of problems with stepwise regression due to the fact that stepwise procedures select the variables to be included on the basis of the analysis of sample data. If a large number of independent variables is considered originally, then by chance some will be statistically significant and entered into the equation. There is little assurance that another sample would yield a stepwise solution with the variables entered in the same order. Furthermore, if two independent variables are highly correlated, only one is likely to get into the regression equation. Any attribution of superiority to the one variable entered would be incorrect because the two variables are redundant (i.e., measure the same construct). Any results obtained from a stepwise regression should be cross-validated (i.e., replicated on another sample) before much faith is placed in the results.

The third commonly used analytic strategy for multiple regression is *hierarchical regression*. In hierarchical regression a researcher chooses the order in which the variables are entered. As opposed to stepwise regression, the order of entry is chosen prior to analysis of the data and should be made on the basis of some rationale. Three alternative rationales have been suggested. First, the order of the variables could be chosen to reflect causal priority such that an independent variable entered at a given step should not be presumed to be a cause of a variable entered at an earlier step. Second, the variables could be entered according to their research relevance, where relevance reflects importance to the researcher or established relationship with the dependent variable. For example, in a study of achievement the independent variables entered first would probably reflect intelligence, because intelligence is an established predictor of achievement, and then independent variables that reflect other constructs, such as creativity, would be added to determine what they added to the prediction of achievement over and above that contributed by intelligence. Finally, the ordering of the variables might be determined by the structural aspects of the design, as we will see subsequently in this chapter.

Because the use of hierarchical regression plays an important role in the applications of multiple regression to a number of contexts discussed in the remainder of this chapter, some additional discussion of this topic follows. As with stepwise regression, in a hierarchical regression the coefficient of interest is the proportion of variance accounted for by a variable over and above the variables entered previously. It is these increments that are tested for statistical significance. To be clear about aspects of hierarchical regression, we summarize the steps of such an analysis schematically in Table 12.1. As can be seen from this table, for hierarchical regression with individual independent variables the increment in predictive ability at each step in the

TABLE 12.1

Hierarchical Regression

Step	Variable or Set Added	R^2	Increment in Predictive Ability	Model I — F	Model I — df	Model II — F	Model II — df
			INDIVIDUAL VARIABLES				
1	X_1	$R^2_{y\cdot1} = r^2_{y1}$	r^2_{y1}	$\dfrac{r^2_{y1}/1}{(1 - r^2_{y1})/(N-2)}$	$1, N-2$	$\dfrac{r^2_{y1}/1}{(1 - R^2_{y\cdot12\ldots K})/(N-K-1)}$	$1, N-K-1$
2	X_2	$R^2_{y\cdot12}$	$r^2_{y(2\cdot1)} = R^2_{y\cdot12} - R^2_{y\cdot1}$	$\dfrac{(R^2_{y\cdot12} - R^2_{y\cdot1})/1}{(1 - R^2_{y\cdot12})/(N-3)}$	$1, N-3$	$\dfrac{(R^2_{y\cdot12} - R^2_{y\cdot1})/1}{(1 - R^2_{y\cdot12\ldots K})/(N-K-1)}$	$1, N-K-1$
3	X_3	$R^2_{y\cdot123}$	$r^2_{y(3\cdot12)} = R^2_{y\cdot123} - R^2_{y\cdot12}$	$\dfrac{(R^2_{y\cdot123} - R^2_{y\cdot12})/1}{(1 - R^2_{y\cdot123})/(N-4)}$	$1, N-4$	$\dfrac{(R^2_{y\cdot123} - R^2_{y\cdot12})/1}{(1 - R^2_{y\cdot12\ldots K})/(N-K-1)}$	$1, N-K-1$
$\vdots$		$\vdots$	$\vdots$	$\vdots$	$\vdots$	$\vdots$	$\vdots$
K	X_K	$R^2_{y\cdot123\ldots K}$	$r^2_{y(K\cdot12\ldots K-1)} =$ $R^2_{y\cdot12\ldots K} - R^2_{y\cdot12\ldots K-1}$	$\dfrac{(R^2_{y\cdot12\ldots K} - R^2_{y\cdot12\ldots K-1})/1}{(1 - R^2_{y\cdot12\ldots K})/(N-K-1)}$	$1, N-K-1$	$\dfrac{(R^2_{y\cdot12\ldots K} - R^2_{y\cdot12\ldots K-1})/1}{(1 - R^2_{y\cdot12\ldots K})/(N-K-1)}$	$1, N-K-1$
			SETS OF VARIABLES				
			(Set A contains G variables, Set B contains H variables, Set C contains I variables)				
1	A	$R^2_{y\cdot A}$	$R^2_{y\cdot A}$	$\dfrac{R^2_{y\cdot A}/G}{(1 - R^2_{y\cdot A})/(N-G-1)}$	$G, N-G-1$	$\dfrac{R^2_{y\cdot A}/G}{(1 - R^2_{y\cdot ABC})/(N-G-H-I-1)}$	$G, N-G-H-I-1$
2	B	$R^2_{y\cdot AB}$	$R^2_{y(B\cdot A)} = R^2_{y\cdot AB} - R^2_{y\cdot A}$	$\dfrac{(R^2_{y\cdot AB} - R^2_{y\cdot A})/H}{(1 - R^2_{y\cdot AB})/(N-G-H-1)}$	$H, N-G-H-1$	$\dfrac{(R^2_{y\cdot AB} - R^2_{y\cdot A})/H}{(1 - R^2_{y\cdot ABC})/(N-G-H-I-1)}$	$H, N-G-H-I-1$
3	C	$R^2_{y\cdot ABC}$	$R^2_{y(C\cdot AB)} = R^2_{y\cdot ABC} - R^2_{y\cdot AB}$	$\dfrac{(R^2_{y\cdot ABC} - R^2_{y\cdot AB})/I}{(1 - R^2_{y\cdot ABC})/(N-G-H-I-1)}$	$I, N-G-H-I-1$	$\dfrac{(R^2_{y\cdot ABC} - R^2_{y\cdot AB})/I}{(1 - R^2_{y\cdot ABC})/(N-G-H-I-1)}$	$I, N-G-H-I-1$

analysis is the proportion of variance accounted for at that step minus the proportion of variance accounted for by the previous step. The total variance accounted for by all K variables in a hierarchical regression is neatly expressed as

$$R_{y \cdot 12 \ldots K}^2 = r_{y1}^2 + r_{y(2 \cdot 1)}^2 + r_{y(3 \cdot 12)}^2 + \cdots + r_{y(K \cdot 12 \ldots K-1)}^2 \qquad (12.43)$$

Hierarchical regression can also be applied to sets of variables, as summarized in the lower portion of Table 12.1.

One final note on hierarchical regression concerns the statistical test of the increments in predictive ability. At each step k the statistical test of the increment involves the F ratio

$$F = \frac{r_{y(k \cdot 12 \ldots k-1)}^2 / 1}{(1 - R_{y \cdot 12 \ldots k}^2 / (N - k - 1)} \qquad (12.44)$$

where k is the step in the hierarchical regression. In this test the error is related to the proportion of variance unaccounted for by the variables entered up to this step (i.e., $1 - R_{y \cdot 12 \ldots k}^2$). This test is referred to as the Model I test. However, a case could be made for using a different error term for the F ratio. When a hierarchical regression is used, the researcher is interested in all the variables and it would be logical to consider the error to be that which remains when all variables are considered. In this case the error term is related to the variance unaccounted for by all K variables (i.e., $1 - R_{y \cdot 12 \ldots K}^2$). Adjusting the degrees of freedom, an alternative test, to be referred to as the Model II test, involves the F ratio

$$F = \frac{r_{y(k \cdot 12 \ldots k-1)}^2 / 1}{(1 - R_{y \cdot 12 \ldots K}^2) / (N - K - 1)} \qquad (12.45)$$

The difference between the Model I and Model II tests is illustrated in Table 12.1 for both individual variables and sets of variables.

Which test should be used? Clearly one would prefer to use the more powerful test. The proportion of variance unaccounted for by the independent variables is always less than or equal to the proportion of the variance unaccounted for by the independent variables up to a particular step; that is

$$1 - R_{y \cdot 12 \ldots K}^2 \leq 1 - R_{y \cdot 12 \ldots k}^2$$

This would tend to make the Model II F ratio greater than the Model I F ratio. However, the degrees of freedom for Model II is less than that of Model I, which would tend to make the Model II F ratio less than that of the Model I F ratio. Generally, when K is relatively small compared to N and it is expected that the remaining variables will account for a sizable proportion of the variance, Model II will result in a more significant statistical test. Because the researcher typically only includes variables in a hierarchical regression that have a purported relationship with the dependent variable, it would seem that Model II would be preferred. In any case the model should be selected prior to the analysis and used throughout the analysis (i.e., do not use Model I for some steps and Model II for other steps).

Applications of Multiple Regression

12.7 Coding Categorical Variables

Up to this point the independent variables in the regression analyses have been limited to quantitative variables. It is, however, not difficult to think of instances for which one or more of the independent variables in a regression analysis would be categorical, for instance, gender, religious affiliation, political party, treatment group, or a host of other attributes. As we will see, accommodating categorical independent variables in a regression analysis is not difficult. Already there have been some hints that the two procedures are not all that different: η^2 is the proportion of the sample variance in the dependent variable accounted for by a categorical variable, the sample covariance can be expressed as a sample comparison, the statistical tests for multiple regression can be developed by partitioning the sum of squares, and so forth. Although we begin the discussion with one categorical independent variable, the ways in which multiple regression can accommodate several categorical independent variables as well as quantitative variables will be mentioned.

Consider a categorical independent variable with J levels or groups and a quantitative dependent variable. To be used in a regression analysis, numerical values need to be assigned to the J levels of the independent variable. One way to do this is to decompose the effects for the independent variable into $(J - 1)$ mutually orthogonal comparisons, as discussed extensively in Section 9.3 (p. 271). Recall that the null hypothesis

$$H_0:\ \mu_1 = \mu_2 = \cdots = \mu_J$$

is equivalent to

$$H_0:\ \psi_1 = 0,\ \psi_2 = 0,\ ...,\ \psi_{(J-1)} = 0$$

where the $(J - 1)$ comparisons are mutually orthogonal. In the context of regression analysis the omnibus F test will be approached by simultaneously testing the $(J - 1)$ mutually orthogonal comparisons. This is accomplished by creating $(J - 1)$ *coded* independent variables such that each represents one of the mutually orthogonal comparisons. This representation is accomplished in the following way. Given the comparison

$$\psi_1:\ c_1\mu_1 + c_2\mu_2 + \cdots + c_J\mu_J$$

the coded independent variable X_1 would be created so that a subject in the jth level of the categorical independent variable would be assigned the value c_j. This process is illustrated for an independent variable with four groups. The following three mutually orthogonal planned comparisons are used, although any three mutually orthogonal comparisons would be suitable:

$$\psi_1:\ 1\mu_1 + (-1)\mu_2 + 0\mu_3 + 0\mu_4$$
$$\psi_2:\ 0\mu_1 + 0\mu_2 + 1\mu_3 + (-1)\mu_4$$
$$\psi_3:\ 1\mu_1 + 1\mu_2 + (-1)\mu_3 + (-1)\mu_4$$

Three independent variables X_1, X_2, and X_3 would be created such that the subjects in each level of the original independent variable would be assigned values for the variables X_1, X_2, and X_3 as follows:

Value Assigned for Coded Variable

	X_1	X_2	X_3
Subject in Group 1	1	0	1
Subject in Group 2	−1	0	1
Subject in Group 3	0	1	−1
Subject in Group 4	0	−1	−1

The test of the omnibus null hypothesis is simply an F test for the proportion of variance accounted for by the $(J - 1)$ coded variables. That is, the value of $R^2_{Y \cdot 12...(J-1)}$ is used to test the null hypothesis that $P^2_{Y \cdot 12...(J-1)} = 0$, which is equivalent to the null hypothesis that the means of the J groups are equal.

The use of orthogonally coded variables is illustrated by considering the data in Table 12.2, where the four groups are represented by three orthogonal variables discussed earlier. The correlations among the three coded variables and the dependent variable are as follows:

	Y	X_1	X_2	X_3
Y		−.344	.115	.811
X_1			.000	.000
X_2				.000
X_3				

For these data $R^2_{y \cdot 123} = .7895$. The corresponding value of F is

$$F = \frac{R^2_{y \cdot 123}/K}{(1 - R^2_{y \cdot 123})/(N - K - 1)} = \frac{.7895/3}{(1 - .7895)/(20 - 3 - 1)} = 20.0$$

which, when compared to an F distribution with 3 and 16 degrees of freedom, leads to rejection of the null hypothesis that $P^2_{Y \cdot 123} = 0$ or equivalently of the null hypothesis that $\mu_1 = \mu_2 = \mu_3 = \mu_4$. The equivalence of the regression procedure and the analysis of variance is illustrated by examining the source table for these data:

Source	Sum of Squares	df	Mean Squares	F
Between	150	3	50.00	20.0
Within	40	16	2.50	
Total	190	19		

The value of F obtained from the analysis of variance is identical to that for regression analysis. Similarly, the sample values for the proportion of variance accounted for are identical. Recall from Section 7.12 (p. 200) that in the context of the analysis of variance, the proportion of variance accounted for (η^2) was equal to SSB/SST, the value of which is equal to .7895 for this example, which illustrates the equivalence of η^2 and R^2.

Further examination of the data in Table 12.2 illustrates a point made earlier about comparisons and leads as well to a discussion of other methods of coding. Note that the correlation among the three coded independent variables are all equal to zero, which should not be surprising given that orthogonal comparisons were used (the sum of the products of the coefficients is zero and the sample sizes are equal). However, when the sample sizes are not equal, the correlations among the independent variables will

TABLE 12.2

Data for Four Independent Groups with Three Orthogonally Coded Variables

	Y	X_1	X_2	X_3
Group 1	4	1	0	1
	5	1	0	1
	6	1	0	1
	7	1	0	1
	8	1	0	1
Group 2	7	-1	0	1
	8	-1	0	1
	9	-1	0	1
	10	-1	0	1
	11	-1	0	1
Group 3	1	0	1	-1
	2	0	1	-1
	3	0	1	-1
	4	0	1	-1
	5	0	1	-1
Group 4	0	0	-1	-1
	1	0	-1	-1
	2	0	-1	-1
	3	0	-1	-1
	4	0	-1	-1

not be zero (as illustrated in Example 12.3, which follows later in this section). However, because the comparisons are not being tested separately, the correlation among the coded variables is not problematic. Indeed, there are several alternative methods for coding that do not rely on orthogonal comparisons, the most popular of which is called *dummy coding*. In dummy coding $(J - 1)$ *dummy variables* are used to create $(J - 1)$ dichotomies, which reflect presence (or absence) in a group. That is, subjects in Group 1 would be assigned the value of 1 on X_1 and zero on all the other dummy variables; subjects in Group 2 would be assigned a value of 1 on X_2 and zero on all the other dummy variables, and so forth, except that the subjects in the Jth group are assigned the value zero for all dummy variables. For the four-group problem discussed in this section the following assignments would be made:

	Value Assigned for Coded Variable		
	X_1	X_2	X_3
Subject in Group 1	1	0	0
Subject in Group 2	0	1	0
Subject in Group 3	0	0	1
Subject in Group 4	0	0	0

Dummy variables X_1, X_2, and X_3 indicate membership in Groups 1, 2, and 3, respectively. A fourth variable is not needed because by the process of elimination, subjects that are not in any of the first three groups are in Group 4. Note that any $(J - 1)$ coded variables are appropriate for regression analysis so long as the variables are not completely linearly redundant. However, it should be emphasized that coding a categorical variable with J groups always requires $(J - 1)$ coded variables. Although the advantages of one coding system over another have not been discussed, for certain purposes particular coding systems are advantageous; a discussion of this topic is beyond the scope of this text (see Cohen & Cohen, 1983, or Pedhazur, 1982).

12.8   Coding Interaction Effects—A Viable Method for Unequal Sample Sizes

Thus far we have discussed coding one categorical independent variable. Inclusion of other categorical independent or quantitative independent variables in the regression analysis presents no special problems. Additional categorical variables require a set of additional coded variables. Suppose that one of the two independent variables contains J levels and the other contains

K levels. Then the first independent variable will be coded with $(J - 1)$ variables and the second independent variable will be coded with $(K - 1)$ independent variables. In such cases the interaction is coded with $(J - 1)(K - 1)$ variables obtained by taking the product of each of the $(J - 1)$ coded variables for the first independent variable with each of the $(K - 1)$ coded variables for the second independent variable. Tests for the main effects and the interaction are conducted by examining the R^2 associated with the $(J - 1)$ coded variables associated with the first independent variable, the R^2 associated with the $(K - 1)$ coded variables associated with the second independent variable, and the R^2 associated with the $(J - 1)(K - 1)$ coded variables associated with the interaction [as expected, the interaction has $(J - 1)(K - 1)$ degrees of freedom]. It should be realized that the two categorical variables and the interaction may not be independent and disentangling the effects of one variable from another may be difficult. The strategy used here is to use a hierarchical regression; if the $(K - 1)$ coded variables for the second independent variable are added to the regression equation after the $(J - 1)$ coded variables for the first independent variable, then the test of the second independent variable is accomplished by removing the effects of the first independent variable. However, it should be realized that the significance levels attached to the independent variables and the interaction are dependent on the order in which these effects are entered into the hierarchical regression. Nevertheless, this approach is particularly useful for nonorthogonal designs, such as typically are found in studies with unequal sample sizes. The following example illustrates a multiple regression for a factorial design with unequal sample sizes.

EXAMPLE 12.3

In this example a factorial design is analyzed with multiple regression. This example differs from previous factorial designs discussed in Chapter 7 in that the sample sizes within cells are unequal (and nonproportional), indicating that the main effects and the interaction effects are not independent.

Suppose that a researcher is interested in determining whether political party and conservative/liberal identification affect attitudes about whether the Soviet Union would comply with an arms reduction treaty. Further, suppose that 52 adults are randomly selected from the population of adults in the United States and asked to indicate their political party (assume all answer Republican, Democrat, or Independent), complete an instrument that is used to classify each subject as either liberal or conservative, and complete an instrument that indicates, on a 9-point scale, belief that the Soviet Union would comply with an arms treaty (high scores indicate a belief in compliance). The mean scores on the compliance instrument as well as the sample sizes are shown here for the three (political parties) by two (conservative/liberal) factorial design.

	Conservative	Liberal	
Democrat	$M_{11} = 5.67$ $N_{11} = 6$	$M_{12} = 6.08$ $N_{12} = 12$	$M_{1.} = 5.94$ $N_{1.} = 18$
Republican	$M_{21} = 5.07$ $N_{21} = 15$	$M_{22} = 5.63$ $N_{22} = 8$	$M_{2.} = 5.26$ $N_{2.} = 23$
Independent	$M_{31} = 6.00$ $N_{31} = 4$	$M_{32} = 6.14$ $N_{32} = 7$	$M_{3.} = 6.08$ $N_{3.} = 11$
	$M_{.1} = 5.36$ $N_{.1} = 25$	$M_{.2} = 5.96$ $N_{.2} = 27$	$M = 5.67$ $N = 52$

To analyze these data, we need to code the two categorical independent variables. For this example orthogonal coded variables are used; two coded variables (denoted by X_1 and X_2) are needed to code political party and one coded variable (denoted by X_3) is needed to code conservative/liberal. The interaction is coded by two variables formed by taking the product of the coded variables for political party with the coded variable for conservative/liberal. The resulting two variables are denoted by X_4 and X_5, where $X_4 = (X_1)(X_3)$ and $X_5 = (X_2)(X_3)$. Using this scheme, we code the subjects in each group in the following way:

Party	Conservative/ Liberal	Party X_1	X_2	Conservative/ Liberal X_3	Interaction $X_4 = (X_1)(X_3)$	$X_5 = (X_2)(X_3)$
Democrat	Conservative	1	1	1	1	1
Democrat	Liberal	1	1	−1	−1	−1
Republican	Conservative	−1	1	1	−1	1
Republican	Liberal	−1	1	−1	1	−1
Independent	Conservative	0	−2	1	0	−2
Independent	Liberal	0	−2	−1	0	2

The correlation matrix for the 52 subjects is as follows:

	Y	Party X_1	X_2	Conservative/ Liberal X_3	Interaction X_4	X_5
Y		.321	−.222	−.309	−.087	.151
X_1			−.056	−.288	.006	.211
X_2				−.121	.152	.167
X_3					.124	−.292
X_4						−.058
X_5						

Before proceeding further with the regression analysis, we must make an important point about the correlation matrix. Although orthogonally coded variables are used, the correlations among the coded variables are not zero, illustrating the point that was made in Section 7.15 (p. 210) that when unequal (and nonproportional) cell sizes are present, statistical independence will not be present. The result of this observation for this example is that the main effects and the interaction are not independent. For example, the correlations between the coded variables for political party and the coded variable for conservative/liberal are nonzero, which is not surprising given the fact that if an individual is a Republican, he or she is likely to be conservative.

Because the main effects and interaction are not independent, the order of entry of the variables in the regression analysis is important, as will become apparent in this example. Suppose that the researcher chooses to enter first the variables for party, next the variable for conservative/liberal, and finally the variables for the interaction. The following results are obtained for this hierarchical regression:

Effect Tested	Variable(s) Added	Variance Accounted for	Increment in Predictive Ability
Party	X_1, X_2	$R^2_{y \cdot 12} = .145$	$R^2_{y \cdot 12} = .145$
Conservative/ Liberal	X_3	$R^2_{y \cdot 123} = .187$	$R^2_{y \cdot 123} - R^2_{y \cdot 12} = .042$
Interaction	X_4, X_5	$R^2_{y \cdot 12345} = .193$	$R^2_{y \cdot 12345} - R^2_{y \cdot 123} = .006$

Suppose that all tests are conducted with Model II error terms and α is set at .05. Then the test for political party involves testing the variance accounted for by X_1 and X_2.

$$F = \frac{R^2_{y \cdot 12}/2}{(1 - R^2_{y \cdot 12345})/(N - K - 1)}$$

$$= \frac{(.145)/2}{(1 - .193)/(52 - 5 - 1)}$$

$$= 4.13$$

which when compared to an F distribution with 2 and 46 degrees of freedom leads to rejection of the null hypothesis related to political party. To test the null hypothesis related to conservative/liberal, the increment in predictive ability of X_3 (over and above X_1 and X_2) is tested:

$$F = \frac{(R^2_{y \cdot 123} - R^2_{y \cdot 12})/1}{(1 - R^2_{y \cdot 12345})/(N - K - 1)} = \frac{(.042)/1}{(1 - .193)/(52 - 5 - 1)} = 2.39$$

which when compared to an F distribution with 1 and 46 degrees of freedom is insufficiently large to reject the null hypothesis. However, if the researcher had decided to enter X_3 before X_1 and X_2, then the test for conservative/liberal would have been statistically significant, illustrating the problems encountered when the main effects are not independent. Finally, to test for interaction effects, we examine the increment in predictive ability for X_4 and X_5 (over and above X_1, X_2, and X_3):

$$F = \frac{(R^2_{y \cdot 12345} - R^2_{y \cdot 123})/2}{(1 - R^2_{y \cdot 12345})/(N - K - 1)} = \frac{(.006)/2}{(1 - .193)/(52 - 5 - 1)} = .17$$

which when compared to an F distribution with 2 and 46 degrees clearly is insufficiently large to reject the null hypothesis.

If two quantitative independent variables X_1 and X_2 are included in the regression analysis, then the variable $X_3 = (X_1)(X_2)$ is included in the regression analysis to test for interaction effects. If one of the independent variables is categorical and the other quantitative, then variables representing the interaction are formed by multiplying each of the $(J - 1)$ coded variables corresponding the categorical variable with the quantitative variable. Again, in each of these cases there is no assurance that the independent variables and the interaction among them will be uncorrelated.

12.9   Analysis of Covariance Revisited

Principles of multiple regression presented in this chapter can lead to an increased understanding of the analysis of covariance as well as an alternative approach to performing such an analysis. First, the material presented in Section 12.5 (p. 387) that utilizes the full versus the restricted model for significance testing is applied to the analysis of covariance, and second, hierarchical regression is used to conduct an analysis of covariance. A review of Section 11.8 (p. 343) on the analysis of covariance will be helpful in appreciating the material in this section.

Recall that the linear model for the analysis of covariance is given by [see Equation (11.54), p. 344]

$$Y_{ij} = \mu + \alpha_j + \beta_{Y \cdot C}(C_{ij} - \mu_C) + e_{ij} \tag{12.46}$$

Given that the null hypothesis for the analysis of covariance is that the treatment effects are all zero (i.e., H_0: $\alpha_j = 0$ for $j = 1$ to $j = J$), the restricted model is the linear model for the analysis of covariance with the condition that $\alpha_j = 0$ for all j. That is, the restricted model is given by

$$Y_{ij} = \mu + \beta_{Y \cdot C}(C_{ij} - \mu_C) + e_{ij} \qquad (12.47)$$

The question is whether the full model explains significantly more of the variance in the dependent variable than does the restricted model. The general form for a test statistic of this type was given by Equation (12.41), which, when adapted to the context of the analysis of covariance, yields

$$\frac{[\text{SSE (restricted)} - \text{SSE (full)}]/(J - 1)}{[\text{SSE (restricted)}]/(N - J - 1)} = F \qquad (12.48)$$

Examination of this test statistic reinforces the need for two different estimators for the regression coefficient $\beta_{Y \cdot C}$ discussed in Section 11.8, p. 343. The restricted model assumes there are no treatment effects, and thus the estimator for $\beta_{Y \cdot C}$ is based on combining all the observations into one sample. This estimator, denoted by B_t, is used to calculate SSE (restricted). On the other hand, the estimator for $\beta_{Y \cdot C}$ for the full model must take into account the treatment effects, and thus is based on pooling the estimates of $\beta_{Y \cdot C}$ calculated for each of the J groups. This estimator, denoted by B_w, is used to calculate SSE (full). (Figure 11.5, p. 346, illustrates the difference between B_t and B_w.) It can be shown that by using the appropriate estimators for $\beta_{Y \cdot C}$, the test statistic given by Equation (12.48) reduces to the test statistic discussed in Section 11.8, p. 343. A significant test statistic indicates that the full model, which contains nonzero treatment effects, explains a significantly greater proportion of the variance in the dependent variable than does the restricted model, for which the treatment effects are all zero.

Although the discussion of the analysis of covariance to this point has not affected the computation of the test statistic, application of multiple regression provides an equivalent and straightforward means to conduct an analysis of covariance. Essentially, as discussed earlier, the analysis of covariance is a test of the proportion of variance accounted for by the independent variable over and above that accounted for by the covariable. Let set A contain the $(J - 1)$ variables needed to code the independent variable with J groups. Then the proportion of variance accounted for by the independent variable over and above that accounted for by the covariable C is given by

$$R^2_{y \cdot AC} - R^2_{y \cdot C}$$

Hence, the test statistic for the hypothesis that the treatment effects are all zero is given by

$$\frac{(R^2_{y \cdot AC} - R^2_{y \cdot C})/(J - 1)}{(1 - R^2_{y \cdot AC})/(N - J - 1)} = F \qquad (12.49)$$

which is compared to an F distribution with $(J - 1)$ and $(N - J - 1)$ degrees of freedom. The equivalence of this test statistic with the one discussed in Section 11.8 (p. 343) is illustrated by reanalyzing the data presented in Example 11.2 (p. 349). Having coded the three groups (two treatments and a control group) with two coded variables, we obtain the following squared multiple correlation coefficients by regression analysis:

$$R^2_{y \cdot C} = .088 \qquad \text{and} \qquad R^2_{y \cdot AC} = .485$$

where C is the pretest and A is the set containing the two coded variables X_1 and X_2. Given that there were 15 subjects, the value of the test statistic is thus

$$\frac{(.485 - .088)/2}{(1 - .485)/(15 - 3 - 1)} = 4.23$$

which is the same value obtained in Section 11.8 using the traditional analysis of covariance formulas.

Recall from Section 11.8 (p. 343) that an assumption of the analysis of covariance is that the population regression coefficient is constant across the J groups (i.e., homogeneity of regression). In the regression context, a direct test of this assumption is provided by examining the interaction of the covariable and the independent variable. Specifically, if the F test for the set of $(J - 1)$ interaction variables [viz., the $(J - 1)$ variables used to code the independent variable multiplied by the covariable—see Section 12.8, p. 401] is statistically significant, then there is persuasive evidence that the assumption is false.

12.10   Curvilinear Regression Revisited

When curvilinear regression was discussed in Section 11.9 (p. 353), the restrictions that the classical regression model with equally spaced values for the independent variable and an equal number of subjects in each group were required. These restrictions limit the usefulness of curvilinear regression, particularly when the experimental paradigm involves sampling from a bivariate distribution. Consider the instance where a researcher is interested in studying the relation between anxiety and performance. An obvious means to study this relation is to have subjects attempt a task and measure performance on the task and anxiety level. Suppose that a researcher conducts such a study where anxiety and performance are measured on a 10-point scale; the results of this hypothetical study are presented in Figure 12.2. There appears to be a quadratic relation with moderately anxious subjects performing the best. The method of orthogonal comparisons discussed earlier is inappropriate because the experimental design violates the restrictions placed on that method. In this section we examine an alternative procedure.

We can rewrite the linear model for multiple regression as

$$Y'_{ij} = \beta_1 X_1 + \beta_2 X_2 + \cdots + \beta_K X_K + \text{regression constant}$$

Now we let $X_1 = X$, $X_2 = X^2$, ..., $X_K = X^K$, so that the model becomes

$$Y'_{ij} = \beta_1 X + \beta_2 X^2 + \cdots + \beta_K X^K + \text{regression constant} \qquad (12.50)$$

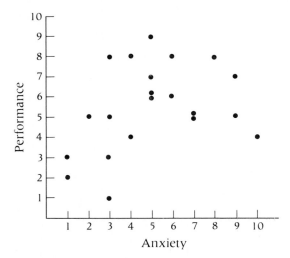

FIGURE 12.2 Scatter Plot for Relation Between Anxiety and Performance

This equation is identical to the polynomial equation discussed in Section 11.9. Clearly the regression model is no longer linear since it contains powers of variables; nevertheless, this equation can be used to test for linear and curvilinear trends. The variable X represents the linear relationship, the variable X^2 the quadratic, X^3 the cubic, and so forth. The presence of linear or curvilinear relations are tested simply by examining the partial coefficients associated with each variable. At this juncture it should be emphasized that the variables will not be independent, and in fact will be highly correlated with each other, as we will see. So, when testing for each relation, we must keep in mind that the contribution of other variables must be partialed out.

One commonly used way to test for linear and curvilinear trends in this context is to perform a hierarchical regression beginning with the linear relation, followed by the quadratic, cubic, and so forth. This method has the appealing feature that priority is given to the simpler relation and thus the emphasis is on parsimony. When performed in this way, the increment in predictive ability of each variable over and above the variables entered previously is tested. Either a Model I or a Model II error term may be used, but should be selected prior to the analysis. Relations up to the $(N-1)$th degree may be tested (where N is the number of bivariate observations), but typically there is little theoretical support in the social sciences for testing relations beyond a third-degree (cubic) relation.

To illustrate the use of hierarchical regression to test for linear and curvilinear relations, the data presented in Figure 12.2 are analyzed using this method where relations are tested up to and including the cubic relation with a Model I error term. The following correlations are obtained:

	Y	$X_1 = X$	$X_2 = X^2$	$X_3 = X^3$
Y		.308	.232	.173
$X_1 = X$			.985	.952
$X_2 = X^2$				.991
$X_3 = X^3$				

This example demonstrates the high correlations among the independent variables that are typical in this type of analysis. The following values for the proportion of variance accounted for are obtained:

$$R^2_{y \cdot 1} = .095 \qquad R^2_{y \cdot 12} = .263, \qquad \text{and} \qquad R^2_{y \cdot 123} = .274$$

To test for the linear relation, the proportion of variance accounted for by $X_1 = X$ is tested:

$$\frac{(R^2_{y \cdot 1})/1}{(1 - R^2_{y \cdot 1})/(N - 2)} = \frac{(.095)/1}{(1 - .095)/18} = 1.89$$

which when compared to the F distribution with 1 and 18 degrees of freedom is not sufficiently large to reject the null hypothesis that there is no linear relation between anxiety and performance. To test for a quadratic relation, the increment in predictive ability of $X_2 = X^2$ over and above $X_1 = X$ is tested:

$$\frac{(R^2_{y \cdot 12} - R^2_{y \cdot 1})/1}{(1 - R^2_{y \cdot 12})/(N - 3)} = \frac{(.263 - .095)/1}{(1 - .263)/17} = 3.88$$

which although close, when compared to an F distribution with 1 and 17 degrees of freedom, is not sufficiently large to reject the null hypothesis that there is no quadratic relation between anxiety and performance. To test for a cubic relation, the increment in predictive ability of $X_3 = X^3$ over and above $X_1 = X$ and $X_2 = X^2$ is tested:

$$\frac{(R^2_{y \cdot 123} - R^2_{y \cdot 12})/1}{(1 - R^2_{y \cdot 123})/(N - 4)} = \frac{(.274 - .263)/1}{(1 - .274)/16} = .24$$

which clearly is not sufficiently large to reject the null hypothesis that there is no cubic relation between anxiety and performance.

NOTES AND SUPPLEMENTARY READINGS

Cohen, J., & Cohen, P. (1983). *Applied multiple regression/correlation analysis for the behavioral sciences* (2nd ed.). Hillsdale, NJ: Lawrence Erlbaum.

This volume is viewed by many as the most readable book on multiple regression. The authors provide a comprehensive examination of this topic, which serves well as an instructional textbook and as a reference.

Darlington, R. B. (1968). Multiple regression in psychological research and practice. *Psychological Bulletin, 69,* 161–182.

In this seminal article Darlington addresses directly the problem with interpreting the results of a multiple regression analysis and the problems associated with selecting a coefficient that reflects the importance of a variable in such an analysis.

Pedhazur, E. J. (1982). *Multiple regression in behavioral research: Explanation and prediction* (2nd ed.). New York: Holt, Rinehart and Winston.

Pedhazur presents a detailed discussion of multiple regression and includes a substantial number of research examples. Some may find this volume mathematically sophisticated.

PROBLEMS

1. Consider three variables Y, X_1, X_2, where Y represents a university professor's salary, X_1 represents the years since the highest degree was obtained, and X_2 represents the number of publications. Suppose that a random sample of 22 professors was selected and the following sample statistics were obtained:

$$M_y = 35{,}000 \qquad M_1 = 15 \qquad M_2 = 12$$
$$S_y = 4000 \qquad S_1 = 5 \qquad S_2 = 4$$

Correlation Coefficients

	Y	X_1	X_2
Y		.60	.50
X_1			.40
X_2			

We wish to predict scores on Y from scores on X_1 and X_2.

a. Calculate the sample partial regression coefficients and the regression equation. Predict the salary for a professor who obtained his or her highest degree ten years ago and who has published six articles.

b. Calculate the multiple correlation coefficient R and R^2. What interpretation can be made about the value of R^2?

c. Test the null hypotheses that the population partial regression coefficients are equal to zero versus the alternative that they are greater than zero ($\alpha = .05$).

d. Test the null hypothesis that the population multiple correlation coefficient is zero ($\alpha = .05$).

e. Calculate the increment in predictive ability of X_2 over and above X_1. Test the null hypothesis H_0: $\rho^2_{Y(2 \cdot 1)} = 0$ ($\alpha = .05$).

    f. Calculate the increment in predictive ability of X_1 over and above X_2. Test the null hypothesis $H_0: \rho^2_{Y(1 \cdot 2)} = 0$ ($\alpha = .05$).

    g. Explain in words the relation among the variables.

2. Consider three variables Y, X_1, X_2, where Y represents the number of unexcused absences during the most recent semester, X_1 represents the students grade point average, and X_2 represents the student's age (in months). Suppose that a random sample of 50 high school students from a large inner city high school were selected and the following sample statistics were obtained:

$$M_y = 28 \qquad M_1 = 2.5 \qquad M_2 = 200$$
$$S_y = 4 \qquad S_1 = .5 \qquad S_2 = 10$$

Correlation Coefficients

	Y	X_1	X_2
Y		$-.40$	.30
X_1			.05
X_2			

We wish to predict scores on Y from scores on X_1 and X_2.

    a. Calculate the sample partial regression coefficients and the regression equation. Predict the number of unexcused absences for a student who has a 3.0 grade point average and is 16 years old.

    b. Calculate the multiple correlation coefficient R and R^2. What interpretation can be made about the value of R^2?

    c. Test the null hypotheses that the population partial regression coefficients are equal to zero versus the alternative that the first (i.e., β_1) is less than zero and the second (i.e., β_2) is greater than zero ($\alpha = .05$).

    d. Test the null hypothesis that the population multiple correlation coefficient is zero ($\alpha = .05$).

    e. Calculate the increment in predictive ability of X_2 over and above X_1. Test the null hypothesis $H_0: \rho^2_{Y(2 \cdot 1)} = 0$ ($\alpha = .05$).

    f. Calculate the increment in predictive ability of X_1 over and above X_2. Test the null hypothesis $H_0: \rho^2_{Y(1 \cdot 2)} = 0$ ($\alpha = .05$).

    g. Explain in words the relation among the variables.

3. Consider the following results from a hierarchical regression analysis for predicting Y from X_1, X_2, and X_3, which resulted from a sample of size 25.

Step	Variable Added	R^2
1	X_1	.03
2	X_2	.20
3	X_3	.50

    a. Indicate the increment in predictive ability for each step.

    b. State the null hypothesis related to the increment in predictive ability for each step.

c. Test the null hypothesis for each step using the Model I test ($\alpha = .05$).
d. Test the null hypothesis for each step using the Model II test ($\alpha = .05$).

4. Consider the following results from a hierarchical regression analysis for predicting Y from X_1, X_2, X_3, and X_4, which resulted from a sample of size 30.

Step	Variable Added	R^2
1	X_1	.15
2	X_2	.23
3	X_3	.26
4	X_4	.27

a. Indicate the increment in predictive ability for each step.
b. State the null hypothesis related to the increment in predictive ability for each step.
c. Test the null hypothesis for each step using the Model I test ($\alpha = .05$).
d. Test the null hypothesis for each step using the Model II test ($\alpha = .05$).

5. A researcher interested in the relation between alcohol and drug use and the amount of time that couples spend together sampled 62 couples, determined whether the husband and the wife were (a) alcohol abusers (but not drug abusers), (b) drug or drug and alcohol abusers, or (c) not abusers of alcohol or drugs, and asked the couples to indicate the number of nonsleeping hours they spent together during the previous weekend. The following means (and sample sizes) were obtained:

| | | HUSBAND | | |
		Alcohol Abuse	Drug/Alcohol Abuse	No Abuse
	Alcohol Abuse	7.17 ($N = 6$)	5.86 ($N = 7$)	3.20 ($N = 5$)
WIFE	Drug/Alcohol Abuse	5.20 ($N = 5$)	5.25 ($N = 8$)	2.20 ($N = 5$)
	No Abuse	5.33 ($N = 9$)	3.50 ($N = 8$)	8.33 ($N = 9$)

Suppose that X_1 and X_2 are the two variables needed to code the three groups related to wife, X_3 and X_4 are the two variables needed to code the three groups related to husband, and $X_5 = (X_1)(X_3)$, $X_6 = (X_1)(X_4)$, $X_7 = (X_2)(X_3)$, and $X_8 = (X_2)(X_4)$. The results of a hierarchical regression are as follows:

Step	Variables Added	R^2
1	X_1, X_2	.045
2	X_3, X_4	.061
3	X_5, X_6, X_7, X_8	.395

Given the results of this hierarchical regression, perform the statistical tests related to wife effects, husband effects, and the interaction of the husband and wife effects using the Model II test ($\alpha = .01$).

6. Suppose that a researcher was interested in the effect of the type of illness and locus of control on a patient's compliance with regard to medication. Twenty subjects were selected from four populations: (a) patients taking medication for an injury, (b) patients taking medication for a chronic illness, (c) patients taking medication for an acute illness, and (d) patients taking medication to prevent some illness. The locus of control for each of the 80 subjects was measured (high scores indicated external locus of control and low scores indicated internal locus of control), as well as the degree to which they were complying with the physician's instructions with regard to taking medication. Suppose that X_1, X_2, and X_3 are the three variables needed to code the four groups related to type of illness, X_4 is the variable related to locus of control, and $X_5 = (X_1)(X_4)$, $X_6 = (X_2)(X_4)$, and $X_7 = (X_3)(X_4)$. The results of a hierarchical regression are as follows:

Step	Variables Added	R^2
1	X_1, X_2, X_3	.10
2	X_4	.15
3	X_5, X_6, X_7	.19

Given the results of this hierarchical regression, perform the statistical tests related to the effects for type of illness, locus of control, and the interaction of type of illness and locus of control using the Model II test ($\alpha = .05$).

7. A researcher thinks that there is a quadratic relation between the adaptability of a family and the marital satisfaction of the mother and father; that is, rigid (low adaptability) as well as overly accommodating (high adaptability) families produce low marital satisfaction. Let X_1 be the variable related to adaptability. From a sample of 15 families the following correlations were obtained:

	Satisfaction	X_1	$X_2 = X_1^2$
Satisfaction		.187	.051
X_1			.972
$X_2 = X_1^2$			

Test the null hypothesis that there is no quadratic relation between adaptability and satisfaction (after having accounted for the linear relation) versus the alternative that there is a quadratic relation ($\alpha = .05$, rounding to a minimum of three decimal points at each step, the calculations give a reasonable approximation here).

8. To investigate the relation between the size of a city and the number of city council members, a sample of 20 cities was selected. Let X_1 represent the number of city council members. The researcher wanted to determine whether the relation between the size of the city and the number of council members was linear or quadratic. The following correlations were obtained:

Size of City	X_1	$X_2 = X_1^2$
Size of City	.623	.585
X_1		.970
$X_2 = X_1^2$		

Test the null hypotheses related to linear and curvilinear relations ($\alpha = .05$, rounding to a minimum of three decimal points at each step, the calculations give a reasonable approximation here).

CHAPTER

13 DISTRIBUTION-FREE STATISTICAL TESTS

F or the most part the statistical tests discussed in this text have been used to make inferences about population parameters. The paradigm for these tests involves random selection from a population with a specified distribution (typically the normal distribution), calculation of a test statistic, and comparison of the test statistic to the sampling distribution of the test statistic, the form of which is established via statistical theory. Generally speaking, statistical tests of this type are referred to as *parametric* statistical procedures. Another statistical paradigm involves statistical tests that do not rely on knowledge of the theoretical population distribution and thus are called *distribution-free* statistical tests. Although called distribution-free tests, as will become clear in this chapter, certain matters related to distributions are needed; the manner in which distributions are used, however, will be quite different from the role of distributions in the traditional parametric tests. Many distribution-free tests are referred to as *nonparametric* tests.

There are a number of reasons that we may prefer to use distribution-free statistical tests. First, there may be insufficient reason to believe that the assumptions of the parametric test are met for the particular population or populations of interest. We have already discussed the need to be cognizant

415

of possible violations of distribution assumptions and their implications for the validity of a decision to reject the null hypothesis. A second reason to use distribution-free statistical methods is that characteristics of the data may preclude the use of parametric methods. For instance, the performance of subjects may be measured on an ordinal scale (e.g., the rank ordering of subjects by judges). Most of the tests described in this chapter require only ordinal measurement. A third reason to use distribution-free statistical tests is that specific applications of these tests can often be made in the case where random selection from a population (or approximations to it) is not realistic. We will elaborate on this point in Section 13.4 (p. 428). A final reason for the use of distribution-free statistical methods involves their simplicity. The statistical rationale for these tests are less complex than parametric tests and more closely tied to basic probability theory.

Several other points should be considered when making a decision to use parametric or distribution-free statistical tests. The hypotheses tested by distribution-free methods are different from those tested by the corresponding parametric test, as we will see. Therefore, even when data are obtained that can be analyzed by either a parametric or a distribution-free statistical test and the decision to reject the null hypothesis is made, the exact conclusion made by the research depends on the choice of statistical methods. However, this difference is often emphasized to differentiate the two methods when it should be kept in mind that the particular statistical hypothesis is rarely of primary interest to the researcher. Rather, the statistical hypotheses are derived from scientific conjectures and typically it makes little difference whether the statistical hypotheses are associated with parametric or with distribution-free methods (see Section 5.1, p. 106 for a discussion of statistical versus scientific hypotheses). Another consideration related to the choice of methods is related to their respective power to reject the null hypothesis. Recall from Section 5.9 (p. 121) that power was calculated for hypotheses based on the particular statistical test used and it was mentioned that one way to change power was to change statistical tests. When the assumptions of the parametric test are true, generally, the parametric test will yield a slightly more powerful test of the null hypothesis than the corresponding best distribution-free statistical test (although this superiority is usually less than 5%). However, when the assumptions of the parametric test are not met, the power of the distribution-free statistical test is often greater than the corresponding parametric test. Because the degree to which assumptions are violated in the population is never known, the greater power attributed to parametric tests may be illusory. In actual practice, parametric tests are typically used because they have been generally accepted in the social sciences and because they are readily accessible in the statistical packages commonly available for computer implementation. Nevertheless, the relative advantages of the two competing paradigms is a topic that is of current interest to statisticians. However, it is safe to say that the distribution-free statistical methods are underutilized by researchers given their many desirable characteristics.

In this chapter distribution-free statistical tests for the more common designs discussed in this text are presented. Because these tests are intimately tied to basic probability theory, this chapter begins with some extensions of probability theory discussed in Chapter 1. After this introduction we present the binomial distribution, an important discrete probability distribution. Finally, commonly used distribution-free statistical tests are discussed. It should be noted that the discussion of distribution-free statistical methods is cursory and the interested reader should consult texts that are devoted entirely to this subject (see references at the end of this chapter).

Properties of Distribution-Free Statistical Tests

13.1 Counting Rules

Counting rules are used to enumerate the number of possible results of a particular experiment. For example, suppose that a die was rolled four times, to calculate the probability of any particular outcome, we need to know the number of possible outcomes that can result from this experiment. Later in this chapter we will want to count such things as the number of ways that N observations can be assigned to J groups. Such counting is instrumental to many distribution-free statistical tests.

To understand counting, it is convenient to speak of *sequences* of observations. If an experiment consists of tossing a die three times, the outcome may be thought of as a sequence of three numbers, each of which corresponds to the spots on each toss; for example, one outcome would be the sequence (2, 5, 4). Essentially, this experiment involves repeating a simple experiment, tossing one die, three times. Often in this chapter we discuss experiments that consist of N repetitions or trials of a simple experiment. Indeed, the first counting rule involves counting the number of outcomes of such an experiment.

Consider again the experiment consisting of tossing a die three times. For this experiment, how many outcomes (sequences) are possible? Clearly, on each of the three trials, there are six possible outcomes, yielding a total of $(6)(6)(6) = 6^3 = 216$ possible outcomes or sequences. Generalization of this strategy results in Counting Rule 1.

> *Counting Rule 1*: **If an experiment consists of N repetitions (i.e., N trials) and there are K mutually exclusive and exhaustive events that can occur on each of the N trials, then there are K^N different outcomes (sequences).**

We will find occasion to use this rule to count the number of outcomes of an experiment when it is impractical to list each of the outcomes, as was the case for a process as simple as tossing a die three times.

The second counting rule involves arranging a set of objects in order. Before discussing the rule, however, we need to introduce the symbol $N!$, which is read as "*N* factorial." Provided *N* is a nonnegative integer, $N!$ is defined in the following way:

$$N! = N(N - 1)(N - 2)\cdots(2)(1) \text{ if } N > 1 \tag{13.1}$$

and

$$N! = 1 \quad \text{ if } N = 0 \text{ (i.e., } 0! = 1) \tag{13.2}$$

For example, $4! = 4(3)(2)(1) = 24$.

Consider the problem of arranging four people in a line. There are four possible people who can be placed in the first position in the line. Given one person was placed in the first position, there are three possible people who can be placed in the second position. Similarly, there are two possible people who can be placed in the third position. Finally, the remaining person is placed in the final position. Thus, there are $4! = 4(3)(2)(1) = 24$ different arrangements. Generalization of this process yields Counting Rule 2.

Counting Rule 2: The number of different ways that *N* distinct objects can be arranged in order is *N*!.

An arrangement of objects in order is called a *permutation*.

The third counting rule involves selecting a subset of objects and arranging the subset of objects in order. Consider a classroom of 30 children where 1 child is selected to be president, 1 child is selected to be secretary, and 1 child is selected to be treasurer. How many different ways can 3 children be assigned to these three offices? (Note that it does not make a difference whether 3 children are selected and then assigned to offices or 1 child is selected for a particular office, a second child is selected for another office, and so forth.) There are 30 possible children who can be assigned to be president, 29 who can be assigned to secretary, and 28 who can be assigned to treasurer. Thus, there are $30(29)(28) = 24,360$ possible arrangements. Although there is no special notation for a product such as this, $30(29)(28)$ can be written in terms of factorial notation:

$$30(29)(28) = \frac{30(29)(28)\cdots(2)(1)}{27(26)\cdots(2)(1)} = \frac{30!}{27!} = \frac{30!}{(30 - 3)!}$$

The general rule in this instance is

Counting Rule 3: The number of ways of selecting and arranging *r* objects from among *N* distinct objects is

$$\frac{N!}{(N - r)!}$$

The next rule discussed here is similar to Counting Rule 3, with the exception that the order is not important. That is, the *r* objects selected are not arranged in order. Suppose that from a group of seven subjects three are assigned to a treatment group. In other words, in how many ways can three objects be selected from seven objects when the order of three objects does not make a difference. Suppose for the moment that the order does make a difference; then, by Counting Rule 3, there would be

$$\frac{7!}{(7-3)!} = 7(6)(5) = 210$$

different arrangements of the three subjects. Further, suppose that the three subjects chosen are Abbey, Bob, and Candy. Because the 210 possible arrangements consider order, the arrangement Abbey, Bob, and Candy is different from the arrangement Bob, Abbey, and Candy. Because Counting Rule 3 considers order, the 210 arrangements obtained by Counting Rule 3 overstates the number of ways that three subjects are selected from a set of seven subjects. All of the different arrangements of Bob, Abbey, and Candy must be considered to be the same, since the order of the subjects makes no difference. By Counting Rule 2, there are 3! = 6 ways that these three subjects can be arranged in order; thus, the 210 arrangements obtained by Counting Rule 3 is overstated by a factor of 3! = 6. Therefore, the number of ways of selecting three subjects from a set of seven subjects is given by

$$\frac{7!}{(7-3)!3!} = \frac{210}{6} = 35$$

Before stating the general rule for this instance, we need to introduce an additional notation:

$$\binom{N}{r} = \frac{N!}{(N-r)!r!} \tag{13.3}$$

The symbol $\binom{N}{r}$ is read as "number of combinations of N things taken r at a time" or simply as "N choose r." The word "combination" is used because each of the possible ways of selecting r objects from N objects, regardless of order, is called a *combination*. We are now ready to state the next counting rule to be used here.

Counting Rule 4: **The number of ways of selecting r distinct combinations of N objects is**

$$\binom{N}{r} = \frac{N}{(N-r)!r!}$$

A few points about this rule and the notation will be helpful. First, returning to the example of selecting three subjects from a set of seven subjects, it should be noted that when the three subjects are selected, a group of four subjects remains. Thus, $\binom{7}{3} = \binom{7}{4}$. In general

$$\binom{N}{r} = \binom{N}{N - r} \qquad (13.4)$$

Therefore, $\binom{N}{r}$ or $\binom{N}{N - r}$ is the number of ways that r subjects can be assigned to one group and $(N - r)$ subjects assigned to another group. Although factorials are cumbersome to calculate, calculation of the number of combinations of N things taken r at a time is facilitated by cancellation. For example, calculate the number of ways that a four-person committee can be formed from a class of 20 graduate students:

$$\binom{20}{4} = \frac{20!}{(20 - 4)!4!} = \frac{20(19)(18)(17)(\cancel{16})(\cancel{15})\cdots(\cancel{2})(\cancel{1})}{[(\cancel{16})(\cancel{15})\cdots(\cancel{2})(\cancel{1})][4(3)(2)(1)]}$$

$$= \frac{20(19)(18)(17)}{(4)(3)(2)(1)}$$

$$= 5(19)(3)(17) = 4845$$

The final rule involves assigning N objects to J groups. Let N_j be the number of objects assigned to Group j such that $\Sigma N_j = N$.

Counting Rule 5: The number of ways that N_1, N_2, ..., N_J objects can be assigned to Groups 1, 2, ..., J is

$$\frac{N!}{N_1!(N_2!)\cdots(N_J!)}$$

For example, the number of ways that 20 subjects can be assigned to four groups such that each group contains five subjects is

$$\frac{20!}{5!5!5!5!}$$

which exceeds five thousand trillion!

13.2   The Binomial Distribution

The binomial distribution is one of the most valuable discrete probability distributions used in statistics. As was the case for many of the continuous distributions discussed previously in this text (e.g., the t, χ^2, and F distributions), the binomial distribution is a family of distributions, each of which is defined by the value of parameters. The characteristics of the binomial distribution will be discussed in the context of repetitions of a simple experiment where the experiment is restricted to have only two possible outcomes.

An experiment that can result in one of two possible outcomes is called a *Bernoulli trial*. One of the outcomes is referred to as a "success" and the other is referred to as a "failure." Designation of success and failure is arbitrary, although it is often convenient to designate a particular outcome as a success. We let p be the probability of a success and q be the probability of a failure; because there are exactly two outcomes, $p = 1 - q$. Tossing a coin is an obvious example of a Bernoulli trial because there are two outcomes, heads and tails. If heads is designated as success, then $p = \frac{1}{2}$ and thus tails is designated failure and $q = \frac{1}{2}$. Rolling a die is a Bernoulli trial if, for instance, obtaining one or two spots is designated as success; thus, $p = \frac{1}{3}$ and $q = \frac{2}{3}$. Other examples include selecting a person at random and noting gender, votes in an election with two candidates, and passing or failing a class. The two outcomes and their associated probabilities are called a *Bernoulli process*.

In most instances the value of p, the probability of a success, is unknown. For example, suppose that in a population, one-third of the voters favors candidate A and the remainder favors candidate B (no undecided voters). If an experiment involves selecting a voter at random and asking whom the voter favors (assuming the voter answers truthfully) and success is defined as favoring candidate A, then the experiment represents a Bernoulli trial with $p = \frac{1}{3}$ and $q = \frac{2}{3}$. However, the value of p is not known unless all voters are queried. That is to say, p is a population parameter. Inferences about p are made by sampling from the Bernoulli process, as we will see in the next section.

Sampling in the context of a Bernoulli process involves performing N independent Bernoulli trials for which the values of p and q remain unchanged over the N trials. If the Bernoulli trial consists of flipping a coin, the probability of heads and tails will not change from one flip to another (unless some dramatic event occurs to the coin). For our voter example the probability of success remains unchanged only so long as sampling with replacement is used. When the values of p and q remain unchanged, the process is said to be *stationary*.

Of primary interest in sampling from a stationary Bernoulli process is the number of successes and failures obtained in the N trials. More specifically, we want to be able to find the probability that a given number of successes is obtained from N trials of a stationary Bernoulli process with the parameter p. For example, what is the probability that if five voters are sampled, four will favor candidate B? To aid in answering this type of question, we form a random variable X, where X is the number of successes in N trials. Thus, we want to be able to calculate the probability that X takes on a certain value, which will be denoted by r, given N independent trials of a stationary Bernoulli process where the probability of a success is p. Notationally, we write

$$P(X = r; N, p)$$

(The N and p are often omitted from this statement if their values are clear from the context.) The probability that four out of five voters sampled will favor candidate A (recall that Candidate A was a success and $p = \frac{1}{3}$) can, using the random variable X, be written as

$$P(X = 4; 5, \frac{1}{3})$$

To calculate the probability of the number of successes, the voter example will be explored in detail by finding $P(X = r; 5, \frac{1}{3})$ for $r = 0$ to $r = 5$; thereafter, a general formula for finding such probabilities will be discussed.

We begin by finding the probability that, of the five voters sampled, all five will favor candidate A. In other words, we want to find the probability of the sequence (A, A, A, A, A). That is

$$P(X = 5) = P(A, A, A, A, A)$$

Because the trials are independent, the probability of the sequence (A, A, A, A, A) is equal to the product of the probabilities of each individual event (see Section 1.6, p. 10).

$$P(X = 5) = P(A, A, A, A, A) = (\frac{1}{3})(\frac{1}{3})(\frac{1}{3})(\frac{1}{3})(\frac{1}{3})$$
$$= (\frac{1}{3})^5 = \frac{1}{243}$$

We next turn to finding the probability of obtaining four voters who favor candidate A. There are many ways that this result may be obtained: the first four candidates may favor A and the last favors B, or the first may favor B and the remaining four A, and so forth. In fact there are five sequences for which four voters favor A:

(A, A, A, A, B)
(A, A, A, B, A)
(A, A, B, A, A)
(A, B, A, A, A)
(B, A, A, A, A)

The probability of each of these sequences is the same; namely

$$(\frac{1}{3})^4(\frac{2}{3})$$

Because there are five of these sequences

$$P(X = 4) = 5(\frac{1}{3})^4(\frac{2}{3}) = \frac{10}{243}$$

We now turn to the probability of obtaining three voters who favor A. One such sequence is (A, A, A, B, B). However, there are nine other sequences that also have three voters favoring A (we could list them but shortly we will have a simple way of counting them). Because the probability of each sequence is $(\frac{1}{3})^3(\frac{2}{3})^2$

$$P(X = 3) = 10(\frac{1}{3})^3(\frac{2}{3})^2 = \frac{40}{243}$$

The probability of obtaining two voters who favor A is found from sequences containing two A's and three B's; for example, (A, A, B, B, B). Because there are 10 such sequences and the probability of each is $(\frac{1}{3})^2(\frac{2}{3})^3$

$$P(X = 2) = 10(\tfrac{1}{3})^2(\tfrac{2}{3})^3 = {}^{80}/_{243}$$

The probability of obtaining one voter who favors A is found from sequences containing only one A, such as (A, B, B, B, B), of which there are five. Because the probability of each is $(\frac{1}{3})(\frac{2}{3})^4$

$$P(X = 1) = 5(\tfrac{1}{3})(\tfrac{2}{3})^4 = {}^{80}/_{243}$$

Finally, the probability of obtaining no voters who favor A is found from the sequence $(B, B, B, B, B,)$; thus

$$P(X = 0) = (\tfrac{2}{3})^5 = {}^{32}/_{243}$$

To summarize

$$P(X = 5; 5, \tfrac{1}{3}) = (\tfrac{1}{3})^5 = {}^{1}/_{243}$$
$$P(X = 4; 5, \tfrac{1}{3}) = 5(\tfrac{1}{3})^4(\tfrac{2}{3}) = {}^{10}/_{243}$$
$$P(X = 3; 5, \tfrac{1}{3}) = 10(\tfrac{1}{3})^3(\tfrac{2}{3})^2 = {}^{40}/_{243}$$
$$P(X = 2; 5, \tfrac{1}{3}) = 10(\tfrac{1}{3})^2(\tfrac{2}{3})^3 = {}^{80}/_{243}$$
$$P(X = 1; 5, \tfrac{1}{3}) = 5(\tfrac{1}{3})(\tfrac{2}{3})^4 = {}^{80}/_{243}$$
$$P(X = 0; 5, \tfrac{1}{3}) = (\tfrac{2}{3})^5 = {}^{32}/_{243}$$

Because each value of X possible is associated with a probability, the preceding discussion describes the probability distribution of X (check to see that the probabilities sum to 1). Examination of these probabilities reveals a pattern expressed by the statement

$$P(X = r; N, p) = (\text{coefficient})p^r q^{N-r} \qquad (13.5)$$

where the coefficient equals the number of sequences that yield r successes.

Further examination of the sequences will reveal that Counting Rule 4 can be used to determine the value of the coefficient. The coefficient of Equation (13.5) is the number of sequences of length N that contains r successes. Now think of the sequences in the following way. Consider N openings; a sequence is constructed by placing the r successes in r of the N openings. The question arises with regard to the number of ways that the r openings can be chosen. This conceptualization suggests that Counting Rule 4 can be invoked. Counting Rule 4 states that the number of ways that r objects can be chosen from a set of N objects is equal to $\binom{N}{r}$. Therefore, the

number of sequences of size N that contain r successes is equal to the number of ways that r objects can be selected from a set of N objects and is equal to $\binom{N}{r}$, the coefficient of interest. To illustrate, the value of the coefficient for

the voting example for three voters for candidate A is equal to

$$\binom{5}{3} = \frac{5!}{(5-3)!3!} = 10$$

Substituting in Equation (13.5), we obtain a formula for finding the number of successes for sampling from a Bernoulli process. **If the random variable X is the number of successes r obtained in N independent trials from a stationary Bernoulli process, then**

$$P(X = r; N, p) = \binom{N}{r}p^r q^{N-r} \tag{13.6}$$

which is referred to as the binomial equation. The distribution of X is called a *binomial distribution* with parameters N and p.

Several examples will illustrate the use of the binomial equation. Suppose that a fair coin (i.e., $p = \frac{1}{2}$ where success is a head) is tossed six times, the probability of obtaining three heads is

$$P(X = 3; 6, \tfrac{1}{2}) = \binom{6}{3}(\tfrac{1}{2})^3(\tfrac{1}{2})^3 = 20(\tfrac{1}{64}) = .313$$

Assuming that the probability of a girl being born is $\frac{1}{2}$ (actually it is slightly lower) and that births are independent events, the probability that a mother, out of four births, will have four girls is

$$P(X = 4; 4, \tfrac{1}{2}) = \binom{4}{4}(\tfrac{1}{2})^4 = 1(\tfrac{1}{16}) = .0625$$

If we define rolling a six as success, the probability of rolling three sixes out of four rolls is

$$P(X = 3; 4, \tfrac{1}{6}) = \binom{4}{3}(\tfrac{1}{6})^3(\tfrac{5}{6}) = 4(\tfrac{1}{216})(\tfrac{5}{6}) = {}^{20}\!/_{1296} = .0154$$

The binomial distribution can also be used to find probabilities for intervals. For example, the probability of obtaining two, three, or four heads when a coin is tossed six times is

$$P(2 \le X \le 4; 6, \tfrac{1}{2}) = \binom{6}{2}(\tfrac{1}{2})^2(\tfrac{1}{2})^4 + \binom{6}{3}(\tfrac{1}{2})^3(\tfrac{1}{2})^3 + \binom{6}{4}(\tfrac{1}{2})^4(\tfrac{1}{2})^2$$
$$= \sum_{i=2}^{4}\binom{6}{i}(\tfrac{1}{2})^i(\tfrac{1}{2})^{6-i}$$
$$= 15(\tfrac{1}{64}) + 20(\tfrac{1}{64}) + 15(\tfrac{1}{64})$$
$$= .78125$$

One observation of calculating probabilities with the binomial distribution should be quite apparent. As N becomes large, the computations are cumbersome, to say the least. Fortunately, the normal distribution can be used to approximate probabilities associated with random variables that have a binomial distribution. It can be shown that if X has a binomial distribution

$$E(X) = Np \tag{13.7}$$

and

$$V(X) = Npq \tag{13.8}$$

The random variable X is standardized in the usual fashion:

$$Z = \frac{X - E(X)}{\sqrt{V(X)}} = \frac{X - Np}{\sqrt{Npq}} \tag{13.9}$$

For a given value of p the random variable Z approaches a standard normal distribution as N becomes large. However, approximating a discrete distribution with a continuous distribution presents a special problem. Essentially, the problem is this. With the binomial distribution, the probability of obtaining a given number of successes can be found; however, when the value of X is transformed to a z score, the probability that the random variable Z assumes that value is zero because Z is treated as a standard normal variable (i.e., a continuous variable). This problem is handled in the following way. Figure 13.1 presents a histogram of the binomial distribution

$$P(X = r;\ 10,\ \tfrac{1}{2}) = \binom{N}{r} (\tfrac{1}{2})^r (\tfrac{1}{2})^{N-r}$$

(which would be the distribution associated with flipping a fair coin ten times) and a continuous distribution. Suppose that we want to approximate the probability that $X = a$. Recall from Section 2.2 (p. 18) that probabilities associated with discrete distributions can be conceptualized as areas in rectangles that extend one-half unit to either side of the value of the random

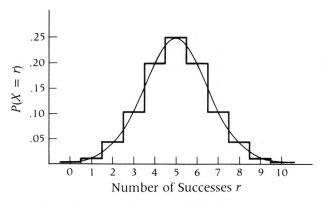

FIGURE 13.1   Binomial Distribution $P(x = r;\ 10,\ \tfrac{1}{2})$ and Approximate Continuous Distribution

variable, as shown in Figure 13.1. To approximate the area in a rectangle using the continuous distribution, we find the area under the curve from $a - .5$ to $a + .5$. However, because the standard normal distribution is used, the approximation of $P(X = a)$ is given by

$$P\left(\frac{a - .5 - Np}{\sqrt{Npq}} \le Z \le \frac{a + .5 - Np}{\sqrt{Npq}}\right) \tag{13.10}$$

where Z has a standard normal distribution. To illustrate, we approximate $P(X = 4; 10, \frac{1}{2}) = .2051$. Using Statement (13.10), we obtain the approximation

$$P\left(\frac{4 - .5 - 10(\frac{1}{2})}{\sqrt{10(\frac{1}{2})(\frac{1}{2})}} \le Z \le \frac{4 + .5 - 10(\frac{1}{2})}{\sqrt{10(\frac{1}{2})(\frac{1}{2})}}\right)$$
$$= P(-.9487 \le Z \le -.3162)$$
$$= .2047$$

which clearly is a good approximation.

Typically, the approximation to the binomial distribution is used to find probabilities in the tails of the distribution, as we will illustrate in the next section. In these instances a half-unit adjustment is also needed to account for the fact that a discrete distribution is being approximated with a continuous distribution. To approximate $P(X \ge a)$ when a is greater than Np, we use

$$P\left(Z \ge \frac{a - .5 - Np}{\sqrt{Npq}}\right) \tag{13.11}$$

and to approximate $P(X \le a)$ when a is less than Np, we use

$$P\left(Z \le \frac{a + .5 - Np}{\sqrt{Npq}}\right) \tag{13.12}$$

The addition or subtraction of .5 is referred to as a *correction for continuity*.

The normal approximation to the binomial distribution improves as N becomes larger for a given p. When $p = \frac{1}{2}$, the binomial distribution is symmetric, and thus the normal approximation is better than when $p \ne \frac{1}{2}$, in which case the binomial distribution is skewed. Generally, if $Np > 5$, the normal distribution is an adequate approximation to the binomial.

13.3   Hypothesis Testing with the Binomial Distribution

The binomial distribution can be used to perform tests in a variety of situations. Typically, the value of p, the probability of success, is unknown to the researcher and the null hypothesis is stated in terms of values of p.

We illustrate this process first with a fabricated example and then apply it to a familiar problem, the dependent group problem.

Consider the problem of testing whether a coin is fair versus the alternative that it is biased in favor of heads. Letting heads be a success, the null hypothesis is H_0: $p = \frac{1}{2}$ versus the alternative H_a: $p > \frac{1}{2}$. An obvious experiment to test these hypotheses is to flip the coin a number of times, say, eight times. Suppose that six heads were obtained, which suggests that the coin is biased in favor of heads. If obtaining six or more heads is an unusual result under the null hypothesis (i.e., this probability is less than α), then the null hypothesis is rejected in favor of the alternative. To calculate the probability that six or more heads are obtained in eight stationary Bernoulli trials, assuming that $p = \frac{1}{2}$, we have

$$P(X \geq 6; 8, \tfrac{1}{2}) = \sum_{i=6}^{8} \binom{8}{i} (\tfrac{1}{2})^i (\tfrac{1}{2})^{8-i}$$

$$= \binom{8}{6}(\tfrac{1}{2})^6(\tfrac{1}{2})^2 + \binom{8}{7}(\tfrac{1}{2})^7(\tfrac{1}{2})^1 + \binom{8}{8}(\tfrac{1}{2})^8$$

$$= 28(\tfrac{1}{256}) + 8(\tfrac{1}{256}) + 1(\tfrac{1}{256}) = .145$$

That is to say, the probability of obtaining six or more heads out of eight flips with a fair coin is .145, which is not sufficiently small to reject the null hypothesis that $p = \frac{1}{2}$.

The preceding test of the hypothesis concerning the probability of success is called an *exact* test because the probability of a particular result is calculated exactly (i.e., not approximated). Although exact tests are elegant, many applications of exact tests involve laborious calculations and approximations are frequently used, such as the normal approximation to the binomial distribution, illustrated in the following example.

Suppose that it is believed that Form A of a test results in higher scores than Form B of the test. The obvious design to determine whether or not this is true is the dependent sample design. Suppose that 20 subjects are selected and administered both forms of the test (10 of the subjects are randomly chosen to receive Form A first; the other 10 receive Form B first). The binomial distribution can be used to provide a test of the null hypothesis that the forms are equivalent. Each subject is assigned a "plus" if his or her score on Form A is higher than the score on Form B and a "minus" if his or her score on Form A is lower than the score on Form B (for now, assume there are no tied scores). The general conjecture that Form A results in higher scores than Form B would imply that for any subject a plus would be more likely than a minus. Under the null hypothesis that the forms were equivalent, it would be expected that for any subject the likelihood of a plus would be equal to the likelihood of a minus (i.e., the probability of each would be $\frac{1}{2}$). If a plus is defined as a success, then the null hypothesis is that $p = \frac{1}{2}$ and the alternative is that $p > \frac{1}{2}$. If the observations are independent, then this design represents 20 trials of a stationary Bernoulli process. Suppose that

16 pluses and 4 minuses resulted from this experiment. The probability of obtaining 16 or more pluses is given by

$$P(X \geq 16;\ 20,\ \tfrac{1}{2}) = \sum_{i=16}^{20} \binom{20}{i} (\tfrac{1}{2})^i (\tfrac{1}{2})^{20-i}$$

Computing these probabilities to conduct the exact test is extremely cumbersome; consequently, the normal approximation is used. Using Statement (13.11)(p. 426), $P(X \geq 16)$ is approximately equal to

$$P\left(Z \geq \frac{16 - .5 - Np}{\sqrt{Npq}}\right) = P\left(Z \geq \frac{15.5 - 20(\tfrac{1}{2})}{\sqrt{20(\tfrac{1}{2})(\tfrac{1}{2})}}\right) = P(Z \geq 2.46) = .007$$

which is sufficiently small to reject the null hypothesis that the probability of obtaining a plus is equal to the probability of obtaining a minus in favor of the alternative that the probability of obtaining a plus is greater than $\tfrac{1}{2}$. (Another way to look at this is to say that $Z = 2.46$ is sufficiently large to reject the null hypothesis.)

Several points should be made about the test involving pluses and minuses. This test, which is called the *sign test*, provides an alternative to the dependent sample t test, although more powerful alternatives for this design will be presented later in this chapter. It should be kept in mind that although the normal distribution was used in this test, the sign test is a distribution-free test. The normal distribution was used to approximate the binomial distribution but no assumptions about the distribution of the population were made. Although both the dependent sample t test or the sign test can be used to test the "equivalence" of the two forms, the statistical hypotheses are different. In the former case H_0: $\mu_1 - \mu_2 = 0$ is tested against H_a: $\mu_1 - \mu_2 > 0$, whereas in the latter case H_0: $p = \tfrac{1}{2}$ is tested against H_a: $p > \tfrac{1}{2}$. In the examples discussed in this section a one-tailed test was used. A two-tailed test is conducted by obtaining a probability in either tail of less than $\alpha/2$. Finally, the issues of ties needs to be mentioned. Although many distribution-free tests, such as those discussed in this section, have ways to accommodate ties, a general, but conservative, method is to resolve ties in a way that favors the null hypothesis. For instance, subjects whose scores were equal on the two forms of the test would be assigned a minus.

13.4   Randomization Tests and the Rationale Behind Distribution-Free Tests

Although called tests, randomization tests is the name given to a procedure used to assign significance to statistics that are sensitive to some aspect of the data. Discussion of randomization tests will help to clarify several mystifying features of distribution-free statistical tests; however, we should keep in mind that the randomization test procedure is a legitimate and powerful procedure in its own right.

We present randomization tests by presenting data from a simple two-group problem. Suppose that six subjects are randomly assigned to two groups, for example, a treatment group and a control group, such that three subjects are in each group and the subjects are measured in some way (this example has exceedingly small numbers of subjects, but it is illustrative of the concepts). Suppose that the following observations were recorded:

Group 1	Group 2
9	3
18	12
15	6

The difference between the groups can be assessed with the t statistic for independent groups:

$$t = \frac{M_1 - M_2}{\sqrt{\dfrac{S_1^2 + S_2^2}{N - 1}}} = \frac{14 - 7}{\sqrt{\dfrac{14 + 14}{2}}} = 2.45$$

The traditional way to assign significance to this value of t would be to compare it to a t distribution with 4 degrees of freedom. However, the validity of this test relies on the assumptions of normality and homogeneity of variance. Given that only a small number of subjects was involved in the study, we should view this approach with trepidation. Furthermore, validity of the parametric approach relies on random selection of the subjects from the population, a condition infrequently satisfied in the social sciences.

Randomization tests provide a procedure for assigning significance that does not rely on making assumptions about the population distributions and, indeed, **does not even require that observations be selected from a population.** The null hypothesis in the randomization test procedure is that the obtained score for each subject will be the same for one arrangement as for any other possible arrangement; that is, all arrangements of the observations are equally likely. For our continuing example, if there is no treatment effect, then all arrangements of the six scores are equally likely. However, if there is a treatment effect, then higher scores will tend to be in one group. Accordingly, assignment of significance to a statistic, such as the t statistic for our example, involves three steps. First, the statistic is computed for the experimental data, as was done earlier. Second, the experimental data are rearranged in all possible ways, as indicated by the random assignment procedure in the experiment, and the statistic is calculated for each of the possible arrangements. In the continuing example there are $\binom{6}{3} = 20$ different ways that six scores can be assigned to two groups such that three scores are in each group. The 20 arrangements, as well as the t statistic for each arrangement, are presented in Table 13.1. The distribution of statistics for all possible arrangement of the data is called the *randomization distribution*. Finally, significance is assigned to the statistic by comparing the obtained statistic to the randomization distribution. Specifically, the significance level of the test statistic is the proportion of test statistics in the randomization distribution that are as large or larger (or as small or smaller, depending on the statistic) than the obtained statistic, and if this significance level is smaller than the value of α set by the researcher, the null hypothesis is rejected. In our continuing example the significance level of the obtained statistic is $2/20 = .10$, because two statistics are as large or larger than the obtained statistic (including the obtained statistic). Because this proportion is not less than .05, the null hypothesis cannot be rejected at the .05 level. That is to say, there is not sufficient evidence to give up the belief that the arrangement of scores obtained in this experiment happened by chance.

TABLE 13.1

Randomization Distribution for t Statistic

Arrangement	Observations in Group 1	Observations in Group 2	M_1	S_1^2	M_2	S_2^2	t	
1	3, 6, 9	12, 15, 18	6	6	15	6	-5.81	
2	3, 6, 12	9, 15, 18	7	14	14	14	-2.96	
3	3, 6, 15	9, 12, 18	8	26	13	14	-1.77	
4	3, 6, 18	9, 12, 15	9	42	12	6	$-.97$	
5	3, 9, 12	6, 15, 18	8	14	13	26	-1.77	
6	3, 9, 15	6, 12, 18	9	24	12	24	$-.97$	
7	3, 9, 18	6, 12, 15	10	38	11	14	$-.31$	
8	3, 12, 15	6, 9, 18	10	26	11	26	$-.31$	
9	3, 12, 18	6, 9, 15	11	38	10	14	.31	
10	3, 15, 18	6, 9, 12	12	42	9	6	.96	
11	6, 9, 12	3, 1, 1	9	6	12	42	$-.96$	
12	6, 9, 15	3, 12, 18	10	14	11	38	$-.31$	
13	6, 9, 18	3, 12, 15	11	26	10	26	.31	
14	6, 12, 15	3, 9, 18	11	14	10	38	.31	
15	6, 12, 18	3, 9, 15	12	24	9	24	.97	
16	6, 15, 18	3, 9, 12	13	26	8	14	1.77	
17	9, 12, 15	3, 6, 18	12	6	9	42	.97	
18	9, 12, 18	3, 6, 15	13	14	8	26	1.77	
19	9, 15, 18	3, 6, 12	14	14	7	14	2.96	obtained statistic
20	12, 15, 18	3, 6, 9	15	6	6	6	5.81	larger than obtained statistic

Several important points about randomization tests need to be made. The most important point is that the randomization test procedure for assigning significance to a statistic is a viable method that could be applied to any research design so long as some aspect of the design involves random assignment. For example, suppose that N subjects are randomly assigned to J groups with N_j subjects in each group. The significance for the F ratio obtained via the analysis of variance could be assigned by comparing the obtained F ratio to the $\dfrac{N}{N_1!(N_2!)\cdots(N_J!)}$ (see Counting Rule 5, p. 420) F ratios obtained for all possible ways that N subjects can be assigned to J groups with N_j subjects in each group. Actually traditional statistics, such as t or F, are not required; any statistic sensitive to the expected effect is suitable. For the example of this section, any statistic sensitive to differences between the two groups (e.g., $M_1 - M_2$) would be appropriate.

To those trained primarily in parametric statistical tests, randomization tests may seem an odd way to assign significance levels to a statistic. Actually randomization tests provide exact probabilities that some result would occur by chance and parametric tests can be thought of as approximations to randomization tests. In fact parametric tests were developed because at the time randomization tests were unfeasible due to the large number of arrangements resulting from experiments with even modest sized samples (recall from Section 13.1, p. 417, that 20 subjects can be assigned to four groups with five subjects in each group in over five thousand trillion ways). However, with methods that reduce the size of the referent randomization distribution and computer applications, assigning significance to statistics with randomization tests is not difficult.

Randomization tests are often conducted on ranks. Observations are ranked in some way and a statistic that is sensitive to departures from the null hypothesis is computed on the ranks. The methods discussed subsequently in this chapter involve computing a statistic from ranks. When randomization tests are applied to ranks, a significance level is assigned by comparing the obtained statistic to the distribution of the statistic for all possible arrangements of the ranks. For a given problem (e.g., assigning six subjects to two groups with three subjects in each group) there will only be one distribution of a statistic based on ranks; that is, the distribution of the statistic is not dependent on the particular scores obtained in the experiment. Consequently, tables exist for the exact probabilities of commonly used statistics based on ranks for small to moderate sized samples (and will be mentioned in subsequent sections).

It is worth mentioning again that randomization tests do not require that the subjects be randomly selected from a population; in fact, as described here, the null hypothesis does not refer to a population. However, when subjects are randomly selected from a population, the model of distribution-free statistical tests can be altered to make inferences about the population. When reference is made to a population, the typical null hypothesis in the distribution-free context is that populations have identical distributions; for example, in the two-group problem the null hypothesis is that the first population distribution is identical to the second population distribution. The alternative hypothesis is that the distributions are not identical (note that in this instance a two-tailed test would always be used). However, we are trained to think in terms of central tendency so that typically alternative hypotheses refer to the location of the distributions; that is, one distribution is to the right or to the left of the other distribution (the average of one distribution is greater than or less than the average of the other distribution). When hypotheses are stated in terms of identical distributions, the alternative hypothesis can refer to central tendency only if it is assumed that the population distributions have identical form. There is a trade-off here: the more informative the alternative hypothesis is, the more stringent the assumptions become. Another point should be made about the null hypotheses in distribution-free tests based on ranks. When reference is made to populations, it is assumed that the distribution of the scores is continuous; observations are sampled and then ranked in some fashion. However, when the null hypothesis refers to all possible arrangements of the data being equally likely, the raw data may be ranks, such as that which might be obtained when judges are asked to rank order subjects on some basis. Although discussion of the remainder of the statistical tests to be described will make reference to populations, keep in mind that inferences can be made about arrangements of the data without regard to sampling from a population.

Commonly Used Distribution-Free Statistical Tests

13.5 The Two-Independent-Sample Design— The Mann–Whitney Test

In the two-independent-sample design N_1 observations are randomly selected from one population and N_2 observations are randomly selected from a second population. In the distribution-free context the null hypothesis is that the distributions of the populations are identical. The distribution-free statistical test to be discussed for two independent groups is the Mann–Whitney test. For this test the $N = N_1 + N_2$ observations are arranged in order and ranked from smallest to largest, as illustrated with the following data (the ranks for each observation are in parentheses):

Sample 1	Sample 2
65 (8)	73 (10)
72 (9)	81 (11)
49 (1)	53 (3)
52 (2)	61 (6)
57 (4)	63 (7)
59 (5)	

The statistic that is used in the Mann–Whitney test is based on the sum of the ranks for either group. Rationale for this choice should be clear: accumulation of low ranks in one group is evidence that the population distributions are not identical. Let T_1 be the sum of the ranks in sample 1. For our example $T_1 = 29$. The U statistic used in the Mann–Whitney test is defined as follows:

$$U = \text{minimum of } U_A \text{ and } U_B \qquad (13.13)$$

where

$$U_A = N_1 N_2 + \frac{N_1(N_1 + 1)}{2} - T_1 \qquad (13.14)$$

and

$$U_B = N_1 N_2 - U_A \qquad (13.15)$$

For our continuing example

$$U_A = (6)(5) + \frac{6(6 + 1)}{2} - 29 = 22$$

and

$$U_B = (6)(5) - 22 = 8$$

and therefore $U = 8$.

In our presentation U was calculated based on the sum of the ranks from the first sample; the same value of U would result from calculations based on the sum of the ranks from the second sample. Clearly, taking the minimum of U_A and U_B indicates that the small values of U are evidence against the null hypothesis. Indeed, sufficiently small values of U lead to rejection of the null hypothesis.

Exact tests for determining the significance of U exist and are based on the fact that there are $\binom{N}{N_1}$ ways to assign the N_1 ranks to sample 1. For our example there are $\binom{11}{6} = 462$ ways to arrange the 11 observations.

When the larger of the two samples is less than 20, significance levels for the U statistic should be assigned by reference to tables based on the randomization distribution of U. However, for larger samples significance can be assigned by noting that the distribution of U is approximately normal with

$$E(U) = \frac{N_1 N_2}{2} \quad \text{and} \quad V(U) = \frac{N_1 N_2 (N_1 + N_2 + 1)}{12} \qquad (13.16)$$

Thus, decisions about the null hypothesis are made by comparing

$$z = \frac{U - E(U)}{\sqrt{V(U)}} \tag{13.17}$$

to the standard normal distribution. Although the normal approximation would not be used for the data presented in this section, for illustration

$$E(U) = \frac{(6)(5)}{2} = 15 \quad \text{and} \quad V(U) = \frac{6(5)(6 + 5 + 1)}{12} = 30$$

and

$$z = \frac{8 - 15}{\sqrt{30}} = -1.27$$

which is insufficient to reject the null hypothesis of identical populations. For a two-tailed test either U_A or U_B can be used when the normal approximation is used and will yield the same results (the z values obtained will have opposite signs). When a one-tailed test is used, we should be cognizant of the direction of the results. When tied ranks exist, ranks can be associated with scores by assigning the average rank (i.e., the midrank). However, when this method is used, the variance of U must be adjusted; the reader is referred to more detailed texts.

13.6   Dependent Sample Design—The Wilcoxon Test

In Section 13.3, (p. 426), the sign test was used to make inferences for the dependent sample design. Although a valid test, the sign test is not very powerful because differences between scores are reduced to a dichotomy—either plus or minus. The Wilcoxon test improves on the sign test by ranking the size of the difference between scores.

The test statistic for the Wilcoxon test is calculated in the following way. First, the difference between the paired scores is calculated. Second, the absolute values of the scores are ranked from smallest to largest. Third, the sign (plus or minus) of the difference is attached to the associated rank. This procedure is illustrated in Table 13.2, which presents the intelligence test scores of kindergarten twins, where one of the twins was randomly assigned to a preschool experience and the other assigned to a control condition. So long as tied ranks do not exist for differences of zero, there is no difficulty in assigning midranks, as illustrated in Table 13.2.

The test statistic T for the Wilcoxon test is the sum of ranks with the less frequent sign. In the twin example positive differences are less frequent, and thus T is the sum of the ranks for positive differences:

$$T = 9.5 + 2.5 + 9.5 + 2.5 + 4 = 28$$

There is an exact test based on the 2^N possible arrangements of the ranks; however, a satisfactory approximation is based on the normal distribution

TABLE 13.2

Scores, Differences, and Signed Ranks
for Two Dependent Groups

Twin	Preschool	Control	Differences	Rank of Absolute Value of Difference	Signed Rank
1	101	87	14	9.5	9.5
2	110	127	−17	11.0	−11.0
3	92	103	−11	8.0	−8.0
4	98	99	−1	1.0	−1.0
5	108	106	2	2.5	2.5
6	115	125	−10	7.0	−7.0
7	96	127	−31	12.0	−12.0
8	101	87	14	9.5	9.5
9	105	103	2	2.5	2.5
10	90	99	−9	6.0	−6.0
11	86	83	3	4.0	4.0
12	97	105	−8	5.0	−5.0

when the number of pairs is greater than eight. Specifically, decisions about the null hypothesis are made by comparing

$$z = \frac{T - E(T)}{\sqrt{V(T)}} \tag{13.18}$$

to the standard normal distribution, where

$$E(T) = \frac{N(N + 1)}{4} \quad \text{and} \quad V(T) = \frac{N(N + 1)(2N + 1)}{24} \tag{13.19}$$

As was the case for the Mann–Whitney test, directional or nondirectional alternative hypotheses may be specified. For the present example

$$E(T) = \frac{12(12 + 1)}{4} = 39 \quad \text{and}$$

$$V(T) = \frac{12(12 + 1)[2(12) + 1]}{24} = 162.5$$

and therefore

$$z = \frac{28 - 39}{\sqrt{162.5}} = -.86$$

and the null hypothesis of identical populations cannot be rejected.

Ties are not problematic in the Wilcoxon test unless they occur for a difference of zero. In that case, if the number of ties is even, midranks are assigned and one-half of the ties is randomly assigned a plus and the other half is assigned a minus. If the number of ties is odd, then one pair of observations is randomly deleted from the data and the test is performed with $(N - 1)$ pairs.

13.7   The J Independent Sample Design— The Kruskal–Wallis Test

One of the best distribution-free alternatives to the fixed-effects one-way analysis of variance is the Kruskal–Wallis test. Essentially, the Kruskal–Wallis test is a generalization of the Mann–Whitney test. Consider J groups with N_j observations in the jth group. The observations are rank ordered from smallest to largest and the sum of the ranks within each group is summed (midranks are assigned to ties). Let T_j be the sum of the ranks in the jth group. Table 13.3 presents the calculations of T_j for three groups.

The test statistic H for the Kruskal–Wallis test is given by

$$H = \frac{12}{N(N + 1)} \sum_j \left(\frac{T_j^2}{N_j} \right) - 3(N + 1) \tag{13.20}$$

TABLE 13.3

Scores and Ranks for Three Independent Groups

	Group	
1	**2**	**3**
83(3)	91(15)	101(26)
91(15)	90(11.5)	100(25)
94(20.5)	81(1)	91(15)
89(9)	83(3)	93(19)
89(9)	84(5.5)	96(23.5)
96(23.5)	83(3)	95(22)
91(15)	88(7)	94(20.5)
92(18)	91(15)	
90(11.5)	89(9)	
	84(5.5)	
$T_1 = 124.5$	$T_2 = 75.5$	$T_3 = 151.0$

Clearly the greater the dispersion of the sum of the ranks, the greater the value of H, and thus, as expected, H is sensitive to differences among the groups. Large values of H are expected when the null hypothesis is not true. Exact tests based on the $\dfrac{N}{N_1!(N_2!)\cdots(N_J!)}$ arrangement of the ranks exist for small N. However, an adequate means of assigning significance levels to H is to compare H to a chi-square distribution with $(J-1)$ degrees of freedom. For the data in Table 13.3

$$H = \frac{12}{26(26+1)}\left[\frac{(124.5)^2}{9} + \frac{(75.5)^2}{10} + \frac{(151)^2}{7}\right] - 3(26+1) = 13.9$$

which when compared to a chi-square distribution with 2 degrees of freedom leads to rejection of the null hypothesis that the three populations from which the samples were drawn have identical distributions. Actually, because of the presence of ties, the value obtained for H needs to be corrected. However, when the number of ties relative to the number of observations is small, the correction makes little difference; nevertheless, in practice, the correction factor should be used and the reader is referred to distribution-free texts.

13.8 The Randomized Block Design or Repeated Measures Design—The Friedman Test

Recall that the randomized block design and the repeated measures design were conceptualized as mixed models without replication (see Sections 8.4, p. 247, and 8.5, p. 250). In the randomized block design each of the J blocks contained K subjects who were randomly assigned to the K treatments. In the repeated measures design, each of the J subjects receives K treatments (in random order) or is measured at K time points. The Friedman test is appropriate for either of these designs.

The Friedman test is conducted in the following way. First, the K scores within each row are rank ordered from smallest to largest. That is, for the randomized block design the scores within a block are ranked; for the repeated measures the scores for each individual are ranked. This ranking is illustrated in Table 13.4. The second step involves calculating the sum of the ranks, T_k, for each of the K treatments, as illustrated in Table 13.4. Finally, the test statistic S is given by the following expression:

$$S = \sum_k \left[T_k - \frac{J(K+1)}{2}\right]^2 \tag{13.21}$$

TABLE 13.4

Scores and Ranks for Randomized Block Design
or Repeated Measures Design

		Treatments			
		1	2	3	4
	1	8.3(4)	7.9(3)	6.1(2)	5.9(1)
	2	9.1(4)	7.5(2)	7.7(3)	6.8(1)
	3	10.9(3)	10.1(1)	10.5(2)	11.0(4)
	4	9.9(3)	6.6(1)	7.7(2)	10.2(4)
	5	10.7(4)	8.1(2)	6.2(1)	8.9(3)
Block or	6	5.7(3)	5.0(1)	5.2(2)	6.2(4)
Subjects	7	7.7(1)	8.5(3)	7.9(2)	10.1(4)
	8	6.6(2)	7.7(4)	6.4(1)	7.2(3)
	9	10.2(3)	8.1(1)	9.7(2)	12.1(4)
	10	8.2(4)	6.0(1)	7.5(3)	7.0(2)
	11	9.3(4)	7.9(2)	8.2(3)	7.0(1)
	12	6.7(3)	6.0(1)	6.5(2)	8.2(4)
		$T_1 = 38$	$T_2 = 22$	$T_3 = 25$	$T_4 = 35$

For the data in Table 13.4

$$S = \left[38 - \frac{12(4 + 1)}{2} \right]^2 + \left[22 - \frac{12(4 + 1)}{2} \right]^2$$

$$+ \left[25 - \frac{12(4 + 1)}{2} \right]^2 + \left[35 - \frac{12(4 + 1)}{2} \right]^2$$

$$= 178$$

Significance levels can be assigned to the statistic S based on $(K!)^J$ arrangements of the ranks ($K!$ arrangements within each row repeated over the J rows). However, for moderately large samples ($J \geq 10$ and $K \geq 4$) the statistic

$$\frac{12S}{JK(K + 1)} \tag{13.22}$$

has approximately a chi-square distribution with $(K - 1)$ degrees of freedom and the null hypothesis is rejected provided this statistic is sufficiently large. For the data in Table 13.4

$$\frac{12S}{JK(K + 1)} = \frac{12(178)}{12(4)(4 + 1)} = 8.9$$

which, when compared to a chi-square distribution with 3 degrees of freedom, leads to rejection of the null hypothesis.

The statistic S can be simplified to

$$S = \sum_k (T_k - M_T)^2 \qquad (13.23)$$

that is, S is the sum of the squared deviations of the sum of the column ranks (viz., the T_j's) from the mean of the sum of the column ranks (M_T). If the null hypothesis is not true, then low ranks will accumulate in some groups and high ranks in others, creating a greater dispersion among the T_j's and a larger value of S. Thus, S is sensitive to departures from the null hypothesis.

Tied scores within each row are accommodated in the Friedman test by breaking the ties in such a way that the dispersion of the T_k's are minimized, yielding a conservative but legitimate test of the null hypothesis.

Contrary to what one might expect, when $K = 2$, the Friedman test reduces to the sign test (rather than to the Wilcoxon test).

NOTES AND SUPPLEMENTARY READINGS

Edgington, E.S. (1980). *Randomization tests*. New York: Marcel Dekker.

Edgington has provided considerable stimulus to methodologists in the behavioral sciences for examining the utility of randomization tests, which is becoming increasingly evident in the literature. This procedure is not widely found in earlier books on nonparametric statistics and worthy of further consideration.

Bradley, J. V. (1968). *Distribution-free statistical tests*. Englewood Cliffs, NJ: Prentice-Hall.

Interested readers may find this volume in their university library. Unfortunately, it is out of print and may not be easily accessed otherwise.

Lehman, E. L. (1975). *Nonparametrics: Statistical methods based on ranks*. San Francisco: Holden-Day.

As suggested by the title, this volume focuses specifically on nonparametric statistics for use on ordinal or rank-type data. It is based heavily on the early work of Wilcoxon. Perhaps the most detailed volume available on rank tests, this book may be somewhat mathematically oriented for readers with a limited mathematical background.

Siegel, S. (1956). *Nonparametric statistics for the behavioral sciences*. New York: McGraw-Hill.

Siegel's volume on nonparametric statistics is perhaps one of the most durable books in the statistics field. It is very readable and a constant source for those in circumstances where nonparametric techniques are advisable. Over the years it has set a record as being the most "borrowed and never returned" book on the shelves of Professor Drew. There are relatively few volumes focusing specifically on nonparametrics, and this is the classic.

PROBLEMS

1. Suppose that the probability that a newborn will be a boy is ½.
   a. Find the probability that only one of the five babies born in a hospital on a certain day will be a boy.
   b. Find the probability that on a day when six babies are born, more than four will be girls.
   c. Find the probability that on a day when eight babies are born, four will be girls.
   d. Using the normal approximation to the binomial distribution (with correction for continuity), find the probability that of the next 20 babies born, 13 or more will be boys.

2. Suppose that the probability that a student will successfully complete an undergraduate degree in four years is ⅔.
   a. Find the probability that four of the starting five on the basketball team will graduate in four years.
   b. Find the probability that one or none of the five starters will graduate in four years.
   c. Find the probability that four of the six students admitted from a certain high school will graduate in four years.
   d. Using the normal approximation to the binomial distribution (with correction for continuity), find the probability that of the 100 students admitted in a given year, 55 or fewer will graduate.

3. Eleven distressed couples and twelve nondistressed couples were recorded during an hour's conversation and the proportion of negative statements to total statements were determined. The following data were obtained:

Distressed	Nondistressed
.32	.18
.21	.33
.27	.16
.45	.12
.29	.10
.17	.42
.34	.06
.36	.22
.52	.19
.07	.23
.15	.11
	.28

Use the Mann–Whitney test to determine if distressed couples emit a greater proportion of negative statements than do nondistressed couples ($\alpha = .05$).

4. To study the efficacy of two treatments for stuttering, subjects with severe stuttering problems were randomly assigned to two treatment groups. At the end of the treatment each subject was asked to read a passage and the number of words pronounced correctly per minute was recorded.

Treatment I	Treatment II
75	85
50	55
23	63
64	48
95	105
21	92
52	110
31	53
45	46
67	83
27	
57	
35	

Determine which, if either, treatment is more effective using the Mann–Whitney test ($\alpha = .05$).

5. In a study of the conversational styles of college men with other men and women a sample of college males was asked to participate in an experiment. When a subject arrived for the experiment, he was seated in a waiting room, shortly thereafter another student (a research assistant) entered and carried on a conversation with the subject. A while

after the assistant left, another student (also a research assistant) of the other gender entered and also carried on a conversation with the subject. The proportion of the time that the subject had eye contact with each research assistant was noted.

Subject	Male Research Assistant	Female Research Assistant
1	.45	.56
2	.33	.44
3	.17	.65
4	.37	.39
5	.61	.57
6	.45	.58
7	.12	.19
8	.34	.32
9	.47	.40
10	.25	.14
11	.45	.65
12	.27	.41

Determine if men students make more eye contact with women students than they do with other men students by using (a) the sign test and (b) the Wilcoxon test ($\alpha = .05$).

6. In an experiment to compare two methods for teaching algebra skills, 20 students were matched based on their score on an algebra aptitude test and one of each pair was randomly assigned to the first treatment. The students' scores on an algebra test after completion of the instruction are as follows:

Pair	Treatment I	Treatment II
1	27	39
2	15	23
3	38	33
4	35	42
5	27	22
6	41	42
7	23	31
8	10	21
9	26	24
10	31	41

Determine which, if either, treatment is more effective using (a) the sign test and (b) the Wilcoxon test ($\alpha = .05$).

7. Experienced typists were randomly assigned to type a passage on four brands of computer keyboards. The following table shows the number of words typed correctly per minute for each subject:

Keyboard A	Keyboard B	Keyboard C	Keyboard D
74	60	54	49
78	77	63	52
84	82	67	57
85	88	66	73
68	93	75	64

Using the Kruskal–Wallis test, determine if typing speed on the four keyboards differ ($\alpha = .05$).

8. In a study in a hospital the length of time (in minutes) that staff spent eating lunch was recorded. The following are the times for three groups of subjects:

Physicians	Nurses	Clerical Staff
12	34	37
23	18	35
38	41	27
31	28	47
19	33	65
	25	24
		29
		22

Using the Kruskal–Wallis test, determine if the time spent eating lunch for the three groups differs ($\alpha = .05$).

9. To determine which of four designs for a dial led to the most accurate settings, ten subjects made ten settings with each dial. Error in each setting was measured to the nearest .1 and the outcome measure for each subject and each dial was the total error for the ten settings. The results are presented as follows:

Subject	Dial 1	Dial 2	Dial 3	Dial 4
1	2.1	2.7	1.9	2.9
2	3.0	2.6	1.8	2.8
3	5.3	4.8	3.5	4.0
4	1.9	2.4	2.6	2.7
5	3.5	3.6	1.7	3.9
6	1.9	2.5	2.0	3.3
7	4.8	4.0	3.7	4.4
8	3.7	4.8	3.6	5.0
9	2.9	2.6	2.0	2.8
10	3.0	3.6	2.6	2.4

Test whether there are differences in the accuracy produced by the four dials using the Friedman test ($\alpha = .05$).

10. To examine the efficacy of three brief treatments for phobic reactions to flying, subjects were matched based on a measure of their pretreatment anxiety related to flying. Specifically, eight blocks of three subjects each were formed so that subjects in each block had equal or nearly equal scores on the pretreatment anxiety measure; the three subjects in each block were then randomly assigned to the three treatments. The outcome measure was the number of sessions needed for the subject to board an airplane. Here are the results.

Block	Treatment I	Treatment II	Treatment III
1	6	4	9
2	8	10	13
3	8	2	6
4	9	5	4
5	10	6	15
6	7	3	8
7	4	8	10
8	3	4	12

Test whether there are differences in the efficacy of the three treatments using the Friedman test ($\alpha = .05$).

COMPUTER DATA SET

T he following table is a data set upon which most of the statistical procedures discussed in this text can be performed with computer assistance.

Assume that the data were collected according to the following procedure. Ninety subjects were randomly selected from a population and randomly assigned to five groups (18 subjects per group): treatments A, B, and C, a wait-list control group (WLC), and a placebo control group (PC). The treatments (A, B, and C) involved different types of instructional programs; each instructional program contained information about 12 independent topics. Furthermore, the subjects in the treatment groups (A, B, C) were randomly assigned to either an enrichment condition or to a no-enrichment condition and to one of three specified orders of presentation of topics, where the three orders were randomly selected from the 12! possible orders. Assignment of subjects to treatments, enrichment factor, and order yielded a balanced design (i.e., three subjects in each combination). As well, the gender, age, and score on an intelligence test were recorded. Three equivalent tests of knowledge were given: before assignment of subjects to groups (pretest), upon completion of treatment (posttest), and again six months later (follow-up).

TABLE A.1

Data Set for Computer Applications

ID	Treatment	Enrichment	Order	Gender	Age	IQ	Pretest	Posttest	Follow-up
1	A	Yes	1	Male	24	93.00	48.00	79.81	66.77
2	A	Yes	1	Female	34	74.00	50.00	52.25	50.07
3	A	Yes	1	Male	23	59.00	50.00	61.60	52.62
4	A	No	1	Male	32	87.00	53.00	50.64	43.28
5	A	No	1	Female	25	84.00	54.00	61.72	51.35
6	A	No	1	Male	18	131.00	74.00	72.49	71.74
7	A	Yes	2	Male	31	124.00	62.00	77.48	79.89
8	A	Yes	2	Male	37	73.00	58.00	77.12	62.67
9	A	Yes	2	Female	24	112.00	63.00	70.90	67.00
10	A	No	2	Male	32	113.00	51.00	64.01	64.16
11	A	No	2	Male	39	119.00	71.00	61.91	68.89
12	A	No	2	Male	37	65.00	51.00	59.75	45.10
13	A	Yes	3	Male	37	87.00	52.00	74.36	70.90
14	A	Yes	3	Female	42	115.00	65.00	80.46	67.62
15	A	Yes	3	Male	26	101.00	44.00	46.41	61.69
16	A	No	3	Male	27	101.00	40.00	51.30	62.09
17	A	No	3	Male	27	86.00	47.00	51.13	50.46
18	A	No	3	Female	31	109.00	64.00	55.94	55.38
19	B	Yes	1	Female	34	124.00	64.00	65.55	77.94
20	B	Yes	1	Male	19	115.00	58.00	66.75	61.38
21	B	Yes	1	Male	21	97.00	50.00	56.42	45.61
22	B	No	1	Female	31	101.00	71.00	60.76	46.18
23	B	No	1	Male	31	109.00	67.00	79.92	74.86
24	B	No	1	Female	29	109.00	51.00	56.24	55.13
25	B	Yes	2	Male	28	83.00	51.00	63.77	50.77
26	B	Yes	2	Male	15	100.00	69.00	69.31	45.31
27	B	Yes	2	Female	32	131.00	71.00	81.41	70.35
28	B	No	2	Male	34	100.00	56.00	43.74	47.91
29	B	No	2	Female	30	109.00	53.00	47.03	52.57
30	B	No	2	Female	30	101.00	71.00	64.01	60.29
31	B	Yes	3	Male	29	92.00	70.00	63.12	51.82
32	B	Yes	3	Female	26	101.00	63.00	63.20	53.00
33	B	Yes	3	Male	30	94.00	50.00	46.14	41.70
34	B	No	3	Male	41	97.00	36.00	48.94	46.24
35	B	No	3	Female	21	95.00	50.00	53.03	56.27
36	B	No	3	Male	22	86.00	28.00	41.72	30.22
37	C	Yes	1	Female	35	106.00	50.00	57.63	42.52
38	C	Yes	1	Male	21	103.00	70.00	67.18	67.84
39	C	Yes	1	Male	31	119.00	62.00	58.90	74.17
40	C	No	1	Male	33	102.00	64.00	64.62	56.13
41	C	No	1	Female	29	110.00	68.00	75.97	62.24
42	C	No	1	Female	14	122.00	60.00	67.75	63.85
43	C	Yes	2	Male	24	66.00	62.00	50.95	57.02
44	C	Yes	2	Female	35	113.00	57.00	68.59	51.29
45	C	Yes	2	Male	27	86.00	49.00	62.48	69.19

TABLE A.1 (*Continued*)

ID	Treatment	Enrichment	Order	Gender	Age	IQ	Pretest	Posttest	Follow-up
46	C	No	2	Male	26	99.00	54.00	57.22	50.07
47	C	No	2	Female	32	100.00	57.00	49.76	60.52
48	C	No	2	Female	24	89.00	54.00	59.14	49.47
49	C	Yes	3	Male	28	109.00	60.00	72.45	56.86
50	C	Yes	3	Female	25	136.00	49.00	65.04	69.28
51	C	Yes	3	Female	25	104.00	47.00	72.03	56.74
52	C	No	3	Female	18	84.00	52.00	51.59	28.42
53	C	No	3	Female	24	86.00	56.00	48.97	48.27
54	C	No	3	Male	17	113.00	68.00	61.56	64.27
55	WLC			Male	29	72.00	53.00	56.42	52.39
56	WLC			Male	27	129.00	74.00	64.42	74.33
57	WLC			Male	33	98.00	61.00	54.22	61.38
58	WLC			Male	19	104.00	48.00	59.03	54.77
59	WLC			Female	28	118.00	48.00	66.53	60.35
60	WLC			Female	23	100.00	63.00	53.68	56.76
61	WLC			Female	25	111.00	72.00	67.37	65.25
62	WLC			Male	35	120.00	61.00	60.35	72.66
63	WLC			Male	18	85.00	47.00	46.07	32.22
64	WLC			Male	25	116.00	65.00	65.73	64.00
65	WLC			Female	28	89.00	50.00	49.87	50.87
66	WLC			Male	11	122.00	59.00	68.35	69.34
67	WLC			Male	26	86.00	35.00	55.09	45.39
68	WLC			Female	31	115.00	76.00	71.65	77.53
69	WLC			Male	21	70.00	54.00	55.82	33.13
70	WLC			Male	26	92.00	57.00	45.79	45.43
71	WLC			Female	24	111.00	58.00	61.41	79.20
72	WLC			Female	29	118.00	37.00	72.63	69.64
73	PC			Male	30	135.00	67.00	72.85	70.36
74	PC			Female	24	85.00	55.00	35.47	41.94
75	PC			Male	25	103.00	50.00	53.48	63.86
76	PC			Male	36	96.00	53.00	51.45	40.30
77	PC			Female	32	99.00	63.00	55.88	61.03
78	PC			Male	25	80.00	43.00	50.27	45.33
79	PC			Female	12	118.00	60.00	67.02	68.84
80	PC			Male	32	92.00	56.00	59.32	57.82
81	PC			Male	29	87.00	36.00	41.65	34.75
82	PC			Male	41	88.00	52.00	50.85	50.93
83	PC			Male	24	94.00	52.00	62.29	51.48
84	PC			Female	27	126.00	43.00	61.79	55.94
85	PC			Female	21	91.00	60.00	54.44	53.25
86	PC			Female	22	112.00	70.00	56.81	63.70
87	PC			Male	36	90.00	37.00	42.45	41.79
88	PC			Female	24	96.00	56.00	53.85	56.08
89	PC			Male	26	99.00	55.00	52.69	61.23
90	PC			Male	22	77.00	54.00	46.94	58.28

APPENDIX B

TABLES

TABLE B.1

Cumulative Normal Probabilities

z	$F(z)$	z	$F(z)$	z	$F(z)$	z	$F(z)$
0.00	.5000000	0.46	.6772419	0.92	.8212136	1.38	.9162067
0.01	.5039894	0.47	.6808225	0.93	.8238145	1.39	.9177356
0.02	.5079783	0.48	.6843863	0.94	.8263912	1.40	.9192433
0.03	.5119665	0.49	.6879331	0.95	.8289439	1.41	.9207302
0.04	.5159534	0.50	.6914625	0.96	.8314724	1.42	.9221962
0.05	.5199388	0.51	.6949743	0.97	.8339768	1.43	.9236415
0.06	.5239222	0.52	.6984682	0.98	.8364569	1.44	.9250663
0.07	.5279032	0.53	.7019440	0.99	.8389129	1.45	.9264707
0.08	.5318814	0.54	.7054015	1.00	.8413447	1.46	.9278550
0.09	.5358564	0.55	.7088403	1.01	.8437524	1.47	.9292191
0.10	.5398278	0.56	.7122603	1.02	.8461358	1.48	.9305634
0.11	.5437953	0.57	.7156612	1.03	.8484950	1.49	.9318879
0.12	.5477584	0.58	.7190427	1.04	.8508300	1.50	.9331928
0.13	.5517168	0.59	.7224047	1.05	.8531409	1.51	.9344783
0.14	.5556700	0.60	.7257469	1.06	.8554277	1.52	.9357445
0.15	.5596177	0.61	.7290691	1.07	.8576903	1.53	.9369916
0.16	.5635595	0.62	.7323711	1.08	.8599289	1.54	.9382198
0.17	.5674949	0.63	.7356527	1.09	.8621434	1.55	.9394292
0.18	.5714237	0.64	.7389137	1.10	.8643339	1.56	.9406201
0.19	.5753454	0.65	.7421539	1.11	.8665005	1.57	.9417924
0.20	.5792597	0.66	.7453731	1.12	.8686431	1.58	.9429466
0.21	.5831662	0.67	.7485711	1.13	.8707619	1.59	.9440826
0.22	.5870604	0.68	.7517478	1.14	.8728568	1.60	.9452007
0.23	.5909541	0.69	.7549029	1.15	.8749281	1.61	.9463011
0.24	.5948349	0.70	.7580363	1.16	.8769756	1.62	.9473839
0.25	.5987063	0.71	.7611479	1.17	.8789995	1.63	.9484493
0.26	.6025681	0.72	.7642375	1.18	.8809999	1.64	.9494974
0.27	.6064199	0.73	.7673049	1.19	.8829768	1.65	.9505285
0.28	.6102612	0.74	.7703500	1.20	.8849303	1.66	.9515428
0.29	.6140919	0.75	.7733726	1.21	.8868606	1.67	.9525403
0.30	.6179114	0.76	.7763727	1.22	.8887676	1.68	.9535213
0.31	.6217195	0.77	.7793501	1.23	.8906514	1.69	.9544860
0.32	.6255158	0.78	.7823046	1.24	.8925123	1.70	.9554345
0.33	.6293000	0.79	.7852361	1.25	.8943502	1.71	.9563671
0.34	.6330717	0.80	.7881446	1.26	.8961653	1.72	.9572838
0.35	.6368307	0.81	.7910299	1.27	.8979577	1.73	.9581849
0.36	.6405764	0.82	.7938919	1.28	.8997274	1.74	.9590705
0.37	.6443088	0.83	.7967306	1.29	.9014747	1.75	.9599408
0.38	.6480273	0.84	.7995458	1.30	.9031995	1.76	.9607961
0.39	.6517317	0.85	.8023375	1.31	.9049021	1.77	.9616364
0.40	.6554217	0.86	.8051055	1.32	.9065825	1.78	.9624620
0.41	.6590970	0.87	.8078498	1.33	.9082409	1.79	.9632730
0.42	.6627573	0.88	.8105703	1.34	.9098773	1.80	.9640697
0.43	.6664022	0.89	.8132671	1.35	.9114920	1.81	.9648521
0.44	.6700314	0.90	.8159399	1.36	.9130850	1.82	.9656205
0.45	.6736448	0.91	.8185887	1.37	.9146565	1.83	.9663750

This table is condensed from Table 1 of the *Biometrika Tables for Statisticians*, Vol. 1 (ed. 3), edited by E. S. Pearson and H. O. Hartley. Reproduced here with the kind permission of E. S. Pearson and the trustees of *Biometrika*.

TABLE B.1   (*Continued*)

z	F(z)	z	F(z)	z	F(z)	z	F(z)
1.84	.9671159	2.07	.9807738	2.29	.9889893	2.51	.9939634
1.85	.9678432	2.08	.9812372	2.30	.9892759	2.52	.9941323
1.86	.9685572	2.09	.9816911	2.31	.9895559	2.53	.9942969
1.87	.9692581	2.10	.9821356	2.32	.9898296	2.54	.9944574
1.88	.9699460	2.11	.9825708	2.33	.9900969	2.55	.9946139
1.89	.9706210	2.12	.9829970	2.34	.9903581	2.56	.9947664
1.90	.9712834	2.13	.9834142	2.35	.9906133	2.57	.9949151
1.91	.9719334	2.14	.9838226	2.36	.9908625	2.58	.9950600
1.92	.9725711	2.15	.9842224	2.37	.9911060	2.59	.9952012
1.93	.9731966	2.16	.9846137	2.38	.9913437	2.60	.9953388
1.94	.9738102	2.17	.9849966	2.39	.9915758	2.70	.9965330
1.95	.9744119	2.18	.9853713	2.40	.9918025	2.80	.9974449
1.96	.9750021	2.19	.9857379	2.41	.9920237	2.90	.9981342
1.97	.9755808	2.20	.9860966	2.42	.9922397	3.00	.9986501
1.98	.9761482	2.21	.9864474	2.43	.9924506	3.20	.9993129
1.99	.9767045	2.22	.9867906	2.44	.9926564	3.40	.9996631
2.00	.9772499	2.23	.9871263	2.45	.9928572	3.60	.9998409
2.01	.9777844	2.24	.9874545	2.46	.9930531	3.80	.9999277
2.02	.9783083	2.25	.9877755	2.47	.9932443	4.00	.9999683
2.03	.9788217	2.26	.9880894	2.48	.9934309	4.50	.9999966
2.04	.9793248	2.27	.9883962	2.49	.9936128	5.00	.9999997
2.05	.9798178	2.28	.9886962	2.50	.9937903	5.50	.9999999
2.06	.9803007						

TABLE B.2

Upper Percentage Points of the *t* Distribution

df	Q = 0.4	0.25	0.1	0.05	0.025	0.01	0.005	0.001
1	0.325	1.000	3.078	6.314	12.704	31.821	63.657	318.31
2	0.289	0.816	1.886	2.920	4.303	6.965	9.925	22.326
3	0.277	0.765	1.638	2.353	3.182	4.541	5.841	10.213
4	0.271	0.741	1.533	2.132	2.776	3.747	4.604	7.173
5	0.267	0.727	1.476	2.015	2.571	3.365	4.032	5.893
6	0.265	0.718	1.440	1.943	2.447	3.143	3.707	5.208
7	0.263	0.711	1.415	1.895	2.365	2.998	3.499	4.785
8	0.262	0.706	1.397	1.860	2.306	2.896	3.355	4.501
9	0.261	0.703	1.383	1.833	2.262	2.821	3.250	4.297
10	0.260	0.700	1.372	1.812	2.228	2.764	3.169	4.144
11	0.260	0.697	1.363	1.796	2.201	2.718	3.106	4.025
12	0.259	0.695	1.356	1.782	2.179	2.681	3.055	3.930
13	0.259	0.694	1.350	1.771	2.160	2.650	3.012	3.852
14	0.258	0.692	1.345	1.761	2.145	2.624	2.977	3.787
15	0.258	0.691	1.341	1.753	2.131	2.602	2.947	3.733
16	0.258	0.690	1.337	1.746	2.120	2.583	2.921	3.686
17	0.257	0.689	1.333	1.740	2.110	2.567	2.898	3.646
18	0.257	0.688	1.330	1.734	2.101	2.552	2.878	3.610
19	0.257	0.688	1.328	1.729	2.093	2.539	2.861	3.579
20	0.257	0.687	1.325	1.725	2.086	2.528	2.845	3.552
21	0.257	0.686	1.323	1.721	2.080	2.518	2.831	3.527
22	0.256	0.686	1.321	1.717	2.074	2.508	2.819	3.505
23	0.256	0.685	1.319	1.714	2.069	2.500	2.807	3.485
24	0.256	0.685	1.318	1.711	2.064	2.492	2.797	3.467
25	0.256	0.684	1.316	1.708	2.060	2.485	2.787	3.450
26	0.256	0.684	1.315	1.706	2.056	2.479	2.779	3.435
27	0.256	0.684	1.314	1.703	2.052	2.473	2.771	3.421
28	0.256	0.683	1.313	1.701	2.048	2.467	2.763	3.408
29	0.256	0.683	1.311	1.699	2.045	2.462	2.756	3.396
30	0.256	0.683	1.310	1.697	2.042	2.457	2.750	3.385
40	0.255	0.681	1.303	1.684	2.021	2.423	2.704	3.307
60	0.254	0.679	1.296	1.671	2.000	2.390	2.660	3.232
120	0.254	0.677	1.289	1.658	1.980	2.358	2.617	3.160
∞	0.253	0.674	1.282	1.645	1.960	2.326	2.576	3.090

This table is condensed from Table 12 of the *Biometrika Tables for Statisticians*, Vol. 1 (ed. 3), edited by E. S. Pearson and H. O. Hartley. Reproduced here with the kind permission of E. S. Pearson and the trustees of *Biometrika*.

TABLE B.3

Upper Percentage Points of the χ^2 Distribution

Q v	0.995	0.990	0.975	0.950	0.900	0.750	0.500
1	$392{,}704.10^{-10}$	$157{,}088.10^{-9}$	$982{,}069.10^{-9}$	$393{,}214.10^{-8}$	0.0157908	0.1015308	0.454937
2	0.0100251	0.0201007	0.0506356	0.102587	0.210720	0.575364	1.38629
3	0.0717212	0.114832	0.215795	0.351846	0.584375	1.212534	2.36597
4	0.206990	0.297110	0.484419	0.710721	1.063623	1.92255	3.35670
5	0.411740	0.554300	0.831211	1.145476	1.61031	2.67460	4.35146
6	0.675727	0.872085	1.237347	1.63539	2.20413	3.45460	5.34812
7	0.989265	1.239043	1.68987	2.16735	2.83311	4.25485	6.34581
8	1.344419	1.646482	2.17973	2.73264	3.48954	5.07064	7.34412
9	1.734926	2.087912	2.70039	3.32511	4.16816	5.89883	8.34283
10	2.15585	2.55821	3.24697	3.94030	4.86518	6.73720	9.34182
11	2.60321	3.05347	3.81575	4.57481	5.57779	7.58412	10.3410
12	3.07382	3.57056	4.40379	5.22603	6.30380	8.43842	11.3403
13	3.56503	4.10691	5.00874	5.89186	7.04150	9.29906	12.3398
14	4.07468	4.66043	5.62872	6.57063	7.78953	10.1653	13.3393
15	4.60094	5.22935	6.26214	7.26094	8.54675	11.0365	14.3389
16	5.14224	5.81221	6.90766	7.96164	9.31223	11.9122	15.3385
17	5.69724	6.40776	7.56418	8.67176	10.0852	12.7919	16.3381
18	6.26481	7.01491	8.23075	9.39046	10.8649	13.6753	17.3379
19	6.84398	7.63273	8.90655	10.1170	11.6509	14.5620	18.3376
20	7.43386	8.26040	9.59083	10.8508	12.4426	15.4518	19.3374
21	8.03366	8.89720	10.28293	11.5913	13.2396	16.3444	20.3372
22	8.64272	9.54249	10.9823	12.3380	14.0415	17.2396	21.3370
23	9.26042	10.19567	11.6885	13.0905	14.8479	18.1373	22.3369
24	9.88623	10.8564	12.4011	13.8484	15.6587	19.0372	23.3367
25	10.5197	11.5240	13.1197	14.6114	16.4734	19.9393	24.3366
26	11.1603	12.1981	13.8439	15.3791	17.2919	20.8434	25.3364
27	11.8076	12.8786	14.5733	16.1513	18.1138	21.7494	26.3363
28	12.4613	13.5648	15.3079	16.9279	18.9392	22.6572	27.3363
29	13.1211	14.2565	16.0471	17.7083	19.7677	23.5666	28.3362
30	13.7867	14.9535	16.7908	18.4926	20.5992	24.4776	29.3360
40	20.7065	22.1643	24.4331	26.5093	29.0505	33.6603	39.3354
50	27.9907	29.7067	32.3574	34.7642	37.6886	42.9421	49.3349
60	35.5346	37.4848	40.4817	43.1879	46.4589	52.2938	59.3347
70	43.2752	45.4418	48.7576	51.7393	55.3290	61.6983	69.3344
80	51.1720	53.5400	57.1532	60.3915	64.2778	71.1445	79.3343
90	59.1963	61.7541	65.6466	69.1260	73.2912	80.6247	89.3342
100	67.3276	70.0648	74.2219	77.9295	82.3581	90.1332	99.3341

This table is taken from Table 8 of the *Biometrika Tables for Statisticians,* Vol. 1 (ed. 3), edited by E. S. Pearson and H. O. Hartley. Reproduced here with the kind permission of E. S. Pearson and the trustees of *Biometrika.*

TABLE B.3 (*Continued*)

v \ Q	0.250	0.100	0.050	0.025	0.010	0.005	0.001
1	1.32330	2.70554	3.84146	5.02389	6.63490	7.87944	10.828
2	2.77259	4.60517	5.99147	7.37776	9.21034	10.5966	13.816
3	4.10835	6.25139	7.81473	9.34840	11.3449	12.8381	16.266
4	5.38527	7.77944	9.48773	11.1433	13.2767	14.8602	18.467
5	6.62568	9.23635	11.0705	12.8325	15.0863	16.7496	20.515
6	7.84080	10.6446	12.5916	14.4494	16.8119	18.5476	22.458
7	9.03715	12.0170	14.0671	16.0128	18.4753	20.2777	24.322
8	10.2188	13.3616	15.5073	17.5346	20.0902	21.9550	26.125
9	11.3887	14.6837	16.9190	19.0228	21.6660	23.5893	27.877
10	12.5489	15.9871	18.3070	20.4831	23.2093	25.1882	29.588
11	13.7007	17.2750	19.6751	21.9200	24.7250	26.7569	31.264
12	14.8454	18.5494	21.0261	23.3367	26.2170	28.2995	32.909
13	15.9839	19.8119	22.3621	24.7356	27.6883	29.8194	34.528
14	17.1170	21.0642	23.6848	26.1190	29.1413	31.3193	36.123
15	18.2451	22.3072	24.9958	27.4884	30.5779	32.8013	37.697
16	19.3688	23.5418	26.2962	28.8454	31.9999	34.2672	39.252
17	20.4887	24.7690	27.5871	30.1910	33.4087	35.7185	40.790
18	21.6049	25.9894	28.8693	31.5264	34.8053	37.1564	42.312
19	22.7178	27.2036	30.1435	32.8523	36.1908	38.5822	43.820
20	23.8277	28.4120	31.4104	34.1696	37.5662	39.9968	45.315
21	24.9348	29.6151	32.6705	35.4789	38.9321	41.4010	46.797
22	26.0393	30.8133	33.9244	36.7807	40.2894	42.7956	48.268
23	27.1413	32.0069	35.1725	38.0757	41.6384	44.1813	49.728
24	28.2412	33.1963	36.4151	39.3641	42.9798	45.5585	51.179
25	29.3389	34.3816	37.6525	40.6465	44.3141	46.9278	52.620
26	30.4345	35.5631	38.8852	41.9232	45.6417	48.2899	54.052
27	31.5284	36.7412	40.1133	43.1944	46.9630	49.6449	55.476
28	32.6205	37.9159	41.3372	44.4607	48.2782	50.9933	56.892
29	33.7109	39.0875	42.5569	45.7222	49.5879	52.3356	58.302
30	34.7998	40.2560	43.7729	46.9792	50.8922	53.6720	59.703
40	45.6160	51.8050	55.7585	59.3417	63.6907	66.7659	73.402
50	56.3336	63.1671	67.5048	71.4202	76.1539	79.4900	86.661
60	66.9814	74.3970	79.0819	83.2976	88.3794	91.9517	99.607
70	77.5766	85.5271	90.5312	95.0231	100.425	104.215	112.317
80	88.1303	96.5782	101.879	106.629	112.329	116.321	124.839
90	98.6499	107.565	113.145	118.136	124.116	128.299	137.208
100	109.141	118.498	124.342	129.561	135.807	140.169	149.449

TABLE B.4

Upper Percentage Points of the F Distribution (5%)

$\nu_2 \backslash \nu_1$	1	2	3	4	5	6	7	8	9	10	12	15	20	24	30	40	60	120	∞
1	161.4	199.5	215.7	224.6	230.2	234.0	236.8	238.9	240.5	241.9	243.9	245.9	248.0	249.1	250.1	251.1	252.2	253.3	254.3
2	18.51	19.00	19.16	19.25	19.30	19.33	19.35	19.37	19.38	19.40	19.41	19.43	19.45	19.45	19.46	19.47	19.48	19.49	19.50
3	10.13	9.55	9.28	9.12	9.01	8.94	8.89	8.85	8.81	8.79	8.74	8.70	8.66	8.64	8.62	8.59	8.57	8.55	8.53
4	7.71	6.94	6.59	6.39	6.26	6.16	6.09	6.04	6.00	5.96	5.91	5.86	5.80	5.77	5.75	5.72	5.69	5.66	5.63
5	6.61	5.79	5.41	5.19	5.05	4.95	4.88	4.82	4.77	4.74	4.68	4.62	4.56	4.53	4.50	4.46	4.43	4.40	4.36
6	5.99	5.14	4.76	4.53	4.39	4.28	4.21	4.15	4.10	4.06	4.00	3.94	3.87	3.84	3.81	3.77	3.74	3.70	3.67
7	5.59	4.74	4.35	4.12	3.97	3.87	3.79	3.73	3.68	3.64	3.57	3.51	3.44	3.41	3.38	3.34	3.30	3.27	3.23
8	5.32	4.46	4.07	3.84	3.69	3.58	3.50	3.44	3.39	3.35	3.28	3.22	3.15	3.12	3.08	3.04	3.01	2.97	2.93
9	5.12	4.26	3.86	3.63	3.48	3.37	3.29	3.23	3.18	3.14	3.07	3.01	2.94	2.90	2.86	2.83	2.79	2.75	2.71
10	4.96	4.10	3.71	3.48	3.33	3.22	3.14	3.07	3.02	2.98	2.91	2.85	2.77	2.74	2.70	2.66	2.62	2.58	2.54
11	4.84	3.98	3.59	3.36	3.20	3.09	3.01	2.95	2.90	2.85	2.79	2.72	2.65	2.61	2.57	2.53	2.49	2.45	2.40
12	4.75	3.89	3.49	3.26	3.11	3.00	2.91	2.85	2.80	2.75	2.69	2.62	2.54	2.51	2.47	2.43	2.38	2.34	2.30
13	4.67	3.81	3.41	3.18	3.03	2.92	2.83	2.77	2.71	2.67	2.60	2.53	2.46	2.42	2.38	2.34	2.30	2.25	2.21
14	4.60	3.74	3.34	3.11	2.96	2.85	2.76	2.70	2.65	2.60	2.53	2.46	2.39	2.35	2.31	2.27	2.22	2.18	2.13
15	4.54	3.68	3.29	3.06	2.90	2.79	2.71	2.64	2.59	2.54	2.48	2.40	2.33	2.29	2.25	2.20	2.16	2.11	2.07
16	4.49	3.63	3.24	3.01	2.85	2.74	2.66	2.59	2.54	2.49	2.42	2.35	2.28	2.24	2.19	2.15	2.11	2.06	2.01
17	4.45	3.59	3.20	2.96	2.81	2.70	2.61	2.55	2.49	2.45	2.38	2.31	2.23	2.19	2.15	2.10	2.06	2.01	1.96
18	4.41	3.55	3.16	2.93	2.77	2.66	2.58	2.51	2.46	2.41	2.34	2.27	2.19	2.15	2.11	2.06	2.02	1.97	1.92
19	4.38	3.52	3.13	2.90	2.74	2.63	2.54	2.48	2.42	2.38	2.31	2.23	2.16	2.11	2.07	2.03	1.98	1.93	1.88
20	4.35	3.49	3.10	2.87	2.71	2.60	2.51	2.45	2.39	2.35	2.28	2.20	2.12	2.08	2.04	1.99	1.95	1.90	1.84
21	4.32	3.47	3.07	2.84	2.68	2.57	2.49	2.42	2.37	2.32	2.25	2.18	2.10	2.05	2.01	1.96	1.92	1.87	1.81
22	4.30	3.44	3.05	2.82	2.66	2.55	2.46	2.40	2.34	2.30	2.23	2.15	2.07	2.03	1.98	1.94	1.89	1.84	1.78
23	4.28	3.42	3.03	2.80	2.64	2.53	2.44	2.37	2.32	2.27	2.20	2.13	2.05	2.01	1.96	1.91	1.86	1.81	1.76
24	4.26	3.40	3.01	2.78	2.62	2.51	2.42	2.36	2.30	2.25	2.18	2.11	2.03	1.98	1.94	1.89	1.84	1.79	1.73
25	4.24	3.39	2.99	2.76	2.60	2.49	2.40	2.34	2.28	2.24	2.16	2.09	2.01	1.96	1.92	1.87	1.82	1.77	1.71
26	4.23	3.37	2.98	2.74	2.59	2.47	2.39	2.32	2.27	2.22	2.15	2.07	1.99	1.95	1.90	1.85	1.80	1.75	1.69
27	4.21	3.35	2.96	2.73	2.57	2.46	2.37	2.31	2.25	2.20	2.13	2.06	1.97	1.93	1.88	1.84	1.79	1.73	1.67
28	4.20	3.34	2.95	2.71	2.56	2.45	2.36	2.29	2.24	2.19	2.12	2.04	1.96	1.91	1.87	1.82	1.77	1.71	1.65
29	4.18	3.33	2.93	2.70	2.55	2.43	2.35	2.28	2.22	2.18	2.10	2.03	1.94	1.90	1.85	1.81	1.75	1.70	1.64
30	4.17	3.32	2.92	2.69	2.53	2.42	2.33	2.27	2.21	2.16	2.09	2.01	1.93	1.89	1.84	1.79	1.74	1.68	1.62
40	4.08	3.23	2.84	2.61	2.45	2.34	2.25	2.18	2.12	2.08	2.00	1.92	1.84	1.79	1.74	1.69	1.64	1.58	1.51
60	4.00	3.15	2.76	2.53	2.37	2.25	2.17	2.10	2.04	1.99	1.92	1.84	1.75	1.70	1.65	1.59	1.53	1.47	1.39
120	3.92	3.07	2.68	2.45	2.29	2.17	2.09	2.02	1.96	1.91	1.83	1.75	1.66	1.61	1.55	1.50	1.43	1.35	1.25
∞	3.84	3.00	2.60	2.37	2.21	2.10	2.01	1.94	1.88	1.83	1.75	1.67	1.57	1.52	1.46	1.39	1.32	1.22	1.00

TABLE B.4   (Continued) (1%)

$v_2 \backslash v_1$	1	2	3	4	5	6	7	8	9	10	12	15	20	24	30	40	60	120	∞
1	4,052	4,999.5	5,403	5,625	5,764	5,859	5,928	5,982	6,022	6,056	6,106	6,157	6,209	6,235	6,261	6,287	6,313	6,339	6,366
2	98.50	99.00	99.17	99.25	99.30	99.33	99.36	99.37	99.39	99.40	99.42	99.43	99.45	99.46	99.47	99.47	99.48	99.49	99.50
3	34.12	30.82	29.46	28.71	28.24	27.91	27.67	27.49	27.35	27.23	27.05	26.87	26.69	26.60	26.50	26.41	26.32	26.22	26.13
4	21.20	18.00	16.69	15.98	15.52	15.21	14.98	14.80	14.66	14.55	14.37	14.20	14.02	13.93	13.84	13.75	13.65	13.56	13.46
5	16.26	13.27	12.06	11.39	10.97	10.67	10.46	10.29	10.16	10.05	9.89	9.72	9.55	9.47	9.38	9.29	9.20	9.11	9.02
6	13.75	10.92	9.78	9.15	8.75	8.47	8.26	8.10	7.98	7.87	7.72	7.56	7.40	7.31	7.23	7.14	7.06	6.97	6.88
7	12.25	9.55	8.45	7.85	7.46	7.19	6.99	6.84	6.72	6.62	6.47	6.31	6.16	6.07	5.99	5.91	5.82	5.74	5.65
8	11.26	8.65	7.59	7.01	6.63	6.37	6.18	6.03	5.91	5.81	5.67	5.52	5.36	5.28	5.20	5.12	5.03	4.95	4.86
9	10.56	8.02	6.99	6.42	6.06	5.80	5.61	5.47	5.35	5.26	5.11	4.96	4.81	4.73	4.65	4.57	4.48	4.40	4.31
10	10.04	7.56	6.55	5.99	5.64	5.39	5.20	5.06	4.94	4.85	4.71	4.56	4.41	4.33	4.25	4.17	4.08	4.00	3.91
11	9.65	7.21	6.22	5.67	5.32	5.07	4.89	4.74	4.63	4.54	4.40	4.25	4.10	4.02	3.94	3.86	3.78	3.69	3.60
12	9.33	6.93	5.95	5.41	5.06	4.82	4.64	4.50	4.39	4.30	4.16	4.01	3.86	3.78	3.70	3.62	3.54	3.45	3.36
13	9.07	6.70	5.74	5.21	4.86	4.62	4.44	4.30	4.19	4.10	3.96	3.82	3.66	3.59	3.51	3.43	3.34	3.25	3.17
14	8.86	6.51	5.56	5.04	4.69	4.46	4.28	4.14	4.03	3.94	3.80	3.66	3.51	3.43	3.35	3.27	3.18	3.09	3.00
15	8.68	6.36	5.42	4.89	4.56	4.32	4.14	4.00	3.89	3.80	3.67	3.52	3.37	3.29	3.21	3.13	3.05	2.96	2.87
16	8.53	6.23	5.29	4.77	4.44	4.20	4.03	3.89	3.78	3.69	3.55	3.41	3.26	3.18	3.10	3.02	2.93	2.84	2.75
17	8.40	6.11	5.18	4.67	4.34	4.10	3.93	3.79	3.68	3.59	3.46	3.31	3.16	3.08	3.00	2.92	2.83	2.75	2.65
18	8.29	6.01	5.09	4.58	4.25	4.01	3.84	3.71	3.60	3.51	3.37	3.23	3.08	3.00	2.92	2.84	2.75	2.66	2.57
19	8.18	5.93	5.01	4.50	4.17	3.94	3.77	3.63	3.52	3.43	3.30	3.15	3.00	2.92	2.84	2.76	2.67	2.58	2.49
20	8.10	5.85	4.94	4.43	4.10	3.87	3.70	3.56	3.46	3.37	3.23	3.09	2.94	2.86	2.78	2.69	2.61	2.52	2.42
21	8.02	5.78	4.87	4.37	4.04	3.81	3.64	3.51	3.40	3.31	3.17	3.03	2.88	2.80	2.72	2.64	2.55	2.46	2.36
22	7.95	5.72	4.82	4.31	3.99	3.76	3.59	3.45	3.35	3.26	3.12	2.98	2.83	2.75	2.67	2.58	2.50	2.40	2.31
23	7.88	5.66	4.76	4.26	3.94	3.71	3.54	3.41	3.30	3.21	3.07	2.93	2.78	2.70	2.62	2.54	2.45	2.35	2.26
24	7.82	5.61	4.72	4.22	3.90	3.67	3.50	3.36	3.26	3.17	3.03	2.89	2.74	2.66	2.58	2.49	2.40	2.31	2.21
25	7.77	5.57	4.68	4.18	3.85	3.63	3.46	3.32	3.22	3.13	2.99	2.85	2.70	2.62	2.54	2.45	2.36	2.27	2.17
26	7.72	5.53	4.64	4.14	3.82	3.59	3.42	3.29	3.18	3.09	2.96	2.81	2.66	2.58	2.50	2.42	2.33	2.23	2.13
27	7.68	5.49	4.60	4.11	3.78	3.56	3.39	3.26	3.15	3.06	2.93	2.78	2.63	2.55	2.47	2.38	2.29	2.20	2.10
28	7.64	5.45	4.57	4.07	3.75	3.53	3.36	3.23	3.12	3.03	2.90	2.75	2.60	2.52	2.44	2.35	2.26	2.17	2.06
29	7.60	5.42	4.54	4.04	3.73	3.50	3.33	3.20	3.09	3.00	2.87	2.73	2.57	2.49	2.41	2.33	2.23	2.14	2.03
30	7.56	5.39	4.51	4.02	3.70	3.47	3.30	3.17	3.07	2.98	2.84	2.70	2.55	2.47	2.39	2.30	2.21	2.11	2.01
40	7.31	5.18	4.31	3.83	3.51	3.29	3.12	2.99	2.89	2.80	2.66	2.52	2.37	2.29	2.20	2.11	2.02	1.92	1.80
60	7.08	4.98	4.13	3.65	3.34	3.12	2.95	2.82	2.72	2.63	2.50	2.35	2.20	2.12	2.03	1.94	1.84	1.73	1.60
120	6.85	4.79	3.95	3.48	3.17	2.96	2.79	2.66	2.56	2.47	2.34	2.19	2.03	1.95	1.86	1.76	1.66	1.53	1.38
∞	6.63	4.61	3.78	3.32	3.02	2.80	2.64	2.51	2.41	2.32	2.18	2.04	1.88	1.79	1.70	1.59	1.47	1.32	1.00

TABLE B.5

Upper Percentage Points for the Studentized Range Statistic q

(5%)

v \ r	2	3	4	5	6	7	8	9	10	11	12	13	14	15	16	17	18	19	20
1	18.0	27.0	32.8	37.1	40.4	43.1	45.4	47.4	49.1	50.6	52.0	53.2	54.3	55.4	56.3	57.2	58.0	58.8	59.6
2	6.09	8.3	9.8	10.9	11.7	12.4	13.0	13.5	14.0	14.4	14.7	15.1	15.4	15.7	15.9	16.1	16.4	16.6	16.8
3	4.50	5.91	6.82	7.50	8.04	8.48	8.85	9.18	9.46	9.72	9.95	10.15	10.35	10.52	10.69	10.84	10.98	11.11	11.24
4	3.93	5.04	5.76	6.29	6.71	7.05	7.35	7.60	7.83	8.03	8.21	8.37	8.52	8.66	8.79	8.91	9.03	9.13	9.23
5	3.64	4.60	5.22	5.67	6.03	6.33	6.58	6.80	6.99	7.17	7.32	7.47	7.60	7.72	7.83	7.93	8.03	8.12	8.21
6	3.46	4.34	4.90	5.31	5.63	5.89	6.12	6.32	6.49	6.65	6.79	6.92	7.03	7.14	7.24	7.34	7.43	7.51	7.59
7	3.34	4.16	4.68	5.06	5.36	5.61	5.82	6.00	6.16	6.30	6.43	6.55	6.66	6.76	6.85	6.94	7.02	7.09	7.17
8	3.26	4.04	4.53	4.89	5.17	5.40	5.60	5.77	5.92	6.05	6.18	6.29	6.39	6.48	6.57	6.65	6.73	6.80	6.87
9	3.20	3.95	4.42	4.76	5.02	5.24	5.43	5.60	5.74	5.87	5.98	6.09	6.19	6.28	6.36	6.44	6.51	6.58	6.64
10	3.15	3.88	4.33	4.65	4.91	5.12	5.30	5.46	5.60	5.72	5.83	5.93	6.03	6.11	6.20	6.27	6.34	6.40	6.47
11	3.11	3.82	4.26	4.57	4.82	5.03	5.20	5.35	5.49	5.61	5.71	5.81	5.90	5.99	6.06	6.14	6.20	6.26	6.33
12	3.08	3.77	4.20	4.51	4.75	4.95	5.12	5.27	5.40	5.51	5.62	5.71	5.80	5.88	5.95	6.03	6.09	6.15	6.21
13	3.06	3.73	4.15	4.45	4.69	4.88	5.05	5.19	5.32	5.43	5.53	5.63	5.71	5.79	5.86	5.93	6.00	6.05	6.11
14	3.03	3.70	4.11	4.41	4.64	4.83	4.99	5.13	5.25	5.36	5.46	5.55	5.64	5.72	5.79	5.85	5.92	5.97	6.03
15	3.01	3.67	4.08	4.37	4.60	4.78	4.94	5.08	5.20	5.31	5.40	5.49	5.58	5.65	5.72	5.79	5.85	5.90	5.96
16	3.00	3.65	4.05	4.33	4.56	4.74	4.90	5.03	5.15	5.26	5.35	5.44	5.52	5.59	5.66	5.72	5.79	5.84	5.90
17	2.98	3.63	4.02	4.30	4.52	4.71	4.86	4.99	5.11	5.21	5.31	5.39	5.47	5.55	5.61	5.68	5.74	5.79	5.84
18	2.97	3.61	4.00	4.28	4.49	4.67	4.82	4.96	5.07	5.17	5.27	5.35	5.43	5.50	5.57	5.63	5.69	5.74	5.79
19	2.96	3.59	3.98	4.25	4.47	4.65	4.79	4.92	5.04	5.14	5.23	5.32	5.39	5.46	5.53	5.59	5.65	5.70	5.75
20	2.95	3.58	3.96	4.23	4.45	4.62	4.77	4.90	5.01	5.11	5.20	5.28	5.36	5.43	5.49	5.55	5.61	5.66	5.71
24	2.92	3.53	3.90	4.17	4.37	4.54	4.68	4.81	4.92	5.01	5.10	5.18	5.25	5.32	5.38	5.44	5.50	5.54	5.59
30	2.89	3.49	3.84	4.10	4.30	4.46	4.60	4.72	4.83	4.92	5.00	5.08	5.15	5.21	5.27	5.33	5.38	5.43	5.48
40	2.86	3.44	3.79	4.04	4.23	4.39	4.52	4.63	4.74	4.82	4.91	4.98	5.05	5.11	5.16	5.22	5.27	5.31	5.36
60	2.83	3.40	3.74	3.98	4.16	4.31	4.44	4.55	4.65	4.73	4.81	4.88	4.94	5.00	5.06	5.11	5.16	5.20	5.24
120	2.80	3.36	3.69	3.92	4.10	4.24	4.36	4.48	4.56	4.64	4.72	4.78	4.84	4.90	4.95	5.00	5.05	5.09	5.13
∞	2.77	3.31	3.63	3.86	4.03	4.17	4.29	4.39	4.47	4.55	4.62	4.68	4.74	4.80	4.85	4.89	4.93	4.97	5.01

TABLE B.5 *(Continued)*

Upper Percentage Points for the Studentized Range Statistic q

(1%)

v \ r	2	3	4	5	6	7	8	9	10	11	12	13	14	15	16	17	18	19	20
1	90.0	135	164	186	202	216	227	237	246	253	260	266	272	277	282	286	290	294	298
2	14.0	19.0	22.3	24.7	26.6	28.2	29.5	30.7	31.7	32.6	33.4	34.1	34.8	35.4	36.0	36.5	37.0	37.5	37.9
3	8.26	10.6	12.2	13.3	14.2	15.0	15.6	16.2	16.7	17.1	17.5	17.9	18.2	18.5	18.8	19.1	19.3	19.5	19.8
4	6.51	8.12	9.17	9.96	10.6	11.1	11.5	11.9	12.3	12.6	12.8	13.1	13.3	13.5	13.7	13.9	14.1	14.2	14.4
5	5.70	6.97	7.80	8.42	8.91	9.32	9.67	9.97	10.24	10.48	10.70	10.89	11.08	11.24	11.40	11.55	11.68	11.81	11.93
6	5.24	6.33	7.03	7.56	7.97	8.32	8.61	8.87	9.10	9.30	9.49	9.65	9.81	9.95	10.08	10.21	10.32	10.43	10.54
7	4.95	5.92	6.54	7.01	7.37	7.68	7.94	8.17	8.37	8.55	8.71	8.86	9.00	9.12	9.24	9.35	9.46	9.55	9.65
8	4.74	5.63	6.20	6.63	6.96	7.24	7.47	7.68	7.87	8.03	8.18	8.31	8.44	8.55	8.66	8.76	8.85	8.94	9.03
9	4.60	5.43	5.96	6.35	6.66	6.91	7.13	7.32	7.49	7.65	7.78	7.91	8.03	8.13	8.23	8.32	8.41	8.49	8.57
10	4.48	5.27	5.77	6.14	6.43	6.67	6.87	7.05	7.21	7.36	7.48	7.60	7.71	7.81	7.91	7.99	8.07	8.15	8.22
11	4.39	5.14	5.62	5.97	6.25	6.48	6.67	6.84	6.99	7.13	7.25	7.36	7.46	7.56	7.65	7.73	7.81	7.88	7.95
12	4.32	5.04	5.50	5.84	6.10	6.32	6.51	6.67	6.81	6.94	7.06	7.17	7.26	7.36	7.44	7.52	7.59	7.66	7.73
13	4.26	4.96	5.40	5.73	5.98	6.19	6.37	6.53	6.67	6.79	6.90	7.01	7.10	7.19	7.27	7.34	7.42	7.48	7.55
14	4.21	4.89	5.32	5.63	5.88	6.08	6.26	6.41	6.54	6.66	6.77	6.87	6.96	7.05	7.12	7.20	7.27	7.33	7.39
15	4.17	4.83	5.25	5.56	5.80	5.99	6.16	6.31	6.44	6.55	6.66	6.76	6.84	6.93	7.00	7.07	7.14	7.20	7.26
16	4.13	4.78	5.19	5.49	5.72	5.92	6.08	6.22	6.35	6.46	6.56	6.66	6.74	6.82	6.90	6.97	7.03	7.09	7.15
17	4.10	4.74	5.14	5.43	5.66	5.85	6.01	6.15	6.27	6.38	6.48	6.57	6.66	6.73	6.80	6.87	6.94	7.00	7.05
18	4.07	4.70	5.09	5.38	5.60	5.79	5.94	6.08	6.20	6.31	6.41	6.50	6.58	6.65	6.72	6.79	6.85	6.91	6.96
19	4.05	4.67	5.05	5.33	5.55	5.73	5.89	6.02	6.14	6.25	6.34	6.43	6.51	6.58	6.65	6.72	6.78	6.84	6.89
20	4.02	4.64	5.02	5.29	5.51	5.69	5.84	5.97	6.09	6.19	6.29	6.37	6.45	6.52	6.59	6.65	6.71	6.76	6.82
24	3.96	4.54	4.91	5.17	5.37	5.54	5.69	5.81	5.92	6.02	6.11	6.19	6.26	6.33	6.39	6.45	6.51	6.56	6.61
30	3.89	4.45	4.80	5.05	5.24	5.40	5.54	5.65	5.76	5.85	5.93	6.01	6.08	6.14	6.20	6.26	6.31	6.36	6.41
40	3.82	4.37	4.70	4.93	5.11	5.27	5.39	5.50	5.60	5.69	5.77	5.84	5.90	5.96	6.02	6.07	6.12	6.17	6.21
60	3.76	4.28	4.60	4.82	4.99	5.13	5.25	5.36	5.45	5.53	5.60	5.67	5.73	5.79	5.84	5.89	5.93	5.98	6.02
120	3.70	4.20	4.50	4.71	4.87	5.01	5.12	5.21	5.30	5.38	5.44	5.51	5.56	5.61	5.66	5.71	5.75	5.79	5.83
∞	3.64	4.12	4.40	4.60	4.76	4.88	4.99	5.08	5.16	5.23	5.29	5.35	5.40	5.45	5.49	5.54	5.57	5.61	5.65

TABLE B.6

r to *Z* Transformation

r	*r* (3rd decimal)					*r*	*r* (3rd decimal)				
	.000	.002	.004	.006	.008		.000	.002	.004	.006	.008
.00	.0000	.0020	.0040	.0060	.0080	.25	.2554	.2575	.2597	.2618	.2640
1	.0100	.0120	.0140	.0160	.0180	6	.2661	.2683	.2704	.2726	.2747
2	.0200	.0220	.0240	.0260	.0280	7	.2769	.2790	.2812	.2833	.2855
3	.0300	.0320	.0340	.0360	.0380	8	.2877	.2899	.2920	.2942	.2964
4	.0400	.0420	.0440	.0460	.0480	9	.2986	.3008	.3029	.3051	.3073
.05	.0500	.0520	.0541	.0561	.0581	.30	.3095	.3117	.3139	.3161	.3183
6	.0601	.0621	.0641	.0661	.0681	1	.3205	.3228	.3250	.3272	.3294
7	.0701	.0721	.0741	.0761	.0782	2	.3316	.3339	.3361	.3383	.3406
8	.0802	.0822	.0842	.0862	.0882	3	.3428	.3451	.3473	.3496	.3518
9	.0902	.0923	.0943	.0963	.0983	4	.3541	.3564	.3586	.3609	.3632
.10	.1003	.1024	.1044	.1064	.1084	.35	.3654	.3677	.3700	.3723	.3746
1	.1104	.1125	.1145	.1165	.1186	6	.3769	.3792	.3815	.3838	.3861
2	.1206	.1226	.1246	.1267	.1287	7	.3884	.3907	.3931	.3954	.3977
3	.1307	.1328	.1348	.1368	.1389	8	.4001	.4024	.4047	.4071	.4094
4	.1409	.1430	.1450	.1471	.1491	9	.4118		.4165	.4189	.4213
.15	.1511	.1532	.1552	.1573	.1593	.40	.4236	.4260	.4284	.4308	.4332
6	.1614	.1634	.1655	.1676	.1696	1	.4356	.4380	.4404	.4428	.4453
7	.1717	.1737	.1758	.1779	.1799	2	.4477	.4501	.4526	.4550	.4574
8	.1820	.1841	.1861	.1882	.1903	3	.4599	.4624	.4648	.4673	.4698
9	.1923	.1944	.1965	.1986	.2007	4	.4722	.4747	.4772	.4797	.4822
.20	.2027	.2048	.2069	.2090	.2111	.45	.4847	.4872	.4897	.4922	.4948
1	.2132	.2153	.2174	.2195	.2216	6	.4973	.4999	.5024	.5049	.5075
2	.2237	.2258	.2279	.2300	.2321	7	.5101	.5126	.5152	.5178	.5204
3	.2342	.2363	.2384	.2405	.2427	8	.5230	.5256	.5282	.5308	.5334
4	.2448	.2469	.2490	.2512	.2533	9	.5361	.5387	.5413	.5440	.5466

TABLE B.6   (*Continued*)

r	r (3rd decimal)					r	r (3rd decimal)				
	.000	.002	.004	.006	.008		.000	.002	.004	.006	.008
.50	.5493	.5520	.5547	.5573	.5600	.75	0.973	0.978	0.982	0.987	0.991
1	.5627	.5654	.5682	.5709	.5736	6	0.996	1.001	1.006	1.011	1.015
2	.5763	.5791	.5818	.5846	.5846	7	1.020	1.025	1.030	1.035	1.040
3	.5901	.5929	.5957	.5985	.5985	8	1.045	1.050	1.056	1.061	1.006
4	.6042	.6070	.6098	.6127	.6127	9	1.071	1.077	1.082	1.088	1.093
.55	.6184	.6213	.6241	.6270	.6299	.80	1.099	1.104	1.110	1.116	1.121
6	.6328	.6358	.6387	.6416	.6446	1	1.127	1.133	1.139	1.145	1.151
7	.6475	.6505	.6535	.6565	.6595	2	1.157	1.163	1.169	1.175	1.182
8	.6625	.6655	.6685	.6716	.6746	3	1.188	1.195	1.201	1.208	1.214
9	.6777	.6807	.6838	.6869	.6900	4	1.221	1.228	1.235	1.242	1.249
.60	.6931	.6963	.6994	.7026	.7057	.85	1.256	1.263	1.271	1.278	1.286
1	.7089	.7121	.7153	.7185	.7218	6	1.293	1.301	1.309	1.317	1.325
2	.7250	.7283	.7315	.7348	.7381	7	1.333	1.341	1.350	1.358	1.367
3	.7414	.7447	.7481	.7514	.7548	8	1.376	1.385	1.394	1.403	1.412
4	.7582	.7616	.7650	.7684	.7718	9	1.422	1.432	1.442	1.452	1.462
.65	.7753	.7788	.7823	.7858	.7893	.90	1.472	1.483	1.494	1.505	1.516
6	.7928	.7964	.7999	.8035	.8071	1	1.528	1.539	1.551	1.564	1.576
7	.8107	.8144	.8180	.8217	.8254	2	1.589	1.602	1.616	1.630	1.644
8	.8291	.8328	.8366	.8404	.8441	3	1.658	1.673	1.689	1.705	1.721
9	.8480	.8518	.8556	.8595	.8634	4	1.738	1.756	1.774	1.792	1.812
.70	.8673	.8712	.8752	.8792	.8832	.95	1.832	1.853	1.874	1.897	1.921
1	.8872	.8912	.8953	.8994	.9035	6	1.946	1.972	2.000	2.029	2.060
2	.9076	.9118	.9160	.9202	.9245	7	2.092	2.127	2.165	2.205	2.249
3	.9287	.9330	.9373	.9417	.9461	8	2.298	2.351	2.410	2.477	2.555
4	.9505	.9549	.9594	.9639	.9684	9	2.647	2.759	2.903	3.106	3.453

This table is abridged from Table 14 of the *Biometrika Tables for Statisticians,* Vol. 1 (ed. 3), edited by E. S. Pearson and H. O. Hartley. Used with the kind permission of E. S. Pearson and the trustees of *Biometrika*.

TABLE B.7

Coefficients of Orthogonal Polynomials

$J = 3$

	c_1	c_2	c_3
Linear	-1	0	1
Quadratic	1	-2	1

$J = 4$

	c_1	c_2	c_3	c_4
Linear	-3	-1	1	3
Quadratic	1	-1	-1	1
Cubic	-1	3	-3	1

$J = 5$

	c_1	c_2	c_3	c_4	c_5
Linear	-2	-1	0	1	2
Quadratic	2	-1	-2	-1	2
Cubic	-1	2	0	-2	1
Quartic	1	-4	6	-4	1

$J = 6$

	c_1	c_2	c_3	c_4	c_5	c_6
Linear	-5	-3	-1	1	3	5
Quadratic	5	-1	-4	-4	-1	5
Cubic	-5	7	4	-4	-7	5
Quartic	1	-3	2	2	-3	1
Fifth-degree	-1	5	-10	10	-5	1

$J = 7$

	c_1	c_2	c_3	c_4	c_5	c_6	c_7
Linear	-3	-2	-1	0	1	2	3
Quadratic	5	0	-3	-4	-3	0	5
Cubic	-1	1	1	0	-1	-1	1
Quartic	3	-7	1	6	1	-7	3
Fifth-degree	-1	4	-5	0	5	-4	1
Sixth-degree	1	-6	15	-20	15	-6	1

SOLUTIONS TO ODD-NUMBERED PROBLEMS AND ANSWERS TO EVEN-NUMBERED PROBLEMS

CHAPTER 1

1. a. $A \cup B = \{1, 2, 3, 4, 5, 6, 7, 8, 9, 10\}$
   b. $A \cap C = \{2\}$
   c. $A \cap B = \emptyset$
   d. $\emptyset, \{1\}, \{2\}, \{5\}, \{1, 2\}, \{1, 5\}, \{2, 5\}, \{1, 2, 5\}$
   e. $\overline{A} = \{1, 3, 5, 7, 9\} = B$

2. a. $A \cup C = \{a, b, c, e, i, o, u\}$
   b. $A \cap C = \{a\}$
   c. $A \cap B = \emptyset$
   d. $\emptyset, \{a\}, \{b\}, \{c\}, \{a, b\}, \{a, c\}, \{b, c\}, \{a, b, c\}$
   e. $\overline{B} = \{a, e, i, o, u\} = A$

3. a. ½   b. ⅙   c. ²⁄₆   d. ⁴⁄₅₂
   e. Let A be the event that an even-numbered card is drawn; then $P(A)$ = ²⁰⁄₅₂. Let B be the event that a heart is drawn; then $P(B)$ = ¹³⁄₅₂. The event $A \cap B$ is the even-numbered heart is drawn; $P(A \cap B)$ = ⁵⁄₅₂. The desired probability is

$$P(A \cup B) = P(A) + P(B) - P(A \cap B) = {}^{20}\!/_{52} + {}^{13}\!/_{52} - {}^{5}\!/_{52}$$
$$= {}^{28}\!/_{52} = {}^{7}\!/_{13}$$

f. Let B be the event that a 10 is drawn and A be the event that an even-numbered card is drawn. The desired probability is

$$P(B|A) = \frac{P(B \cap A)}{P(A)} = \frac{4/52}{20/52} = \frac{1}{5}$$

4. a. $\frac{1}{2}$ b. $\frac{1}{6}$ c. $\frac{3}{6}$ d. $\frac{11}{26}$ e. $\frac{1}{3}$

5. a. i. $\frac{1}{2}$ ii. $\frac{2}{6} = \frac{1}{3}$ iii.   0   iv. $\frac{2}{6} = \frac{1}{3}$

   b. $P(\text{cube}|\text{white}) = \dfrac{P(\text{cube} \cap \text{white})}{P(\text{white})} = \dfrac{0/6}{2/6} = 0$

   c. Because $P(A) = \frac{3}{6}$, $P(B) = \frac{3}{6}$, and $P(A \cap B) = \frac{2}{6}$, $P(A)P(B) \neq P(A \cap B)$, and therefore A and B are not independent.

   d. Because $P(A) = \frac{3}{6}$, $P(C) = \frac{2}{6}$, and $P(A \cap C) = \frac{1}{6}$, $P(A)P(C) = P(A \cap C)$, and therefore A and C are independent.

6. a. i. $\frac{1}{2}$ ii. $\frac{1}{4}$ iii. $\frac{1}{8}$ iv. $\frac{3}{8}$
   b. 1   c. Yes, independent   d. Yes, independent

7. Let A be the event that a widget is defective and B be the event that the test is positive. Then from the information given in the problem

$$P(A) = .05 \qquad P(\overline{A}) = .95 \qquad P(B|A) = .99 \quad \text{and} \quad P(B|\overline{A}) = .03$$

By Bayes' rule the desired probability is given by

$$P(A|B) = \frac{P(B|A)P(A)}{P(B|A)P(A) + P(B|\overline{A})P(\overline{A})} = \frac{.99(.05)}{.99(.05) + .03(.95)} = .635$$

8. .083

CHAPTER 2

1. a.

b.

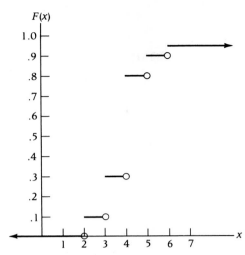

c. (i) $P(2 \le X < 5) = P(X = 2) + P(X = 3) + P(X = 4) = {}^8/_{10} = {}^4/_5$
   (ii) $F(3) = P(X \le 3) = P(X = 2) + P(X = 3) = {}^3/_{10}$
   (iii) $P(X > 4) = P(X = 5) + P(X = 6) = {}^2/_{10}$
   (iv) $1 - F(5) = 1 - P(X \le 5) = P(X = 6) = {}^1/_{10}$
   (v) $P(X \le 4) = P(X = 2) + P(X = 3) + P(X = 4) = {}^8/_{10} = {}^4/_5$

2. c. (i) $\frac{1}{2}$ (ii) $\frac{1}{3}$ (iii) $\frac{1}{4}$ (iv) $\frac{1}{12}$ (v) $\frac{3}{4}$

3. a.

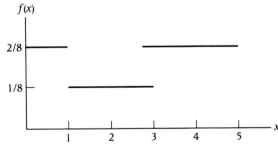

b.

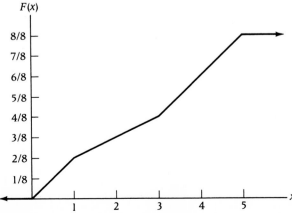

   c. (i) $P(X > 3) = \frac{1}{2}$
     (ii) $F(2) = P(X \leq 2) = \frac{3}{8}$
     (iii) $P(3 \leq X \leq 4) = \frac{1}{4}$
     (iv) $F(3.5) = P(X \leq 3.5) = \frac{5}{8}$

4. c. (i) $\frac{5}{8}$ (ii) $\frac{1}{4}$ (iii) $\frac{1}{8}$ (iv) $\frac{7}{16}$

5. a. (i) $P(Z \leq 2.5) = F(2.5) = .9938$
     (ii) $P(Z > 2.25) = 1 - F(2.25) = 1 - .9878 = .0122$
     (iii) $P(-1.00 < Z < 1.00) = F(1) - F(-1) = .8413 - .1587$
                                           $= .6826$
     (iv) $P(Z < -3.00) = 1 - F(3) = 1 - .9987 = .0013$
     (v) $P(Z > -1.50) = F(1.5) = .9332$
   b. (i) $P(Z < a) = F(a) = .850$; therefore, $a = 1.04$
     (ii) $P(Z > a) = .10$ implies that $P(Z \leq a) = F(a) = .90$, therefore
       $a = 1.28$.
     (iii) $P(Z < a) = .025$ implies that $P(Z < -a) = F(-a) = .975$;
       therefore, $-a = 1.96$ and $a = -1.96$.
     (iv) $P(-a < Z < a) = .900$ implies that $P(Z > a) = .05$ and thus
       $P(Z < a) = F(a) = .95$; therefore, $a = 1.65$ (and $-a = -1.65$).

6. a. i. .997   ii. .0082   iii. .8664   iv. .0062   v. .8413
  b. i. .67   ii. 1.65   iii. -1.65 iv. 1.28

CHAPTER 3

1. a. $E(X) = \sum x_i P(X = x_i) = 9(\frac{1}{3}) + 4(\frac{1}{4}) + 0(\frac{1}{6}) + 8(\frac{1}{4}) = 6$
  b. $V(X) = \sum P(X = x_i)[x_i - E(X)]^2$
       $= \frac{1}{3}(9 - 6)^2 + \frac{1}{4}(4 - 6)^2 + \frac{1}{6}(0 - 6)^2 + \frac{1}{4}(8 - 6)^2$
       $= \frac{1}{3}(9) + \frac{1}{4}(4) + \frac{1}{6}(36) + \frac{1}{4}(4)$
       $= 11$

  c. By the rules of expectation

$$E(2X + 3) = E(2X) + 3 = 2E(X) + 3 = 2(6) + 3 = 15$$

    By the rules of variance

$$V(2X + 3) = 2^2 V(X) = 4(11) = 44$$

  d. $E(X^2 - 1) = \sum(x_i^2 - 1)P(X = x_i)$
                $= (9^2 - 1)(\frac{1}{3}) + (4^2 - 1)(\frac{1}{4}) + (0^2 - 1)(\frac{1}{6})$
                   $+ (8^2 - 1)(\frac{1}{4})$
                $= 46$

2. a. 6   b. 11   c. 22, 99   d. 20

3. a. $M = (1/N) \sum x_i = (\frac{1}{6})(36) = 6$

b. $S^2 = \dfrac{1}{N} \Sigma (x_i - M)^2$

$\quad = \dfrac{1}{6}[(5 - 6)^2 + (3 - 6)^2 + (6 - 6)^2 + (9 - 6)^2 + (2 - 6)^2$
$\quad\quad + (11 - 6)^2]$

$\quad = \dfrac{1}{6}(1 + 9 + 0 + 9 + 16 + 25)$

$\quad = 10$

and $S = \sqrt{S^2} = \sqrt{10} = 3.16$

4. a. 9 b. 14, 3.74

5. a. Marginal distribution for X is

$P(X = 4) = \tfrac{1}{4}$
$P(X = 6) = \tfrac{1}{2}$
$P(X = 8) = \tfrac{1}{4}$

Marginal distribution for Y is

$P(Y = 1) = \tfrac{1}{3}$
$P(Y = 2) = \tfrac{1}{3}$
$P(Y = 3) = \tfrac{1}{3}$

b. $E(X) = \Sigma xP(X = x) = 4(\tfrac{1}{4}) + 6(\tfrac{1}{2}) + 8(\tfrac{1}{4}) = 6$
$\quad V(X) = \Sigma P(X = x)[x - E(X)]^2 = (\tfrac{1}{4})(4) + (\tfrac{1}{2})(0) + (\tfrac{1}{4})(4) = 2$

c. $E(Y) = \Sigma yP(Y = y) = 1(\tfrac{1}{3}) + 2(\tfrac{1}{3}) + 3(\tfrac{1}{3}) = 2$
$\quad V(X) = \Sigma P(Y = y)[y - E(Y)]^2 = (\tfrac{1}{3})(1) + (\tfrac{1}{3})(0) + \tfrac{1}{3}(1) = \tfrac{2}{3}$

d. X and Y are not independent because

$P(X = a, Y = b) \neq P(X = a)P(Y = b)$

in at least one instance. For example

$P(X = 4, Y = 1) \neq P(X = 4)P(Y = 1)$

e. $\text{Cov}(X, Y) = \Sigma P(X = x, Y = y)[(x - E(X))(y - E(Y))]$
$\quad\quad = (\tfrac{1}{9})(-2)(-1) + (\tfrac{1}{12})(-2)(0) + (\tfrac{1}{18})(-2)(1)$
$\quad\quad + (\tfrac{1}{9})(0)(-1) + (\tfrac{1}{6})(0)(0) + (\tfrac{2}{9})(0)(1)$
$\quad\quad + (\tfrac{1}{9})(2)(-1) + (\tfrac{1}{12})(2)(0) + (\tfrac{1}{18})(2)(1)$
$\quad\quad = 0$

This result shows that $\text{Cov}(X, Y) = 0$ does not imply that X and Y are independent.

f. The distribution of $U = X + Y$ is as follows:

u	$P(U = u)$	
5	$\tfrac{1}{9}$	$= \tfrac{4}{36}$
6	$\tfrac{1}{12}$	$= \tfrac{3}{36}$
7	$\tfrac{3}{18}$	$= \tfrac{6}{36}$
8	$\tfrac{1}{6}$	$= \tfrac{6}{36}$
9	$\tfrac{1}{3}$	$= \tfrac{12}{36}$
10	$\tfrac{1}{12}$	$= \tfrac{3}{36}$
11	$\tfrac{1}{18}$	$= \tfrac{2}{36}$

g. $E(X + Y) = E(U) = \Sigma uP(U = u)$
$$= {}^{20}\!/_{36} + {}^{18}\!/_{36} + {}^{42}\!/_{36} + {}^{48}\!/_{36} + {}^{108}\!/_{36}$$
$$+ {}^{30}\!/_{36} + {}^{22}\!/_{36}$$
$$= 8$$

$V(X + Y) = V(U) = \Sigma P(U = u)[u - E(U)]^2$
$$= ({}^{4}\!/_{36})(9) + ({}^{3}\!/_{36})(4) + ({}^{6}\!/_{36})(1) + ({}^{6}\!/_{36})(0)$$
$$+ ({}^{12}\!/_{36})(1) + ({}^{3}\!/_{36})(4) + ({}^{2}\!/_{36})(9)$$
$$= 2{}^{2}\!/_{3}$$

h. By the answer to (g)

$V(X + Y) = 2{}^{2}\!/_{3}$

and by the answers to (b), (c), and (e)

$V(X) + V(Y) + 2\text{Cov}(X, Y) = 2 + {}^{2}\!/_{3} + 2(0) = 2{}^{2}\!/_{3}$

6. a. Marginal distribution for X is
$P(X = 0) = {}^{1}\!/_{3}$
$P(X = 3) = {}^{2}\!/_{3}$

Marginal distribution for Y is
$P(Y = 0) = {}^{1}\!/_{8}$
$P(Y = 1) = {}^{1}\!/_{4}$
$P(Y = 2) = {}^{1}\!/_{8}$
$P(Y = 3) = {}^{1}\!/_{2}$

b. $E(X) = 2, V(X) = 2$ c. $E(Y) = 2, V(Y) = {}^{30}\!/_{24} = {}^{5}\!/_{4}$ d. No   e. ${}^{1}\!/_{8}$
f. The distribution of $U = X + Y$ as follows:

u	$P(U = u)$
0	0
1	${}^{1}\!/_{8} = {}^{3}\!/_{24}$
2	${}^{1}\!/_{8} = {}^{3}\!/_{24}$
3	${}^{5}\!/_{24} = {}^{5}\!/_{24}$
4	${}^{1}\!/_{8} = {}^{3}\!/_{24}$
5	0
6	${}^{5}\!/_{12} = {}^{10}\!/_{24}$

g. $E(X + Y) = 4, V(X + Y) = 3{}^{1}\!/_{2}$

7. $M_x = 5$ and $M_y = 10$

$S_{xy} = \dfrac{1}{N} \Sigma (x - M_x)(y - M_y)$

$$= \frac{1}{6}[(-2)(-5) + (2)(3) + (3)(1) + (-4)(2) + (0)(-1) + (1)(0)]$$

$$= \frac{11}{6} = 1{}^{5}\!/_{6}$$

8. -4.43

9. a.
$$P(X < 1) = P\left(\frac{X - 9}{4} < \frac{1 - 9}{4}\right) = P(Z < -2) = .0228$$

b. $P(X > 25) = P\left(\frac{X - 10}{5} > \frac{25 - 10}{5}\right) = P(Z > 3) = .0013$

c. $P(X < 6) = P\left(\frac{X - 0}{3} < \frac{6 - 0}{3}\right) = P(Z < 2) = .9772$

d. $P(2 < X < 8) = P\left(\frac{2 - 5}{3} < \frac{X - 5}{3} < \frac{8 - 5}{3}\right) = P(-1 < Z < 1)$

$$= .6827$$

10. a. .1587   b. .0228   c. .9987   d. .8664

11. a. $P(X < a) = .01$ implies that
$$P\left(\frac{X - 14}{3} < \frac{a - 14}{3}\right) = .01 \quad \text{and} \quad P\left(Z < \frac{a - 14}{3}\right) = .01$$

and thus $(a - 14)/3 = -2.33$ and $a = 7.01$

b. $P(X > a) = .025$ implies that
$$P\left(\frac{X - 3}{4} > \frac{a - 3}{4}\right) = .025 \quad \text{and} \quad P\left(Z > \frac{a - 3}{4}\right) = .025$$

and thus $(a - 3)/4 = 1.96$ and $a = 10.84$

c. $P(X > a) = .99$ implies that
$$P\left(\frac{X - 24}{3} > \frac{a - 24}{3}\right) = .99 \quad \text{and} \quad P\left(Z > \frac{a - 24}{3}\right) = .99$$

and thus $(a - 24)/3 = -2.33$ and $a = 17.01$

12. a. 16.08   b. 17.65   c. 18.84

CHAPTER 4

1. a. $\hat{\mu} = M = (1/N)\Sigma x_i = 10$
   b. $s^2 = \hat{\sigma}^2 = 1/(N - 1) \Sigma(x_i - M)^2 = 19.2$
   c. If the population is normally distributed, $M \sim N(\mu, \sigma^2/N)$. Considering the estimates made in (a) and (b), the best guess about the distribution of M is that it is normal with mean 10 and variance 19.2/6; that is, $M \sim N (10, 3.2)$

2. a. 9   b. 16.67   c. $M \sim N(9, 2.38)$

3. a. $E(X) = \Sigma\, xP(X = x) = 2(\frac{1}{4}) + 4(\frac{1}{2}) + 6(\frac{1}{4}) = 4$
   $V(X) = \Sigma\, P(X = x)[x - E(X)]^2 = (\frac{1}{4})(4) + (\frac{1}{2})(0) + (\frac{1}{4})(4) = 2$
   b. (i) $\mu_M = \mu = E(X) = 4$ 　　and 　　$\sigma_M^2 = \sigma^2/N = V(X)/N = \frac{2}{3}$
   (ii) All possible samples of size 3, their associated probabilities, and the value of M are given here.

Sample	Probability of Sample	Value of M
2, 2, 2	$\frac{1}{64}$	2
2, 2, 4	$\frac{2}{64}$	$\frac{8}{3}$
2, 2, 6	$\frac{1}{64}$	$\frac{10}{3}$
2, 4, 2	$\frac{2}{64}$	$\frac{8}{3}$
2, 4, 4	$\frac{4}{64}$	$\frac{10}{3}$
2, 4, 6	$\frac{2}{64}$	4
2, 6, 2	$\frac{1}{64}$	$\frac{10}{3}$
2, 6, 4	$\frac{2}{64}$	4
2, 6, 6	$\frac{1}{64}$	$\frac{14}{3}$
4, 2, 2	$\frac{2}{64}$	$\frac{8}{3}$
4, 2, 4	$\frac{4}{64}$	$\frac{10}{3}$
4, 2, 6	$\frac{2}{64}$	4
4, 4, 2	$\frac{4}{64}$	$\frac{10}{3}$
4, 4, 4	$\frac{8}{64}$	4
4, 4, 6	$\frac{4}{64}$	$\frac{14}{3}$
4, 6, 2	$\frac{2}{64}$	4
4, 6, 4	$\frac{4}{64}$	$\frac{14}{3}$
4, 6, 6	$\frac{2}{64}$	$\frac{16}{3}$
6, 2, 2	$\frac{1}{64}$	$\frac{10}{3}$
6, 2, 4	$\frac{2}{64}$	4
6, 2, 6	$\frac{1}{64}$	$\frac{14}{3}$
6, 4, 2	$\frac{2}{64}$	4
6, 4, 4	$\frac{4}{64}$	$\frac{14}{3}$
6, 4, 6	$\frac{2}{64}$	$\frac{16}{3}$
6, 6, 2	$\frac{1}{64}$	$\frac{14}{3}$
6, 6, 4	$\frac{2}{64}$	$\frac{16}{3}$
6, 6, 6	$\frac{1}{64}$	6

Summing the probabilities over the common values of M, we obtain the distribution of M as follows:

m	$P(M = m)$
2	$\frac{1}{64}$
$\frac{8}{3}$	$\frac{6}{64}$
$\frac{10}{3}$	$\frac{15}{64}$
4	$\frac{20}{64}$
$\frac{14}{3}$	$\frac{15}{64}$
$\frac{16}{3}$	$\frac{6}{64}$
6	$\frac{1}{64}$

The parameters of the sampling distribution are

$$\mu_M = E(M) = \sum m P(M = m)$$
$$= 2(\tfrac{1}{64}) + (\tfrac{8}{3})(\tfrac{6}{64}) + (\tfrac{10}{3})(\tfrac{15}{64})$$
$$+ 4(\tfrac{20}{64}) + (\tfrac{14}{3})(\tfrac{15}{64}) + (\tfrac{16}{3})(\tfrac{6}{64}) + 6(\tfrac{1}{64})$$
$$= 4$$

and

$$\sigma_M^2 = V(M) = \sum P(M = m)(m - \mu_M)^2$$
$$= (\tfrac{1}{64})(4) + (\tfrac{6}{64})(\tfrac{16}{9}) + (\tfrac{15}{64})(\tfrac{4}{9}) + (\tfrac{20}{64})(0)$$
$$+ (\tfrac{15}{64})(\tfrac{4}{9}) + (\tfrac{6}{64})(\tfrac{16}{9}) + (\tfrac{1}{64})(4)$$
$$= \tfrac{2}{3}$$

4. a. $E(X) = \mu = 3$, $V(X) = \sigma^2 = 4$ b. $\mu_M = 3$, $\sigma_M^2 = 2$

CHAPTER 5

1. a. The critical value c is such that

$$P(M < c) = .05$$

when the null hypothesis that $\mu = 12$ is true. Thus, given that M has a normal distribution with mean $\mu_M = 12$ and variance $\sigma_M^2 = \sigma^2/N = {}^{16}/_{25}$

$$P\left(\frac{M - 12}{4/5} < \frac{c - 12}{4/5}\right) = .05$$

and $P\left(Z < \dfrac{c - 12}{4/5}\right) = .05$. So, $(c - 12)/(4/5) = -1.65$ and $c = 10.68$.

b. The critical value c in the right-hand tail is such that

$$P(M > c) = .005$$

when the null hypothesis that $\mu = 0$ is true. Thus, given that M has a normal distribution with mean $\mu_M = 0$ and variance $\sigma_M^2 = \sigma^2/N = {}^{8}/_{16} = \tfrac{1}{2}$,

$$P\left(\frac{M - 0}{\sqrt{\tfrac{1}{2}}} > \frac{c - 0}{\sqrt{\tfrac{1}{2}}}\right) = .005$$

and $P\left(Z > \dfrac{c}{\sqrt{\tfrac{1}{2}}}\right) = .005$. So, $c/\sqrt{\tfrac{1}{2}} = 2.58$ and $c = 1.82$.

Because the normal distribution is symmetric, the critical value in the left tail is -1.82.

c. The critical value c is such that

$$P(M > c) = .05$$

when the null hypothesis that $\mu = 10$ is true. Thus, given that M has a normal distribution with mean $\mu_M = 10$ and variance $\sigma_M^2 = \sigma^2/N = {}^{25}/_{100} = {}^1/_4$

$$P\left(\frac{M - 10}{{}^1/_2} > \frac{c - 10}{{}^1/_2}\right) = .05$$

and $P\left(Z > \frac{c - 10}{{}^1/_2}\right) = .05$.

So, $(c - 10)/({}^1/_2) = 1.65$ and $c = 10.83$.
Recall that

Power $= P(\text{rejecting } H_0 | H_a \text{ is true})$

In this case

$$\text{Power} = P(M > 10.83 | M \sim N(12, {}^1/_4))$$
$$= P\left(\frac{M - 12}{{}^1/_2} > \frac{10.83 - 12}{{}^1/_2}\right) = P(Z > -2.34) = .9904$$

2. a. 8.65   b. 102.19 and 97.81   c. $c = 2.7$ and power $= .64$

3. a. $z = \dfrac{M - \mu_0}{\sigma/\sqrt{N}} = \dfrac{(98.5 - 100)}{{}^5/_8} = -2.4$, which is sufficiently large to

reject H_0.

b. $z = \dfrac{M - \mu_0}{\sigma/\sqrt{N}} = \dfrac{13.8 - 14}{2/\sqrt{30}} = -0.55$, which is not sufficiently large to
reject H_0.

c. The result is in the opposite direction from that predicted, and therefore the null hypothesis cannot be rejected.

4. a. $z = 1.72$; do not reject H_0.   b. $z = -3.11$, reject H_0.
c. $z = 2.30$, reject H_0.

5. For (a), (b), and (c) the confidence interval is given by

$$P(M - c_{\alpha/2}\, \sigma/\sqrt{N} \le \mu \le M + c_{\alpha/2}\, \sigma/\sqrt{N}) = 1 - \alpha$$

a. $P[23.5 - 1.96(\sqrt{20/10}) \le \mu \le 23.5 + 1.96(\sqrt{20/10})] = .95$

$P(23.5 - .877 \le \mu \le 23.5 + .877) = .95$
$P(22.62 \le \mu \le 24.38) = .95$

b. $P[3 - 2.58({}^{10}/_5) \le \mu \le 3 + 2.58({}^{10}/_5)] = .99$
$P(3 - 5.16 \le \mu \le 3 + 5.16) = .99$
$P(-2.16 \le \mu \le 8.16) = .99$

c. $P(27.5 - 1.65(\sqrt{8/10}) \leq \mu \leq 27.5 + 1.65(\sqrt{8/10})) = .90$
$P(27.5 - 1.48 \leq \mu \leq 27.5 + 1.48) = .90$
$P(26.02 \leq \mu \leq 28.98) = .90$

6. a. $P(150.80 \leq \mu \leq 155.60) = .95$
   b. $P(10.35 \leq \mu \leq 17.65) = .99$
   c. $P(-3.44 \leq \mu \leq -1.16) = .90$

CHAPTER 6

1. a. For these data $M = 4$ and $s = 2$; thus

$$t = \frac{M - \mu_0}{s/\sqrt{N}} = \frac{4 - 5}{2/\sqrt{6}} = -1.22$$

which when compared to a t distribution with 5 degrees of freedom does not lead to rejection of the null hypothesis.

   b. Given these summary statistics

$$t = \frac{M - \mu_0}{s/\sqrt{N}} = \frac{14.1 - 12}{3/\sqrt{20}} = \frac{2.1}{.671} = 3.13$$

which when compared to a t distribution with 19 degrees of freedom is sufficiently large to reject the null hypothesis.

The limits of the 99% confidence interval for μ are given by

$$M \pm t_{\alpha/2; N-1} \frac{s}{\sqrt{N}} = 14.1 \pm 2.86 \left(\frac{3}{\sqrt{20}} \right)$$
$$= 14.1 \pm 1.92$$

So, $P(12.18 \leq \mu \leq 16.02) = .99$.

   c. Given these summary statistics

$$t = \frac{M - \mu_0}{s/\sqrt{N}} = \frac{98.9 - 100}{\sqrt{10/30}} = -1.91$$

which when compared to a t distribution with 29 degrees of freedom is sufficiently large to reject the null hypothesis.

2. a. 2.93, do not reject H_0.
   b. 2.12, do not reject H_0, limits of 99% confidence interval are $-.70$ and 3.70.
   c. Do not reject H_0 (obtained mean not in direction predicted).

3. a. For sample 1 $N_1 = 5$, $M_1 = 4$, and $S_1^2 = 4$ and for sample 2, $N_2 = 6$, $M_2 = 6$, and $^2_2 = 9.67$. Then

$$t = \frac{(M_1 - M_2) - k_0}{\sqrt{\left(\dfrac{N_1 S_1^2 + N_2 S_2^2}{N_1 + N_2 - 2}\right)\left(\dfrac{N_1 + N_2}{N_1 N_2}\right)}} = \frac{4 - 6}{\sqrt{\left(\dfrac{5(4) + 6(9.67)}{5 + 6 - 2}\right)\left(\dfrac{5 + 6}{5(6)}\right)}}$$

$$= \frac{-2}{1.78} = -1.12$$

which, when compared to a t distribution with 9 degrees of freedom, is not sufficiently large to reject the null hypothesis.

The limits of the 95% confidence interval for $\mu_1 - \mu_2$ are given by

$$(M_1 - M_2) \pm t_{\alpha/2; N_1 + N_2 - 2}\hat{\sigma}_{M_1 - M_2} = -2 \pm 2.26(1.78)$$

Thus, the limits of the confidence interval are -6.02 and 2.02.

b. Given the summary statistics

$$t = \frac{(M_1 - M_2) - k_0}{\sqrt{\dfrac{S_1^2 + S_2^2}{N - 1}}} = \frac{18 - 12}{\sqrt{\dfrac{16 + 11}{11 - 1}}} = \frac{6}{1.64} = 3.65$$

which, when compared to a t distribution with 20 degrees of freedom, is sufficiently large to reject the null hypothesis.

The limits of the 99% confidence interval for $\mu_1 - \mu_2$ are given by

$$(M_1 - M_2) \pm t_{\alpha/2; N_1 + N_2 - 2}\hat{\sigma}_{M_1 - M_2} = 6 \pm 2.845(1.64)$$

Thus, the limits of the confidence interval are 1.33 and 10.67.

c. Given the summary statistics

$$t = \frac{(M_1 - M_2) - k_0}{\sqrt{\left(\dfrac{(N_1 - 1)s_1^2 + (N_2 - 1)s_2^2}{N_1 + N_2 - 2}\right)\left(\dfrac{N_1 + N_2}{N_1 N_2}\right)}}$$

$$= \frac{13.3 - 15.1}{\sqrt{\left(\dfrac{19(25.3) + 21(17.9)}{20 + 22 - 2}\right)\left(\dfrac{20 + 22}{20(22)}\right)}} = -1.26$$

which, when compared to a t distribution with 40 degrees of freedom, is not sufficiently large to reject the null hypothesis.

4. a. $t = 1.98$, do not reject H_0. The limits of the 99% confidence interval are -2.10 to 8.10.

b. $t = -.471$, do not reject H_0. The limits of the 95% confidence interval are $(-2.38$ and $1.48)$.

c. $t = 2.398$, do not reject H_0.

5. The difference scores for the ten subjects are

Subject	d
1	-1
2	-1
3	0
4	-3
5	1
6	-1
7	-1
8	-3
9	0
10	-1

Based on these difference scores $M_D = -1$ and $S_D^2 = 1.4$. Thus

$$t = \frac{M_D}{\sqrt{S_D^2/(N-1)}} = \frac{-1}{\sqrt{1.4/9}} = -2.54$$

which, when compared to a t distribution with 9 degrees of freedom, the null hypothesis is rejected.

The limits of the 95% confidence interval for $\mu_1 - \mu_2$ are given by

$$(M_1 - M_2) \pm t_{\alpha/2; N-1} \sqrt{\frac{S_D^2}{(N-1)}} = -1 \pm 2.26 \sqrt{\frac{1.4}{9}}$$

Thus, the limits of the confidence interval are -1.89 and $-.11$.

6. $t = .98$, do not reject H_0. The limits of the 99% confidence interval are -2.57 and 4.57.

CHAPTER 7

1. a. $T \sim \chi_8^2$; therefore, $P(T > 20.09) = .01$ by Table B.3
   b. $T \sim \chi_8^2$; therefore, $P(2.73 < T < 5.07) = .950 - .750 = .200$ by Table B.3
   c. $T \sim \chi_8^2$; therefore, $P(T > 2.18) = .975$ by Table B.3

2. a. .05   b. .99   c. .50

3. a. F has an F distribution with 8 and 10 degrees of freedom; therefore, by Table B.4 $a = 3.07$.
   b. F has an F distribution with 1 and 12 degrees of freedom; therefore, by Table B.4 $a = 9.33$.
   c. F has an F distribution with 7 and 15 degrees of freedom. That is,

$P(F_{7,15} < a) = .05$ implies that $P(F_{15,7} > 1/a) = .05$. Therefore, by Table B.4 $1/a = 3.51$ and $a = .285$.

    d. F has an F distribution with 20 and 30 degrees of freedom; therefore, by Table B.4 $a = 1.93$.

4. a. 5.06   b. 2.60   c. .350   d. 2.00

5. a. $H_0: \mu_1 = \mu_2 = \mu_3$ or $H_0: \alpha_1 = \alpha_2 = \alpha_3 = 0$
   b. $M_1 = 7$, $M_2 = 13$, $M_3 = 10$, $M = 10$

$$
\begin{aligned}
\text{SST} &= \sum\sum(y_{ij} - M)^2 = (3 - 10)^2 + (5 - 10)^2 + (8 - 10)^2 + \cdots \\
&\qquad + (13 - 10)^2 + (16 - 10)^2 \\
&= 478 \\
\text{SSB} &= \sum N_j(M_j - M)^2 = 10(7 - 10)^2 + 10(13 - 10)^2 \\
&\qquad + 10(10 - 10)^2 \\
&= 180 \\
\text{SSE} &= \sum\sum(y_{ij} - M_j)^2 = (3 - 7)^2 + \cdots + (11 - 7)^2 + (12 - 13)^2 \\
&\qquad + \cdots + (20 - 13)^2 + (7 - 10)^2 \\
&\qquad + \cdots + (16 - 10)^2 \\
&= 298
\end{aligned}
$$

Source Table

Source	Sum of Squares	Degrees of Freedom	Mean Square	F
Between	180	2	90.00	8.15
Error	298	27	11.04	
Total	478	29		

When compared to an F distribution with 2 and 27 degrees of freedom, the obtained value of F is sufficiently large to reject the null hypothesis.

   c. $\hat{\alpha}_1 = M_1 - M = 7 - 10 = -3$, $\hat{\alpha}_2 = M_2 - M = 13 - 10 = 3$, $\hat{\alpha}_3 = M_3 - M = 10 - 10 = 0$

   d. $\eta^2 = \dfrac{\text{SSB}}{\text{SST}} = \dfrac{180}{478} = .377$

$$\hat{\omega}^2 = \frac{\text{SSB} - (J - 1)\text{MSE}}{\text{SST} + \text{MSE}} = \frac{180 - (3 - 1)11.04}{478 + 11.04} = .323$$

6. a. $H_0: \mu_1 = \mu_2 = \mu_3 = \mu_4$ or $H_0: \alpha_1 = \alpha_2 = \alpha_3 = \alpha_4 = 0$
   b.

Source Table

Source	Sum of Squares	Degrees of Freedom	Mean Square	F
Between	134	3	44.67	6.70
Error	120	18	6.67	
Total	254	21		

When compared to an F distribution with 3 and 18 degrees of freedom, the obtained value of F is sufficiently large to reject the null hypothesis.

c. $\hat{\alpha}_1 = -1$, $\hat{\alpha}_2 = -2$, $\hat{\alpha}_3 = 0$, $\hat{\alpha}_4 = 5$

d. $\eta^2 = .528$, $\hat{\omega}^2 = .437$

7. a. For rows: H_0: $\alpha_1 = \alpha_2 = 0$

For columns: H_0: $\beta_1 = \beta_2 = 0$

For interaction: H_0: $\gamma_{11} = \gamma_{21} = \gamma_{12} = \gamma_{22} = 0$

b.
$$SST = \sum\sum\sum (y_{ijk} - M)^2 = (12 - 12)^2 + \cdots + (11 - 12)^2$$
$$+ (12 - 12)^2 + \cdots + (15 - 12)^2$$
$$+ (6 - 12)^2 + \cdots + (3 - 12)^2$$
$$+ (14 - 12)^2 + \cdots + (17 - 12)^2$$
$$= 414$$
$$SSE = \sum\sum\sum (y_{ijk} - M_{jk})^2 = (12 - 12)^2 + \cdots + (11 - 12)^2$$
$$+ (12 - 14)^2 + \cdots + (15 - 14)^2$$
$$+ (6 - 6)^2 + \cdots + (3 - 6)^2$$
$$+ (14 - 16)^2 + \cdots + (17 - 16)^2$$
$$= 78$$
$$SSR = \sum N_{j\cdot}(M_{j\cdot} - M)^2 = 12(9 - 12)^2 + 12(15 - 12)^2 = 216$$
$$SSC = \sum N_{\cdot k}(M_{\cdot k} - M)^2 = 12(13 - 12)^2 + 12(11 - 12)^2 = 24$$
$$SS(R \times C) = \sum\sum n(M_{jk} - M_{j\cdot} - M_{\cdot k} + M)^2$$
$$= 6(12 - 9 - 13 + 12)^2 + 6(14 - 15 - 13 + 12)^2$$
$$+ 6(6 - 9 - 11 + 12)^2 + 6(16 - 15 - 11 + 12)^2$$
$$= 96$$

Source Table

Source	Sum of Squares	Degrees of Freedom	Mean Squares	F
Rows (material)	216	1	216.00	55.38
Columns (media)	24	1	24.00	6.15
Interaction	96	1	96.00	24.62
Error	78	20	3.90	
Total	414	23		

The values of F for rows (material) and interaction are sufficiently large to reject the respective null hypotheses.

c. $\hat{\alpha}_1 = M_{1\cdot} - M = 9 - 12 = -3$

$\hat{\alpha}_2 = M_{2\cdot} - M = 15 - 12 = 3$

$\hat{\beta}_1 = M_{\cdot 1} - M = 13 - 12 = 1$

$\hat{\beta}_2 = M_{\cdot 2} - M = 11 - 12 = -1$

$$\hat{\gamma}_{11} = M_{11} - M_{1.} - M_{.1} + M$$
$$= 12 - 9 - 13 + 12 = 2$$
$$\hat{\gamma}_{21} = M_{21} - M_{2.} - M_{.1} + M$$
$$= 14 - 15 - 13 + 12 = -2$$
$$\hat{\gamma}_{12} = M_{12} - M_{1.} - M_{.2} + M$$
$$= 6 - 9 - 11 + 12 = -2$$
$$\hat{\gamma}_{22} = M_{22} - M_{2.} - M_{.2} + M$$
$$= 16 - 15 - 11 + 12 = 2$$

d.

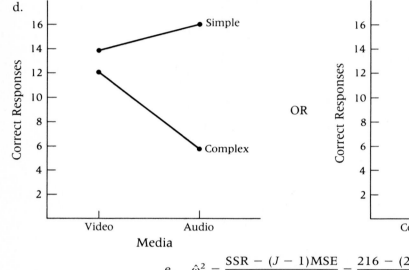

OR

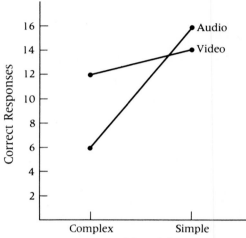

e. $\hat{\omega}_R^2 = \dfrac{SSR - (J - 1)MSE}{MSE + SST} = \dfrac{216 - (2 - 1)3.90}{3.90 + 414} = .508$

$\hat{\omega}_C^2 = \dfrac{SSC - (K - 1)MSE}{MSE + SST} = \dfrac{24 - (2 - 1)3.90}{3.90 + 414} = .048$

$\hat{\omega}_{(R \times C)}^2 = \dfrac{SS(R \times C) - (J - 1)(K - 1)MSE}{MSE + SST} = \dfrac{96 - (1)(1)3.90}{3.90 + 414} = .220$

8. a. For rows: $H_0: \alpha_1 = \alpha_2 = 0$
    For columns: $H_0: \beta_1 = \beta_2 = \beta_3 = 0$
    For interaction: $H_0: \gamma_{11} = \gamma_{21} = \gamma_{12} = \gamma_{22} = \gamma_{13} = \gamma_{23} = 0$

b.   **Source Table**

Source	Sum of Squares	Degrees of Freedom	Mean Squares	F
Rows (gender)	30	1	30.00	4.67
Columns (treatment)	20	2	10.00	1.56
Interaction	0	2	0.00	0.00
Error	154	24	6.42	
Total	204	29		

The values of F for rows (gender) is sufficiently large to reject the null hypothesis that the effects for rows are zero.

c. $\hat{\alpha}_1 = 1$, $\hat{\alpha}_2 = -1$, $\hat{\beta}_1 = 0$, $\hat{\beta}_2 = 1$, $\hat{\beta}_3 = -1$, $\hat{\gamma}_{ij} = 0$ for all i and j.

d.

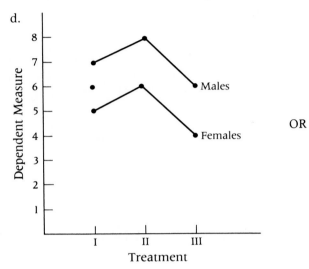   OR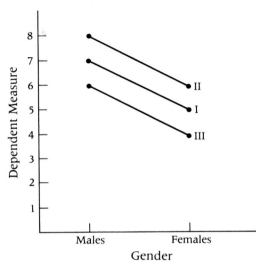

e. $\hat{\omega}_R^2 = .112$, $\hat{\omega}_C^2 = .034$, $\hat{\omega}_{(R \times C)}^2 = 0.00$ (because the obtained value is less than zero, set to zero).

CHAPTER 8

1. a. $H_0: \sigma_{\text{order}}^2 = 0$
   b.

Source Table

Source	Sum of Squares	df	Mean Squares	F
Between (order)	100	4	25.00	2.82
Error	621	70	8.87	
Total	721	74		

When compared to an F distribution with 4 and 74 degrees of freedom, the obtained value of F is sufficiently large to reject the null hypothesis that the order of the items does not make a difference.

c. $\hat{\rho}_I = \dfrac{\text{MSB} - \text{MSE}}{\text{MSB} + (n-1)\text{MSE}} = \dfrac{25.00 - 8.87}{25 + (15-1)8.87} = .108$

2. a. $H_0: \sigma_{\text{tutors}}^2 = 0$

b.

Source Table

Source	Sum of Squares	df	Mean Squares	F
Between (tutors)	150	3	50	7.14
Error	140	20	7	
Total	290	23		

When compared to an F distribution with 3 and 20 degrees of freedom, the obtained value of F is sufficiently large to reject the null hypothesis that the tutor does not make a difference.

c. $\hat{\rho}_I = .506$

3. a. Order: $H_0: \sigma^2_{order} = 0$
   Proctor: $H_0: \sigma^2_{proctor} = 0$
   Interaction: $H_0: \sigma^2_{(R \times C)} = 0$

b.

Source Table

Source	Sum of Squares	df	Mean Squares	F
Rows (order)	280	4	70.00	10.50
Columns (proctor)	60	3	20.00	3.00
Interaction	80	12	6.67	.83
Error	1440	180	8.00	
Total	1860	199		

The value of F for rows (order) is sufficiently large to reject the null hypothesis that order makes a difference.

c. The estimates of the variances are as follows:

$$\hat{\sigma}^2_R = \frac{MSR - MS(R \times C)}{Kn} = \frac{70 - 6.67}{4(10)} = 1.583$$

$$\hat{\sigma}^2_C = \frac{MSC - MS(R \times C)}{Jn} = \frac{20 - 6.67}{5(10)} = .267$$

$$\hat{\sigma}^2_{(R \times C)} = \frac{MS(R \times C) - MSE}{n} = \frac{6.67 - 8.00}{10} = -.133, \text{ set equal to zero}$$

$$\hat{\sigma}^2_e = MSE = 8.00$$

Therefore, the estimates of the intraclass correlation coefficients are

$$\hat{\rho}^R_I = \frac{\hat{\sigma}^2_R}{\hat{\sigma}^2_R + \hat{\sigma}^2_C + \hat{\sigma}^2_{(R \times C)} + \hat{\sigma}^2_e}$$

$$= \frac{1.583}{1.583 + .267 + 0.00 + 8.00} = .161$$

$$\hat{\rho}^C_I = \frac{\hat{\sigma}^2_C}{\hat{\sigma}^2_R + \hat{\sigma}^2_C + \hat{\sigma}^2_{(R \times C)} + \hat{\sigma}^2_e}$$

$$= \frac{.267}{1.583 + .267 + 0.00 + 8.00} = .027$$

$$\hat{\rho}_I^{(R \times C)} = \frac{\hat{\sigma}_{(R \times C)}^2}{\hat{\sigma}_R^2 + \hat{\sigma}_C^2 + \hat{\sigma}_{(R \times C)}^2 + \hat{\sigma}_e^2}$$

$$= \frac{0.00}{1.583 + .267 + 0.00 + 8.00} = 0.00$$

d. The criteria for pooling are satisfied because $(J - 1)(K - 1) = 12 > 6$, $JK(n - 1) = 180 > 6$, and $MS(R \times C)/MSE = .83 < 2.00$.
The pooled MS is given by

$$\text{Pooled MS} = \frac{SSE + SS(R \times C)}{JK(n - 1) + (J - 1)(K - 1)} = \frac{1440 + 80}{5(4)(9) + 4(3)} = 7.92$$

The F value for rows is

$$\frac{MSR}{\text{Pooled MS}} = \frac{70}{7.92} = 8.84$$

which is sufficiently large when compared to an F distribution with 4 and 192 degrees of freedom to reject the null hypothesis related to rows. The F value for columns is

$$\frac{MSC}{\text{Pooled MS}} = \frac{20}{7.92} = 2.53$$

which when compared to an F distribution with 3 and 192 degrees of freedom remains insufficiently large to reject the null hypothesis related to columns.

4. a. Diet: $H_0: \sigma_{\text{diet}}^2 = 0$
   Age: $H_0: \sigma_{\text{age}}^2 = 0$
   Interaction: $H_0: \sigma_{(R \times C)}^2 = 0$

   b.

Source Table

Source	Sum of Squares	df	Mean Squares	F
Rows (diets)	1950	3	650.00	1.81
Columns (age)	4500	5	900.00	2.50
Interaction	5400	15	360.00	3.60
Error	26400	264	100.00	
Total	38250	287		

The F value for the interaction is sufficiently large to reject the null hypothesis related to the interaction.

   c. $\hat{\rho}_I^R = .029$, $\hat{\rho}_I^C = .082$, $\hat{\rho}_I^{(R \times C)} = .158$
   d. Pooling is not appropriate because

$$MS(R \times C)/MSE = 3.60 > 2$$

5. a. The null hypothesis for the random factor, salesperson, is

$H_0: \sigma^2_{salesperson} = 0$

The null hypothesis for the fixed factor, geographical type, is

$H_0: \beta_1 = \beta_2 = \beta_3 = 0$

The null hypothesis for the interaction between salesperson and geographical type is

$H_0: \sigma^2_{interaction} = 0$

b.

Source Table

Source	Sum of Squares	df	Mean Squares	F
Rows—salesperson (random)	9600	4	2400.00	10.00
Columns—geo. type (fixed)	6720	2	3360.00	5.51
Interaction	4880	8	610.00	2.54
Error	32400	135	240.00	
Total	53600	149		

The values of F are sufficiently large that all three hypotheses can be rejected.

6. a. The null hypothesis for the random factor, schools, is

$H_0: \sigma^2_{schools} = 0$

The null hypothesis for the fixed factor, grade level, is

$H_0: \beta_1 = \beta_2 = \beta_3 = \beta_4 = 0$

The null hypothesis for the interaction between schools and grade level is

$H_0: \sigma^2_{interaction} = 0$

b.

Source Table

Source	Sum of Squares	df	Mean Squares	F
Rows—schools (random)	1263	3	421.00	2.17
Columns—grade level (fixed)	8088	3	2696.00	1.98
Interaction	12267	9	1363.00	7.01
Error	15552	80	194.40	
Total	37170	95		

The value of F for interaction is sufficiently large to reject the null hypothesis related to the interaction.

7.

Source Table

Source	Sum of Squares	df	Mean Squares	F
Blocks	45.00	9	5.00	
Treatments	48.80	2	24.40	4.03
Residual	109.00	18	6.06	
Total	202.80	29		

When compared to an F distribution with 2 and 18 degrees of freedom, the value of F is sufficiently large to reject the null hypothesis that the treatments are equally effective.

8.

Source Table

Source	Sum of Squares	df	Mean Squares	F
Blocks	812	11	73.82	
Treatments	216	1	216.00	2.10
Residual	1133	11	103.00	
Total	2161	23		

When compared to an F distribution with 1 and 11 degrees of freedom, the value of F is not sufficiently large to reject the null hypothesis that the intelligence of children who are adopted does not differ from the intelligence of children who are raised by foster parents.

9. a.

Source Table

Source	Sum of Squares	df	Mean Squares	F
Subjects	350	14	25.00	
Treatments (type of partner)	390	2	195.00	8.13
Residual	672	28	24.00	
Total	1412	44		

When compared to an F distribution with 2 and 28 degrees of freedom, the value of F is sufficiently large to reject the null hypothesis that the partner does not make a difference in the number of aggressive behaviors displayed.

b. When the value of F is compared to an F distribution with 1 and $(J - 1) = 14$ degrees of freedom, it is not sufficiently large to reject the null hypothesis.

10. a.

Source Table

Source	Sum of Squares	df	Mean Squares	F
Subjects	38.0	19	2.00	
Repeated measures	17.0	5	3.40	6.80
Residual	47.5	95	.50	
Total	102.5	119		

When compared to an F distribution with 5 and 95 degrees of freedom, the value of F is sufficiently large to reject the null hypothesis that the time since completion of therapy does not affect the number of hours away from the house.

b. When the value of F is compared to an F distribution with 1 and $(J - 1) = 19$ degrees of freedom, it remains sufficiently large to reject the null hypothesis.

CHAPTER 9

1. a. The three sample comparisons are

$$\hat{\psi}_1: 1M_1 - 1M_2 + 0M_3 + 0M_4$$
$$\hat{\psi}_2: 0M_1 + 0M_2 + 1M_3 - 1M_4$$
$$\hat{\psi}_3: 1M_1 + 1M_1 - 1M_3 - 1M_4$$

Because the sample sizes are equal and $\Sigma c_{1j}c_{2j} = 0$ for all pairs of comparisons, the three comparisons are mutually orthogonal.

b. The test of the three comparisons will be conducted via the analysis of variance. First, the sample value for each comparison is calculated.

$$\hat{\psi}_1: 1(6) - 1(8) = -2$$
$$\hat{\psi}_2: 1(10) - 1(12) = -2$$
$$\hat{\psi}_3: 1(6) + 1(8) - 1(10) - 1(12) = -8$$

Next, the sum of squares for each comparison are calculated, noting that

$$SS(\hat{\psi}) = \frac{(\hat{\psi})^2}{\Sigma(c_j^2/N_j)}:$$
$$SS(\hat{\psi}_1) = \frac{4}{(1/20 + 1/20)} = 40$$

$$SS(\hat{\psi}_2) = \frac{4}{(\frac{1}{20} + \frac{1}{20})} = 40$$

$$SS(\hat{\psi}_3) = \frac{64}{(\frac{1}{20} + \frac{1}{20} + \frac{1}{20} + \frac{1}{20})} = 320$$

The tests of the three comparisons are conducted by constructing the source table.

Source Table

Source	Sum of Squares	df	Mean Squares	F
(Between	400	3)		
$\hat{\psi}_1$	40	1	40.00	4.00
$\hat{\psi}_2$	40	1	40.00	4.00
$\hat{\psi}_3$	320	1	320.00	32.00
Error	760	76	10.00	
Total	1160	79		

When compared to an F distribution with 1 and 76 degrees of freedom, the null hypotheses related to the three comparisons can be rejected.

c. The $(1 - \alpha)$ 100% confidence interval is given by

$$P[\hat{\psi} - t_{\alpha/2,N-J}\sqrt{MSE\ \Sigma(c_j^2/N_j)} \le \psi \le \hat{\psi} + t_{\alpha/2,N-J}\sqrt{MSE\ \Sigma(c_j^2/N_j)}] =$$

$$1 - \alpha$$

Thus, the 95% confidence interval for the three comparisons listed in a. are

$$P[-2 - 1.99\sqrt{10(1/10)} \le \psi_1 \le -2 + 1.99\sqrt{10(1/10)}] = .95$$

$$P[-3.99 \le \psi_1 \le -.01] = .95$$

$$P[-2 - 1.99\sqrt{10(1/10)} \le \psi_2 \le -2 + 1.99\sqrt{10(1/10)}] = .95$$

$$P[-3.99 \le \psi_2 \le -.01] = .95$$

and

$$P[-8 - 1.99\sqrt{10(1/5)} \le \psi_3 \le -8 + 1.99\sqrt{10(1/5)}] = .95$$

$$P[-10.81 \le \psi_3 \le -5.19] = .95$$

2. a. $\hat{\psi}_1$: $2M_1 + 2M_2 + 2M_3 - 3M_4 - 3M_5$

$\hat{\psi}_2$: $1M_1 + 1M_2 - 2M_3 + 0M_4 + 0M_5$

$\hat{\psi}_3$: $0M_1 + 0M_2 + 0M_3 + 1M_4 - 1M_5$

Because the sample sizes are equal and $\Sigma c_{1j}c_{2j} = 0$ for all pairs of comparisons, the three comparisons are mutually orthogonal.

b.

Source Table

Source	Sum of Squares	df	Mean Squares	F
(Between	690.0	4)		
$\hat{\psi}_1$	312.5	1	312.5	15.63
$\hat{\psi}_2$	160.0	1	160.0	8.00
$\hat{\psi}_3$	187.5	1	187.5	9.38
Residual	30.0	1	30.0	1.50
Error	1400.0	70	20.0	
Total	2090.0	74		

When compared to an F distribution with 1 and 70 degrees of freedom, the null hypotheses related to the three comparisons can be rejected.

c. $P(12.35 \le \psi_1 \le 37.65) = .95$
$P(\ 2.34 \le \psi_2 \le 13.66) = .95$
$P(\ 1.73 \le \psi_3 \le \ 8.27) = .95$ (Given the coefficients in a.)

3. a. The F ratio for the Scheffé tests is

$$\frac{SS(\psi)/(J-1)}{MSE}$$

For the first comparison

$$F = \frac{40/3}{10} = 1.33$$

which when compared to an F distribution with 3 and 76 degrees of freedom is not sufficiently large to reject the null hypothesis. For the second comparison

$$F = \frac{40/3}{10} = 1.33$$

which, when compared to an F distribution with 3 and 76 degrees of freedom, is not sufficiently large to reject the null hypothesis. For the third comparison

$$F = \frac{320/3}{10} = 10.67$$

which, when compared to an F distribution with 3 and 76 degrees of freedom, is sufficiently large to reject the null hypothesis.

b. The limits of the $(1 - \alpha)100\%$ confidence interval are given by

$$\hat{\psi} \pm \sqrt{(J - 1)F_{\alpha;J-1,N-J}} \sqrt{\text{MSE}\sum\frac{c_j^2}{N_j}}$$

Thus, the limits for the first comparison are (given the coefficients in 1a.)

$$-2 - \sqrt{(3)2.74}\sqrt{10(1/10)} = -4.87$$

and

$$-2 + \sqrt{(3)2.74}\sqrt{10(1/10)} = .87$$

The limits for the second comparison are (given the coefficients in 1a.)

$$-2 - \sqrt{(3)2.74}\sqrt{10(1/10)} = -4.87$$

and

$$-2 + \sqrt{(3)2.74}\sqrt{10(1/10)} = .87$$

Finally, the limits for the third comparison are (given the coefficients in 1a.)

$$-8 - \sqrt{(3)2.74}\sqrt{10(1/5)} = -12.05$$

and

$$-8 + \sqrt{(3)2.74}\sqrt{10(1/5)} = -3.95$$

c. The Neuman–Keuls method is summarized in the following table:

	Lecture	Lecture + Visual Aids	PI with Text	Computer Assisted PI	r	$q_\alpha\sqrt{\text{MSE}/n}$
	I	II	III	IV		
	6.0	8.0	10.0	12.0		
I		2.0----	4.0*----	6.0*	4	$3.74\sqrt{10/20} = 2.64$
II			----2.0----	----4.0*	3	$3.40\sqrt{10/20} = 2.40$
III				----2.0	2	$2.83\sqrt{10/20} = 2.00$
IV						

Those groups that do not differ have a common underline

I   II   III   IV

4. a. The values of F for the three comparisons are 3.91, 2.00, and 2.34, respectively; when compared to an F distribution with 4 and 70

degrees of freedom, only the value of F related to the first comparison leads to rejection of the null hypothesis.

b. $P(4.92 \leq \psi_1 \leq 45.08) = .95$, given the coefficients in 2a.
$P(-.98 \leq \psi_2 \leq 16.98) = .95$, given the coefficients in 2a.
$P(-.18 \leq \psi_3 \leq 10.18) = .95$, given the coefficients in 2a.

c. The Neuman–Keuls method is summarized in the following table:

Normal Controls	Normal Weight Bulemics	Clinical Controls	Bulemic/ Anorexics	Anorexics		
V	III	IV	II	I		
19	23	24	26	28	r	$q\sqrt{MSE/n}$
V	4*⋯	5*⋯	7*⋯	9*------5		4.60
III		⋯1⋯	3 ⋯	5*------4		4.32
IV			⋯2 ⋯	4*------3		3.93
II				2-------2		3.27
I						

Those groups that do not differ have a common underline

V   III   IV   II   I

CHAPTER 10

1. $r_{xy} = \dfrac{\text{sample covariance}}{S_x S_y} = \dfrac{S_{xy}}{S_x S_y}$

where S_{xy} is defined as follows:

$$S_{xy} = \frac{1}{N} \sum (x_i - M_x)(y_i - M_y)$$

The sample statistics are calculated as follows:

$M_x = \dfrac{1}{N}\sum x = 7 \qquad S_x = \sqrt{\dfrac{1}{N}\sum (x_i - M)^2} = 3.16$

and

$M_y = \dfrac{1}{N}\sum y = 11 \qquad S_y = \sqrt{\dfrac{1}{N}\sum (y_i - M)^2} = 2.00$

$S_{xy} = \dfrac{1}{N}\sum (x_i - M_x)(y_i - M_y)$

$\quad = \dfrac{1}{5}[(3-7)(8-11) + (5-7)(12-11) + (6-7)(10-11)$

$\qquad + (9-7)(11-11) + (12-7)(14-11)]$

$\quad = \dfrac{26}{5} = 5.2$

Thus, $r_{xy} = \dfrac{5.2}{(3.16)(2.00)} = .82$.

2. .54

3. The test statistic for problems (a), (b), and (c) is

$$t = \frac{r_{xy}\sqrt{N-2}}{\sqrt{1-r_{xy}^2}}$$

a.

$$t = \frac{.30\sqrt{17-2}}{\sqrt{1-(.30)^2}} = 1.22$$

which, when compared to a t distribution with 15 degrees of freedom, is not sufficiently large to reject the null hypothesis.

b. $$t = \frac{-.50\sqrt{42-2}}{\sqrt{1-(-.50)^2}} = -3.65$$

which, when compared to a t distribution with 40 degrees of freedom, is sufficiently large to reject the null hypothesis.

The 99% confidence interval for ρ_{XY} is found in the following way:

$$r_{xy} = -.50 \qquad \text{corresponds to} \qquad Z = -.5493$$

We know

$$P\left(Z - z_{\alpha/2}\sqrt{\frac{1}{N-3}} \leq E(Z) \leq Z + z_{\alpha/2}\sqrt{\frac{1}{N-3}}\right) = 1 - \alpha$$

and thus is our case

$$P\left(-.5493 - 2.58\sqrt{\frac{1}{39}} \leq E(Z) \leq -.5493 + 2.58\sqrt{\frac{1}{39}}\right) = .99$$

which simplifies to

$$P(-.9624 \leq E(Z) \leq -.1362) = .99$$

To find the corresponding confidence interval for ρ_{XY}, the limits for the confidence interval for $E(Z)$ are transformed to correlation coefficients:

$$Z = -.9624 \qquad \text{corresponds to} \qquad r_{xy} = -.746$$

and

$$Z = -.1362 \qquad \text{corresponds to} \qquad r_{xy} = -.136$$

Therefore

$$P(-.746 \leq \rho_{XY} \leq -.136) = .99$$

c. $$t = \frac{.26\sqrt{29-2}}{\sqrt{1-(.26)^2}} = 1.40$$

which, when compared to a t distribution with 27 degrees of freedom, is not sufficiently large to reject the null hypothesis.

The 95% confidence interval for ρ_{XY} is found in the following way:

$$r_{xy} = .26 \qquad \text{corresponds to} \qquad Z = .2661$$

We know

$$P\left(Z - z_{\alpha/2}\sqrt{\frac{1}{N-3}} \leq E(Z) \leq Z + z_{\alpha/2}\sqrt{\frac{1}{N-3}}\right) = 1 - \alpha$$

and thus is our case

$$P\left(.2661 - 1.96\sqrt{\frac{1}{26}} \leq E(Z) \leq .2661 + 1.96\sqrt{\frac{1}{26}}\right) = .95$$

which simplifies to

$$P(-.1183 \leq E(Z) \leq .6505) = .95$$

To find the corresponding confidence interval for ρ_{XY}, the limits for the confidence interval for $E(Z)$ are transformed to correlation coefficients:

$$Z = -.1183 \qquad \text{corresponds to} \qquad r_{xy} = -.118$$

and

$$Z = .6505 \qquad \text{corresponds to} \qquad r_{xy} = .572$$

Therefore

$$P(-.118 \leq \rho_{XY} \leq .572) = .95$$

d. The test statistic for the null hypothesis that $\rho_{XY} = c$ is given by

$$\frac{Z - E(Z)}{\sqrt{V(Z)}}$$

where Z is the transformed value of the obtained sample correlation coefficient, $V(Z) = 1/(N-3)$, and $E(Z)$ is the expectation of Z when the null hypothesis is true. The value of Z is found by transforming the obtained correlation coefficient:

$$r_{xy} = .80 \qquad \text{corresponds to} \qquad Z = 1.099$$

$E(Z)$ is found by transforming the hypothesized value of .50:

$$\rho_{XY} = .50 \qquad \text{corresponds to} \qquad E(Z) = .5493$$

Finally

$$V(Z) = \frac{1}{35 - 3} = .0313$$

Therefore, the value of the test statistic is given by

$$\frac{1.099 - .5493}{\sqrt{.0313}} = 3.11$$

which when compared to the standard normal distribution leads to rejection of the null hypothesis.

4. a. $t = -1.41$, do not reject H_0
   b. $t = 2.05$, do not reject H_0. $P(-.113 \le \rho_{xy} \le .688) = .99$
   c. $t = -1.22$, do not reject H_0. $P(-.715 \le \rho_{xy} \le .230) = .95$
   d. test statistic $= 1.05$, do not reject H_0

5. The difference in the ranks, d_i, and the squares of the differences, d_i^2, are as follows:

Contestant	Judge 1	Judge 2	d_i	d_i^2
1	1	3	-2	4
2	4	2	2	4
3	2	5	-3	9
4	6	1	5	25
5	3	4	-1	1
6	10	7	3	9
7	5	8	-3	9
8	9	6	3	9
9	7	10	-3	9
10	8	9	-1	1

Thus

$$r_s = 1 - \frac{6\sum d_i^2}{N(N^2 - 1)} = 1 - \frac{6(80)}{10(10^2 - 1)} = .515$$

The test statistic related to this null hypothesis is

$$t = \frac{r_s\sqrt{N - 2}}{\sqrt{1 - r_s^2}} = \frac{.515\sqrt{10 - 2}}{\sqrt{1 - (.515)^2}} = 1.70$$

which when compared to a t distribution with 8 degrees of freedom is not sufficiently large to reject the null hypothesis.

6. $r_s = .245$, $t = 0.80$, do not reject H_0.

7. a. For each cell of the contingency table, the expected value f_{ejk} is given by

$$f_{ejk} = \frac{(f_{oj\cdot})(f_{o\cdot k})}{N}$$

The marginal frequencies and the expected frequencies (in parentheses) for each cell are given as follows:

		Fair	Not Fair	No Opinion	
Acceptance Decision	Rejected	25 (30)	20 (18)	15 (12)	$f_{o1.} = 60$
	Accepted	25 (20)	10 (12)	5 (8)	$f_{o2.} = 40$
		$f_{o.1} = 50$	$f_{o.2} = 30$	$f_{o.3} = 20$	$N = 100$

Opinion of Process

The test statistic

$$\chi^2 = \sum\sum \frac{(f_{ojk} - f_{ejk})^2}{f_{ejk}}$$

$$= \frac{(25 - 30)^2}{30} + \frac{(25 - 20)^2}{20} + \frac{(20 - 18)^2}{18}$$

$$+ \frac{(10 - 12)^2}{12} + \frac{(15 - 12)^2}{12} + \frac{(5 - 8)^2}{8}$$

$$= .833 + 1.250 + .222 + .333 + .750 + 1.125$$

$$= 4.513$$

which, when compared to a χ^2 distribution with 2 degrees of freedom, is not sufficiently large to reject the null hypothesis that acceptance decision and opinion of process are independent.

b. The value of Cramer's statistic for this problem is given by

$$\phi' = \sqrt{\frac{\phi^2}{L}} = \sqrt{\frac{\chi^2}{NL}} = \sqrt{\frac{4.513}{100(1)}} = .212$$

8. a. $\chi^2 = 16.09$, reject H_0
   b. $\phi' = .201$

CHAPTER 11

1. a. The regression coefficient is given by

$$B_{y \cdot x} = r_{xy} \frac{S_y}{S_x} = (-.785) \frac{\sqrt{59.12}}{\sqrt{5}} = -2.70$$

The regression equation is given by

$$y_i' = B_{y \cdot x}(x_i - M_x) + M_y = -2.70(x_i - 5) + 19$$

or

$$y_i' = -2.70x_i + 32.5$$

b. $r^2 = (-.785)^2 = .616$; 61.6% of the variance in depression scores is due to hours of sunlight.

c. To calculate the sum of squares, we need the following statistics:

$$M_{y_1} = 26 \qquad M_{y_2} = 24 \qquad M_{y_3} = 15 \qquad M_{y_4} = 11$$

and

$$y_1' = -2.70x_1 + 32.5 = -2.7(2) + 32.5 = 27.1$$
$$y_2' = -2.70x_2 + 32.5 = -2.7(4) + 32.5 = 21.7$$
$$y_3' = -2.70x_3 + 32.5 = -2.7(6) + 32.5 = 16.3$$
$$y_4' = -2.70x_4 + 32.5 = -2.7(8) + 32.5 = 10.9$$

Then

$$\begin{aligned}
\text{SS error} &= \sum\sum (y_{ij} - M_{y_j})^2 \\
&= (34 - 26)^2 + (20 - 26)^2 + \cdots + (8 - 11)^2 + (12 - 11)^2 \\
&= 330
\end{aligned}$$

$$\begin{aligned}
\text{SS deviations} &= \sum N_j (M_{y_j} - y_j')^2 \\
&= 4(26 - 27.1)^2 + 4(24 - 21.7)^2 + 4(15 - 16.3)^2 \\
&\quad + 4(11 - 10.9)^2 \\
&= 32.8
\end{aligned}$$

$$\begin{aligned}
\text{SS linear regression} &= \sum N_j (y_j' - M_y)^2 \\
&= 4(27.1 - 19)^2 + 4(21.7 - 19)^2 \\
&\quad + 4(16.3 - 19)^2 + 4(10.9 - 19)^2 \\
&= 583.2
\end{aligned}$$

d. To test the null hypothesis that $\beta_{Y \cdot X} = 0$, the following F ratio is calculated:

$$F = \frac{\text{SS linear regression}}{\text{SS (deviations + error)}/(N - 2)} = \frac{583.2}{(32.8 + 330)/(16 - 2)}$$
$$= 22.50$$

When compared to an F distribution with 1 and 14 degrees of freedom, the value of F is sufficiently large to reject the null hypothesis.

e. The limits of the $(1 - \alpha)100\%$ confidence interval for $\beta_{Y \cdot X}$ are

$$B_{y \cdot x} \pm t_{\alpha/2; N-2} \frac{\sqrt{\text{MS(error + deviations)}}}{S_x \sqrt{N}}$$

which in this case is equal to

$$-2.7 \pm 2.145 \frac{\sqrt{(32.8 + 330)/(16 - 2)}}{\sqrt{5}\sqrt{16}} = -2.7 \pm 1.22$$

Thus

$$P(-3.92 \leq \beta_{Y \cdot X} \leq -1.48) = .95$$

The limits of the $(1 - \alpha)100\%$ confidence interval for the true score are

$$y'_j \pm t_{\alpha/2;N-2}\sqrt{\text{MS}(\text{error} + \text{deviations})}\sqrt{1 + \frac{1}{N} + \frac{(x_j - M_x)^2}{NS_x^2}}$$

which in this case is equal to

$$16.3 \pm 2.145\sqrt{(32.8 + 330)/(16 - 2)}\sqrt{1 + \frac{1}{16} + \frac{(6 - 5)^2}{16(5)}}$$

$$= 16.3 \pm 11.32$$

Thus, the limits of the 95% confidence interval for the true value of the dependent variable given that a person was exposed to six hours of sunlight are 4.98 and 27.62.

2. a. $B_{y \cdot x} = -1.7$,　$y'_i = -1.7x_i + 13.1$
   b. 37.7%
   c. SS error = 128, SS deviations = 15.3, SS linear regression = 86.7
   d. $F = 7.87$, do not reject H_0
   e. Limits for confidence interval for $B_{Y \cdot X}$ are -3.53 and .13. Limits for confidence interval for the true score are .45 and 22.35.

3. a. The value of the regression coefficient is given by

$$B_{y \cdot x} = r_{xy}\frac{S_y}{S_x} = .819\frac{\sqrt{5.25}}{\sqrt{3}} = 1.083$$

   The regression equation is given by

$$y'_i = B_{y \cdot x}(x_i - M_x) + M_y = 1.083(x_i - 4) + 6$$

   or

$$y'_i = 1.083x_i + 1.668$$

   b. For the first student $y'_1 = 1.083x_1 + 1.668 = 2.75$. The remaining predictions are summarized in the following table:

Student	Hours of Study x	Problems Solved y	y'	y − y'
1	1	4	2.75	1.25
2	3	6	4.92	1.08
3	6	10	8.17	1.83
4	5	5	7.08	−2.08
5	2	2	3.83	−1.83
6	4	6	6.00	.00
7	6	8	8.17	−.17
8	5	7	7.08	−.08

c. For each student the error in prediction is shown in the table of (b).

The value of the average squared errors $= \dfrac{1}{N}\sum(y_i - y_i')^2$

$$= \frac{1}{8}(13.79)$$

$$= 1.72$$

d. $r^2 = (.819)^2 = .671$; 67.1% of the variance in problems solved is accounted for by the number of hours studied.

e. The standard error of estimate is given by

$$S_{y \cdot x} = S_y\sqrt{1 - r_{xy}^2} = \sqrt{5.25}\sqrt{1 - (.819)^2} = 1.31$$

The standard error of estimate is the square root of the sample variance of estimate, which is equal to the average squared errors. Thus, the standard error of estimate is equal to the square root of the answer to (c).

f. Using the test statistic discussed in Statement (11.45), we have

$$t = \frac{r_{xy}\sqrt{N - 2}}{\sqrt{1 - r_{xy}^2}} = \frac{.819\sqrt{8 - 2}}{\sqrt{1 - (.819)^2}} = 3.50$$

which, when compared to a t distribution with six degrees of freedom, is not sufficiently large to reject the null hypothesis.

4. a. $B_{y \cdot x} = -.168$, $y_i' = -.168x_i + 24.585$.
   b. See the following table:

Subject	Age	Words Recalled		
	x	y	y'	$y - y'$
1	55	18	15.34	2.66
2	60	15	14.50	.50
3	62	12	14.17	−2.17
4	58	14	14.84	−.84
5	73	15	12.32	2.68
6	79	10	11.31	−1.31
7	58	13	14.84	−1.84
8	61	12	14.34	−2.34
9	64	15	13.83	1.17
10	60	16	14.50	1.50

   c. See the table in (b); average squared error $= 3.41$
   d. 29.1%
   e. 1.85
   f. $t = -1.81$, do not reject H_0

5. a. The value of the partial regression is

$$r_{12\cdot3} = \frac{r_{12} - r_{13}r_{23}}{\sqrt{(1 - r_{13}^2)(1 - r_{23}^2)}} = \frac{.40 - (.30)(.50)}{\sqrt{(1 - .09)(1 - .25)}} = .303$$

b. Using the test statistic,

$$\frac{Z}{\sqrt{1/(N - 4)}}$$

where Z is the transformed value corresponding to $r_{12\cdot3}$,

$r_{12\cdot3} = .303$ corresponds to $Z = .3128$

and thus, the value of the test statistic is

$$\frac{.3128}{\sqrt{1/21}} = 1.43$$

which, when compared to a standard normal distribution, is not sufficiently large to reject the null hypothesis.

6. a. .355   b. test statistic = 3.13, reject H_0

7. a. To calculate the adjusted sums of squares, we need the following statistics:

$M_y = 10$ $M_{y_1} = 12$ $M_{y_2} = 8$ $M_{y_3} = 10$
$M_c = 500$ $M_{c_1} = 520$ $M_{c_2} = 500$ $M_{c_3} = 480$

$$\begin{aligned}
\text{CP total} &= \sum\sum(y_{ij} - M_y)(C_{ij} - M_c) \\
&= (10 - 10)(500 - 500) + \cdots + (8 - 10)(500 - 500) \\
&\quad + (11 - 10)(550 - 500) + \cdots + (6 - 10)(400 - 500) \\
&\quad + (12 - 10)(600 - 500) + \cdots + (10 - 10)(400 - 500) \\
&= 2250
\end{aligned}$$

$$\begin{aligned}
\text{CP between} &= \sum N_j(M_{y_j} - M_y)(M_{c_j} - M_c) \\
&= 5(12 - 10)(520 - 500) + 5(8 - 10)(500 - 500) \\
&\quad + 5(10 - 10)(480 - 500) \\
&= 200
\end{aligned}$$

$$\begin{aligned}
\text{CP error} &= \sum\sum(y_{ij} - M_{y_j})(c_{ij} - M_{c_j}) \\
&= (10 - 12)(500 - 520) + \cdots + (8 - 12)(500 - 520) \\
&\quad + (11 - 8)(550 - 500) + \cdots + (6 - 8)(500 - 400) \\
&\quad + (12 - 10)(600 - 480) + \cdots + (10 - 10)(400 - 480) \\
&= 2050
\end{aligned}$$

$$SS_y \text{ total} = \sum\sum (y_{ij} - M_y)^2 = 146$$
$$SS_y \text{ between} = \sum N_j(M_{y_j} - M_y)^2 = 40$$
$$SS_y \text{ error} = \sum\sum (y_{ij} - M_{y_j})^2 = 106$$
$$SS_c \text{ total} = \sum\sum (c_{ij} - M_c)^2 = 95{,}000$$
$$SS_c \text{ between} = \sum N_j(M_{c_j} - M_c)^2 = 4000$$
$$SS_c \text{ error} = \sum\sum (c_{ij} - M_{c_j})^2 = 91{,}000$$

The adjusted sums of squares are now calculated:

$$SS(\text{adj}) \text{ total} = SS_y \text{ total} - \frac{(CP \text{ total})^2}{SS_c \text{ total}} = 146 - \frac{(2250)^2}{95000} = 92.71$$

$$SS(\text{adj}) \text{ between} = SS_y \text{ between} + \frac{(CP \text{ error})^2}{SS_c \text{ error}} - \frac{(CP \text{ total})^2}{SS_c \text{ total}}$$

$$= 40 + \frac{(2050)^2}{91000} - \frac{(2250)^2}{95000}$$

$$= 32.89$$

$$SS(\text{adj}) \text{ error} = SS_y \text{ error} - \frac{(CP \text{ error})^2}{SS_c \text{ error}} = 106 - \frac{(2050)^2}{91000} = 59.82$$

The analysis of covariance is presented in the following table:

Source Table

Source	Sum of Squares	df	Mean Squares	F
Between(adj)	32.89	2	16.45	3.03
Error(adj)	59.82	11	5.44	
Total(adj)	92.71	13		

The value of F, when compared to an F distribution with 2 and 11 degrees of freedom, is not sufficient to reject the null hypothesis that the treatments are equally effective.

b. To calculate the adjusted means, we need the value of B_w:

$$B_w = \frac{CP \text{ error}}{SS_c \text{ error}} = \frac{2050}{91000} = .023$$

The adjusted means are thus

$$M_{y_1}(\text{adj}) = M_{y_1} - B_w(M_{c_1} - M_c) = 12 - .023(520 - 500) = 11.54$$
$$M_{y_2}(\text{adj}) = M_{y_2} - B_w(M_{c_2} - M_c) = 8 - .023(500 - 500) = 8.00$$
$$M_{y_3}(\text{adj}) = M_{y_3} - B_w(M_{c_3} - M_c) = 10 - .023(480 - 500) = 10.46$$

8. a.

Source Table

Source	Sum of Squares	df	Mean Squares	F
Between(adj)	299.18	3	99.73	9.31
Error(adj)	117.82	11	10.71	
Total(adj)	417.00	14		

When compared to an F distribution with 3 and 11 degrees of freedom, the value of the F ratio is sufficiently large to reject the null hypothesis that the reinforcers are equally potent.

b. $M_{y_1}(\text{adj}) = 27.82$, $M_{y_2}(\text{adj}) = 20.00$, $M_{y_3}(\text{adj}) = 21.38$, $M_{y_4}(\text{adj}) = 30.82$

9. This problem is approached using orthogonal comparisons for the linear and quadratic relations between the number of problems assigned and the acquisition of knowledge. First, the sums of squares for the data are calculated:

$$\text{SS between} = \sum N_j (M_j - M)^2 = 64$$
$$\text{SS error} = \sum \sum (y_{ij} - M_j)^2 = 66$$
$$\text{SS total} = \sum \sum (y_{ij} - M)^2 = 130$$

The coefficients for the comparisons are obtained from Table B.7 and are as follows:

	c_1	c_2	c_3	c_4	c_5	c_6
Linear:	−5	−3	−1	1	3	5
Quadratic:	5	−1	−4	−4	−1	5

Given that $M_1 = 2$, $M_2 = 6$, $M_3 = 7$, $M_4 = 6$, $M_5 = 4$, $M_6 = 5$

$$\hat{\psi}_1 = \sum c_j M_{y_j} = 8 \quad \text{and} \quad \hat{\psi}_2 = \sum c_j M_{y_j} = -27$$

Further

$$\text{SS}(\hat{\psi}_1) = \frac{(\hat{\psi}_1)^2}{\sum (c_j^2 / N_j)} = \frac{64}{(70/4)} = 3.66$$

and

$$\text{SS}(\hat{\psi}_2) = \frac{(\hat{\psi}_2)^2}{\sum (c_j^2 / N_j)} = \frac{729}{(84/4)} = 34.7$$

The source table for conducting the tests is as follows:

Source	Sum of Squares	df	Mean Squares	F
(Between	64.00	5)		
$\hat{\psi}_1$: linear	3.66	1	3.66	1.00
$\hat{\psi}_2$: quadratic	34.70	1	34.70	9.45
Residual	25.64	3	8.55	2.33
Error	66.00	18	3.67	
Total	130.00			

When compared to an F distribution with 1 and 18 degrees of freedom, the null hypothesis that there is no quadratic relation between the number of problems solved and the acquisition of knowledge is rejected; the null hypothesis that there is no linear relation cannot be rejected. Furthermore, when compared to an F distribution with 3 and 18 degrees of freedom, the value of F related to the residual is not sufficiently large to believe that there are other curvilinear relations.

10. The source table for conducting the tests is as follows:

Source	Sum of Squares	df	Mean Squares	F
(Between	70.00	4)		
$\hat{\psi}_1$: linear	32.00	1	32.00	7.44
$\hat{\psi}_2$: quadratic	35.71	1	35.71	8.30
$\hat{\psi}_3$: cubic	.50	1	.50	.01
$\hat{\psi}_4$: quartic	1.79	1	1.79	.42
Error	86.00	20	4.30	
Total	156.00	24		

When compared to an F distribution with 1 and 20 degrees of freedom, the null hypotheses that there are no linear and quadratic relations between the number of hours of respite care and the number of attempts to institutionalize can be rejected.

CHAPTER 12

1. a. First, the standardized partial regression coefficients are calculated.

$$b_1 = \frac{r_{y1} - r_{y2}r_{12}}{1 - r_{12}^2} = \frac{.60 - (.50)(.40)}{1 - (.40)^2} = .4762$$

and

$$b_2 = \frac{r_{y2} - r_{y1}r_{12}}{1 - r_{12}^2} = \frac{.50 - (.60)(.40)}{1 - (.40)^2} = .3095$$

The partial regression coefficients are thus

$$B_1 = b_2\left(\frac{S_y}{S_1}\right) = .4762\left(\frac{4000}{5}\right) = 381.0$$

$$B_2 = b_2\left(\frac{S_y}{S_2}\right) = .3095\left(\frac{4000}{4}\right) = 309.5$$

The regression equation is

$$y' = M_y + B_1(x_1 - M_1) + B_2(x_2 - M_2)$$
$$= 35,000 + 381(x_1 - 15) + 309.5(x_2 - 12)$$
$$= 381x_1 + 309.5x_2 + 25,571$$

The predicted salary for a professor who obtained his or her highest degree ten years ago and who has published six articles is

$$y' = 381(10) + 309.5(6) + 25,571 = 31,238 \text{ (i.e., } \$31,238)$$

b. $R_{y \cdot 12} = \sqrt{b_1 r_{y1} + b_2 r_{y2}} = \sqrt{.4762(.60) + (.3095)(.50)} = .6637$
$R_{y \cdot 12}^2 = (.6637)^2 = .4405$

44.05% of the variance in salaries is accounted for by years since the highest degree and the number of publications.

c. The test statistics for the null hypotheses related to the regression coefficients are

$$t = \frac{b_j}{\sqrt{\dfrac{1 - R_{y \cdot AB}^2}{(1 - R_{j \cdot A}^2)(N - K - 1)}}} = \frac{.4762}{\sqrt{\dfrac{1 - .4405}{[1 - (.40)^2](22 - 2 - 1)}}} = 2.54$$

and

$$t = \frac{b_j}{\sqrt{\dfrac{1 - R_{y \cdot AB}^2}{(1 - R_{j \cdot A}^2)(N - K - 1)}}} = \frac{.3095}{\sqrt{\dfrac{1 - .4405}{[1 - (.40)^2](22 - 2 - 1)}}} = 1.65$$

When compared to a t distribution with 19 degrees of freedom, the value of t for the first regression coefficient is sufficiently large so that the null hypothesis that the population value is zero can be rejected in favor of the alternative that it is greater than zero.

d. The value of F related to the test for the multiple correlation coefficient is as follows:

$$F = \frac{R_{y \cdot 12...K}^2 / K}{(1 - R_{y \cdot 12...K}^2)/(N - K - 1)} = \frac{.4405/2}{(1 - .4405)/(22 - 2 - 1)} = 7.48$$

which, when compared to an F distribution with 2 and 19 degrees of freedom, is sufficiently large to reject the null hypothesis that the population multiple correlation coefficient is zero.

e. $r_{y(2\cdot1)}^2 = R_{y\cdot12}^2 - r_{y1}^2 = .4405 - .3600 = .0805$

The F value associated with this test is

$$F = \frac{r_{y(j\cdot A)}^2/1}{(1 - R_{y\cdot AB}^2)/(N - K - 1)} = \frac{.0805}{(1 - .4405)/(22 - 2 - 1)} = 2.74$$

which, when compared to an F distribution with 1 and 19 degrees of freedom, is not sufficiently large to reject the null hypothesis.

f. $r_{y(1\cdot2)}^2 = R_{y\cdot12}^2 - r_{y2}^2 = .4405 - .2500 = .1905$

The F value associated with this test is

$$F = \frac{r_{y(j\cdot A)}^2/1}{(1 - R_{y\cdot AB}^2)/(N - K - 1)} = \frac{.1905}{(1 - .4405)/(22 - 2 - 1)} = 6.47$$

which, when compared to an F distribution with 1 and 19 degrees of freedom, is sufficiently large to reject the null hypothesis.

2. a. $B_1 = -3.328$, $B_2 = .128$, $y' = -3.328x_1 + .128x_2 - 10.72$
   The predicted number of absences is 2.78.

   b. $R_{y\cdot12} = .5125$, $R_{y\cdot12}^2 = .2627$
   c. For B_1: $t = -3.32$, reject H_0
      For B_2: $t = 2.56$, reject H_0
   d. $F = 8.37$, reject H_0
   e. $r_{y(2\cdot1)}^2 = .1027$, $F = 6.55$, reject H_0
   f. $r_{y(1\cdot2)}^2 = .1727$, $F = 11.01$, reject H_0

3. a. $r_{y1}^2 = .03$ $r_{y(2\cdot1)}^2 = .17$, $r_{y(3\cdot12)}^2 = .30$
   b. For step 1: H_0: $\rho_{Y1}^2 = 0$
      For step 2: H_0: $\rho_{Y(2\cdot1)}^2 = 0$
      For step 3: H_0: $\rho_{Y(3\cdot12)}^2 = 0$
   c. For step 1

$$F = \frac{r_{y(k\cdot12...k-1)}^2/1}{(1 - R_{y\cdot12...k}^2)/(N - k - 1)} = \frac{.03}{[1 - .03]/(25 - 1 - 1)} = .71$$

which, when compared to an F distribution with 1 and 23 degrees of freedom, is not sufficiently large to reject the null hypothesis. For step 2

$$F = \frac{r_{y(k\cdot12...k-1)}^2/1}{(1 - R_{y\cdot12...k}^2)/(N - k - 1)} = \frac{.17}{[1 - .20]/(25 - 2 - 1)} = 4.68$$

which, when compared to an F distribution with 1 and 22 degrees of freedom, is sufficiently large to reject the null hypothesis. For step 3

$$F = \frac{r^2_{y(k \cdot 12 \ldots k-1)}/1}{(1 - R^2_{y \cdot 12 \ldots k})/(N - k - 1)} = \frac{.30}{[1 - .50]/(25 - 3 - 1)} = 12.60$$

which, when compared to an F distribution with 1 and 21 degrees of freedom, is sufficiently large to reject the null hypothesis.

d. For step 1

$$F = \frac{r^2_{y(k \cdot 12 \ldots k-1)}/1}{(1 - R^2_{y \cdot 12 \ldots K})/(N - K - 1)} = \frac{.03}{[1 - .50]/(25 - 3 - 1)} = 1.26$$

which, when compared to an F distribution with 1 and 21 degrees of freedom, is not sufficiently large to reject the null hypothesis. For step 2

$$F = \frac{r^2_{y(k \cdot 12 \ldots k-1)}/1}{(1 - R^2_{y \cdot 12 \ldots K})/(N - K - 1)} = \frac{.17}{[1 - .50]/(25 - 3 - 1)} = 7.14$$

which, when compared to an F distribution with 1 and 21 degrees of freedom, is sufficiently large to reject the null hypothesis. For step 3

$$F = \frac{r^2_{y(k \cdot 12 \ldots k-1)}/1}{(1 - R^2_{y \cdot 12 \ldots K})/(N - K - 1)} = \frac{.30}{[1 - .50]/(25 - 3 - 1)} = 12.60$$

which, when compared to an F distribution with 1 and 21 degrees of freedom, is sufficiently large to reject the null hypothesis.

4. a. $r^2_{y1} = .15$, $r^2_{y(2 \cdot 1)} = .08$, $r^2_{y(3 \cdot 12)} = .03$, $r^2_{y(4 \cdot 123)} = .01$

   b. For step 1: H_0: $\rho^2_{Y1} = 0$

   For step 2: H_0: $\rho^2_{Y(2 \cdot 1)} = 0$

   For step 3: H_0: $\rho^2_{Y(3 \cdot 12)} = 0$

   For step 4: H_0: $\rho^2_{Y(4 \cdot 123)} = 0$

   c. For step 1: $F = 4.94$, reject H_0
   For step 2: $F = 2.81$, do not reject H_0
   For step 3: $F = 1.05$, do not reject H_0
   For step 4: $F = .34$, do not reject H_0

   d. For step 1: $F = 5.14$, reject H_0
   For step 2: $F = 2.73$, do not reject H_0
   For step 3: $F = 1.03$, do not reject H_0
   For step 4: $F = .34$, do not reject H_0

5. Let the set of variables entered in step 1 be denoted by A, the variables entered in step 2 by B, and the variables in step 3 by C. The increment in predictive ability at each step is summarized in the following table:

Step	Set	Variables Added	Number of Variables in Set	R^2	Increment in Predictive Ability
1	A	X_1, X_2	$G = 2$	.045	$R^2_{y \cdot A} = .045$
2	B	X_3, X_4	$H = 2$	.061	$R^2_{y(B \cdot A)} = .016$
3	C	X_5, X_6, X_7, X_8	$I = 4$	.395	$R^2_{y(C \cdot AB)} = .334$

The test related to wife effects is

$$F = \frac{R^2_{y \cdot A}/G}{(1 - R^2_{y \cdot ABC})/(N - G - H - I - 1)}$$

$$= \frac{.045/2}{(1 - .395)/(62 - 2 - 2 - 4 - 1)}$$

$$= 1.97$$

which, when compared to an F distribution with 2 and 53 degrees of freedom, is not sufficiently large to reject the null hypothesis related to wife effects. The test related to husband effects is

$$F = \frac{R^2_{y(B \cdot A)}/H}{(1 - R^2_{y \cdot ABC})/(N - G - H - I - 1)}$$

$$= \frac{.016/2}{(1 - .395)/(62 - 2 - 2 - 4 - 1)}$$

$$= .70$$

which, when compared to an F distribution with 2 and 53 degrees of freedom, is not sufficiently large to reject the null hypothesis related to husband effects. With regard to the interaction effects

$$F = \frac{R^2_{y(C \cdot AB)}/I}{(1 - R^2_{y \cdot ABC})/(N - G - H - I - 1)}$$

$$= \frac{.334/4}{(1 - .395)/(62 - 2 - 2 - 4 - 1)}$$

$$= 7.31$$

which, when compared to an F distribution with 2 and 53 degrees of freedom, is sufficiently large to reject the null hypothesis related to interaction effects.

6. For type of illness $F = 2.96$, reject null hypothesis.
   For locus of control $F = 4.44$, reject null hypothesis.
   For interaction $F = 1.19$, do not reject null hypothesis.

7. The test for the quadratic relation, we use a hierarchical regression with the contribution of the linear relation removed first. To find the increment in predictive ability, we begin by calculating the standardized partial regression coefficients:

$$b_1 = \frac{r_{y1} - r_{y2}r_{12}}{1 - r_{12}^2} = \frac{.187 - (.051)(.972)}{1 - (.972)^2} = 2.489$$

and

$$b_2 = \frac{r_{y2} - r_{y1}r_{12}}{1 - r_{12}^2} = \frac{.051 - (.187)(.972)}{1 - (.972)^2} = -2.368$$

Thus, $R_{y \cdot 12} = b_1 r_{y1} + b_2 r_{y2} = 2.489(.187) + (-2.368)(.051) = .345$ and $r_{y(2 \cdot 1)}^2 = R_{y \cdot 12}^2 - r_{y1}^2 = .345 - (.187)^2 = .310$. The F value for the null hypothesis that $\rho_{y(2 \cdot 1)}^2 = 0$ is

$$F = \frac{r_{y(2 \cdot 1)}^2/1}{(1 - R_{y \cdot 12}^2)/(N - K - 1)} = \frac{.310}{(1 - .345)/(15 - 2 - 1)} = 5.68$$

which, when compared to an F distribution with 1 and 12 degrees of freedom, is sufficiently large to reject the null hypothesis that there is no quadratic relation between marital satisfaction and adaptability.

8. The F value for X_1 (linear relation) is 10.89; hence, reject H_0 that there is no linear relation. The F value for the increment in predictive ability of X_2 is .16; hence, do not reject the null H_0 that there is no quadratic relation (after accounting for the linear relation).

CHAPTER 13

1. Given a stationary Bernoulli process p = probability of a boy = ½, the binomial distribution (or the normal approximation) can be used to provide the answers to this problem. Recall that the binomial distribution is given by

$$P(X = r; N, p) = \binom{N}{r} p^r q^{N-r} = \frac{N!}{r!(N-r)!} p^r q^{N-r}$$

a. $P\left(X = 1; 5, \frac{1}{2}\right) = \binom{5}{1}\left(\frac{1}{2}\right)^1\left(\frac{1}{2}\right)^4 = \frac{5!}{1!4!}\left(\frac{1}{2}\right)^1\left(\frac{1}{2}\right)^4 = .156$

b. $P\left(X > 4; 6, \frac{1}{2}\right) = \sum_{i=5}^{6} \binom{6}{i}\left(\frac{1}{2}\right)^i\left(\frac{1}{2}\right)^{6-i} = \binom{6}{5}\left(\frac{1}{2}\right)^1\left(\frac{1}{2}\right)^5 + \binom{6}{6}\left(\frac{1}{2}\right)^6$
   $= .109$

c. $P\left(X = 4; 8, \frac{1}{2}\right) = \binom{8}{4}\left(\frac{1}{2}\right)^4\left(\frac{1}{2}\right)^4 = \frac{8!}{4!4!}\left(\frac{1}{2}\right)^4\left(\frac{1}{2}\right)^4 = .273$

d. $P(X \geq 13) = P\left(Z \geq \frac{13 - .5 - Np}{\sqrt{Npq}}\right) = P\left(Z \geq \frac{13 - .5 - 20(½)}{\sqrt{20(½)(½)}}\right)$
   $= P(Z \geq 1.12) = .131$

2. a. .329   b. .045   c. .329   d. .009

3. The rankings from smallest to largest are given in parentheses.

Distressed	Nondistressed
.32 (17)	.18 (9)
.21 (11)	.33 (18)
.27 (14)	.16 (7)
.45 (22)	.12 (5)
.29 (16)	.10 (3)
.17 (8)	.42 (21)
.34 (19)	.06 (1)
.36 (20)	.22 (12)
.52 (23)	.19 (10)
.07 (2)	.23 (13)
.15 (6)	.11 (4)
	.28 (15)

$T_1 = 158$

$$U_A = N_1 N_2 + \frac{N_1(N_1 + 1)}{2} - T_1 = 11(12) + \frac{11(11 + 1)}{2} - 158 = 40$$

and

$$U_B = N_1 N_2 - U_A = 11(12) - 40 = 92$$

So, U = minimum of U_A and U_B = 40. The normal approximation is used by noting that

$$E(U) = \frac{N_1 N_2}{2} = \frac{(11)(12)}{2} = 66$$

and

$$V(U) = \frac{N_1 N_2 (N_1 + N_2 + 1)}{12} = \frac{11(12)(11 + 12 + 1)}{12} = 264$$

Therefore

$$z = \frac{U - E(U)}{\sqrt{V(U)}} = \frac{40 - 66}{\sqrt{264}} = -1.60$$

which is not sufficiently large to reject the null hypothesis and conclude that distressed couples emit a greater proportion of negative statements than nondistressed couples.

4. $U = 31$, $z = -2.11$, reject null hypothesis

5. The differences between scores, ranks of absolute values of differences, and the signed ranks are shown as follows:

Subject	Male Research Assistant	Female Research Assistant	Difference	Rank of Absolute Value of Difference	Signed Rank
1	.45	.56	−.11	7	−7
2	.33	.44	−.11	7	−7
3	.17	.65	−.48	12	−12
4	.37	.39	−.02	1.5	−1.5
5	.61	.57	.04	3	3
6	.45	.58	−.13	9	−9
7	.12	.19	−.07	4.5	−4.5
8	.34	.32	.02	1.5	1.5
9	.47	.40	.07	4.5	4.5
10	.25	.14	.11	7	7
11	.45	.65	−.20	11	−11
12	.27	.41	−.14	10	−10

a. For the sign test the probability of obtaining eight or more minuses for 12 trials is given by the binomial distribution:

$$P\left(X \ge 8; 12, \frac{1}{2}\right) = \sum_{i=8}^{12} \binom{12}{i}\left(\frac{1}{2}\right)^{i}\left(\frac{1}{2}\right)^{12-i}$$

Using the normal approximation for the binomial, we get

$$P(X \ge 8) = P\left(Z \ge \frac{8 - .5 - Np}{\sqrt{Npq}}\right)$$

$$= P\left(Z \ge \frac{7.5 - (12)(1/2)}{\sqrt{12(1/2)(1/2)}}\right)$$

$$= P(Z \ge .87)$$

$$= .19$$

which is not sufficiently large to conclude that the gender of the student makes a difference in the amount of contact.

b. For the Wilcoxon test the sum of the ranks with the least frequent sign is calculated; in this case $T = 16$. Using the normal approximation, we first calculate the mean and variance of T:

$$E(T) = \frac{N(N + 1)}{4} = \frac{12(13)}{4} = 39$$

and

$$V(T) = \frac{N(N + 1)(2N + 1)}{24} = \frac{12(13)(25)}{24} = 162.5$$

Thus

$$z = \frac{T - E(T)}{\sqrt{V(T)}} = \frac{16 - 39}{\sqrt{162.5}} = -1.80$$

which is sufficiently large to reject the null hypothesis in favor of the alternative that men students make more contact with women students than they do with other men students.

6. a. Using the normal approximation, $z = .95$, do not reject the null hypothesis that the treatments are equally effective.

 b. $T = 9$, using the normal approximation, $z = -1.89$, do not reject the null hypothesis that the treatments are equally effective.

7. The following are the ranks of the observations, in parentheses, as well as the sum of the ranks for each keyboard:

Keyboard A	Keyboard B	Keyboard C	Keyboard D
74 (12)	60 (5)	54 (3)	49 (1)
78 (15)	77 (14)	63 (6)	52 (2)
84 (17)	82 (16)	67 (9)	57 (4)
85 (18)	88 (19)	66 (8)	73 (11)
68 (10)	93 (20)	75 (13)	64 (7)
$T_A = 72$	$T_B = 74$	$T_C = 39$	$T_D = 25$

The value of the test statistic H is given by

$$H = \frac{12}{N(N + 1)} \sum \left(\frac{T_j^2}{N_j} \right) - 3(N + 1)$$

$$= \frac{12}{20(21)} \left[\frac{(72)^2}{5} + \frac{(74)^2}{5} + \frac{(39)^2}{5} + \frac{(25)^2}{5} \right] - 3(21)$$

$$= 10.18$$

When compared to a chi-squared distribution with 3 degrees of freedom, the value of H is sufficiently large to reject the null hypothesis in favor of the alternative that typing speed on the four keyboards differs.

8. $H = 2.01$, when compared to a chi-squared distribution with 2 degrees of freedom, the null hypothesis that physicians, nurses, and clerical staff do not spend different amounts of time eating lunch is not rejected.

9. The following table shows the ranks for each subject as well as the sum of the ranks for each dial:

Subject	Dial 1	Dial 2	Dial 3	Dial 4
1	2.1 (2)	2.7 (3)	1.9 (1)	2.9 (4)
2	3.0 (4)	2.6 (2)	1.8 (1)	2.8 (3)
3	5.3 (4)	4.8 (3)	3.5 (1)	4.0 (2)
4	1.9 (1)	2.4 (2)	2.6 (3)	2.7 (4)
5	3.5 (2)	3.6 (3)	1.7 (1)	3.9 (4)
6	1.9 (1)	2.5 (3)	2.0 (2)	3.3 (4)
7	4.8 (4)	4.0 (2)	3.7 (1)	4.4 (3)
8	3.7 (2)	4.8 (3)	3.6 (1)	5.0 (4)
9	2.9 (4)	2.6 (2)	2.0 (1)	2.8 (3)
10	3.0 (3)	3.6 (4)	2.6 (2)	2.4 (1)
	$T_1 = 27$	$T_2 = 27$	$T_3 = 14$	$T_4 = 32$

The value of the statistic S is

$$S = \sum \left[T_k - \frac{J(K + 1)}{2} \right]^2$$

$$= \left[27 - \frac{10(4 + 1)}{2} \right]^2 + \left[27 - \frac{10(4 + 1)}{2} \right]^2$$

$$+ \left[14 - \frac{10(4 + 1)}{2} \right]^2 + \left[32 - \frac{10(4 + 1)}{2} \right]^2$$

$$= 178$$

Using the chi-square approximation, we obtain

$$\frac{12S}{JK(K + 1)} = \frac{12(178)}{10(4)(5)} = 10.68$$

which is compared to a chi-squared distribution with 3 degrees of freedom; in this case the null hypothesis is rejected in favor of the alternative that the four dials do not produce the same levels of accuracy.

10. $S = 42$, do not reject the null hypothesis.

INDEX

ABOUT THE AUTHORS

BRUCE E WAMPOLD

Bruce E. Wampold received his Ph.D. in Education, with emphases in counseling psychology and applied research methods, from the University of California at Santa Barbara in 1981. Currently, he is the Associate Dean for the Division of Counseling and Educational Psychology at the University of Oregon. He also serves on the board of editors of the *Journal of Consulting and Clinical Psychology* and the *Journal of Counseling Psychology*, and is Associate Editor of *Behavioral Assessment*.

Dr. Wampold's main area of interest is the analysis of social interactions. He has developed statistical methods for examining sequences of behaviors and with colleagues has used these methods to study marital and family interactions, psychotherapy and supervision process, and small group interactions. Articles on these and related topics have appeared in *Psychological Bulletin, Journal of Consulting and Clinical Psychology, Journal of Counseling Psychology, Behavioral Assessment*, among others.

CLIFFORD J. DREW

Clifford J. Drew received his Ph.D. from the University of Oregon in 1968 with emphases in Special Education and Educational Psychology. He has published 15 books and chapters, and over 60 articles and reviews in professional journals. His work has included topics such as research design, statistics, measurement, mental retardation, learning disabilities and others.

Dr. Drew has taught at Kent State University, the University of Texas at Austin, and the University of Utah where he is currently a Professor in both the Special Education and Educational Psychology departments. He has had assignments in both research and academic administration, serving for six years as an Associate Dean in the Graduate School of Education at the University of Utah.

Major Parametric Statistical Tests (*Continued*)

Name or Design	Section and Page Number	Null Hypothesis
Scheffé post hoc comparisons	9.4, p. 279	$\psi = 0$
Newman–Keuls pairwise post hoc comparisons	9.5, p. 281	$\mu_{\text{larger}} - \mu_{\text{smaller}} = 0$
Correlation coefficient	10.3, p. 291	$\rho_{XY} = 0$
Chi-square test of association	10.5, p. 299	A and B are independent
Regression coefficient	11.4, p. 324	$\beta_{Y \cdot X} = 0$
Analysis of covariance	11.8, p. 343 or 12.9, p. 405	$\alpha_1 = \alpha_2 = \cdots = \alpha_J = 0$
Multiple correlation coefficient or total proportion of variance accounted for by independent variables	12.5, p. 387	$P_{Y \cdot 12 \ldots K} = 0$ or $P^2_{Y \cdot 12 \ldots K} = 0$
Increment in predictive ability for an additional variable	12.5, p. 387	$\rho^2_{Y(j \cdot A)} = 0$
Increment in predictive ability for an additional set of H variables	12.5, p. 387	$P^2_{Y(B \cdot A)} = 0$
Partial regression coefficient	12.5, p. 387	$\beta_j = 0$